D0705896

An Introduction to Theoretical and Computational Aerodynamics

Jack Moran
University of Minnesota

Dover Publications, Inc.
Mineola, New York

Copyright

Bibliographical Note

This Dover edition, first published in 2003, is an unabridged republication of the work originally published by John Wiley & Sons, New York, in 1984.

Library of Congress Cataloging-in-Publication Data

Moran, Jack.
An introduction to theoretical and computational aerodynamics / Jack Moran.
p. cm.
Originally published: New York : Wiley, c1984.
Includes bibliographical references and index.
ISBN-13: 978-0-486-42879-6 (pbk.)
ISBN-10: 0-486-4287-6 (pbk.)
1. Aerodynamics. I. Title.

TL570.M587 2003
629.132'3—dc21

2003043969

Manufactured in the United States by Courier Corporation
42879602
www.doverpublications.com

Dedicated to

WILLIAM R. SEARS
Director of the Graduate School of
Aeronautical Engineering at
Cornell University

&

WILLIAM H. THOMSEN
My Father-in-Law

Without whose help I never would
have had the chance to learn what I have

PREFACE

This textbook on aerodynamics is based on courses I have given at the University of Minnesota for the past several years. Perhaps the most useful way of describing what is and is not in the text is to outline the courses and their prerequisites.

Chapters 1 through 5 are the basis for a course called "Aerodynamics of Lifting Surfaces," which deals with the properties of wings and airfoils in incompressible (and mainly inviscid) flow. It is taken by our juniors on completion of a one-quarter course in fluid mechanics. The prerequisite course takes the students up to potential flow; the material in the present Chapters 2 and 3 (up through Rankine ovals) is actually a review for them.

Chapters 6 and 7 are used in our students' second one-quarter course in fluid mechanics, which deals with viscous flows. They are not the sole basis for that course, since they lack any discussion of pipe flows. However, I find most fluid mechanics texts do a good job on that topic and saw no need to compete with them here.

The material in Chapters 2 through 7 is thus required of all aerospace engineering undergraduates at the University of Minnesota. The emphasis on computational methods in Chapters 4, 5, and 7 is a departure from existing texts, but one that we feel is necessary if our students are to move smoothly into the aerospace industry on their graduation. The specific programs developed in depth are sufficiently close to those used in practice to ease this transition. In my opinion, the emphasis on computer usage also reinforces otherwise esoteric concepts like source and vortex flows.

No prior knowledge of computational methods is required, but students are expected to be able to read and make functional changes in FORTRAN programs. The theory behind the methods is not presented rigorously but is intended to be detailed enough that the programs are not just black boxes.

The last four chapters have been used in elective courses for seniors and graduate students interested in a deeper understanding of computational aerodynamics. Chapter 11 introduces the student to the computation of transonic flows. Although it touches briefly on sub- and supersonic flows, it really should be (and is, at the University of Minnesota) preceded by a full one-quarter course on compressible fluid mechanics, which in turn requires a prior course in thermodynamics.

The text includes listings of 14 FORTRAN computer programs that implement methods developed in the text. For the most part, these should run on any FORTRAN compiler. Few of the programs even use the IF-THEN-ELSE construction; they are largely "standard" FORTRAN IV. Exceptions are the printer-plotting routines PLOTXY and CONTUR, which may have to be revised or discarded. Also, program TSDE in Chapter 11 calls subroutines GETPF and REPLACE, available on the University of Minnesota system, which attach and replace, respectively, a named data file. Finally, most of the programs make liberal use of the Cyber-FORTRAN commands READ and PRINT for interactive input and output.

References are made in Chapters 3 and 4 to programs with graphic input and output capabilities on a microcomputer. For further information on this software, contact the author, who will also be glad to receive suggestions and comments on the text.

Jack Moran
Minneapolis, Minnesota
June 1983

CONTENTS

An Introduction to THEORETICAL AND COMPUTATIONAL AERODYNAMICS

CHAPTER 1

WINGS

Most of this book is about the aerodynamics of wings. Therefore, we start by looking at wings, what they do, and how they do it.

1.1. FUNCTION

The main purpose of a wing is to provide a *lift force*, the force that gets the airplane off the ground, raises it to an efficient and safe cruising altitude, lets it maneuver about, and allows a safe landing. The same results can be obtained by pointing a rocket or other engine in the right direction, and that, in fact, is how you fly in the absence of an atmosphere. However, when an atmosphere is available, a lot of energy can be saved by using wings. The wing acts as a *thrust amplifier*, giving you a lifting force many times the force it takes to keep you going.

To make this precise, consider an airplane in straight and level flight at speed V, as shown in Fig. 1.1. The forces on the airplane are its weight W and the *aerodynamic forces*, the forces exerted by the air on its surfaces. *Thrust* is the aerodynamic force

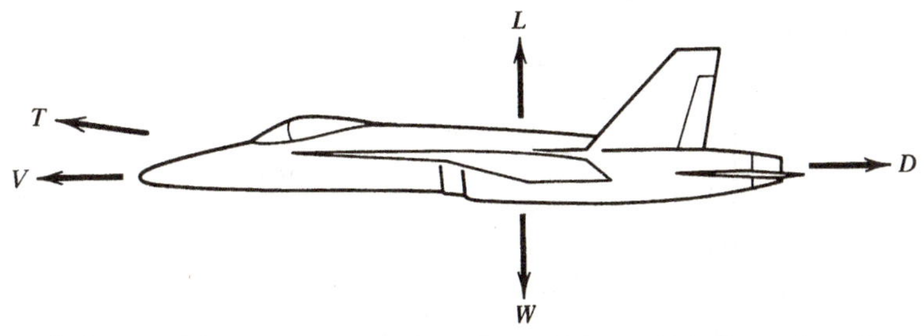

FIG. 1.1. Balance of forces on an airplane in straight and level constant-speed flight. T = thrust, L = lift, D = drag, W = weight.

on the surfaces associated with the propulsive system. Its separation from the other aerodynamic forces is often somewhat arbitrary. For example, in the case of a turbojet-powered aircraft, the usual definition of the engine thrust is the aerodynamic force on all surfaces within the volume enclosed by the engine and surfaces S_1 and S_2 that cover its inlet and exit, respectively, as shown in Fig. 1.2. However, shouldn't the propulsion-system specialist also be held accountable for the drag on the exterior of the engine nacelle? With a little imagination, you can visualize the pain various definitions of thrust could cause different members of the design "team."

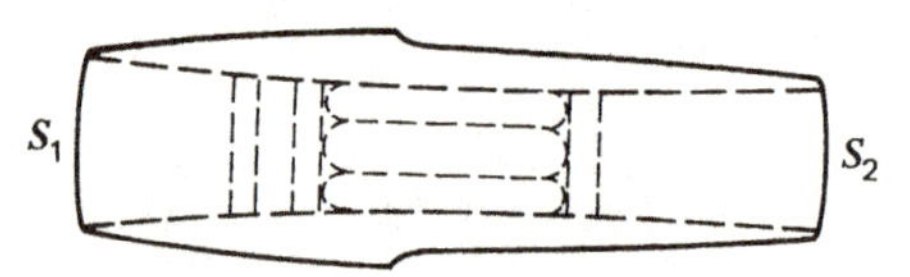

FIG. 1.2. Control volume for analyzing the drag of a propulsion system.

Once the thrust is defined, the aerodynamic forces on the remaining surfaces of the aircraft are resolved into components along and parallel to the direction of flight. *Lift* is the component perpendicular to the velocity vector **V**, whereas *drag* is the component along (but opposite to) **V**.

For flight at constant speed and altitude, the propulsive system thrust is in rough balance with the drag, and the lift with the weight. Of course, to change direction, the lift must exceed the weight, and the thrust must exceed the drag when you need to accelerate. However, what I meant when I called the wing a "thrust amplifier" is that its lift, which must at least equal the airplane's weight in order for it to get off the ground, can be many times as large as its drag, which must be overcome by the engine's thrust.

This can be seen from Fig. 1.3, taken from Chuprun [1], who examined data on the performance of various modern (as of 1980) fighter and transport aircraft. The

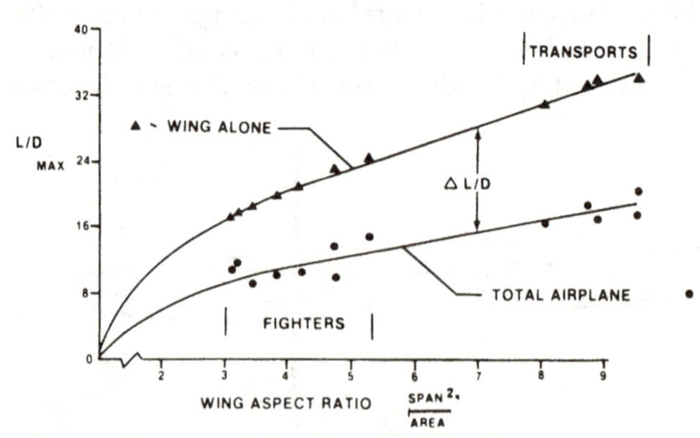

FIG. 1.3. Maximum lift to drag ratios for fighters and transports in subsonic flight. Copyright AIAA; from *The AIAA Evolution of Aircraft Wing Design Symposium*, 1980 [1].

wings of transports are seen to be capable of providing a lift force up to about 35 times their drag, whereas those of fighter aircraft provide lift to drag ratios from 16 to 24. Of course, as is also shown in Fig. 1.3, the lift to drag ratio of the complete airplane is considerably less than that of the wing alone, but it is still more than eight for fighters and about twice that for transports. Thus, for example, the C5A lifts over 750,000 lb with four 41,000-lb thrust engines.

The differences evident in Fig. 1.3 between the performances of the wings of fighters and transports can be understood in terms of their different missions. Since range is directly proportional to the lift/drag ratio,[1] transport wings are designed to give as high a ratio of lift to drag as possible. Fighter aircraft, on the other hand, must be able to change direction rapidly and so must have a high lift/weight ratio. Figure 1.4, also Chuprun's, shows that fighter wings can exert lift forces of up to 90 times their weight, whereas the lift/weight ratio of transport wings is about 22.

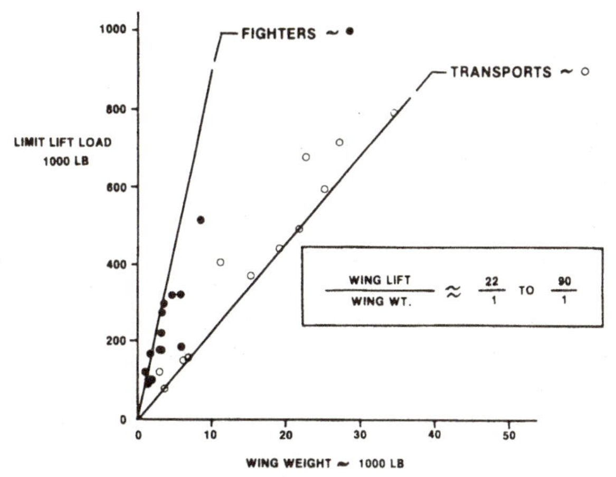

FIG. 1.4. Maximum lift/weight ratio for fighters and transports in subsonic flight. Copyright AIAA; from *The AIAA Evolution of Aircraft Wing Design Symposium*, 1980 [1].

Lift and drag are not the only parameters of interest to the aerodynamicist. He or she must also know where the lift acts, what is called the *center of pressure* of the wing. To fly straight and level, for example, the airplane must be in equilibrium with respect to the moments about its mass center of the forces acting on it; see Fig. 1.5. Also, although our focus in this book is on the *aerodynamic performance* of wings, it should be noted that a wing has nonaerodynamic design parameters, too. In the first place, it must be strong enough to bear the loads the aerodynamicist wants to impose on it. This demands a thicker wing than the aerodynamicist might like. Think of the wing as a beam. Then its stresses and deflection are seen to go inversely with the moment of inertia of the wing section, which is proportional to the cube of the wing thickness. And the storage function of wings must not be overlooked;

[1] See the Breguet range formula in, for example, McCormick [2], p. 440.

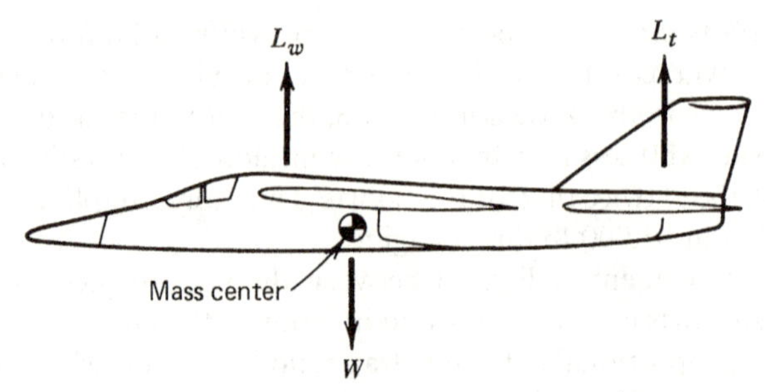

FIG. 1.5. Balance of moments about mass center on a trimmed aircraft.

most aircraft use the wings to store fuel and also the landing gear. Thus, the aerodynamicist will be well received if she or he can design a wing that not only flies the airplane but is nice and thick.

1.2. GEOMETRY

Having briefly outlined the objectives of wing design, let us now look at the geometric variables at the designer's disposal.

The first is *planform*; the shape of the wing as viewed from above. As shown in Fig. 1.6, the *span* of a wing is the distance between its tips. The *chord* is the dimension of the wing from its *leading edge* to its *trailing edge*; generally it varies along the span. The *taper* of a wing is the ratio of its tip chord to its root chord, the chord at midspan. Another planform variable is the *sweep angle* Λ, defined as the angle between a line one-quarter of the chord behind the leading edge (the "quarter-chord line") and the spanwise direction.

Probably the most important planform parameter is the *aspect ratio* of the wing,

$$\mathcal{R} \equiv \frac{\text{span}^2}{\text{area}} \tag{1-1}$$

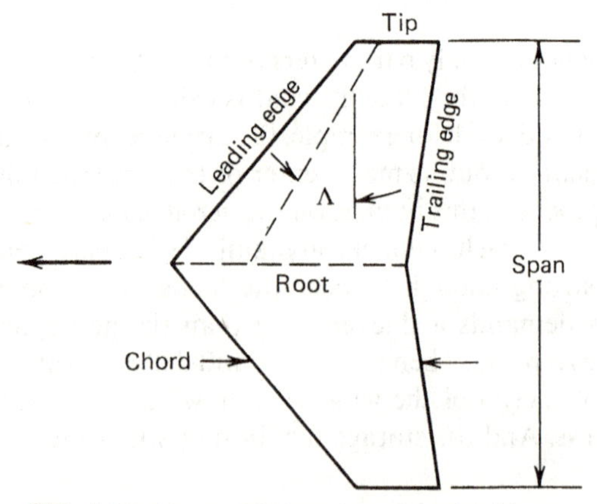

FIG. 1.6. Nomenclature for wing planforms.

The aspect ratio can be regarded as the ratio of the span to an average chord, which is in turn defined as the ratio of the planform area to the span. This can be seen from Fig. 1.3 of the preceding section to correlate very well the maximum lift to drag ratios of a variety of real wings, and even complete airplanes. Generally, for reasons to be discussed in Chapter 5, the higher the aspect ratio, the higher the ratio of lift to drag. Sailplanes, for which a high lift to drag ratio is of paramount importance, are therefore characterized by very high aspect ratios, limited mainly by structural and weight considerations.[2]

The sweep angle of a wing is usually selected with its design speed in mind. For flight at low speeds, little or no sweep is best. Roughly speaking, it just reduces the aspect ratio for a given weight, while introducing some serious problems in getting performance from all of the wing. As the speed gets close to and beyond the speed of sound, sweep becomes more desirable. Some military aircraft must perform over such a wide speed range that their wings are designed to pivot in flight to change the sweep angle.

Some of the variety of planform shapes that have been used on one airplane or another are shown in Fig. 1.7 [1]. I hope the preceding discussion sheds some light on the rationale for their design, although, as Chuprun points out, some geometries "may simply represent the artistic flair of the designer." However, most of the designs shown are rational. In particular, the forward-swept wing has a number of aerodynamic advantages over conventionally swept-rearward designs but has not been used because of an aeroelastic instability called "divergence." The recent advent of composite materials, with their very high strength/weight ratio, may make this design more attractive.

Much work has been done on the design of *wing sections* or *airfoils*; that is, on the shape of the wing in planes perpendicular to the span. The analysis and design of airfoils are generally conducted under the assumption of a two-dimensional flow; that is, the wing section is taken to be constant along the span, and the span to be infinite. As will be shown in Chapter 5, results obtained under this assumption can be used for real wings of large aspect ratio and little or no sweep. However, the main reason for the emphasis of this book on two-dimensional problems is that they are easier to deal with than three-dimensional problems and so more suitable for an introductory textbook. On the other hand, emphasis is given to approaches that have been or could be carried over to the three-dimensional world.

It is convenient to describe an airfoil shape in terms of its *thickness* and *camber* distributions. As shown in Fig. 1.8, we define a *chord line* to run from the airfoil's leading edge to its trailing edge and a *camber line*, which is midway between the airfoil's upper and lower surfaces. The *camber* of the airfoil is the distance between the camber and chord lines. The *thickness* may be defined as the dimension of the airfoil in the direction perpendicular to the chord line, or, as in Fig. 1.8, to the camber line; usually it doesn't make too much difference.

As will be seen in Chapter 4, the lift and moment per unit width of a thin airfoil depend mainly on the shape of its camber line and on the *angle of attack* of the

[2] In the case of the Gossamer Condor and Gossamer Albatross, the famous human-powered aircraft, the limiting factor was the dimensions of the hangar available [3].

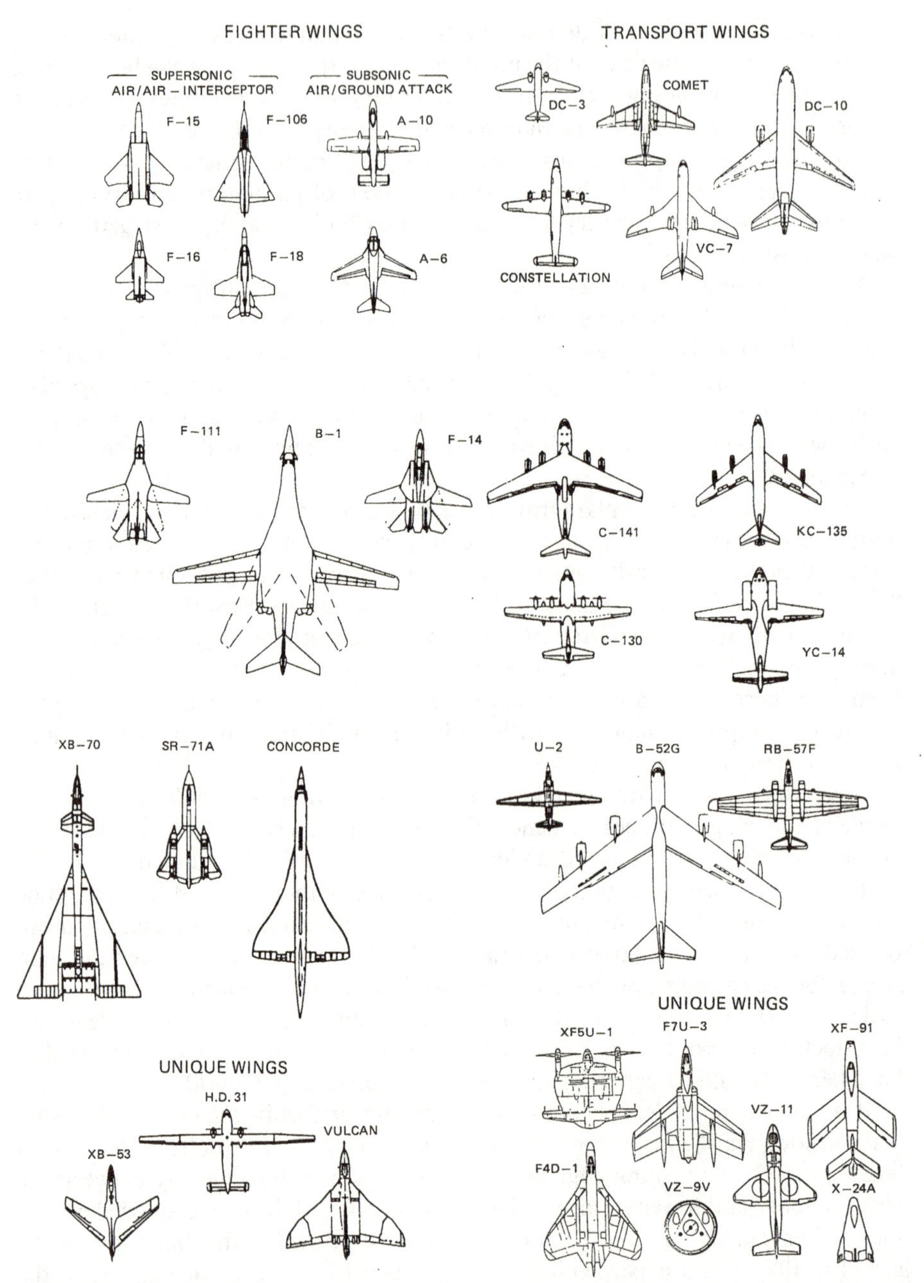

FIG. 1.7. Some of the variety of planforms that have been used. Copyright AIAA; from *The AIAA Evolution of Aircraft Wing Design Symposium*, 1980 [1].

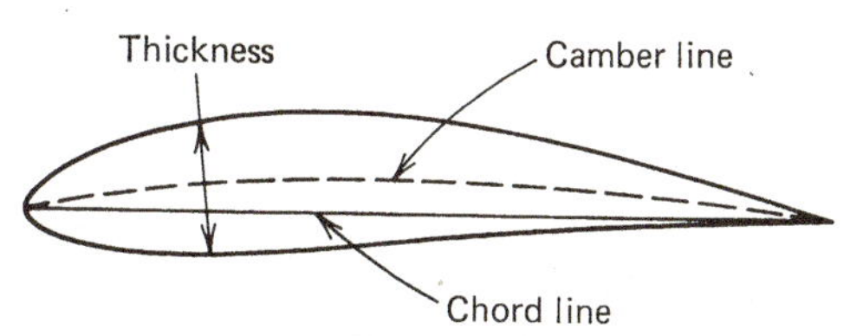

FIG. 1.8. Nomenclature for airfoils.

airfoil, the angle between its chord line and the velocity vector of the undisturbed flow relative to the airfoil. The thickness distribution does influence the pressure distribution along the airfoil surface, which, as described in Chapter 7, controls the behavior of the boundary layer on the airfoil, including the phenomenon of stall, which limits the maximum lift an airfoil can generate.

Most airfoil shapes are defined by giving the coordinates of 50 or more points on its surface. In some cases, the camber line is described analytically and the thickness distribution by giving its value at selected points. Two rather famous series of airfoils, the *NACA four- and five-digit series*, are defined completely by formulas. In both cases, the thickness distribution is

$$T(x) = 10\tau c\left[0.2969\sqrt{\frac{x}{c}} - 0.126\frac{x}{c} - 0.3537\left(\frac{x}{c}\right)^2 + 0.2843\left(\frac{x}{c}\right)^3 - 0.1015\left(\frac{x}{c}\right)^4\right] \tag{1-2}$$

Here c is the airfoil *chord* (length of its chord line) and x the distance along the chord line from the leading edge. The parameter τ is the *thickness ratio* of the airfoil (maximum thickness/chord). This thickness distribution was derived in 1930 by Eastman Jacobs, of the National Advisory Committee for Aeronautics' Langley Laboratory, on the basis of examination of airfoils known to be efficient (in particular, the airfoils known as the Gottingen 398 and the Clark Y) [4].

The camber line of the four-digit airfoils consists of two parabolas joined at the maximum camber point, as shown in Fig. 1.9. If εc is the maximum camber, and pc is the distance between the leading edge and the maximum camber point, we then have

$$\begin{aligned} \bar{Y}(x) &= \frac{\varepsilon x}{p^2}\left(2p - \frac{x}{c}\right) && \text{for } 0 < \frac{x}{c} < p \\ &= \frac{\varepsilon(c - x)}{(1 - p)^2}\left(1 + \frac{x}{c} - 2p\right) && \text{for } p < \frac{x}{c} < 1 \end{aligned} \tag{1-3}$$

The first digit of the four-digit airfoil's designation is the maximum camber ratio ε times 100, while the second digit is the chordwise position of the maximum camber p times 10. The last two digits are 100 times the thickness ratio τ. Thus, a NACA 2412 airfoil is 12% thick ($\tau = 0.12$) and has maximum camber $\varepsilon = 0.02$ at $x = 0.4c$. As indicated in Fig. 1.9, the thickness of the NACA four-digit airfoil is defined in the direction perpendicular to the camber line.

The four-digit airfoils are thus defined by three parameters: τ, p, and ε. Airfoil models with systematic variations in these parameters were tested in the Langley

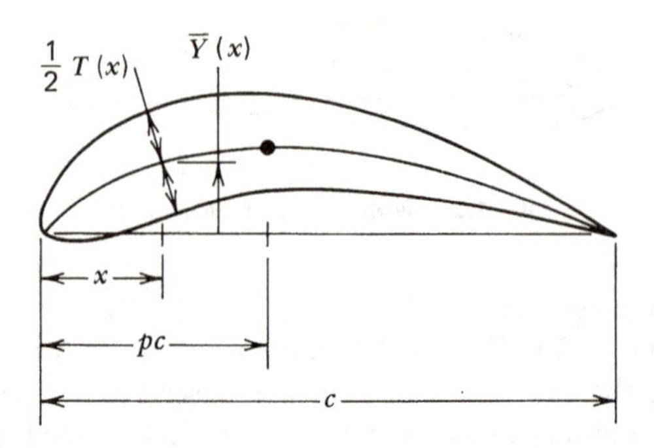

FIG. 1.9. Geometry of NACA four-digit airfoils.

Laboratory's variable-density wind tunnel (which, for the first time, allowed model testing at full-scale Reynolds numbers, by working at pressures of up to 20 atm) in the 1930s. Because their performance was so well documented [5] (and good, too), these airfoils became very popular. When it was found that the two-parabola camber line was not so useful for far-forward positions of the maximum camber, NACA engineers developed the five-digit series, which used the same thickness distribution but a camber line comprised of a cubic polynomial in x up to some point past the maximum camber, followed by a straight line. Modern airfoils have much more complex camber lines. As noted above, their shape is described, not by explicit formulas, but by giving the coordinates of points on their surface. Criteria for airfoil design will be discussed in Chapter 7.

Aside from fighter aircraft, whose maneuverability requirements place an extra demand on the aircraft designer, the lift force required of an airplane wing does not change much over its flight profile. However, as we shall show in Chapter 4, the aerodynamic force on a particular geometry is proportional to the density of the atmosphere and to the square of the airplane's velocity, factors that vary considerably during the course of a flight. Although it seems possible to meet the varying demands on lift by changing the incidence of the aircraft (and its wing) to the direction of flight, some variation in the geometry of the airplane's lifting surfaces (tailplane as well as wing) is required to keep the aircraft "trimmed" (in moment equilibrium) and to control its direction of flight. One of the Wright brothers' major contributions was the invention of a system for varying the camber of their wings by warping them, by tugging on wires attached to the corners of the wing tips. Glenn Curtiss was the first to use hinged ailerons to provide differential lift on the wings, a system that has been used by most aircraft since then (1908), including later products of the Wrights.[3] From the aileron it is but a small step[4] to the flapped wing, which increases the total lift without requiring the whole aircraft to change its inclination to the direction of flight. Flaps are therefore very nice to have during takeoff and

[3] They had the poor grace to sue Curtiss for infringement on their patent for lateral control by warping the wing and won. After 1915, even Wright airplanes used ailerons.

[4] In logic, if not in time; flaps did not come into widespread use until the 1930s.

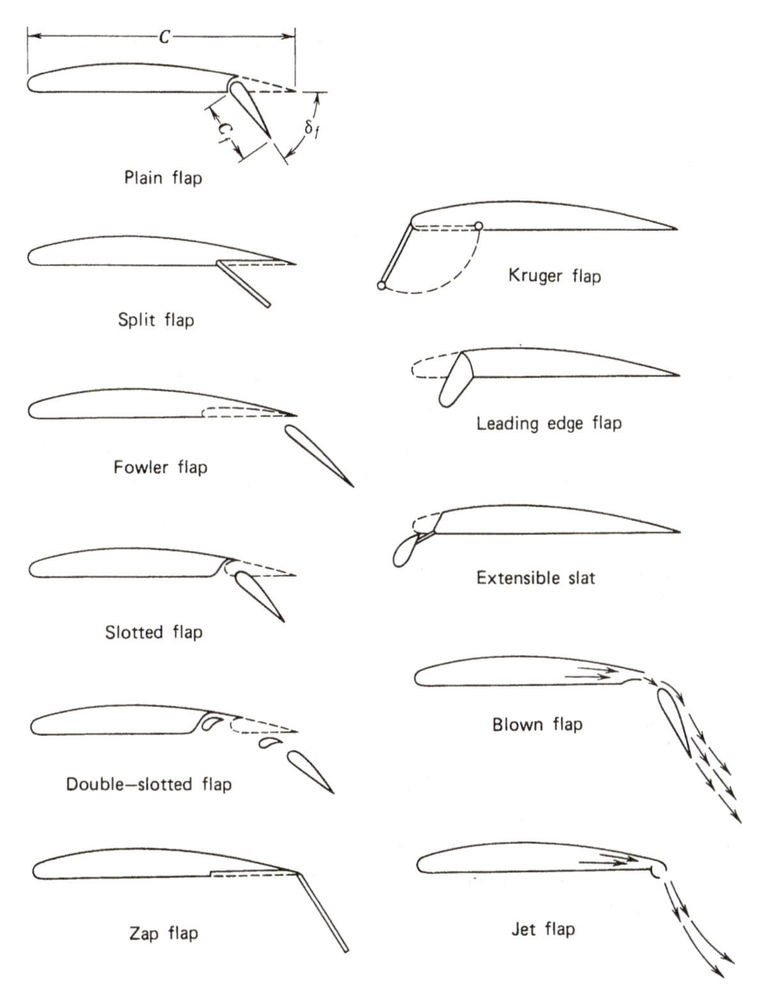

FIG. 1.10. Some flap designs. Reprinted with permission from Ref. [2].

landing. At low speeds you need extra performance from the wings to get a given lift force, and you don't want to fly at a large angle from horizontal.

Some of the variety of flap designs that have been and are used are shown in Fig. 1.10, taken from McCormick [2], who also gives an excellent summary of design data for flaps. Although the man-powered Gossamer Condor and Gossamer Albatross achieved their goals by warping the wings, multielement airfoils are here to stay. Their proper design remains an important and interesting challenge to the aerodynamicist.

1.3. REFERENCES

1. Chuprun, J., "Wings," in *The AIAA Evolution of Aircraft Wing Design Symposium.* Air Force Museum, Dayton (March 1980).

2. McCormick, B. W., *Aerodynamics, Aeronautics, and Flight Mechanics.* Wiley, New York (1979).
3. Lissaman, P. B. S., "Wings for Human-Powered Flight," in *The Evolution of Aircraft Wing Design, AIAA Symposium.* Air Force Museum, Dayton (March 1980).
4. Abbott, I., "Airfoils," in *The Evolution of Aircraft Wing Design, AIAA Symposium.* Air Force Museum, Dayton (March 1980).
5. Abbott, I. H. and von Doenhoff, A. E., *Theory of Wing Sections.* Dover, New York (1959).

1.4. PROBLEMS

1. Compare any three of the planforms shown in Fig. 1.7. Explain their design in terms of the mission of the aircraft to which they belong.
2. If an airplane's engine fails at an altitude of 5000 ft, how far can it glide at constant speed before reaching the ground if its lift to drag ratio is 16? What if the ratio is 12?
3. Compute the aspect ratio of a wing with the trapezoidal planform shown.

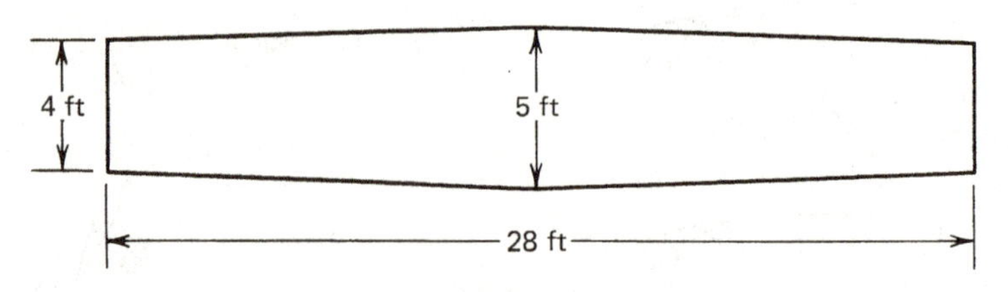

PROBLEM 1.3.

4. Make a scale drawing of an NACA four-digit airfoil. Use a computer or programmable calculator, and find the coordinates of at least 15 points each on the upper and lower surfaces.

CHAPTER 2

REVIEW OF BASIC FLUID DYNAMICS

This chapter contains in outline form results and derivations you should have seen in previous courses in fluid mechanics and/or mathematics. For more details, see the books listed in the bibliography at the end of the chapter.

2.1. FORCES AND MOMENTS DUE TO PRESSURE

In Chapters 3 and 4, we shall exaggerate the smallness of shear stresses in flows of interest to the aerodynamicist by ignoring them entirely. In this approximation, the forces and moments on a body are due entirely to the pressure exerted by the fluid at the body surface. Since the magnitudes of this force and moment are among the prime objectives of an aerodynamic analysis, let us see how we would calculate them if we knew the pressure distribution on the body.

Specifically, consider a two-dimensional airfoil to be in motion with constant velocity $-\mathbf{V}_\infty$, as shown in Fig. 2.1. The flow is *steady* in coordinates (x, y) fixed to the airfoil; that is, the fluid velocity and pressure depend only on the position coordinates (x, y) and not on time. In this frame of reference, the airfoil appears to be immersed in an *onset flow* whose velocity is $\mathbf{V}_\infty$; that is, far from the body, the flow velocity is $\mathbf{V}_\infty$. Let the *angle of attack* of the airfoil—the angle between its chord line (the x axis) and the direction of the onset flow—be α.

The shape of the airfoil can be described by functions $Y_u(x)$ and $Y_l(x)$, which give the contours of its upper and lower surfaces, respectively, as shown in Fig. 2.2. Let $\hat{\mathbf{n}}$ be a unit vector normal to the body surface and directed from the body into the fluid, and $\hat{\mathbf{t}}$ be a unit vector tangential to the surface. If $\hat{\mathbf{k}}$ is the unit vector out of the paper, the direction of $\hat{\mathbf{t}}$ is made definite by writing

$$\hat{\mathbf{t}} = \hat{\mathbf{n}} \times \hat{\mathbf{k}}$$

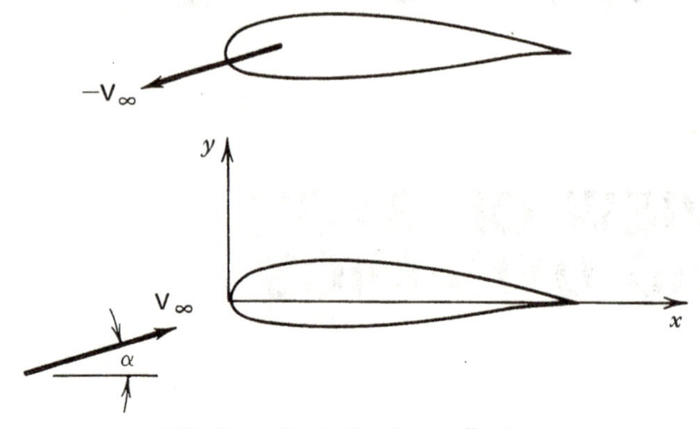

FIG. 2.1. Body-fixed coordinates.

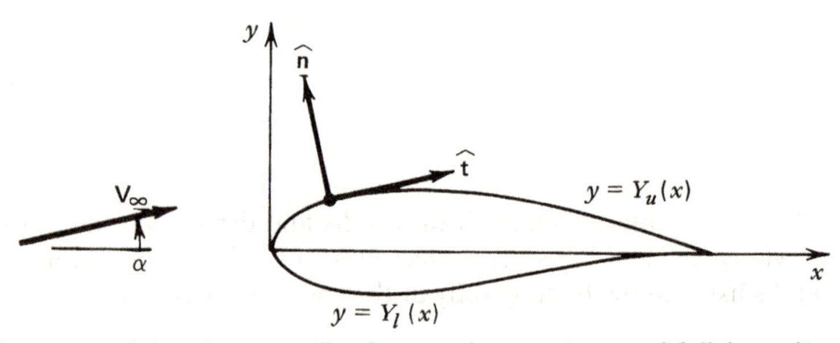

FIG. 2.2. Nomenclature for computing force and moment on an airfoil due to its pressure distribution.

Let θ be the inclination of $\hat{\mathbf{t}}$ to the x axis. Note from Fig. 2.3 how $\hat{\mathbf{n}}$, $\hat{\mathbf{t}}$, and θ behave on the upper and lower surfaces. In either case,

$$\hat{\mathbf{n}} = -\sin\theta\hat{\mathbf{i}} + \cos\theta\hat{\mathbf{j}}$$

$$\hat{\mathbf{t}} = \cos\theta\hat{\mathbf{i}} + \sin\theta\hat{\mathbf{j}} \tag{2-1}$$

where $\hat{\mathbf{i}}$ and $\hat{\mathbf{j}}$ are unit vectors in the x and y directions, respectively.

The net force on the body can be found by dividing the airfoil contour into a large number of small segments dl and summing the forces on all such segments. In the limit, as the number of such segments approaches infinity, the sum becomes an integral. If the unit normal to dl is $\hat{\mathbf{n}}$, and the fluid pressure at dl is p, the force felt by dl (per unit width of the airfoil) is

$$-p\hat{\mathbf{n}}\,dl$$

and so the net force is

$$\mathbf{F} = -\oint_{\text{airfoil}} p\hat{\mathbf{n}}\,dl \tag{2-2}$$

the circle on the integral sign meaning that we integrate around the closed contour of the airfoil. Note that both $\hat{\mathbf{n}}$ and p generally vary around the airfoil and so cannot

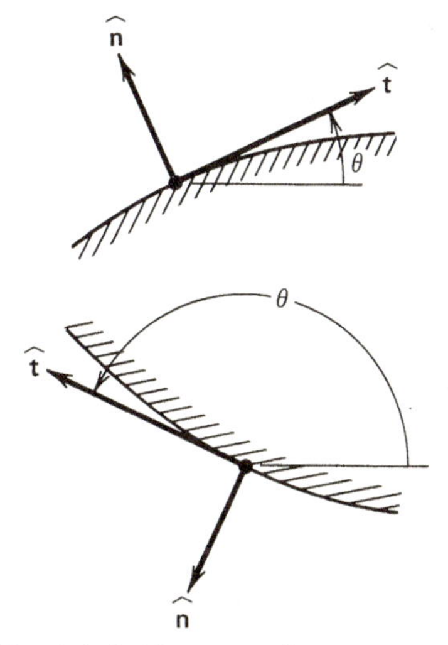

FIG. 2.3. Details of definitions of unit vectors and surface slope.

be taken outside the integral. Thus, to evaluate **F**, we need to know the distribution of pressure around the airfoil contour. For now, we shall suppose that it is known and proceed to put equation 2-2 into a form more convenient for computation.

Observe that, with θ defined as shown in Fig. 2.3,

$$dx = \cos\theta\, dl$$

$$dY = \sin\theta\, dl$$

on both upper and lower surfaces. Then the force associated with the segment dl is, from equations 2-1 and 2-2,

$$\begin{aligned}\mathbf{dF} &= -p\hat{\mathbf{n}}\, dl \\ &= p(\sin\theta\hat{\mathbf{i}} - \cos\theta\hat{\mathbf{j}})\, dl \\ &= p\left(\frac{dY}{dx}\hat{\mathbf{i}} - \hat{\mathbf{j}}\right) dx \end{aligned} \tag{2-3}$$

so the force can be calculated by integrating over x, if the pressure p on the airfoil and its slope dY/dx are known functions of x:

$$\mathbf{F} = \oint_{\text{airfoil}} p\left(\frac{dY}{dx}\hat{\mathbf{i}} - \hat{\mathbf{j}}\right) dx$$

Since we integrate clockwise around the airfoil, x runs from c (the chord) to 0 on the lower surface and from 0 to c on the upper surface. Letting the pressure on the upper and lower surfaces be $p_u(x)$ and $p_l(x)$, respectively, we then have

$$\mathbf{F} = \int_c^0 p_l\left(\frac{dY_l}{dx}\hat{\mathbf{i}} - \hat{\mathbf{j}}\right) dx + \int_0^c p_u\left(\frac{dY_u}{dx}\hat{\mathbf{i}} - \hat{\mathbf{j}}\right) dx$$

Thus, the x and y components of the force on the airfoil are

$$F_x = \int_0^c \left(p_u \frac{dY_u}{dx} - p_l \frac{dY_l}{dx} \right) dx$$

$$F_y = \int_0^c (p_l - p_u)\, dx \tag{2-4}$$

Aerodynamicists define the moment on an airfoil as being positive if clockwise ("nose up"). From equation 2-3 and Fig. 2.4, the moment about the origin (leading edge) of the force on the element dl is

$$\circlearrowright\!\!+\; dM_{\text{l.e.}} = xp\, dx + Yp \frac{dY}{dx}\, dx$$

and the total moment is found to be

$$\circlearrowright\!\!+\; M_{\text{l.e.}} = \int_0^c \left[p_u \left(x + Y_u \frac{dY_u}{dx} \right) - p_l \left(x + Y_l \frac{dY_l}{dx} \right) \right] dx \tag{2-5}$$

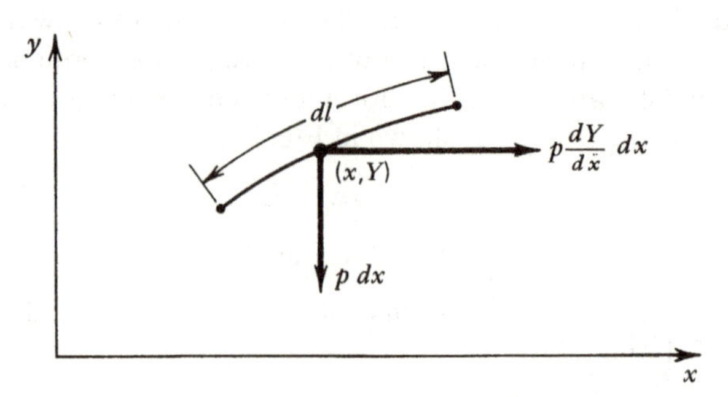

FIG. 2.4. Forces on an element of the surface.

For airfoils of practical interest, the slope dY/dx is small except, perhaps, near the leading edge, where Y itself is small. In such cases, equation 2-5 is well approximated by

$$\circlearrowright\!\!+\; M_{\text{l.e.}} \approx \int_0^c (p_u - p_l) x\, dx \tag{2-6}$$

2.2. THE BASIC CONSERVATION LAWS OF FLUID MECHANICS

The opening section of this chapter should demonstrate the importance of knowing the variation of pressure over the surfaces of bodies on which you want to know the aerodynamic force and moment. Now we will review the conservation laws that connect the pressure and velocity fields.

Consider an arbitrary volume $\mathscr{V}$ within the fluid. Let S be its bounding surface and $\hat{\mathbf{n}}$ a unit vector normal to S and directed out of[1] $\mathscr{V}$, as shown in Fig. 2.5. The mass, momentum, and internal energy of the fluid within $\mathscr{V}$ are, respectively,

$$\int_{\mathscr{V}} \rho \, d\mathscr{V}$$

$$\int_{\mathscr{V}} \rho \mathbf{V} \, d\mathscr{V}$$

$$\int_{\mathscr{V}} \rho(e + \tfrac{1}{2}V^2) \, d\mathscr{V}$$

Here ρ is the fluid density, $\mathbf{V}$ the fluid velocity, and e the internal energy per unit mass.

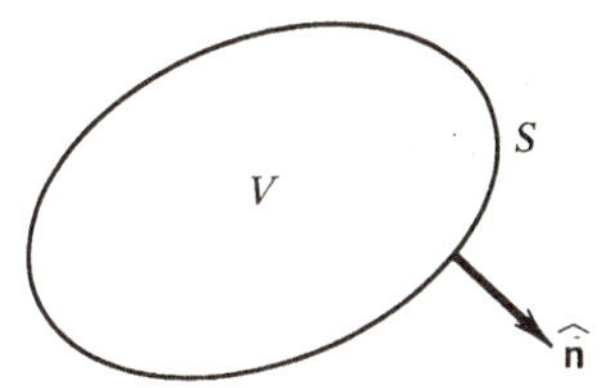

FIG. 2.5. Control volume for application of conservation laws.

The mass within $\mathscr{V}$ changes with time if there is an imbalance in the flux of mass through the bounding surface S:

$$\frac{d}{dt} \int_{\mathscr{V}} \rho \, d\mathscr{V} = - \int_S \rho \mathbf{V} \cdot \hat{\mathbf{n}} \, dS \tag{2-7}$$

The flow of fluid across S may also change the momentum and energy within $\mathscr{V}$, at the rates

$$- \int_S \rho \mathbf{V}(\mathbf{V} \cdot \hat{\mathbf{n}}) \, dS \qquad \text{and} \qquad - \int_S \rho(e + \tfrac{1}{2}V^2)\mathbf{V} \cdot \hat{\mathbf{n}} \, dS$$

respectively. From Newton's second law, the rate of change of momentum of a *fixed portion of matter*—matter whose composition does not change with time—equals the net external force acting on that matter. Assuming the force on the fluid in $\mathscr{V}$ is

[1] Note that this normal will point into any bodies on the boundary of $\mathscr{V}$ and so is directed oppositely from the unit normal of the preceding section. That section was for aerodynamicists; this one, for mathematicians. Unfortunately, these groups often disagree on sign conventions. Wait till you see how aerodynamicists plot pressure coefficients! In the meantime, be prepared to worry about the sign of any equation containing a unit normal.

due only to the pressure exerted at its boundary, we can write

$$\frac{d}{dt}\int_{\mathscr{V}} \rho \mathbf{V}\, d\mathscr{V} = -\int_S [\rho \mathbf{V}(\mathbf{V}\cdot\hat{\mathbf{n}}) + p\hat{\mathbf{n}}]\, dS \tag{2-8}$$

The pressure force also does work on the fluid, which, from the first law of thermodynamics, increases its energy. In the absence of heat addition, we then have

$$\frac{d}{dt}\int_{\mathscr{V}} \rho(e + \tfrac{1}{2}V^2)\, d\mathscr{V} = -\int_S [\rho(e + \tfrac{1}{2}V^2)\mathbf{V}\cdot\hat{\mathbf{n}} + p\hat{\mathbf{n}}\cdot\mathbf{V}]\, dS \tag{2-9}$$

Equations 2-7 to 2-9 are the fundamental principles that govern every fluid motion. They are usually referred to as *conservation laws* for mass, momentum, and energy, although, as was noted above, the momentum and energy of any system change when it is subjected to external forces. Thus, cases in which energy and momentum are actually conserved are not very interesting.

The mass conservation law always has the form of equation 2-7, but you should know that the forms (2-8) and (2-9) given for the momentum and energy conservation laws are based on approximations that cannot always be justified. In particular, they must often be modified to include the effects of shear stresses, heat conduction, and (if the environment is hot enough) radiation. The inclusion of shear stresses in the momentum conservation principle is the subject of Chapter 6.

In any case, the conservation laws must be supplemented by connections among the fluid properties called *constitutive equations.* Most often in this book we shall consider the special case of *incompressible fluid motion*, for which

$$\rho = \text{constant} \tag{2-10}$$

In this case, the mass and momentum conservation laws (2-7) and (2-8) decouple from the energy equation 2-9, with the result that the energy equation is not needed to determine the pressure field. Values of the density of common fluids at room temperature are listed in Table 2.1.

TABLE 2.1
Densities of Common Liquids at 68°F

Liquid	*Density* ($slugs/ft^3$)
Water	1.94
Benzene	1.71
Glycerine	2.46
Mercury	26.3
Engine oil	1.73
Gasoline	1.30
Kerosene	1.59
Castor oil	1.88
Seawater	1.99

TABLE 2.2
Gas Constants for Common Gases

Gas	*R* (*ft lb/slug °R*)
Air	1,715
Carbon dioxide	1,130
Carbon monoxide	1,776
Helium	12,430
Hydrogen	24,660
Nitrogen	1,776
Oxygen	1,555
Water vapor	2,762

Another constitutive equation important to aerodynamicists is the *perfect gas law*

$$p = \rho RT \tag{2-11}$$

T being the absolute temperature and R a constant (the *gas constant*). From thermodynamics, equation 2-11 implies that the internal energy e depends only on T. The relation between e and T is a second constitutive equation that must be supplied to study the motion of a perfect gas. Air behaves very much like a perfect gas up to temperatures of about 1000 °R, with a gas constant R of 1715 ft lb/slug °R. Gas constants for some other gases are listed in Table 2.2. As is shown in courses on compressible fluid flow, a perfect gas may be treated as an incompressible fluid when the flow speed (in a frame of reference in which the flow is steady) is small compared with the speed of sound waves (Chapter 11 will deal with compressible flows, especially in the transonic regime).

An important result concerning the force on a solid body immersed in a steady flow may be derived directly from the momentum principle (2-8). Take $\mathscr{V}$ to be the fluid between the body surface S_B and some control surface S_C (that is often taken to be far from the body); see Fig. 2.6. Because the flow is steady, the left side of equation 2-8 is zero. The surface S on the right side now has two pieces, S_B and S_C, for which separate integrals of the same form may be written. Since the flow velocity is tangential to the surface of a solid body, $\mathbf{V} \cdot \hat{\mathbf{n}} = 0$ on S_B, whereas the pressure

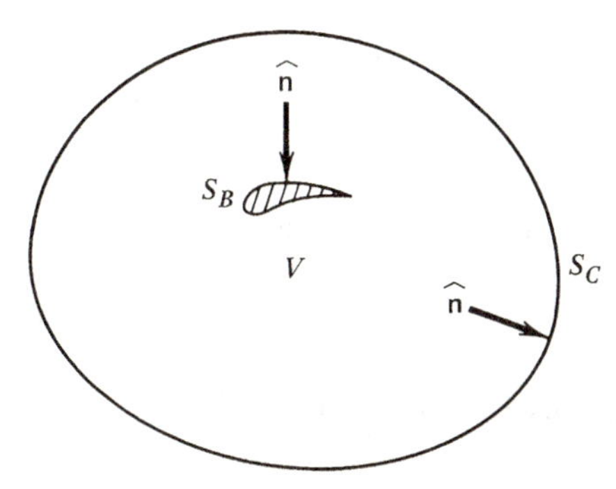

FIG. 2.6. Control surfaces for analysis of forces by application of momentum theorem.

integral over S_B is, from equation 2-2,[2] just the force $\mathbf{F}_B$ on the body. Thus

$$\int_{S_B} p\hat{\mathbf{n}}\, dS = \mathbf{F}_B = -\int_{S_C} [\rho\mathbf{V}(\mathbf{V}\cdot\hat{\mathbf{n}}) + p\hat{\mathbf{n}}]\, dS \tag{2-12}$$

and the force may be deduced by measuring (or analyzing) the flow on the control surface S_C.

2.3. VECTOR CALCULUS

From time to time in this book, you will need some familiarity with vector calculus. One of those times is coming up shortly, and so we shall review here the basic results that will be required.

First, we must define the differential operators grad, div, and curl. The *gradient* operator "transforms" a scalar f into a vector ∇f (read "del f") or grad f. Its value at a point P is

$$\nabla f \equiv \lim_{\nabla\mathscr{V}\to 0} \left(\int_{\Delta S} f\hat{\mathbf{n}}\, dS\right)\Big/\Delta\mathscr{V} \tag{2-13}$$

in which, as shown in Fig. 2.7, $\Delta\mathscr{V}$ is a small volume containing P, ΔS is the surface of $\Delta\mathscr{V}$, and $\hat{\mathbf{n}}$ is a unit normal to ΔS, directed out of $\Delta\mathscr{V}$.

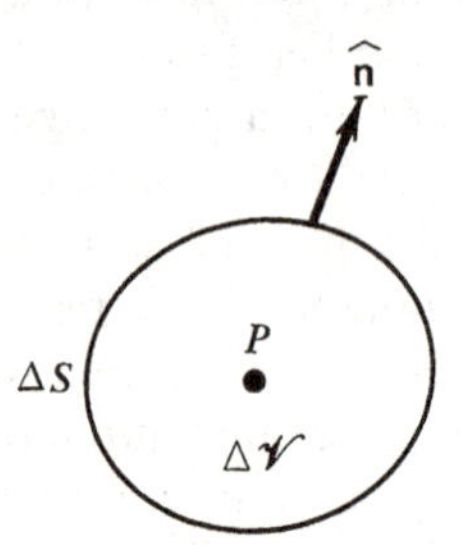

FIG. 2.7. Nomenclature for definition of gradient and divergence operators.

If we take $\Delta\mathscr{V}$ to be a rectangular parallelepiped, with surfaces normal to the Cartesian coordinate directions, we get a formula for the gradient in Cartesian coordinates,

$$\nabla f = \frac{\partial f}{\partial x}\hat{\mathbf{i}} + \frac{\partial f}{\partial y}\hat{\mathbf{j}} + \frac{\partial f}{\partial z}\hat{\mathbf{k}} \tag{2-14}$$

This formula may be used to get a useful interpretation of ∇f: it is a vector whose magnitude and direction are those of the maximum rate of increase of f. Thus the gradient is perpendicular to surfaces on which f is constant; see Fig. 2.8.

The *divergence* of a vector has a definition in terms similar to those of ∇f:

$$\operatorname{div}\mathbf{U} \equiv \lim_{\Delta\mathscr{V}\to 0} \left(\int_{\Delta S} \hat{\mathbf{n}}\cdot\mathbf{U}\, dS\right)\Big/\Delta\mathscr{V} \tag{2-15}$$

[2] With the sign of $\hat{\mathbf{n}}$ reversed.

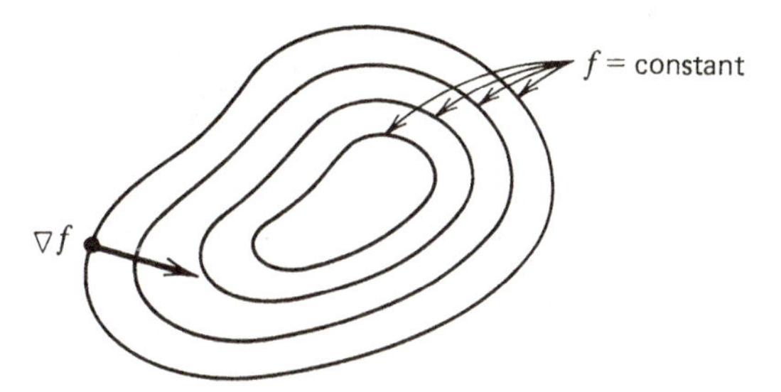

FIG. 2.8. Geometric interpretation of the gradient.

If, as above, we take $\Delta\mathscr{V}$ to be a rectangular parallelepiped whose sides are aligned with the Cartesian axes, we get

$$\operatorname{div} \mathbf{U} = \frac{\partial U_x}{\partial x} + \frac{\partial U_y}{\partial y} + \frac{\partial U_z}{\partial z} \tag{2-16}$$

in which U_x, U_y, and U_z are the Cartesian components of $\mathbf{U}$.

The *divergence theorem* can be derived directly from the definition (2-15). If $\mathscr{V}$ is an arbitrary volume, with surface S and outward directed normal $\hat{\mathbf{n}}$, and div $\mathbf{U}$ exists everywhere in $\mathscr{V}$ (meaning, as can be seen from equation 2-16, that $\mathbf{U}$ has finite—not necessarily continuous—derivatives in all of $\mathscr{V}$), then

$$\int_{\mathscr{V}} \operatorname{div} \mathbf{U}\, d\mathscr{V} = \int_S \hat{\mathbf{n}} \cdot \mathbf{U}\, dS \tag{2-17}$$

A similar (but nameless) formula can be derived from the definition (2-13) of the gradient

$$\int_{\mathscr{V}} \nabla f\, d\mathscr{V} = \int_S \hat{\mathbf{n}} f\, dS \tag{2-18}$$

Equations 2-17 and 2-18 are useful for "integration by parts" in two and three dimensions.

The curl of a vector has a rather more cumbersome definition. Its component in a particular direction—say, in the direction of the unit vector $\hat{\mathbf{e}}$—is determined by constructing a small surface ΔS around the point at which curl $\mathbf{U}$ is sought, such that ΔS is perpendicular to $\hat{\mathbf{e}}$; see Fig. 2.9. Then

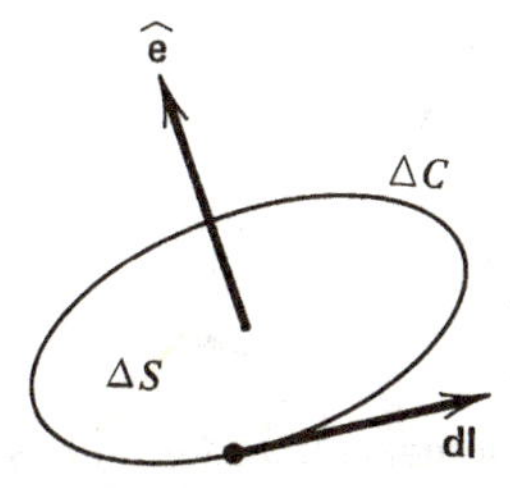

FIG. 2.9. Nomenclature for definition of the curl.

$$\hat{\mathbf{e}} \cdot \operatorname{curl} \mathbf{U} \equiv \lim_{\Delta S \to 0} \left(\int_{\Delta C} \mathbf{U} \cdot \mathbf{dl} \right) \Big/ \Delta S \tag{2-19}$$

in which ΔC is the perimeter of ΔS and we integrate around ΔS in the direction associated with $\hat{\mathbf{e}}$ by a right-hand rule. In Cartesian coordinates,

$$\operatorname{curl} \mathbf{U} = \hat{\mathbf{i}}\left(\frac{\partial U_z}{\partial y} - \frac{\partial U_y}{\partial z}\right) + \hat{\mathbf{j}}\left(\frac{\partial U_x}{\partial z} - \frac{\partial U_z}{\partial x}\right) + \hat{\mathbf{k}}\left(\frac{\partial U_y}{\partial x} - \frac{\partial U_x}{\partial y}\right) \tag{2-20}$$

a formula that is remembered in the form of a determinant

$$\operatorname{curl} \mathbf{U} = \begin{vmatrix} \hat{\mathbf{i}} & \hat{\mathbf{j}} & \hat{\mathbf{k}} \\ \dfrac{\partial}{\partial x} & \dfrac{\partial}{\partial y} & \dfrac{\partial}{\partial z} \\ U_x & U_y & U_z \end{vmatrix} \tag{2-21}$$

which form leads to the representation

$$\operatorname{curl} \mathbf{U} = \nabla \times \mathbf{U} \tag{2-22}$$

∇ ("del") being a differential operator (see equation 2-14) whose form in Cartesian coordinates is

$$\nabla = \hat{\mathbf{i}}\frac{\partial}{\partial x} + \hat{\mathbf{j}}\frac{\partial}{\partial y} + \hat{\mathbf{k}}\frac{\partial}{\partial z} \tag{2-23}$$

Equation 2-16 can be written in similar terms:

$$\operatorname{div} \mathbf{U} = \nabla \cdot \mathbf{U} \tag{2-24}$$

Stokes's theorem follows from the definition (2-19):

$$\int_S \hat{\mathbf{n}} \cdot \operatorname{curl} \mathbf{U}\, dS = \int_C \mathbf{U} \cdot \mathbf{dl} \tag{2-25}$$

Here $\hat{\mathbf{n}}$ is the unit normal to an arbitrary surface S, and the integration along C, the boundary of S, is again associated with that of $\hat{\mathbf{n}}$ by a right-hand rule, as shown in Fig. 2.10.

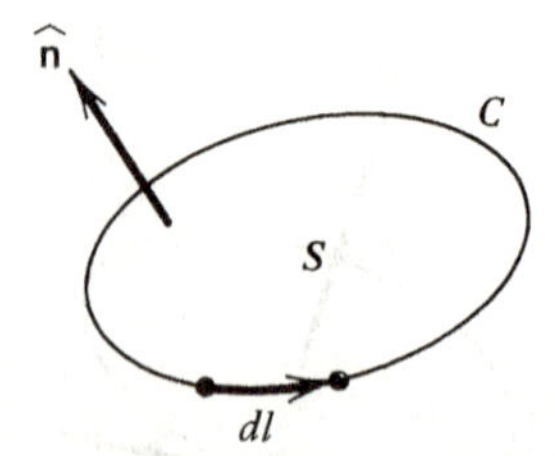

FIG. 2.10. Nomenclature for Stokes's theorem.

Example 1 If $f = x^2z - 2y/z$, find ∇f.

Solution

$$\nabla f = \hat{\mathbf{i}}\frac{\partial f}{\partial x} + \hat{\mathbf{j}}\frac{\partial f}{\partial y} + \hat{\mathbf{k}}\frac{\partial f}{\partial z}$$

$$= \hat{\mathbf{i}}\,2xz + \hat{\mathbf{j}}\left(-\frac{2}{z}\right) + \hat{\mathbf{k}}\left(x^2 + \frac{2y}{z^2}\right)$$

Example 2 If $\mathbf{V} = 3xy\hat{\mathbf{i}} + 2y^2z\hat{\mathbf{j}} - z^3/x\hat{\mathbf{k}}$, find div $\mathbf{V}$ and curl $\mathbf{V}$.

Solution

$$\text{div}\,\mathbf{V} = \frac{\partial}{\partial x}(3xy) + \frac{\partial}{\partial y}(2y^2z) + \frac{\partial}{\partial z}\left(-\frac{z^3}{x}\right)$$

$$= 3y + 4yz - \frac{3z^2}{x}$$

$$\text{curl}\,\mathbf{V} = \begin{vmatrix} \hat{\mathbf{i}} & \hat{\mathbf{j}} & \hat{\mathbf{k}} \\ \dfrac{\partial}{\partial x} & \dfrac{\partial}{\partial y} & \dfrac{\partial}{\partial z} \\ 3xy & 2y^2z & -\dfrac{z^3}{x} \end{vmatrix} = \begin{aligned} &\hat{\mathbf{i}}\left[\frac{\partial}{\partial y}\left(-\frac{z^3}{x}\right) - \frac{\partial}{\partial z}(2y^2z)\right] \\ &+ \hat{\mathbf{j}}\left[\frac{\partial}{\partial z}(3xy) - \frac{\partial}{\partial x}\left(-\frac{z^3}{x}\right)\right] \\ &+ \hat{\mathbf{k}}\left[\frac{\partial}{\partial x}(2y^2z) - \frac{\partial}{\partial y}(3xy)\right] \end{aligned}$$

$$= \hat{\mathbf{i}}(-2y^2) + \hat{\mathbf{j}}\left(-\frac{z^3}{x^2}\right) + \hat{\mathbf{k}}(-3x)$$

2.4. DIFFERENTIAL FORMS OF THE CONSERVATION LAWS

If ρ is continuous in $\mathscr{V}$, we can interchange the order of integration and differentiation in the mass conservation law (2-7) and write it

$$\int_{\mathscr{V}} \frac{\partial \rho}{\partial t}\, d\mathscr{V} = -\int_S \rho\mathbf{V}\cdot\hat{\mathbf{n}}\, dS$$

If $\mathbf{V}$ is also continuous in $\mathscr{V}$, the divergence theorem (2-17) can be used to turn the surface integral into a volume integral, and we get

$$\int_{\mathscr{V}} \left(\frac{\partial \rho}{\partial t} + \text{div}\,\rho\mathbf{V}\right) d\mathscr{V} = 0$$

This can hold for an arbitrary volume $\mathscr{V}$ within a flow only if the integral is zero everywhere within the flow, and so we get the *continuity equation*, a partial differential equation that expresses the law of mass conservation for a flow in which

ρ and $\mathbf{V}$ are continuous:

$$\frac{\partial \rho}{\partial t} + \operatorname{div} \rho \mathbf{V} = 0 \tag{2-26}$$

A similar treatment may be given the momentum equation 2-8, with results most easily obtained in Cartesian coordinates, one component at a time. Letting (u, v, w) be the Cartesian components of the velocity $\mathbf{V}$, we get

$$\frac{\partial \rho u}{\partial t} + \operatorname{div} \rho \mathbf{V} u + \frac{\partial p}{\partial x} = 0$$

$$\frac{\partial \rho v}{\partial t} + \operatorname{div} \rho \mathbf{V} v + \frac{\partial p}{\partial y} = 0$$

$$\frac{\partial \rho w}{\partial t} + \operatorname{div} \rho \mathbf{V} w + \frac{\partial p}{\partial z} = 0 \tag{2-27}$$

These are the *Euler equations*. The differential form of the energy equation 2-9 is

$$\frac{\partial}{\partial t}[\rho(e + \tfrac{1}{2}V^2)] + \operatorname{div}[\rho \mathbf{V}(e + \tfrac{1}{2}V^2) + p\mathbf{V}] = 0 \tag{2-28}$$

By subtracting various multiples of the continuity equation 2-26 from the Euler and energy equations, we may transform them into

$$\rho\left(\frac{\partial u}{\partial t} + \mathbf{V} \cdot \nabla u\right) + \frac{\partial p}{\partial x} = 0$$

$$\rho\left(\frac{\partial v}{\partial t} + \mathbf{V} \cdot \nabla v\right) + \frac{\partial p}{\partial y} = 0$$

$$\rho\left(\frac{\partial w}{\partial t} + \mathbf{V} \cdot \nabla w\right) + \frac{\partial p}{\partial z} = 0$$

$$\rho\left[\frac{\partial}{\partial t}(e + \tfrac{1}{2}V^2) + \mathbf{V} \cdot \nabla(e + \tfrac{1}{2}V^2)\right] + \operatorname{div} p\mathbf{V} = 0$$

where, in Cartesian coordinates,

$$\mathbf{V} \cdot \nabla f = u\frac{\partial f}{\partial x} + v\frac{\partial f}{\partial y} + w\frac{\partial f}{\partial z} \tag{2-29}$$

The same combination of derivatives may be made to appear in the continuity equation itself, by expanding the divergence of the product $\rho\mathbf{V}$:

$$\frac{\partial \rho}{\partial t} + \mathbf{V} \cdot \nabla \rho + \rho \operatorname{div} \mathbf{V} = 0$$

Introduce the notation

$$\frac{Df}{Dt} \equiv \frac{\partial f}{\partial t} + \mathbf{V} \cdot \nabla f \tag{2-30}$$

Called variously the *convective derivative, total derivative, particle derivative,* or *substantive derivative,* Df/Dt has the significance of being the time rate of change of f as observed while moving through space with the local fluid velocity. In terms of this derivative, the differential conservation laws may be written

$$\frac{D\rho}{Dt} + \rho \operatorname{div} \mathbf{V} = 0 \tag{2-31}$$

$$\rho \frac{D\mathbf{V}}{Dt} + \nabla p = 0 \tag{2-32}$$

$$\rho \frac{D}{Dt}(e + \tfrac{1}{2}V^2) + \operatorname{div} p\mathbf{V} = 0 \tag{2-33}$$

Example 3 If $\mathbf{V}$ is as in Example 2 and f as in Example 1, find Df/Dt.

Solution

$$\begin{aligned}\frac{Df}{Dt} &= 3xy\frac{\partial}{\partial x}\left(x^2z - \frac{2y}{z}\right) + 2y^2z\frac{\partial}{\partial y}\left(x^2z - \frac{2y}{z}\right) + \left(-\frac{z^3}{x}\right)\frac{\partial}{\partial z}\left(x^2z - \frac{2y}{z}\right)\\ &= 3xy \cdot 2xz + 2y^2z \cdot \left(-\frac{2}{z}\right) + \left(-\frac{z^3}{x}\right)\left(x^2 + \frac{2y}{z^2}\right)\\ &= 6x^2yz - 4y^2 - z^3x - \frac{2yz}{x}\end{aligned}$$

2.5. ROTATIONAL VELOCITY AND IRROTATIONAL FLOW

The *angular velocity* of a fluid particle ω can be shown to be half the *vorticity*, which is what the curl of the fluid velocity $\mathbf{V}$ is called

$$\omega = \frac{1}{2} \operatorname{curl} \mathbf{V} \tag{2-34}$$

A conservation law for angular velocity (or vorticity) can be derived by taking the curl of the momentum equation 2-32. For a two-dimensional incompressible inviscid flow, this equation takes the form (see problem 3)

$$\frac{D\omega}{Dt} = 0 \tag{2-35}$$

Thus, the angular velocity of a fluid particle in two-dimensional incompressible inviscid flow is constant. This is not generally true in three-dimensional and/or variable-density flows, but it is still the case, even then, that a fluid particle that is initially nonrotating ($\omega = 0$) stays that way. Physically, the forces on a particle in a nonviscous flow are normal to its surface and so exert no net moment about its mass center.

In this course, we always consider flows that are *uniform* (invariant with position) far upstream of the body with whose aerodynamics we are concerned. Thus, from equation 2-34, $\omega = 0$ for upstream. Since ω remains zero for every fluid particle in the field, $\omega = 0$ everywhere (to the extent that the flow is nonviscous). Such a flow is called *irrotational.*

For an irrotational flow, equation 2-34 implies curl $\mathbf{V} = 0$. This, in turn, implies the existence of a *velocity potential* ϕ, in terms of which the velocity can be derived from

$$\mathbf{V} = \nabla\phi \tag{2-36}$$

This is very helpful; it reduces the problem of finding the velocity field from a vector problem to a scalar one. Also, it allows the momentum equation to be integrated into a formula for the pressure in terms of derivatives of ϕ. Consider the case of steady incompressible inviscid flow, for which equation 2-32 reduces to

$$\rho\mathbf{V}\cdot\nabla u + \frac{\partial p}{\partial x} = 0$$

$$\rho\mathbf{V}\cdot\nabla v + \frac{\partial p}{\partial y} = 0$$

$$\rho\mathbf{V}\cdot\nabla w + \frac{\partial p}{\partial z} = 0$$

When equation 2-36 is used to eliminate $\mathbf{V}$, the first of these equations can be written

$$\begin{aligned}\rho\,\nabla\phi\cdot\nabla\frac{\partial\phi}{\partial x} = -\frac{\partial p}{\partial x} &\\ &= \rho\,\nabla\phi\cdot\frac{\partial}{\partial x}(\nabla\phi)\\ &= \frac{\rho}{2}\frac{\partial}{\partial x}(\nabla\phi\cdot\nabla\phi)\\ &= \frac{\rho}{2}\frac{\partial}{\partial x}V^2\end{aligned}$$

$$\frac{\partial}{\partial x}\left(p + \frac{\rho}{2}V^2\right) = 0$$

Similarly treating the other two equations, we find $p + \frac{1}{2}\rho V^2$ to be independent of y and z, too. Thus it must be constant throughout the flow field, which is *Bernoulli's equation*:

$$p + \tfrac{1}{2}\rho V^2 = \text{constant for steady, incompressible irrotational flow}^3 \tag{2-37}$$

Under the same assumptions, the continuity equation 2-31 becomes

[3] It can be shown that Bernoulli's equation is also valid along the streamlines of a steady inviscid rotational flow, but then the value of the constant can change from one streamline to another.

$$\begin{aligned}\operatorname{div} \mathbf{V} &= \operatorname{div} \nabla\phi \\ &= \nabla \cdot \nabla\phi \\ &= \frac{\partial^2\phi}{\partial x^2} + \frac{\partial^2\phi}{\partial y^2} + \frac{\partial^2\phi}{\partial z^2} = 0 \end{aligned} \tag{2-38}$$

This partial differential equation is called the *Laplace equation.*

For an irrotational flow, the central problem is to find the velocity potential ϕ. In the incompressible case, ϕ satisfies the Laplace equation 2-38. Once ϕ is determined, the velocity field can be found by differentiating it according to the definition (2-36). Then the pressure can be found from an integral of the momentum equation, which, for incompressible steady flow,[4] is Bernoulli's equation 2-37.

2.6. TWO-DIMENSIONAL INCOMPRESSIBLE FLOW

For this case, the continuity equation 2-31 reduces to (see equation 2-16)

$$\frac{\partial u}{\partial x} + \frac{\partial v}{\partial y} = 0 \tag{2-39}$$

in Cartesian coordinates, while the irrotationality condition $\omega = 0$ becomes (see equations 2-34 and 2-21)

$$\frac{\partial v}{\partial x} - \frac{\partial u}{\partial y} = 0 \tag{2-40}$$

The definition of the velocity potential (2-36) reduces to (see equation 2-14)

$$u = \frac{\partial\phi}{\partial x}, \qquad v = \frac{\partial\phi}{\partial y} \tag{2-41}$$

That the existence of a velocity potential follows from the irrotationality condition is easy to see in the two-dimensional case; equations 2-41 assure the satisfaction of equation 2-40 for any twice differentiable ϕ.

Similarly, the continuity equation 2-39 is automatically satisfied by the definition of the *stream function* ψ

$$u = \frac{\partial\psi}{\partial y}, \qquad v = -\frac{\partial\psi}{\partial x} \tag{2-42}$$

It has the significance that lines of constant ψ are *streamlines*, that is, lines parallel to the velocity $\mathbf{V}$. Also, the difference in ψ between any two points equals the volume rate of flow (per unit depth) across any curve C connecting those points; see Fig. 2.11:

$$\psi_2 - \psi_1 = \int_{P_1}^{P_2} \mathbf{V} \cdot \hat{\mathbf{n}}\, dl$$

[4] Counterparts exist for unsteady compressible inviscid flows.

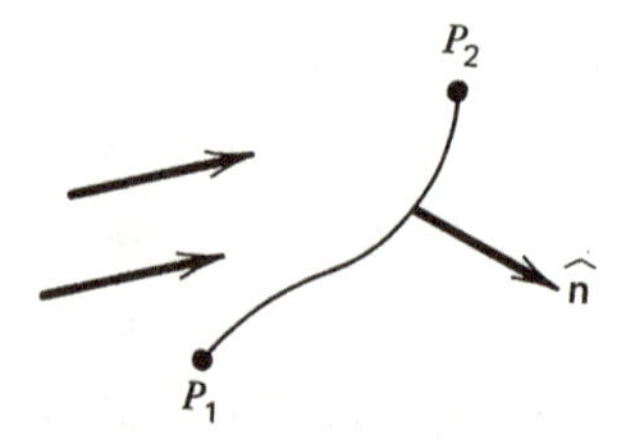

FIG. 2.11. Computation of the rate of flow between two points.

Example 4 If $\psi = 2x + 5y - 7 \tan^{-1} y/x$, find $\mathbf{V}(1, 2)$.

Solution

$$u = \frac{\partial \psi}{\partial y} = 5 - 7\frac{1}{1 + (y/x)^2} \cdot \frac{1}{x}$$

$$= 5 - \frac{7x}{x^2 + y^2} = 3.6 \text{ at } (1, 2)$$

$$v = -\frac{\partial \psi}{\partial x} = -2 + 7\frac{1}{1 + (y/x)^2}\left(-\frac{y}{x^2}\right)$$

$$= -2 - \frac{7y}{x^2 + y^2} = -4.8 \text{ at } (1, 2)$$

Example 5 Show that the stream function of Example 4 describes an irrotational flow.

Solution

$$\frac{\partial v}{\partial x} - \frac{\partial u}{\partial y} = \frac{\partial}{\partial x}\left(-2 - \frac{7y}{x^2 + y^2}\right) - \frac{\partial}{\partial y}\left(5 - \frac{7x}{x^2 + y^2}\right)$$

$$= + \frac{7y \cdot 2x}{(x^2 + y^2)^2} - \frac{7x \cdot 2y}{(x^2 + y^2)^2}$$

$$= 0 \qquad \text{QED}$$

Example 6 If the pressure far from the origin is $p_\infty = 6$, and the density $\rho = 2$, and the flow field is described by the stream function of Example 4, find $p(1, 2)$.

Solution Since the flow field is irrotational, Bernoulli's equation 2-37 applies. Evaluate the constant in equation 2-37 far from the origin, where the velocity is, as can be seen from Example 4,

$$\mathbf{V}_\infty = 5\hat{\imath} - 2\hat{\jmath}$$

Then

$$p + \tfrac{1}{2}\rho V^2 = p_\infty + \tfrac{1}{2}\rho V_\infty^2$$
$$= 6 + \tfrac{1}{2} \cdot 2 \cdot (5^2 + 2^2) = 35$$

and

$$p(1, 2) = 35 - \tfrac{1}{2} \cdot 2 \cdot (3.6^2 + 4.8^2)$$
$$= -1.0$$

When the definition of the velocity potential (2-41) is substituted into the continuity equation 2-39, we again see that ϕ satisfies the Laplace equation, the two-dimensional version of which is

$$\nabla^2\phi = \frac{\partial^2\phi}{\partial x^2} + \frac{\partial^2\phi}{\partial y^2} = 0 \tag{2-43}$$

For an irrotational flow, substitution of equations 2-42 into equation 2-40 shows that the stream function also satisfies the Laplace equation:

$$\nabla^2\psi = \frac{\partial^2\psi}{\partial x^2} + \frac{\partial^2\psi}{\partial y^2} = 0 \tag{2-44}$$

The solution of either equation 2-43 or 2-44 can be used to obtain, via equations 2-41 or 2-42, a velocity field that satisfies the conditions of irrotationality and continuity. These are not the only conditions that the velocity field must meet; it must, for example, be tangent to the surface of any solid bodies in the flow. However, as we shall see in the next two chapters, solutions satisfying all the conditions for an irrotational incompressible flow can be built up efficiently from certain simple solutions of the Laplace equation, which we shall now describe.

2.6.1. Uniform Flow

The velocity components u and v do not vary with position.

$$u = A = \text{constant}$$

$$v = B = \text{constant} \tag{2-45}$$

Then

$$\phi = Ax + By$$

$$\psi = Ay - Bx \tag{2-46}$$

2.6.2. Source Flow

The velocity field is radial from the origin and symmetric about the origin. As can be seen from Fig. 2.12, this implies

$$\frac{v}{u} = \frac{y}{x} \tag{2-47}$$

If Q is the volume rate of flow from the origin (*the source strength*),

$$V = \frac{Q}{2\pi r} \tag{2-48}$$

and

$$\phi = \frac{Q}{2\pi} \ln r$$

$$\psi = \frac{Q}{2\pi}\theta \tag{2-49}$$

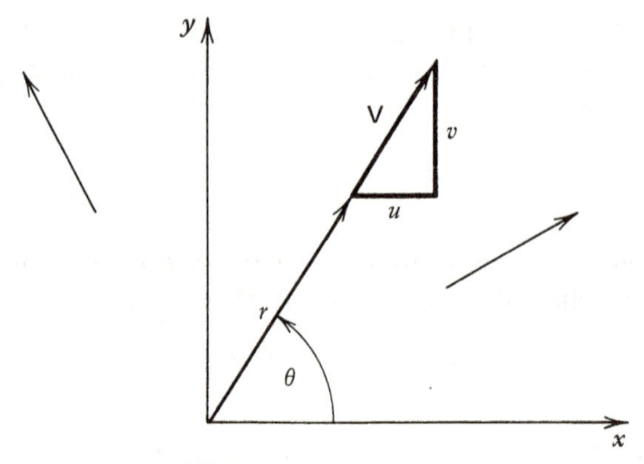

FIG. 2.12. Source flow.

2.6.3. Vortex Flow

The streamlines are circles about the origin. Again, the magnitude of the velocity depends only on radial distance from the origin. With the help of Fig. 2.13, you can show that this means

$$\frac{v}{u} = -\frac{x}{y} \tag{2-50}$$

If Γ is the *circulation* of the vortex (or *vortex strength*), the line integral of the velocity on a curve encircling the origin,

$$\Gamma = \oint_C \mathbf{V} \cdot \mathbf{dl} \tag{2-51}$$

then

$$V = \frac{\Gamma}{2\pi r}$$

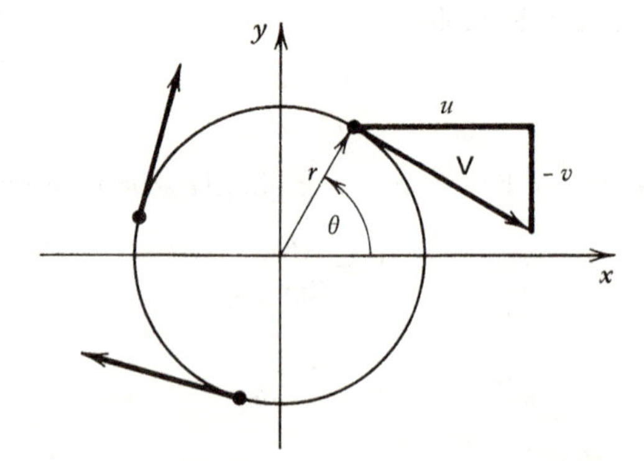

FIG. 2.13. Vortex flow.

and

$$\phi = -\frac{\Gamma}{2\pi}\theta$$

$$\psi = \frac{\Gamma}{2\pi}\ln r \tag{2-52}$$

Note that Γ is defined positive if the circular flow is clockwise.

2.7. BIBLIOGRAPHY

1. Kuethe, A. M. and Chow, C.-Y., *Foundations of Aerodynamics*, 3rd ed., J. Wiley, New York (1976).
2. Li, W. H. and Lam, S. H., *Principles of Fluid Mechanics*. Addison-Wesley, Reading, Massachusetts (1964).
3. White, F. M., *Fluid Mechanics*. McGraw-Hill, New York (1979).

2.8. PROBLEMS

1. In a real (viscous) flow, the force per unit area $\boldsymbol{\tau}$ on the surface of a fluid is not normal to the surface. Thus, the force on a solid body immersed in the flow must be written

$$\mathbf{F} = -\int_S \boldsymbol{\tau}\, dS$$

instead of equation 2-2. For the case in which the flow is steady, find $\mathbf{F}$ in terms of an integral over a control surface S_C removed from the body (along the lines of equation 2-12).

2. Using the fact that

$$\operatorname{div} p\mathbf{V} = \nabla p \cdot \mathbf{V} + p \operatorname{div} \mathbf{V}$$

which you can verify from the definitions of div and ∇ in Cartesian coordinates, equations 2-14 and 2-16, show from equations 2-31–2-33 that, in steady flow

$$\mathbf{V} \cdot \nabla\left(e + \frac{1}{2}V^2 + \frac{p}{\rho}\right) = 0$$

3. Derive equation 2-35 by following the procedure outlined in the paragraph preceding it. *Hint*: First specialize the momentum equation 2-32 to the case of two-dimensional incompressible flow, and write out its x and y components. Then, as is appropriate in taking the curl of a two-dimensional vector, subtract the y derivative of the x component from the x derivative of the y component. You will also have to use the continuity equation 2-39.

4. Find the potential and stream function for the *doublet* that results by positioning sources of equal and opposite strength $\pm Q$ at $(\pm\Delta x/2, 0)$ and letting $Q \to \infty$ as $\Delta x \to 0$ in such a way that $Q\Delta x = M$, the "doublet strength."

5. Show that the velocity potential

$$\phi = V_\infty x\left(1 + \frac{a^2}{r^2}\right) - \frac{\Gamma}{2\pi}\theta$$

describes a two-dimensional incompressible flow that has the uniform velocity $V_\infty \hat{\mathbf{i}}$ far from the origin and is tangent to the circle $r = a$.

6. Calculate the velocity on the circle of radius $r = a$ when the potential is as in problem 5, and show that the circulation about the circle is Γ.

7. Calculate the pressure distribution on the circle of problems 5 and 6, and use equation 2-2 to calculate the force on the circle. *Hint*: Take x and y components of equation 2-2.

8. Also for the case where ϕ is as in problem 5, show that, far from the circle,

$$\mathbf{V} \approx \left(V_\infty - \frac{\Gamma}{2\pi r}\cos\theta\right)\hat{\mathbf{i}} - \frac{\Gamma}{2\pi r}\sin\theta\hat{\mathbf{j}}$$

$$p \approx p_\infty + \frac{\rho V_\infty \Gamma}{2\pi r}\cos\theta$$

Then take the surface S_C of equation 2-12 to be a circle of radius $R \gg a$, and recompute the force on the circle of radius a.

9. Sketch some of the streamlines (lines on which ψ = constant) and equipotentials (lines on which ϕ is constant) for the uniform flow, source flow, and vortex flow (three separate sketches). You should find the streamlines to be orthogonal to the equipotentials. Why?

10. For a point source in three dimensions, the velocity should be

$$V = \frac{Q}{4\pi\rho^2} \tag{2-53}$$

where ρ is the distance from the source; in cylindrical coordinates, if the source is located at x_0 on the x axis,

$$\rho^2 = (x - x_0)^2 + r^2$$

Show that

$$\phi = -\frac{Q}{4\pi}[(x - x_0)^2 + r^2]^{-\frac{1}{2}}$$

is the velocity potential of such a source. That is, show that the velocity field

$$\mathbf{V} = \frac{\partial\phi}{\partial x}\hat{\mathbf{i}} + \frac{\partial\phi}{\partial r}\hat{\mathbf{e}}_r$$

satisfies equation 2-53 and that ϕ satisfies the Laplace equation 2-34, which in cylindrical coordinates can be written

$$\frac{\partial^2\phi}{\partial r^2} + \frac{1}{r}\frac{\partial\phi}{\partial r} + \frac{1}{r^2}\frac{\partial^2\phi}{\partial\theta^2} + \frac{\partial^2\phi}{\partial x^2} = 0$$

11. For axisymmetric incompressible flow, there exists a *Stokes stream function* ψ, defined so that the axial and radial velocity components are, respectively,

$$u = \frac{1}{r}\frac{\partial \psi}{\partial r}, \qquad v = -\frac{1}{r}\frac{\partial \psi}{\partial x}$$

Show that this definition identically satisfies the continuity equation, which, in cylindrical coordinates, is

$$\frac{\partial u}{\partial x} + \frac{\partial v}{\partial r} + \frac{v}{r} = 0$$

if the flow is axisymmetric (no θ dependence). Show also that ψ is constant on streamlines and that the difference between the values of ψ on any two streamlines, times 2π, is the volume flux through the annular region defined by rotating the streamlines about the x axis.

12. Show that the Stokes stream function of a uniform flow of speed V_∞ in the x direction is

$$\psi = \tfrac{1}{2} V_\infty r^2$$

13. Show that the Stokes stream function of the point source discussed in problem 10 is

$$\psi = -\frac{Q}{4\pi}\frac{x}{\rho}$$

CHAPTER 3

INCOMPRESSIBLE IRROTATIONAL FLOW ABOUT SYMMETRIC AIRFOILS AT ZERO LIFT

Let us now collect from Chapter 2 the equations that govern the solution of the problem stated in the title above. The *continuity equation* takes the forms

$$\frac{\partial u}{\partial x} + \frac{\partial v}{\partial y} = 0 \tag{3-1a}$$

$$\nabla^2 \phi = \frac{\partial^2 \phi}{\partial x^2} + \frac{\partial^2 \phi}{\partial y^2} = 0 \tag{3-1b}$$

$$u = \frac{\partial \psi}{\partial y}, \qquad v = -\frac{\partial \psi}{\partial x} \tag{3-1c}$$

depending on what variable ($\mathbf{V} = u\hat{\mathbf{i}} + v\hat{\mathbf{j}}, \phi$, or ψ) we choose to work with. The motion equations are replaced by *Bernoulli's equation*

$$p + \tfrac{1}{2}\rho(u^2 + v^2) = \text{constant} \tag{3-2}$$

and by the condition of *irrotationality*, which may be written

$$\frac{\partial u}{\partial y} = +\frac{\partial v}{\partial x} \tag{3-3a}$$

$$u = \frac{\partial \phi}{\partial x}, \qquad v = \frac{\partial \phi}{\partial y} \tag{3-3b}$$

$$\nabla^2 \psi = \frac{\partial^2 \psi}{\partial x^2} + \frac{\partial^2 \psi}{\partial y^2} = 0 \tag{3-3c}$$

3.1. UNIFORM TWO-DIMENSIONAL IRROTATIONAL INCOMPRESSIBLE FLOW ABOUT AN ISOLATED BODY

Equations 3-1 to 3-3 apply to any two-dimensional incompressible irrotational flow. What distinguishes one flow from another are the *boundary conditions*. *Far from the body*, the velocity takes on a prescribed value, so

$$(u, v) \to (V_\infty \cos \alpha, V_\infty \sin \alpha) \qquad \text{as } x^2 + y^2 \to \infty \tag{3-4a}$$

$$\phi \to V_\infty x \cos \alpha + V_\infty y \sin \alpha \qquad \text{as } x^2 + y^2 \to \infty \tag{3-4b}$$

$$\psi \to V_\infty y \cos \alpha - V_\infty x \sin \alpha \qquad \text{as } x^2 + y^2 \to \infty \tag{3-4c}$$

Here α, the *angle of attack*, is defined as the inclination of the velocity vector at infinity to a reference line fixed in the body. For an airfoil, this reference line is the *chord line*, which runs from the nose $(0, 0)$ to the tail $(c, 0)$; see Fig. 3.1.

Finally, the flow must be tangent to the body surface. If the body is located at $y = Y(x)$, as in Fig. 3.1, the *flow-tangency condition* may be written

$$\frac{v}{u} = \frac{dY}{dx} \qquad \text{at } y = Y(x) \tag{3-5a}$$

$$\hat{\mathbf{n}} \cdot \nabla\phi = \frac{\partial \phi}{\partial n} = 0 \qquad \text{at } y = Y(x) \tag{3-5b}$$

$$\psi = \text{constant} \qquad \text{at } y = Y(x) \tag{3-5c}$$

In equation 3-5b, $\hat{\mathbf{n}}$ is a unit vector normal to the body surface.

Note that, in a nonviscous flow, the only requirement at a solid surface is that the velocity have no component normal to the surface. In a real (i.e., viscous) flow, the tangential component also vanishes on the body surface (see Chapter 6). In most aerodynamic situations (to be precise, if the *Reynolds number* is large enough), the flow adjusts from a no-slip condition at the wall to a nearly parallel-flow condition through a thin boundary layer, without change in pressure, as will be discussed in Chapter 7.

We will see in Chapter 4 that the conditions of continuity and irrotationality and the boundary conditions given above are not always sufficient to determine the flow uniquely. In fact, they do not even determine the lift of an airfoil at an angle of attack. To make them suffice, the discussion in this chapter will be restricted to flows past airfoils that are symmetric about the x axis and at zero angle of attack. Such

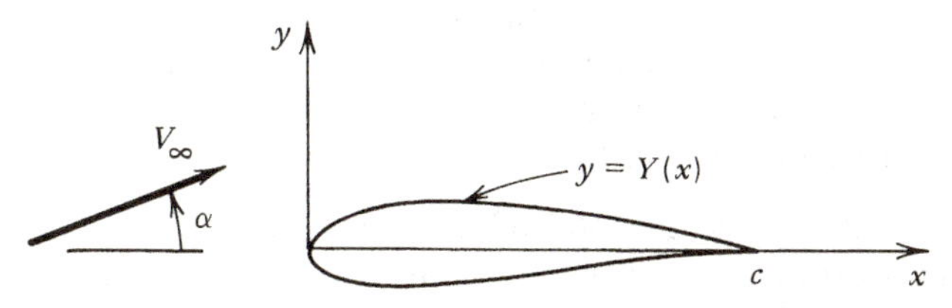

FIG. 3.1. Nomenclature for flow past an isolated airfoil.

nonlifting flows are not, to be sure, of great practical interest. This disadvantage is outweighed by the nice introduction they give you to the superposition method, which will be extended to the lifting case in Chapter 4.

3.2. SUPERPOSITION OF FUNDAMENTAL SOLUTIONS

One of the beauties of incompressible irrotational flow is that the differential equations that govern the velocity, potential, and stream function are linear and homogeneous. This allows us to superpose simple solutions of those equations so as to create more complicated (even realistic) ones. For example, if ψ_1 and ψ_2 satisfy the Laplace equation

$$\nabla^2\psi_1 = \nabla^2\psi_2 = 0$$

then so does their sum:

$$\nabla^2(\psi_1 + \psi_2) = 0$$

A powerful procedure for solving irrotational flow problems is to represent ϕ and ψ by linear combinations of known solutions of Laplace's equation,

$$\phi = \sum_i c_i\phi_i, \qquad \psi = \sum_i c_i\psi_i$$

finding the coefficients c_i so that the boundary conditions are satisfied both far from the body and on the body surface. Since, by their construction, ϕ and ψ also satisfy the Laplace equation, they then meet all the conditions set forth above in equations 3-1 and 3-3 to 3-5.

To begin with, we will consider a few examples of the so-called *inverse problem*, in which ψ is specified and we seek the related body shape. In particular, consider the flow generated by placing a source of strength Q in a uniform stream in the $+x$ direction. From equations 2-46 and 2-49, the velocity potential and stream function of this flow are

$$\phi = +V_\infty x + \frac{Q}{2\pi}\ln r \tag{3-6a}$$

$$\psi = V_\infty y + \frac{Q}{2\pi}\theta \tag{3-6b}$$

The net velocity field can be constructed geometrically by adding vectors that represent the uniform flow $V_\infty\hat{\mathbf{i}}$ and the source flow $(Q/2\pi r)\hat{\mathbf{e}}_r$. Such a process is illustrated in Fig. 3.2, the resultant velocity vectors being indicated by relatively heavy arrows. Because the velocity field due to the source varies inversely with distance from the source, we see that

1. Far from the source, the total velocity becomes $V_\infty\hat{\mathbf{i}}$.
2. On the x axis, downstream of the source, the flow is directed downstream.
3. On the x axis, just upstream of the source, the flow is directed upstream.
4. Further upstream on the x axis, the flow is directed downstream.

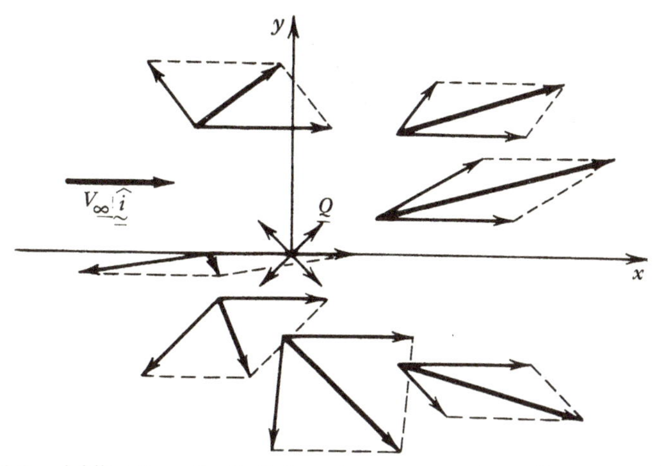

FIG. 3.2. Adding the velocity fields of a source and a uniform onset flow.

Thus, we see that there must be one point on the x axis, upstream of the source, at which $\mathbf{V} = 0$. Such a point is called a *stagnation point*, and any streamline that passes through such a point is called a *stagnation streamline*, whose special significance we will get to shortly.

The idea of a streamline going through a stagnation point may puzzle you, but look at the flow field just above or below the stagnation point, say on the circle centered at the source that passes through the stagnation point, as sketched in Fig. 3.3. Remember that a streamline is just a curve that is everywhere tangential to the velocity vector. You should be able to see there are *two streamlines going through the stagnation point*, at right angles to one another, as is shown in Fig. 3.4. To put it another way, the streamline that goes through the stagnation point comes along the x axis from far upstream to the stagnation point and then splits into three pieces, one of which continues along the x direction, while the other two go off in the positive and negative y directions.

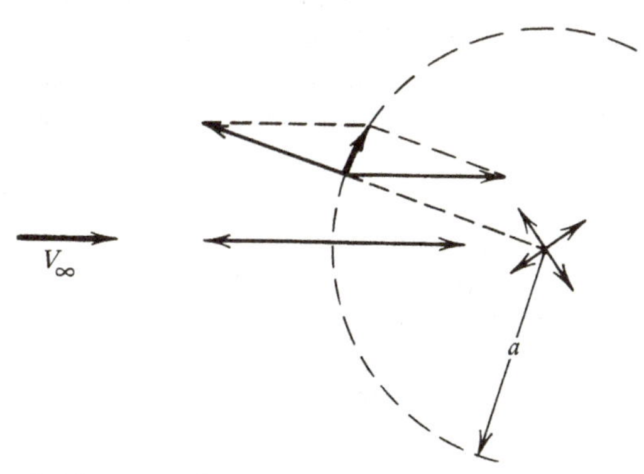

FIG. 3.3. Velocity field near the stagnation point of the source/uniform flow combination.

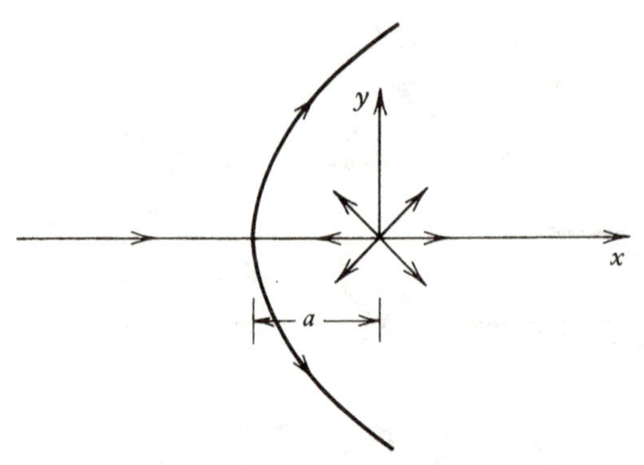

FIG. 3.4. Details of the streamlines near the stagnation point of the source/uniform flow combination.

The reason we are so interested in the streamlines of this very artificially generated flow is that any one of them could be replaced by the surface of a solid, and the streamlines on either side of it can be regarded as the streamlines of a flow past the solid. That is, the only boundary condition we can apply at a solid surface in nonviscous flow is that the flow velocity be parallel to the surface. Thus, not only is any solid surface a streamline, but *any streamline is potentially a solid surface.* Since we started by writing down the stream function of the flow, if any of the streamlines looks like an interesting body contour, we can find the velocity and pressure fields of the flow on and about that body simply by differentiating the stream function and using equations 3-1c and 3-2.

For the case at hand, with ψ given by equation 3-6b, some of the streamlines are sketched in Fig. 3.5*a*. One of the more interesting possibilities is to replace the streamline C_1SC_2 by a solid surface, as shown in Fig. 3.5*b*. The stream function given in equation 3-6b then describes a flow, uniform far upstream, past a semi-infinite board with a rounded nose. We will look at this case in detail shortly. However, first a caveat: in selecting the streamline that will be regarded as the surface of a solid body, *keep the sources inside the body*. For example, replacing the streamline FQI by a solid surface, as in Fig. 3.5*c*, does not yield results as interesting as they may look, since the stream function of equation 3-6b describes a flow that is singular (blows up) at the source point Q. That is, it does not describe the flow past an ordinary solid having the shape of FQI, but rather past one that is squirting fluid out of the cavity in front.

Returning to the case where the streamline C_1SC_2 is regarded as a solid surface, let the coordinates of the stagnation point S be $(-a, 0)$. The velocity fields of a uniform flow and a point source can be obtained from equations 2-45, 2-47, and 2-48 (or by differentiating equations 3-6 and using equations 3-1c or 3-3b), and so we find

$$u(x, y) = V_\infty + \frac{Q}{2\pi}\frac{x}{x^2 + y^2}$$

$$v(x, y) = +\frac{Q}{2\pi}\frac{y}{x^2 + y^2} \qquad (3\text{-}7)$$

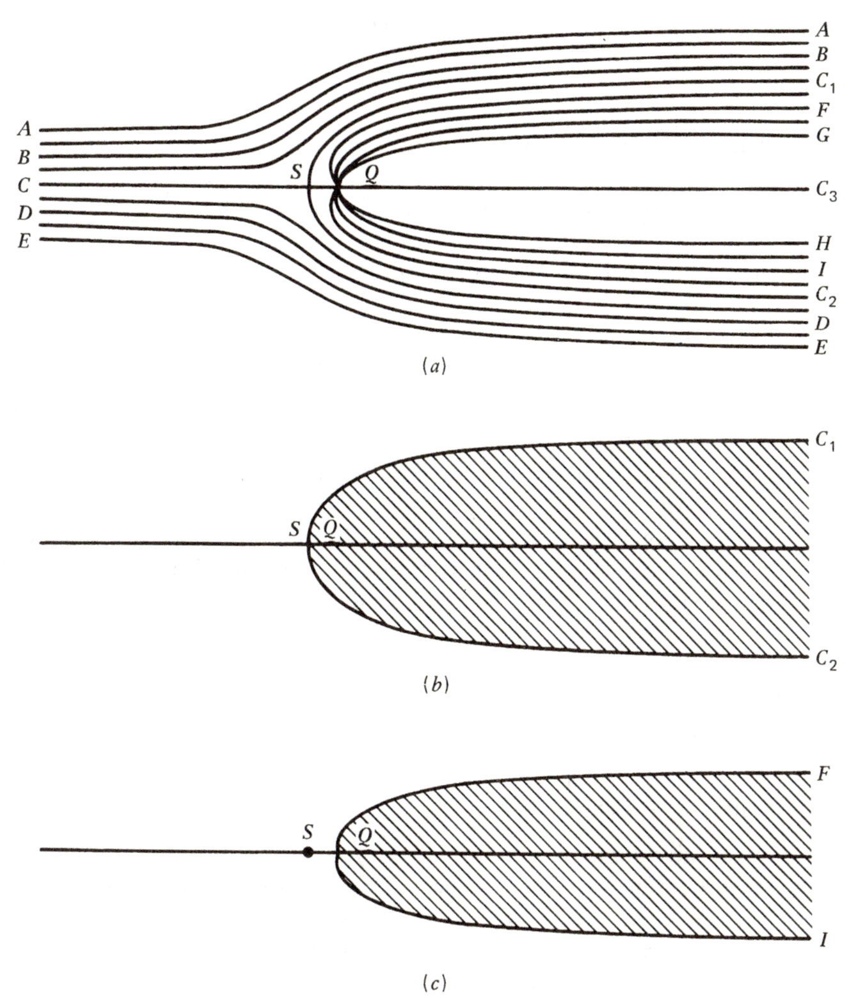

FIG. 3.5. Some streamlines of the flow past a source immersed in a uniform flow. (*a*) Global picture. (*b*) The stagnation streamline. (*c*) A streamline emanating from the source.

Requiring

$$u(-a, 0) = 0$$

gives

$$0 = V_\infty - \frac{Q}{2\pi}\frac{1}{a} \Rightarrow a = \frac{Q}{2\pi V_\infty} \tag{3-8}$$

The stream function is constant on any streamline. At the stagnation point, $\theta = +\pi$ if we approach S through positive values of θ and $-\pi$ if we go through the lower half-plane. Thus, at S,

$$\begin{aligned}\psi &= V_\infty \cdot 0 + \frac{Q}{2\pi} \cdot (\pm\pi) \\ &= \pm\tfrac{1}{2}Q\end{aligned}$$

and the shape of the stagnation streamline, $y = Y(x)$, can be found from

$$\psi[x, Y(x)] = \pm\tfrac{1}{2}Q = V_\infty Y(x) + \frac{Q}{2\pi}\tan^{-1}\left[\frac{Y(x)}{x}\right]$$

which can be rearranged for convenience into

$$\begin{aligned}x &= Y/\tan\left(\pm\pi - \frac{V_\infty Y 2\pi}{Q}\right) \\ &= Y/\tan(\pm\pi - Y/a) \qquad (3\text{-}9)\end{aligned}$$

in which we have used equation 3-8. The upper sign in these equations is associated with $Y > 0$, and so with the upper surface of the body, whereas the lower sign goes with the lower surface.

In problem 2, you'll be asked to show that the asymptotic width of the body is $2\pi a$, and then to plot the pressure distribution on the body as a function of x. The following procedure can be used for the second part of the problem.

1. Pick $Y < \pi a$.
2. Calculate x from equation 3-9.
3. Then calculate u and v from equations 3-7.
4. Finally, calculate the pressure distribution from Bernoulli's equation 3-2.

3.3. DIMENSIONLESS VARIABLES

Unless you have some more information, you will not be able to execute this procedure, straightforward though it seems. How can you pick $Y < \pi a$ unless a is specified? Similarly, what is V_∞, which must be known before u and v are calculated?

I could, of course, specify a and V_∞ numerically, but it is more useful to introduce *dimensionless variables* that will allow you to scale the results with any choice of input data. Specifically, since a and V_∞ are natural choices for length and velocity scales, let

$$x^* \equiv x/a, \qquad y^* \equiv y/a, \qquad Y^* \equiv Y/a$$

$$u^* \equiv u/V_\infty, \qquad v^* \equiv v/V_\infty$$

Then equation 3-9 becomes

$$x^* = Y^*/\tan(\pi - Y^*) \qquad (3\text{-}9')$$

whereas, with the aid of equation 3-8, equations 3-7 may be written in dimensionless form as

$$u^* = 1 + \frac{x^*}{x^{*2} + y^{*2}}$$

$$v^* = \frac{y^*}{x^{*2} + y^{*2}} \qquad (3\text{-}7')$$

We also need a dimensionless version of Bernoulli's equation 3-2. Evaluating the constant far from the source, where $u^* = 1$, $v^* = 0$, and $p = p_\infty$, say, we may write the equation in terms of the dimensionless velocity components as

$$p + \tfrac{1}{2}\rho V_\infty^2(u^{*2} + v^{*2}) = p_\infty + \tfrac{1}{2}\rho V_\infty^2$$

which suggests the definition of the *pressure coefficient* C_p:

$$C_p \equiv \frac{p - p_\infty}{\frac{1}{2}\rho V_\infty^2} \tag{3-10}$$

According to the Bernoulli equation, the pressure coefficient can be calculated from

$$C_p = 1 - u^{*2} - v^{*2} = 1 - \left(\frac{u}{V_\infty}\right)^2 - \left(\frac{v}{V_\infty}\right)^2 \tag{3-2'}$$

The procedure described in the preceding section may now be implemented, substituting equations 3-2′, 3-7′, and 3-9′ for their unprimed versions, without having to specify any extraneous data. This reflects the fact, to be elaborated in Section 4.8, that the inviscid velocity fields of flows past bodies of the same shape but different size L, say, can be scaled:

$$\frac{\mathbf{V}}{V_\infty} = f\left(\frac{x}{L}, \frac{y}{L}\right)$$

3.4. RANKINE OVALS

A similar approach may be employed in more complicated inverse problems. For a second example, put a source of strength Q at $(-x_0, 0)$, one of opposite strength $(-Q)$ at $(+x_0, 0)$, and superpose a uniform flow on them, as shown in Fig. 3.6. Then

$$\psi = +V_\infty y + \frac{Q}{2\pi}\theta_2 - \frac{Q}{2\pi}\theta_1 \tag{3-11}$$

where

$$\theta_2 = \tan^{-1}\frac{y}{x + x_0}$$

$$\theta_1 = \tan^{-1}\frac{y}{x - x_0} \tag{3-12}$$

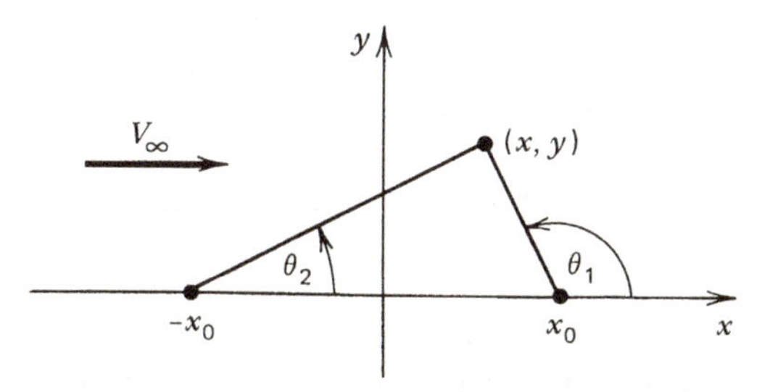

FIG. 3.6. Nomenclature for the source/sink/uniform flow combination.

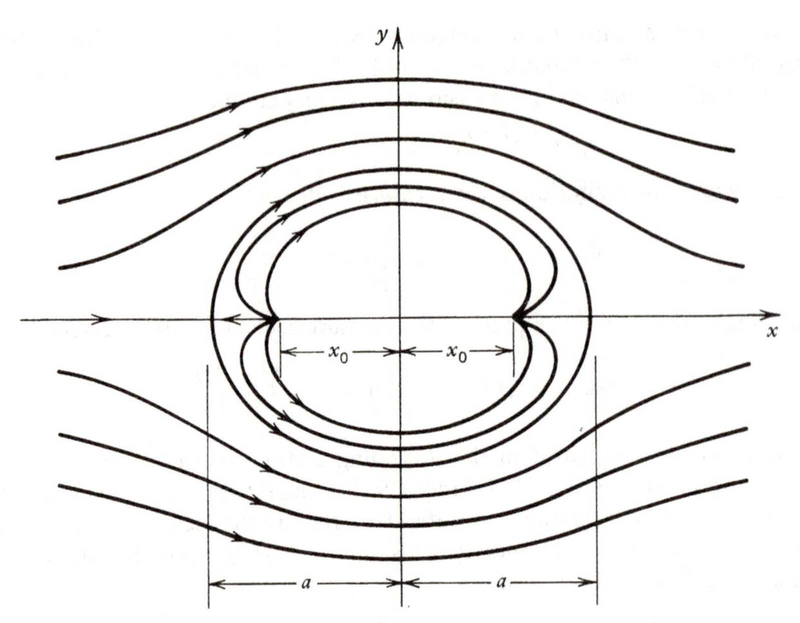

FIG. 3.7. Some streamlines of the flow within and outside the Rankine oval.

The velocity at any point now has three contributions: one from the uniform flow, one from the source, and one from the "sink" (source of negative strength).

Again we start by examining the flow along its axis of symmetry. When $y = 0$, and $|x|$ is a little bigger than x_0, the contribution of the nearest singularity dominates, and the net velocity is upstream (to the left). As $|x|$ increases, the contributions of both singularities diminish, so that V becomes more positive, finally reaching 0. Thus we see there are now *two stagnation points* and that, by symmetry, they are both at the same distance from the origin (call it a).

As in the single-source case, we also see that there are two streamlines going through each stagnation point, one in the x direction and one in the y direction, as is sketched in Fig. 3.7. The stagnation streamline, which comes in along the negative x axis, splits in two at the left stagnation point, rejoins its two halves at the right stagnation point, and then goes off along the positive x axis, is again of great interest. The body that conforms to the stagnation streamline is called a *Rankine oval*, after William John Macquorn Rankine.

Note that this time the stagnation streamline is closed, that is, of finite length. What makes this case different from the previous one in this respect is that the net source strength is zero; whatever fluid is emitted by the source is sucked up by the sink.[1]

A study of the flow about the Rankine oval proceeds exactly as in the single-source-plus-uniform-flow example.

[1] However, not every combination of sources and sinks with zero total strength yields a closed stagnation streamline. Even putting the flow past a source–sink pair at an angle of attack is enough to put the two stagnation points on different streamlines.

1. Locate the stagnation points. The x component of velocity corresponding to equation 3-11 is found from equation 3-1c to be

$$u = V_\infty + \frac{Q}{2\pi}\left[\frac{x + x_0}{(x + x_0)^2 + y^2} - \frac{x - x_0}{(x - x_0)^2 + y^2}\right] \tag{3-13}$$

Letting the coordinates of the stagnation points be $(\pm a, 0)$, we thus require

$$0 = u(\pm a, 0) = V_\infty + \frac{Q}{2\pi}\left(\frac{1}{x_0 \pm a} - \frac{1}{\pm a - x_0}\right)$$

$$= V_\infty - \frac{Q}{\pi} x_0/(a^2 - x_0^2)$$

which yields

$$a^2 = x_0^2 + \frac{x_0 Q}{\pi V_\infty} \tag{3-14}$$

2. Find the value of ψ on the stagnation streamline. From equation 3-11,

$$\psi(-a, 0) = V_\infty \cdot 0 + \frac{Q}{2\pi} \cdot \pi - \frac{Q}{2\pi} \cdot \pi = 0$$

3. Find the body shape $Y(x)$, as follows. A dimensionless measure of the source strength is

$$Q/2\pi V_\infty a \equiv q$$

For a fixed value of this parameter,

(a) Choose a value of Y/a. Since it is not so easy now to calculate Y_{max}/a, it is necessary to check the value of ψ at the point finally located, if any, to see if it really lies on the stagnation streamline.

(b) Since $\psi = 0$ on the body surface, equation 3-11 shows

$$0 = \frac{Y}{a} + q(\theta_2 - \theta_1) \tag{3-15}$$

so

$$\theta_2 - \theta_1 = -\frac{Y}{qa}$$

Take the tangent of both sides:

$$-\tan\left(\frac{Y}{qa}\right) = \tan(\theta_2 - \theta_1)$$

$$= \frac{\tan\theta_2 - \tan\theta_1}{1 + \tan\theta_2 \tan\theta_1} = \frac{Y/(x + x_0) - Y/(x - x_0)}{1 + y^2/(x^2 - x_0^2)}$$

$$= -\frac{2x_0 Y}{x^2 - x_0^2 + Y^2}$$

and so

$$x^2 = x_0^2 - Y^2 + \frac{2x_0 Y}{\tan(Y/aq)} \tag{3-16}$$

Provided the chosen value of Y/a does not make the right side of equation 3-16 negative, it can be used to determine x/a, once you solve equation 3-14 for x_0/a:

$$\frac{x_0}{a} = -q + \sqrt{q^2 + 1} \tag{3-17}$$

4. Calculate u/V_∞ and v/V_∞ at the point whose coordinates were just determined from

$$\frac{u}{V_\infty} = 1 + q\left[\frac{(x + x_0)a}{(x + x_0)^2 + Y^2} - \frac{(x - x_0)a}{(x - x_0)^2 + Y^2}\right]$$
$$\frac{v}{V_\infty} = q\frac{Y}{a}\left[\frac{a^2}{(x + x_0)^2 + Y^2} - \frac{a^2}{(x - x_0)^2 + Y^2}\right] \tag{3-18}$$

The pressure coefficient at this point on the body may then be calculated from Bernoulli's equation 3-2′.

Problems 2 and 4 will give you a chance to implement the procedures outlined above for determining the pressure distributions on bodies generated by positioning either a source or source–sink pair in a uniform flow. These procedures are easily programmed for automatic execution, even on a hand-held calculator. A FORTRAN program called RANKIN that can be used for the Rankine-oval study is listed at the end of the problem section. I urge you to do both problems. They should help you gain some appreciation of the connection between a body shape and its pressure distribution, and also of the detail required to describe them.

Certain limiting cases of the Rankine oval are of special interest. First, suppose the dimensionless source strength $q \ll 1$. Then the oval is quite thin. Let $Y(0) = \tau a$, so that τ is the ratio of the maximum thickness of the oval to its length $2a$. At $x = 0$, where $Y = \tau a$, we see from Fig. 3.8 that $\theta_1 \approx \pi$ and $\theta_2 \approx 0$. Thus, from equation 3-15

$$\tau \approx \pi q \qquad \text{for } q \ll 1$$

The thickness ratio is proportional to the source strength.

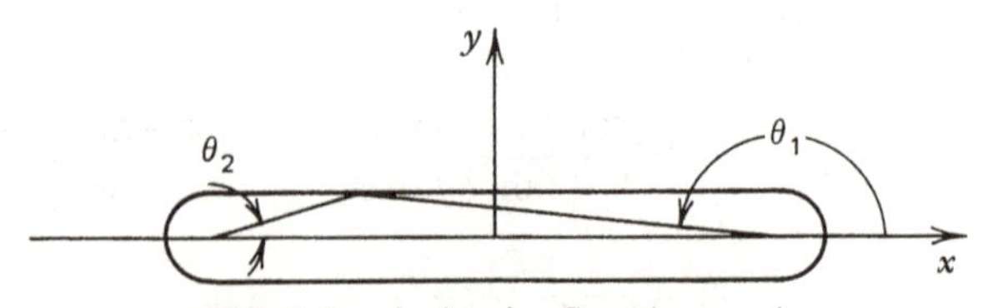

FIG. 3.8. A slender Rankine oval.

At the other extreme, let $q \to \infty$ with a fixed. From equation 3-17,

$$\frac{x_0}{a} \approx \frac{1}{2q} \qquad \text{for } q \gg 1$$

Thus the distance between the sources, $2x_0$, vanishes in this limit, with the product of source strengths and that distance approaching a constant:

$$\begin{aligned} 2x_0 Q &= 2x_0 \cdot 2\pi V_\infty aq \\ &\to 2\pi V_\infty a^2 \qquad \text{as } q \to \infty \end{aligned}$$

From problem 4 of Chapter 2, the source–sink pair is coalescing into a *doublet* of strength $2\pi V_\infty a^2$, so that the stream function given by equation 3-11 approaches

$$\psi = V_\infty y \left(1 - \frac{a^2}{x^2 + y^2}\right) \tag{3-19}$$

As you should know already,[2] but as is also easily seen by setting the ψ given by equation 3-19 to zero, the stagnation streamline in this case becomes a circle of radius a.

You can learn a lot of aerodynamics by studying the streamline patterns and velocity distributions associated with prescribed distributions of sources, vortices, and doublets in a uniform flow. To facilitate this process, a program called STRACE has been prepared that does just that on a microcomputer. Specific exercises in the use of STRACE are contained in problem 5.

3.5. LINE SOURCE DISTRIBUTIONS

We could go on with this inverse approach, constructing any number of exact solutions of the problem stated in equations 3-1 through 3-5. It is time to move on, however, to the more practically interesting *direct problem*, in which the body shape is specified rather than determined as part of the solution.

For this purpose, consider a string of sources along the x axis. If q_i is the strength of the source at x_i, and we put the string in a uniform flow in the $+x$ direction,

$$\psi = +V_\infty y + \sum_i \frac{q_i}{2\pi} \tan^{-1} \frac{y}{x - x_i}$$

More generally, let the source distribution be continuous, and

$$q(t)\, dt \equiv \text{net strength of sources between } x = t \text{ and } x = t + dt$$

Then

$$\psi(x, y) = V_\infty y + \psi_s(x, y) \tag{3-20a}$$

[2] In problem 5 of Chapter 2 you considered the flow due to a coincident doublet–vortex pair in a uniform flow, which also describes uniform flow past a cylinder, but one that carries circulation.

and

$$\phi(x, y) = +V_\infty x + \phi_s(x, y) \tag{3-20b}$$

where ψ_s and ϕ_s are the stream function and velocity potential, respectively, due to a continuous source distribution of strength q per unit length along the x axis. With the help of Fig. 3.9, they may be written

$$\psi_s = \frac{1}{2\pi} \int q(t) \tan^{-1}\left(\frac{y}{x - t}\right) dt \tag{3-21a}$$

$$\phi_s = \frac{1}{2\pi} \int q(t) \ln[(x - t)^2 + y^2]^{\frac{1}{2}}\, dt \tag{3-21b}$$

Because of the superposition principle, equations 3-20 and 3-21 satisfy the conditions of irrotationality and continuity throughout the flow field and the boundary condition at infinity, for any function $q(t)$. It remains to find the strength of the source distribution that allows satisfaction of the flow tangency condition (3-5).

Let the body shape be given as $y = Y(x)$, and let it be closed, that is, of finite length c. Clearly, it ought to be symmetric about the x axis if we are to represent it with sources along the x axis. Also, the sources ought to be confined within the body. Otherwise, the velocity $\mathbf{V}$ blows up in the flow field.

To get an equation that connects the source strength with the body shape, we set ψ equal to its value at the stagnation points. As in the inverse problem, we find this value simply by evaluating ψ at the stagnation points, which are clearly on the x axis ($y = 0$). You can see from Fig. 3.9 that, at the upstream stagnation point,

$$\tan^{-1} \frac{y}{x - t} = \pi$$

while at the one downstream of the source distribution,

$$\tan^{-1} \frac{y}{x - t} = 0$$

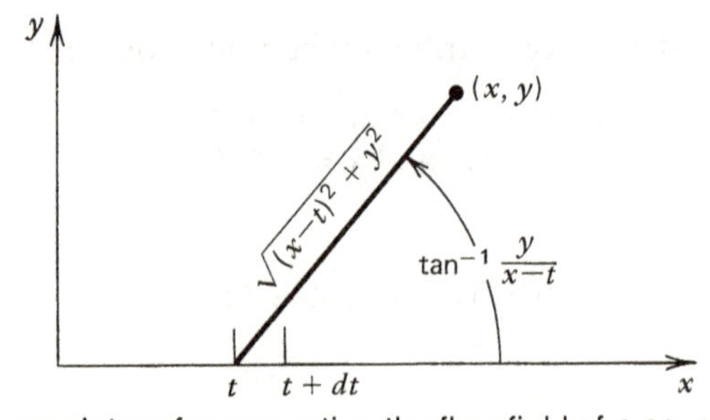

FIG. 3.9. Nomenclature for computing the flow field of a source distribution.

Thus,

$$\psi = 0 - \int_0^c \frac{q(t)\,dt}{2\pi} \cdot \pi \tag{3-22a}$$

at the upstream stagnation point, and

$$\psi = 0 - \int_0^c \frac{q(t)\,dt}{2\pi} \cdot 0 \tag{3-22b}$$

at the downstream stagnation point.

Equation 3-22b tells us that the stream function is zero at the downstream stagnation point, and so all along the body contour,[3]

$$\psi[x, Y(x)] = 0$$

Thus, on setting $y = Y(x)$ in equation 3-20a, we get what is called an *integral equation* for the source strength q:

$$0 = V_\infty Y(x) + \frac{1}{2\pi} \int_0^c q(t) \tan^{-1} \frac{Y(x)}{x - t}\,dt \qquad \text{for } 0 < x < c \tag{3-23}$$

Equation 3-22a is also informative; since $\psi = 0$ at the upstream stagnation point, it tells us that the net source strength is zero:

$$\int_0^c q(t)\,dt = 0 \tag{3-24}$$

Again, this is a consequence of the body's being closed and so having a downstream stagnation point.

Before we continue, be sure you understand the difference between x and t in equations 3-21 and 3-23. Both quantities represent distances along the x axis. However, x is the x coordinate of a point at which we want to evaluate ϕ, ψ, or $\mathbf{V}$, whereas t is the variable x coordinate of points located along the source distribution. Thus the integral in equation 3-23 must be evaluated at each x in the domain of the body-shape function $Y(x)$ and equated to $-V_\infty Y(x)$. In evaluating the integral, x is treated as a constant, whereas t is the variable of integration. The value of the integral is the area under a curve like those shown in Fig. 3.10 and is (generally) a function of x. We sometimes call the point (x, y) at which integrals like those in equations 3-21 are evaluated a *field point*, and $(t, 0)$, a *source point*.

Shortly, we will meet a doublet distribution function, and in the next chapter we will start working with vortex distribution functions. It may help to consider yet another kind of distribution function, one that is useful in statistics and probability. Suppose we wanted to know how many people in a certain group (say, in your

[3] CAUTION: We are getting $\psi = 0$ on the body only because it is closed and because the flow is symmetric. Generally, ψ is some nonzero constant on the stagnation streamline, which must be determined by evaluating ψ at the stagnation point(s) as we did here.

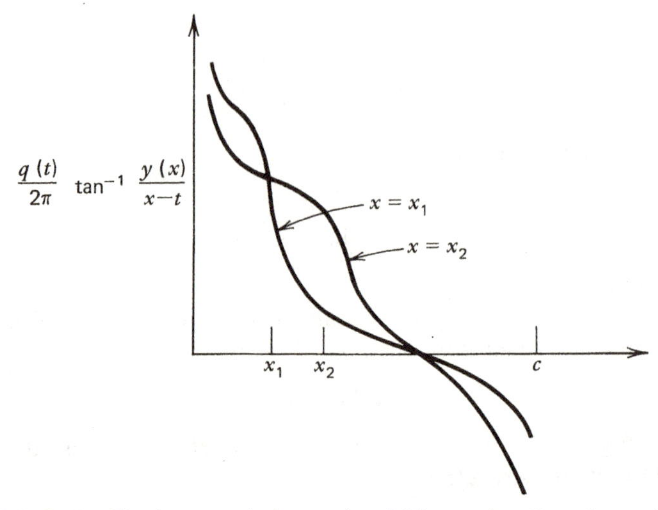

FIG. 3.10. The integrand of equation 3-23, as a function of x and t.

classroom) were a certain height. If you could measure heights to sufficiently high accuracy, then, no matter what height you picked (5 ft 11.5 in., for example), there would almost certainly be no one with exactly that height. Maybe someone would be 5 ft 11.502 in., but no one would have exactly the height you picked. Therefore, in analyzing the distribution of heights, it makes sense to ask questions like, "How many have heights between 5 ft 11 in. and 6 ft?" Clearly, the number with heights between h and $h + \Delta h$ is proportional to Δh. Thus, we define a "height distribution function" $f(h)$ by

$$f(h)\,dh = \text{the limit as } dh \to 0 \text{ of the number of persons with heights between } h \text{ and } h + dh$$

The analogy of $f(h)$ with the strength per unit length $q(x)$ of the source distribution is exact, as you can see by referring back to the beginning of this section for the definition of q.

3.6. FLOW PAST THIN SYMMETRIC AIRFOILS

Equation 3-23 is, as noted, an integral equation for the source strength $q(x)$. Given the body-shape function $Y(x)$, we must determine the source distribution function $q(t)$, whose domain is the interval $0 < t < c$, so that equation 3-23 is satisfied for every x between 0 and c.

To determine the source strength associated with an arbitrary body shape is hard (and sometimes impossible; not every symmetric flow can be represented by a line source distribution). Later on we will see how it may be done numerically. For now, we seek an approximate analytic solution appropriate for the practically interesting case of a thin airfoil:

$$Y(x) \ll c$$

In this case, it is convenient to use, instead of equation 3-23, the velocity-based form of the flow tangency condition, equation 3-5a. As was the case with the stream function and velocity potential in equation 3-20, we let u_s and v_s be the velocity components due to the source distribution:

$$u = V_\infty + u_s \tag{3-25}$$

$$u_s = \frac{1}{2\pi}\int_0^c q(t)\frac{x-t}{(x-t)^2+y^2}\,dt \tag{3-26}$$

$$v = v_s = \frac{1}{2\pi}\int_0^c q(t)\frac{y}{(x-t)^2+y^2}\,dt \tag{3-27}$$

Then equation 3-5a is written

$$\frac{v_s}{V_\infty + u_s} = \frac{dY}{dx} \qquad \text{at } y = Y(x)$$

In the thin-airfoil approximation, we assume the source-induced velocity to be small compared to the velocity of the onset flow:

$$u_s,\, v_s \ll V_\infty \tag{3-28}$$

Then the boundary condition above is approximated by

$$v_s \approx V_\infty \frac{dY}{dx} \tag{3-29}$$

Further, since $Y(x) \ll c$, equation 3-29 is applied at $y = 0$ rather than on the actual body contour. Then we need to evaluate v_s, as given by equation 3-27, on the x axis.

As $y \to 0$, the integrand of equation 3-27 vanishes, except near $t = x$, where, as shown in Fig. 3.11, it peaks sharply. Therefore, we can alter the integrand away from $t = x$ without affecting the result as $y \to 0$. In particular, we can approximate $q(t)$ by its value at x, namely $q(x)$ and thus take the source strength out of the integral (remember that x—and hence $q(x)$—is a constant in evaluating the integral). Also, in the limit $y \to 0$, the value of the integral should not be affected if we let the limits

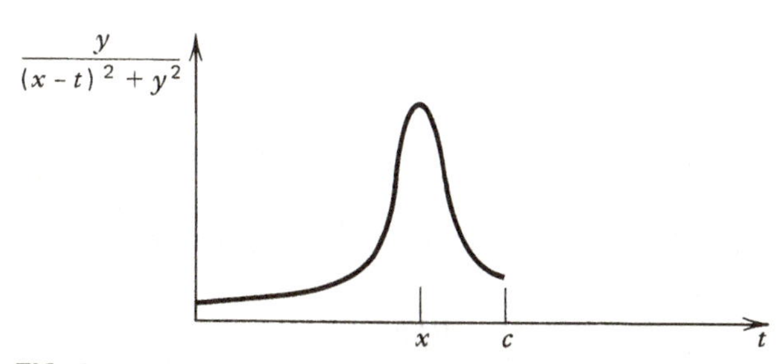

FIG. 3.11. The integrand of equation 3-27 near the x axis.

extend to $-\infty$ and $+\infty$, and so we write

$$\begin{aligned}\lim_{y\to 0} v_s(x, y) &= \frac{1}{2\pi} q(x) \int_{-\infty}^{\infty} \frac{y}{(x-t)^2 + y^2}\, dt \\ &= \frac{1}{2\pi} q(x) \tan^{-1} \frac{t-x}{y} \bigg|_{t=-\infty}^{t=\infty} \\ &= \frac{1}{2\pi} q(x) \left[\frac{\pi}{2} - \left(-\frac{\pi}{2}\right)\right] \\ &= \tfrac{1}{2} q(x) \end{aligned} \tag{3-30}$$

Thus, from equations 3-29 and 3-30, we obtain an approximation for the source strength associated with a thin airfoil directly in terms of its slope:

$$q(x) \approx 2V_\infty \frac{dY}{dx} \tag{3-31}$$

Note that the source strength is then positive in the front of the body and negative near the trailing edge, as should be expected. The sources of positive strength push the streamlines of the onset flow apart, and those of negative strength (sinks) pull them back in.

Once the source strength is known, we can calculate the velocity field from equations 3-25 to 3-27 and then the pressure field from the Bernoulli equation 3-2 or 3-2′. We are particularly interested in the velocity and pressure distribution on the body surface. We already know the y component of velocity on the body surface, from the simplified boundary condition (3-29). As for the x component, note that, if $y \ll c$, the integrand of equation 3-26 may be approximated by $q(t)/(x-t)$, so long as t is not too close to x. Thus we break up the range of integration into three pieces: $0 < t < x - \varepsilon$, $x - \varepsilon < t < x + \varepsilon$, and $x + \varepsilon < t < c$, where ε is supposed to be small. In the middle piece, we can let $q(t) \to q(x)$. Then

$$\begin{aligned} 2\pi u_s(x, Y(x)) \approx{}& \int_0^{x-\varepsilon} q(t) \frac{1}{x-t}\, dt + q(x) \int_{x-\varepsilon}^{x+\varepsilon} \frac{x-t}{(x-t)^2 + Y^2(x)}\, dt \\ &+ \int_{x+\varepsilon}^{c} q(t) \frac{1}{x-t}\, dt \end{aligned}$$

Let $z = x - t$ in the middle integral, which then becomes

$$q(x) \int_{-\varepsilon}^{\varepsilon} \frac{z\, dz}{z^2 + Y^2(x)}$$

Since the integrand is odd and the limits are even, this integral vanishes. Thus

$$\begin{aligned} 2\pi u_s[x, Y(x)] &\approx \lim_{\varepsilon\to 0} \left[\int_0^{x-\varepsilon} \frac{q(t)\, dt}{x-t} + \int_{x+\varepsilon}^{c} \frac{q(t)\, dt}{x-t}\right] \\ &\equiv \mathrm{P}\!\!\int_0^c \frac{q(t)\, dt}{x-t} \end{aligned} \tag{3-32}$$

or, after introducing the thin-airfoil result (3-31) for the source strength,

$$u_s[x, Y(x)] \approx \frac{V_\infty}{\pi} P\int_0^c \frac{Y'(t)\,dt}{x - t} \tag{3-33}$$

Here the P indicates that the *Cauchy principal value* of the integral is to be taken, which is what the limiting process on the first line of equation 3-32 is called. As you can see from Appendix B, this limiting process is largely transparent in the evaluation of the only integrals of this form you could reasonably be expected to work out with elementary methods.

Now the pressure formula (3-2′) can be written

$$\begin{aligned} C_p &= 1 - \left(1 + \frac{u_s}{V_\infty}\right)^2 - \left(\frac{v_s}{V_\infty}\right)^2 \\ &= -2\frac{u_s}{V_\infty} - \left(\frac{u_s}{V_\infty}\right)^2 - \left(\frac{v_s}{V_\infty}\right)^2 \end{aligned} \tag{3-34}$$

But, according to equation 3-28, both u_s and v_s are small compared to V_∞. In particular, equations 3-30 to 3-32 show that, on the body surface, u_s and v_s are roughly proportional to the thickness of the body. Thus, if the thickness of the body is halved, the first term on the second line of equation 3-34 is also halved, but the other two are quartered. For sufficiently thin airfoils, therefore, the first term is bound to be much larger than the last two, and so

$$C_p \approx -2\frac{u_s}{V_\infty} \tag{3-35}$$

Example 1: Flow Past an Ellipse For an ellipse of chord c and maximum thickness τc,[4] see Fig. 3.12,

$$\left(x - \frac{c}{2}\right)^2 + \frac{Y^2(x)}{\tau^2} = \frac{c^2}{4} \Rightarrow Y(x) = \tau\sqrt{x(c - x)} \tag{3-36}$$

and

$$Y'(x) = \frac{\tau}{2}\frac{c - 2x}{\sqrt{x(c - x)}} \tag{3-37}$$

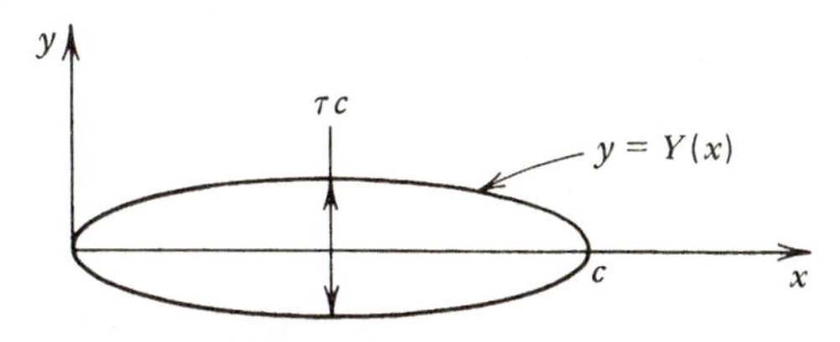

FIG. 3.12. An ellipse.

[4] τ is thus the *thickness ratio*, the ratio of the maximum thickness to the chord.

so, from equation 3-31, the source strength associated by thin airfoil theory with an ellipse is

$$q(x) = \tau V_\infty \frac{c - 2x}{\sqrt{x(c - x)}} \tag{3-38}$$

All we are really interested in is the evaluation of u_s from equation 3-33. Because of the square-root terms in Y', it is convenient to introduce a trigonometric substitution:

$$x = \frac{c}{2}(1 - \cos\theta_0)$$

$$t = \frac{c}{2}(1 - \cos\theta) \tag{3-39}$$

Then

$$t = 0 \Rightarrow \theta = 0$$

$$t = c \Rightarrow \theta = \pi$$

$$dt = \frac{c}{2}\sin\theta\, d\theta$$

From equations 3-37 and 3-39,

$$Y'(t) = \frac{\tau}{2}\frac{c - c(1 - \cos\theta)}{\sqrt{\frac{c}{2}(1 - \cos\theta)\left[c - \frac{c}{2}(1 - \cos\theta)\right]}} = \tau\frac{\cos\theta}{\sin\theta}$$

and so equation 3-33 becomes

$$u_s[x, Y(x)] \simeq \frac{\tau V_\infty}{\pi} \mathrm{P}\!\int_0^\pi \frac{\cos\theta}{\cos\theta - \cos\theta_0}\, d\theta$$

In Appendix A, it is shown that

$$\mathrm{P}\!\int_0^\pi \frac{\cos n\theta}{\cos\theta - \cos\theta_0}\, d\theta = \pi\frac{\sin n\theta_0}{\sin\theta_0} \tag{3-40}$$

Here we have the case $n = 1$, for which the result is simply π. Thus, in the thin-airfoil approximation, the x component of the perturbation velocity on the surface of an ellipse is

$$u_s[x, Y(x)] \approx \tau V_\infty \tag{3-41}$$

a constant! From equation 3-35, the pressure distribution on an ellipse is also constant in this approximation. As will be seen later from a numerical solution of this problem, these results of thin-airfoil theory are not bad, except near the ends of the body. There the flow stagnates, and $C_p = 1.0$.

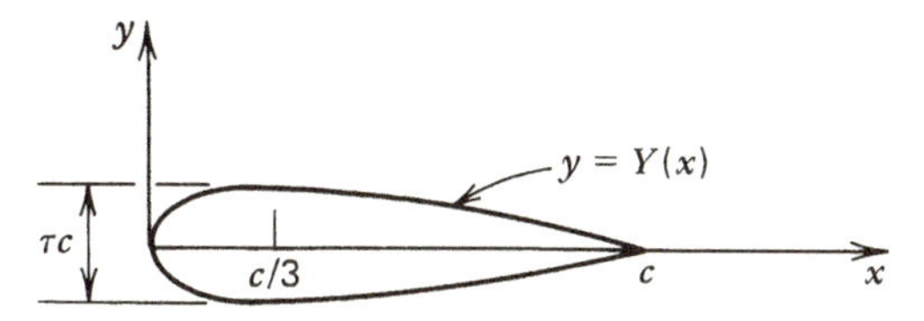

FIG. 3.13. An airfoil-like shape (see equation 3-42).

Example 2 A contour that at least looks like an airfoil (see Fig. 3.13), though it would not be a very good one, can be created by setting

$$Y(x) = A\sqrt{\frac{x}{c}}(c - x) \tag{3-42}$$

Like an ellipse, this shape has a rounded nose, but its tail is sharp. Since

$$Y'(x) = \frac{A}{2}\left(\sqrt{\frac{c}{x}} - 3\sqrt{\frac{x}{c}}\right)$$

the maximum thickness of the airfoil is at $x/c = \frac{1}{3}$. If τ is the thickness ratio, we then have $Y = \tau c/2$ at $x = \frac{1}{3}$, so

$$A = \sqrt{\frac{27}{16}}\,\tau$$

Thus, according to thin-airfoil theory, the source strength associated with uniform flow past this body is, from equation 3-31,

$$q(x) = \sqrt{\frac{27}{16}}\,V_\infty \tau\left(\sqrt{\frac{c}{x}} - 3\sqrt{\frac{x}{c}}\right) \tag{3-43}$$

and, on the body surface, the perturbation in the x component of velocity is, from equation 3-33,

$$u_s = \sqrt{\frac{27}{16}}\,\frac{V_\infty \tau}{2\pi} \oint_0^c \frac{\sqrt{c/t} - 3\sqrt{t/c}}{x - t}\,dt$$

The integral can be evaluated using the list compiled in Appendix B, so

$$\frac{u_s}{V_\infty} = \frac{\tau}{2\pi}\sqrt{\frac{27}{16}}\left[\left(\sqrt{\frac{c}{x}} - 3\sqrt{\frac{x}{c}}\right)\ln\frac{\sqrt{x} + \sqrt{c}}{\sqrt{c} - \sqrt{x}} + 6\right] \tag{3-44}$$

Plugging this into equation 3-35 yields the thin-airfoil approximation for the pressure distribution on the body surface, as sketched in Fig. 3.14.[5]

[5] Note that aerodynamicists plot negative pressure coefficients above the axis. From equation 3-2′, you can, if you like, think of such graphs as plots of $(V/V_\infty)^2 - 1$.

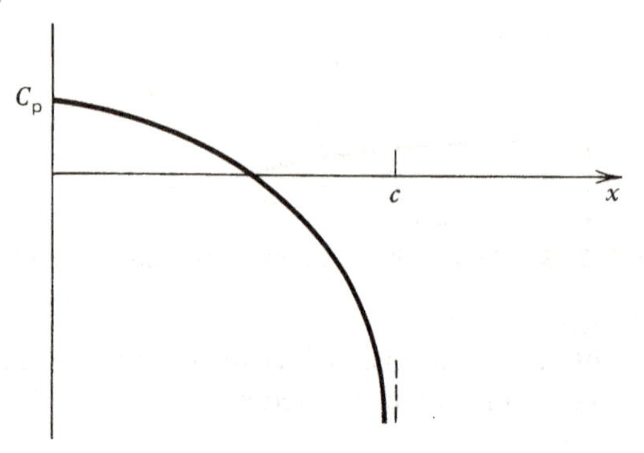

FIG. 3.14. Pressure distribution on the body pictured in Fig. 3.13 according to thin-airfoil theory.

As in the case of the ellipse, the results of thin-airfoil theory shown in Fig. 3.14 are considerably in error at the airfoil's stagnation points.[6] In particular, note from equations 3-35 and 3-44 that

$$C_p = -\frac{4\tau}{\pi}\sqrt{\frac{27}{16}}$$

at the nose of the airfoil ($x = 0$), whereas

$$C_p \to \infty$$

at the trailing edge ($x = c$).

3.7. ERRORS NEAR THE STAGNATION POINTS

In both examples presented above, thin-airfoil theory failed near the stagnation points. This is always the case. The approximation of the pressure formula (3-34) by equation 3-35 is clearly invalid at the stagnation points. So is the approximation of the flow tangency condition (3-5a) by equation 3-29, which was the basis for the thin-airfoil formula for the source strength (3-31). Less obvious are errors in equations 3-30 and 3-32, the thin-airfoil results for the velocity on the airfoil surface in terms of $q(x)$. These errors are due to the implicit assumption that the source distribution extends to the ends of the body. Recognizing that it does not is important to the success of the numerical method to be described in the next section. Thus we shall now estimate the proper distance between the ends of the source distribution and the stagnation points.

[6] That the leading edge is a stagnation point should be clear; $v = 0$ on $y = 0$ by symmetry, whereas $u = 0$ at the origin because of the flow-tangency condition. As for the trailing edge, note that as you approach $(c, 0)$ along the upper and lower body surfaces, the flow must be tangent to those surfaces. As shown in Fig. 3.15, if the velocity were finite at the trailing edge, its direction there could not be tangent to both surfaces. Therefore, it must be zero. See also Appendix C, in which it is shown that at any concave corner in an incompressible irrotational flow, the velocity must be zero.

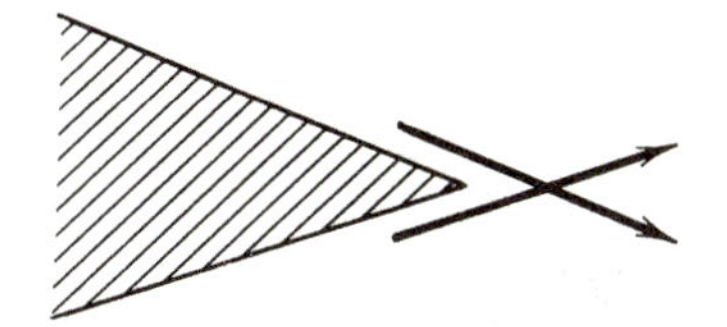

FIG. 3.15. On the need for a stagnation point at a sharp trailing edge.

To do so, we shall use a mixture of thin-airfoil theory and the inverse approach to source distributions. Using thin-airfoil theory to determine the source distribution associated with a given body, we shall locate the stagnation points of the flow that results when that source distribution is immersed in a uniform onset flow. Since the stagnation points of interest are on the x axis,[7] let the forward one be located at $(-\varepsilon, 0)$, where $\varepsilon > 0$ is to be determined. From the exact formula (3-26) for u_s,

$$\begin{aligned} u_s(-\varepsilon, 0) &= -\frac{1}{2\pi}\int_0^c \frac{q(t)\,dt}{\varepsilon + t} \\ &= -V_\infty \end{aligned} \tag{3-45}$$

since the source-induced velocity must cancel that of the onset flow at the stagnation point.

If, as seems reasonable, ε is small, the denominator of the integrand in the formula for $u_s(-\varepsilon, 0)$ emphasizes the values of the source strength $q(t)$ for t near zero. Thus, in equation 3-31, the thin-airfoil formula for the source strength, we will approximate the body-shape function Y for small values of its argument. The shape of a blunt-nosed airfoil near its nose $x = 0$ is approximately parabolic:

$$Y^2(x) \approx Cx \text{ for small } x$$

See, for example, equations 3-36 and 3-42. In particular, if the radius of curvature of the nose is ρ,

$$(x - \rho)^2 + Y^2 \approx \rho^2$$

as can be seen from Fig. 3.16, and so

$$Y \approx \sqrt{2\rho x} \tag{3-46}$$

Then

$$Y'(x) \approx \sqrt{\frac{\rho}{2x}}$$

and so, from equation 3-31, the source strength associated by thin-airfoil theory with an airfoil whose nose has radius of curvature ρ is

$$q(t) \approx V_\infty \sqrt{\frac{2\rho}{t}}$$

near the leading edge of the source distribution.

[7] You can create stagnation points off the x axis by carefully juggling the source strength, but the stagnation points of interest are at the ends of the body which, of course, are on the x axis.

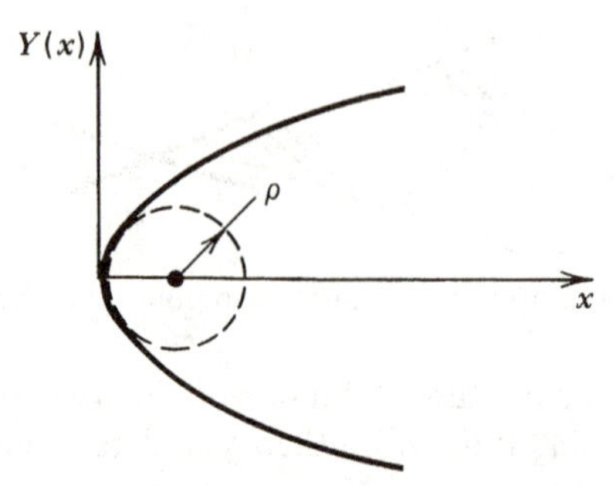

FIG. 3.16. The radius of curvature of a leading edge.

Substituting this approximation into equation 3-45, we get

$$V_\infty \approx \frac{1}{2\pi} V_\infty \sqrt{2\rho} \int_0^c \frac{dt}{\sqrt{t(\varepsilon + t)}}$$

To help evaluate the integral, let $z = (t/c)^{1/2}$. Then

$$V_\infty \approx \frac{1}{2\pi} V_\infty \sqrt{2\rho} \int_0^1 \frac{2\sqrt{cz}\, dz}{z(\varepsilon + cz^2)}$$

$$= \frac{1}{\pi} V_\infty \sqrt{2\rho} \frac{\sqrt{c}}{\sqrt{c\varepsilon}} \tan^{-1} \sqrt{\frac{c}{\varepsilon}} z \Big|_0^1$$

If $\varepsilon \ll c$,

$$\tan^{-1} \sqrt{c/\varepsilon} \approx \pi/2$$

and so

$$V_\infty \approx \frac{1}{2} V_\infty \sqrt{\frac{2\rho}{\varepsilon}}$$

from which we get

$$\varepsilon \approx \frac{\rho}{2} \tag{3-47}$$

That is, the gap between the forward stagnation point and the leading edge of the source distribution is approximately half the radius of curvature of the nose of the body. A similar result can be obtained, of course, at the other end of the body.

Example 3 Estimate the gap between the ends of an ellipse of thickness ratio 0.1 and the ends of the associated source distribution.

Solution From equation 3-36,

$$Y(x) \approx \tau\sqrt{xc}$$

near the leading edge of an ellipse. Comparing this with equation 3-46, we see that the radius of curvature of the leading edge (and, by symmetry, of the trailing edge, too) is

$$\rho = \tfrac{1}{2}\tau^2 c$$

For $\tau = 0.1$, then, equation 3-47 yields

$$\varepsilon = 0.0025c$$

Note how extremely small this is!

The preceding analysis can be performed with fewer approximations when the body under study is an ellipse, in which case thin-airfoil theory predicts the source strength to be given by equation 3-38. In fact it can be shown that this source distribution does, when immersed in a uniform onset flow, generate a stagnation streamline whose shape is exactly elliptical.[8] However, the thickness ratio of the ellipse differs from the τ of equation 3-38, and as we have shown, the source distribution does not extend to the ends of the ellipse.

3.8. NUMERICAL SOLUTION BASED ON LINE DOUBLET DISTRIBUTIONS

As an alternative to the thin-airfoil approximations we have been using, we can solve the integral equation 3-23 by a numerical method. From the discussion of the last section, the source distribution should not extend to the ends of the body but must be displaced from those points by approximately half their radii of curvature. Thus, we rewrite equation 3-23 as

$$0 = V_\infty Y(x) + \frac{1}{2\pi}\int_{x_s}^{x_f} q(t)\tan^{-1}\left[\frac{Y(x)}{x-t}\right] dt \qquad \text{for } 0 < x < c \qquad (3\text{-}23')$$

with

$$x_s \approx \tfrac{1}{2}\rho_0, \qquad x_f \approx c - \tfrac{1}{2}\rho_c \qquad (3\text{-}48)$$

A standard numerical procedure for solving integral equations like (3-23′) is to

1. Assume a functional form for the unknown function [here, the source strength $q(t)$] that involves N, say, parameters. To simplify the determination of those parameters, it is best to assume a functional form that is linear in the parameters. For example, you could use a polynomial of degree $N - 1$ (though that would turn out to be a rather poor choice), in which case the parameters would be the N coefficients of the polynomial.
2. Evaluate the integral in terms of x and the unknown parameters.
3. Stipulate that the integral equation be satisfied at N different values of x. This leads to a system of N linear algebraic equations for the N parameters. Solve them, and substitute the results back into the functional form assumed for q.

[8] To prove this directly is difficult. It is easier to obtain an exact solution for flow past an ellipse by complex-variable methods (see equation 3-82 below) and then to use equation 3-30 to determine the associated source strength.

In principle, if the functional form chosen in step 1 is sufficiently general, the solution may be made as accurate as desired, just by picking N large enough. One of the simplest and most reliable functional forms is the piecewise-constant function. That is, we divide the interval $x_s < x < x_f$ into N subintervals (not necessarily of equal length) and assume the source strength to be constant in each subinterval:

$$q(t) = q_i \qquad \text{for } t_i < t < t_{i+1}, \qquad i = 1, \ldots, N$$

Here $t_1 = x_s$, $t_{N+1} = x_f$, and the other t_i's are also specified, but the q_i's are not; they are the N unknown parameters referred to in the procedure outlined above. The integral equation 3-23′ is then approximated by

$$\sum_{j=1}^{N} A_j(x) q_j = b(x) \qquad \text{for } 0 < x < c \tag{3-49}$$

where

$$A_j(x) = \frac{1}{2\pi} \int_{t_j}^{t_{j+1}} \tan^{-1}\left[\frac{Y(x)}{x - t}\right] dt$$

$$b(x) = -V_\infty Y(x)$$

This can be turned into a system of N linear algebraic equations for the N unknowns q_j by evaluating it at any N values of x.

However, as we saw in equation 3-31, thin-airfoil theory predicts that the source strength is proportional to the slope of the airfoil, which blows up at the usual rounded leading edge. This makes it desirable to transform the integral equation through an integration by parts. Let

$$m(x) \equiv \int_{x_s}^{x} q(t)\, dt \qquad \text{for } x_s < x < x_f \tag{3-50}$$

Then the stream function of the flow (see equations 3-20a and 3-21a) can be written

$$\psi(x, y) = V_\infty y + \frac{1}{2\pi} m(t) \tan^{-1} \frac{y}{x - t}\Bigg|_{t=x_s}^{t=x_f} - \frac{1}{2\pi} \int_{x_s}^{x_f} m(t) \frac{y}{y^2 + (x - t)^2}\, dt$$

Since, for a closed body, the net source strength is zero, equation 3-50 shows that

$$m(x_f) = 0$$

Equation 3-50 also implies $m = 0$ at x_s. Thus the part integrated out vanishes, and

$$\psi = V_\infty y - \frac{1}{2\pi} \int_{x_s}^{x_f} \frac{m(t) y}{y^2 + (x - t)^2}\, dt \tag{3-51}$$

The integral in equation 3-51 is the stream function of a *line doublet distribution*, as you may recognize from your solution to problem 4 of Chapter 2. As was the case in the derivation of equation 3-23, an integral equation for the *doublet strength* $m(t)$ is found by setting $\psi = 0$ on the body surface $y = Y(x)$. After dividing equation 3-51 through by $V_\infty Y(x)$, we obtain

$$1 = \frac{1}{2\pi V_\infty} \int_{x_s}^{x_f} \frac{m(t)}{Y^2(x) + (x - t)^2}\, dt \qquad \text{for } 0 < x < c \tag{3-52}$$

By assuming a piecewise-constant doublet strength

$$m(t) = 2\pi V_\infty m_j \qquad \text{for } t_j < t < t_{j+1}, \qquad j = 1, \ldots, N \tag{3-53}$$

we reduce equation 3-52 to

$$1 = \sum_{j=1}^{N} A_j(x)_{m_j} \qquad \text{for } 0 < x < c \tag{3-54}$$

where

$$\begin{aligned} A_j(x) &= \int_{t_j}^{t_{j+1}} \frac{dt}{Y^2(x) + (x - t)^2} \\ &= \frac{1}{Y(x)} \tan^{-1} \left[\frac{t - x}{Y(x)} \right] \Bigg|_{t=t_j}^{t=t_{j+1}} \end{aligned} \tag{3-55}$$

An algebraic system for the N m_j's is now set up in the form

$$\sum_{j=1}^{N} A_{ij} m_j = 1 \qquad \text{for } i = 1, \ldots, N \tag{3-56}$$

by evaluating the approximating equation 3-54 at N different values of x:

$$A_{ij} = A_j(x_i) \qquad \text{for } i = 1, \ldots, N \tag{3-57}$$

It is at this point that the method becomes "numerical." If you have ever tried solving more than three simultaneous algebraic equations analytically (whether by determinants or by "Gaussian elimination," that is, by subtracting multiples of one equation from the others), you will appreciate the ability of a computer to solve a system of any reasonable size with no algebraic errors.[9] Of course, the numerical solution is good only for the particular body whose profile shape $Y(x)$ appears in the coefficients A_{ij} defined by equations 3-55 and 3-57. Although an analytical solution would enable direct parametric studies of the effect of body shape on the pressure distribution, such a solution is simply not available, except for certain special shapes, or for values of N so low that the solution would be insufficiently accurate. The specificity of a numerical method's results—it only solves the problem you ask it to solve—is a price one is usually willing to pay for wide applicability and controllable accuracy.

Once equations 3-56 are solved for the m_j's, we can calculate the stream function by substituting equations 3-53 into 3-51:

$$\psi = V_\infty y - V_\infty \sum_{j=1}^{N} m_j \tan^{-1} \left(\frac{t - x}{y} \right) \Bigg|_{t=t_j}^{t=t_{j+1}} \tag{3-58}$$

The velocity components can then be found by differentiating equation 3-58 in

[9] Not counting the "roundoff errors" that result from the computer's restriction to a finite number of significant figures. These can be disastrous in their accumulated effect, in which case the algebraic system is called *ill conditioned*.

accordance with equations 3-1c:

$$u = V_\infty + V_\infty \sum_{j=1}^{N} m_j \frac{t - x}{(t - x)^2 + y^2}\Bigg|_{t=t_j}^{t=t_{j+1}} \tag{3-59}$$

$$v = -V_\infty \sum_{j=1}^{N} m_j \frac{y}{(t - x)^2 + y^2}\Bigg|_{t=t_j}^{t=t_{j+1}} \tag{3-60}$$

Finally, the pressure coefficient can be calculated from equation 3-2′.

A FORTRAN computer program for carrying out this procedure is listed at the end of the problem section. In this program, called DUBLET, the points t_j are concentrated near the ends of the distribution, where the variation of the doublet strength is expected to be most rapid. The abscissas x_i of the points at which the integral equation is satisfied are chosen as the midpoints of the subintervals on which the doublet strength is constant

$$x_i = \frac{t_i + t_{i+1}}{2}$$

The parameters x_s and x_f, which are only approximately predicted by equations 3-48, are adjusted by trial and error to ensure satisfaction of the boundary conditions at the stagnation points. That is, after you input N (the number of subintervals) and XS and XF (DUBLET's nomenclature for x_s and x_f), the program, in subroutine FINDM, distributes the nodes and sets up the coefficients of the matrix A of equations 3-56. These equations are solved by Gauss elimination in subroutine GAUSS. Then subroutine PRESS is called to compute the values of u at both ends of the body. If you are satisfied that they are both sufficiently close to zero, you accept the trial values of XS and XF. If not, you try new values.

Once you are satisfied with your guesses for XS and XF, DUBLET computes and outputs the pressure at selected points on the body surface. To emphasize the separation of the procedure into two steps—first find the doublet distribution and then compute the velocity distribution—the number of points at which the body surface pressure is output (NPRINT) is set separately from the number of subintervals (N), and the output points are spaced evenly along the length of the body. There is no reason for NPRINT to bear any relation to N. Note that the accuracy of the solution should be improved by increasing N, the number of subintervals, but that it is unaffected by NPRINT.

3.9. RELATIONS OF NUMERICAL TO ANALYTICAL SOLUTIONS

We have described three different ways of solving the potential-flow problem for symmetric bodies aligned with the onset flow. All are based on superposition of elementary solutions of the Laplace equation, which ensures satisfaction of the conditions of irrotationality, continuity, and the boundary condition at infinity. The solutions differ in the way they connect the source strength to the body shape:

1. In the *inverse* method, the source distribution is specified and the body shape determined by regarding some particular streamline as the surface of a solid body.

2. In the *thin-airfoil theory*, the body shape is specified, but it must be slender enough that the flow can be approximated as a small perturbation on the onset flow. An approximate version of the flow-tangency condition yields an analytic formula for the source strength directly in terms of the body shape.
3. The *numerical method* makes no obvious restrictions on the body shape but approximates the singularity[10] strength parametrically and determines the parameters numerically so as to satisfy flow tangency at specified points on the (numerically) prescribed body contour.

Of the three methods, only the first gives an exact connection between the body shape and the singularity strength and, hence, exact results for the velocity and pressure distributions on the body. However, practical problems require that the body shape be input. Thin-airfoil theory offers that possibility, but its accuracy depends on the body shape and on how far you are from stagnation points, at which it is doomed to fail. Numerical methods are also approximate but can often be made as accurate as desired for a *given* body shape, just by increasing the number of parameters employed in approximating the singularity strength. Therefore, since the advent of programmable computers made it possible to obtain cheap error-free numerical solutions, most aerodynamic analysis has relied heavily on numerical methods.

FIG. 3.17. A body that cannot be modeled by a line source (or doublet) distribution.

The numerical method implemented in program DUBLET is not actually as generally applicable as the preceding comparison may imply. Being based on a distribution of doublets on the axis of symmetry, it does not apply to problems whose solution requires distribution of singularities off the axis. For a trivial example, it will not work for the body whose shape is generated by positioning sources and sinks in the symmetrical pattern shown in Fig. 3.17. There is no simple way of determining, just from the body shape, whether the line singularity distribution underlying DUBLET is sufficient. However, as will be shown in Chapter 8, we can guarantee a solution of the incompressible inviscid flow problem in terms of singularities distributed on the body surface. This is the basis for the panel method to be described in Chapter 4. Such methods, which will be seen to be

[10] "Singularity" is used here as a generic term for sources, vortices, doublets, and other fundamental solutions of the Laplace equation that blow up—are "singular"—at some point, necessarily outside the flow field.

very similar in structure to the line-singularity method, have been widely used in the aircraft industry since about 1970.

3.10. COMPLEX–VARIABLE METHODS

Powerful methods for solving two-dimensional irrotational flow problems can be based on certain properties of functions of complex variables. We shall treat them here from the inverse point of view, using them to generate exact solutions, including some which really look like flows past airfoils.

Here is what I assume you know about the algebra of complex numbers.

1. Define

$$i = \sqrt{-1}$$

That is, $i^2 = -1$. Any number that can be represented by a point on the number line is a *real* number. If a is a real number, ai is an *imaginary* number. If a and b are real numbers, $a + ib$ is a *complex* number. A complex number is neither real nor imaginary; if you add a real number to an imaginary number, you get a complex number. We call a the *real part* of $a + ib$, and b (not ib) the *imaginary part*.

2. An important fact about complex numbers is that they are equal if and only if their real and imaginary parts are equal:

$$a + ib = c + id \Leftrightarrow a = b \quad \text{and} \quad c = d$$

Otherwise, the algebra of complex numbers is the same as the algebra of numbers like $a + bx$, except that $i^2 = -1$:

$$(a + ib) \pm (c + id) = a \pm c + i(b \pm d)$$

$$\begin{aligned}(a + ib)\cdot(c + id) &= ac + aid + ibc + i^2bd \\ &= ac - bd + i(ad + bc)\end{aligned}$$

Note, then, that

$$(a + ib)(a - ib) = a^2 + b^2$$

3. If

$$z = a + ib$$

we call

$$\bar{z} \equiv a - ib$$

the *complex conjugate* of z, and

$$|z| = |\bar{z}| = \sqrt{a^2 + b^2}$$

the *magnitude* or *modulus* of z.

4. The *Argand diagram* shown in Fig. 3.18 is a geometric representation of a complex number. Its real part becomes the abscissa, and its imaginary part, the ordinate. Thus, conversely, we can use a complex number as a kind of vector; to locate a point in the x–y plane, just let $z = x + iy$. Analogous to the polar representation of position is the expression

$$z = |z|e^{i\theta}$$

where θ is the *argument* of z, the angle between the "vector" z and the x axis, and

$$e^{i\theta} = \cos\theta + i\sin\theta$$

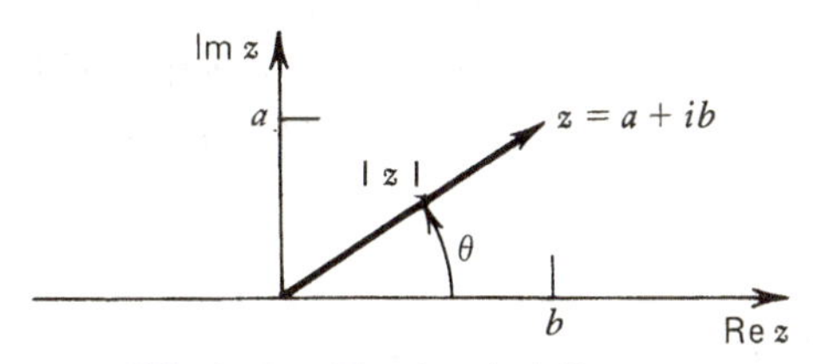

FIG. 3.18. The Argand diagram.

Now let $W(z)$ be a complex-valued function of the complex variable $z = x + iy$. We say that W is differentiable (*analytic* is the technical term) if and only if the limit

$$W'(z) \equiv \lim_{\Delta z \to 0} \frac{W(z + \Delta z) - W(z)}{\Delta z}$$

is independent of the way $\Delta z \to 0$. This has important ramifications. Let

$$W = \phi + i\psi \tag{3-61}$$

(Yes, we will eventually set the real and imaginary parts of W to be the velocity potential and stream function.) If $\Delta z = \Delta x$,

$$\begin{aligned} W'(z) &= \lim_{\Delta x \to 0} \frac{[\phi(x + \Delta x, y) + i\psi(x + \Delta x, y) - \phi(x, y) - i\psi(x, y)]}{\Delta x} \\ &= \frac{\partial \phi}{\partial x} + i\frac{\partial \psi}{\partial x} \end{aligned} \tag{3-62}$$

On the other hand, if $\Delta z = i\Delta y$,

$$\begin{aligned} W'(z) &= \lim_{\Delta y \to 0} \frac{[\phi(x, y + \Delta y) + i\psi(x, y + \Delta y) - \phi(x, y) - i\psi(x, y)]}{i\Delta y} \\ &= \frac{1}{i}\left(\frac{\partial \phi}{\partial y} + i\frac{\partial \psi}{\partial y}\right) \\ &= -i\frac{\partial \phi}{\partial y} + \frac{\partial \psi}{\partial y} \end{aligned} \tag{3-63}$$

Therefore, if W is analytic, we can equate the two expressions (3-62) and (3-63)

$$\frac{\partial \phi}{\partial x} + i\frac{\partial \psi}{\partial x} = -i\frac{\partial \phi}{\partial y} + \frac{\partial \psi}{\partial y}$$

which implies

$$\frac{\partial \phi}{\partial x} = \frac{\partial \psi}{\partial y}$$

$$\frac{\partial \psi}{\partial x} = -\frac{\partial \psi}{\partial y} \tag{3-64}$$

These are called the *Cauchy–Riemann equations*; they are satisfied by the real and imaginary parts of any analytic (i.e., differentiable) function of $z = x + iy$. Note their consistency with equations 3-1c and 3-3b, which, together with equations 3-62 and 3-63, suggest that we call the real and imaginary parts of $W'(z)$ u and $-v$, respectively,

$$W'(z) = u - iv \tag{3-65}$$

Now assume[11] that $W'(z)$ is analytic. Then the same proof shows that u and $-v$ satisfy the Cauchy–Riemann equations:

$$\frac{\partial u}{\partial x} = -\frac{\partial v}{\partial y}$$

$$-\frac{\partial v}{\partial x} = -\frac{\partial u}{\partial y}$$

which are exactly the continuity and irrotationality conditions (3-1a) and (3-3a), if u and v are, as is suggested by equations 3-62 to 3-65, the x and y components of the velocity vector. We conclude the following: *if W is any analytic function of z, its real and imaginary parts are the velocity potential and stream function, respectively, of some irrotational incompressible plane flow.* Therefore, we can generate exact inverse solutions simply by writing down specific forms of $W(z)$, which we now call the *complex potential.*

Example 4 Discuss the flow whose complex potential is

$$W = V_\infty e^{-i\alpha} z$$

Solution Resolve both sides into their real and imaginary parts:

$$\begin{aligned}\phi + i\psi &= V_\infty(\cos\alpha - i\sin\alpha)(x + iy)\\ &= V_\infty(x\cos\alpha + y\sin\alpha) + iV_\infty(y\cos\alpha - x\sin\alpha)\end{aligned}$$

so

$$\phi = V_\infty(x\cos\alpha + y\sin\alpha)$$

$$\psi = V_\infty(y\cos\alpha - x\sin\alpha)$$

[11] Actually, no further assumption is necessary; if $W(z)$ is analytic, so is $W'(z)$, $W''(z)$, and so on.

Comparing these formulas with equations 3-4b and 3-4c, we see that they are, respectively, the velocity potential and stream function of a uniform flow whose speed is V_∞ and which is inclined at the angle α to the x axis. Thus the given W is the complex potential of such a flow.

Example 5 Discuss the flow whose complex potential is

$$W = z + \frac{a^2}{z}, \qquad a > 0 \tag{3-66}$$

Solution Resolving W and z into their real and imaginary parts, we have

$$\phi + i\psi = x + iy + \frac{a^2}{x + iy} \cdot \frac{x - iy}{x - iy}$$

$$= x + iy + \frac{a^2(x - iy)}{x^2 + y^2}$$

so, with $r^2 = x^2 + y^2$,

$$\phi = x\left(1 + \frac{a^2}{r^2}\right)$$

$$\psi = y\left(1 - \frac{a^2}{r^2}\right) \tag{3-67}$$

You should recognize ψ as the stream function of flow at unit speed past a circle of radius a (see equation 3-19). It is, and ϕ is the velocity potential of the flow, and W its complex potential.

In problem 17, you will show that

$$W = \frac{1}{2\pi}(Q + i\Gamma) \ln z$$

is the complex potential of a source of strength Q and vortex of strength Γ at the origin.

3.10.1. Flow Past an Ellipse

So much for putting what you already know in a new form. Now let's make some progress. Consider the complex potential

$$W = V_\infty\left(\tilde{z} + \frac{a^2}{\tilde{z}}\right), \qquad a > 0 \tag{3-68}$$

You could say, from the last example, that this is the complex potential of a uniform flow of speed V_∞ past a circle of radius a, in the $\tilde{z} = \tilde{x} + i\tilde{y}$ plane. Now, however, we regard the $\tilde{z}$ plane as some intermediate mathematical entity and relate it to the physical $z = x + iy$ plane by a transformation of form similar to that of equation 3-68:

$$z = \tilde{z} + \frac{b^2}{\tilde{z}}, \qquad b > 0 \tag{3-69}$$

Then, just as equation 3-66 led to 3-67, we find

$$\phi = V_\infty \tilde{x}\left(1 + \frac{a^2}{\tilde{r}^2}\right) \tag{3-70}$$

$$\psi = V_\infty \tilde{y}\left(1 - \frac{a^2}{\tilde{r}^2}\right) \tag{3-71}$$

$$x = \tilde{x}\left(1 + \frac{b^2}{\tilde{r}^2}\right) \tag{3-72}$$

$$y = \tilde{y}\left(1 - \frac{b^2}{\tilde{r}^2}\right) \tag{3-73}$$

where

$$\tilde{r}^2 \equiv \tilde{x}^2 + \tilde{y}^2 \tag{3-74}$$

As is usual in inverse approaches, we now try to see what problem we have solved, however inadvertently. As $\tilde{z} \to \infty$, equation 3-69 shows $z \to \tilde{z}$, so $W \to V_\infty z$. Thus, from the first example of this section, W describes a flow that becomes uniform and has speed V_∞ in the $+x$ direction far from the origin of the physical z plane.

To locate the body, we can look for curves in the z plane on which the stream function is constant. Since $\psi = 0$ on the circle $\tilde{r} = a$ in the $\tilde{z}$ plane, let us look at the "image," as it is called, of that circle in the z plane. That is, let $\tilde{r} = a$ in equations 3-72 and 3-73, and so obtain

$$x = \tilde{x}\left(1 + \frac{b^2}{a^2}\right), \qquad y = \tilde{y}\left(1 - \frac{b^2}{a^2}\right) \tag{3-75}$$

when $\tilde{r} = a$ and $\psi = 0$. On substituting for $\tilde{x}$ and $\tilde{y}$ in equation 3-74, we find

$$\left[\frac{x}{1 + (b^2/a^2)}\right]^2 + \left[\frac{y}{1 - (b^2/a^2)}\right]^2 = \tilde{r}^2 = a^2 \tag{3-76}$$

when $\tilde{r} = a$ and $\psi = 0$. This is the equation of an ellipse. Thus the streamline $\psi = 0$, a circle in the $\tilde{z}$ plane, is "mapped," as they say, onto an ellipse in the z plane by the transformation (3-69). Since ψ is constant (zero) on that ellipse, the W given by equations 3-68 and 3-69 is the complex potential of a uniform flow past the ellipse in the physical z plane.

To fix the parameters a and b, let the ends of the ellipse be at $x = \pm 1$, and its thickness ratio be τ, as shown in Fig. 3.19. Evaluating equation 3-76 at $x = \pm 1$, $y = 0$ then yields

$$a = \frac{1}{1 + (b^2/a^2)}$$

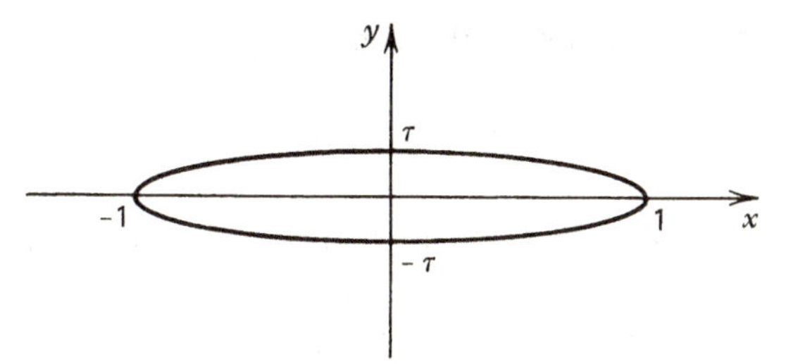

FIG. 3.19. An ellipse.

whereas setting $x = 0$, $y = \pm\tau$, in the same equation gives

$$a = \frac{\tau}{1 - (b^2/a^2)}$$

These equations may be solved simultaneously for a and b:

$$a = \frac{1 + \tau}{2}, \qquad b = \frac{\sqrt{1 - \tau^2}}{2} \tag{3-77}$$

To find the velocity field, we use equation 3-65 and the chain rule:

$$\begin{aligned} u - iv &= \frac{dW}{dz} \\ &= \frac{dW}{d\tilde{z}} \frac{d\tilde{z}}{dz} \\ &= \frac{dW}{d\tilde{z}} \bigg/ \frac{dz}{d\tilde{z}} \end{aligned} \tag{3-78}$$

With $W(\tilde{z})$ and $z(\tilde{z})$ given by equations 3-68 and 3-69, respectively, we obtain

$$u - iv = V_\infty \left(1 - \frac{a^2}{\tilde{z}^2}\right) \bigg/ \left(1 - \frac{b^2}{\tilde{z}^2}\right) \tag{3-79}$$

In general, the easiest way to use these results is to determine u, v, x, and y parametrically in terms of $\tilde{x}$ and $\tilde{y}$ from equations 3-72, 3-73, and 3-79. However, on the ellipse (which is where the results are most interesting anyway), we can proceed more directly. The circle of radius a in the $\tilde{z}$ plane of which the ellipse is the image can be described by

$$\tilde{z} = ae^{i\theta}$$

so that

$$\tilde{x} = a \cos \theta, \qquad \tilde{y} = a \sin \theta \tag{3-80}$$

Then, from equations 3-75, 3-77, and 3-80

$$\begin{aligned} x &= \cos \theta \\ y &= \tau \sin \theta \end{aligned} \tag{3-81}$$

on the ellipse, whereas equation 3-79 becomes

$$u - iv = V_\infty \frac{1 - e^{-2i\theta}}{1 - (b^2/a^2)e^{-2i\theta}}$$

on the ellipse. Using equations 3-77 for a and b, we can manipulate this into

$$u - iv = V_\infty \frac{(1 + \tau)(1 - x^2 + ixy)}{1 - (1 - \tau^2)x^2} \tag{3-82}$$

Note that, at $x = 0$ and $y = \tau$, this yields

$$u = V_\infty(1 + \tau), \qquad v = 0$$

in exact agreement with the thin-airfoil result, equation 3-41! However, equation 3-82 also yields the stagnation conditions expected at $x = \pm 1$, $y = 0$, whereas thin-airfoil theory continues to predict $u = V_\infty(1 + \tau)$ there.

3.10.2. Joukowsky Airfoils

Equation 3-69 is called the *Joukowsky transformation.* It ought to be called "airfoil transformation," since it shows up in virtually every complex-variable solution of an airfoil problem. For a particularly important example, consider what happens when it is used in conjunction with the complex potential

$$W(\tilde{z}) = V_\infty\left(\tilde{z} - c + \frac{a^2}{\tilde{z} - c}\right), \qquad a > 0 \tag{3-83}$$

which describes the uniform flow in the $\tilde{z}$ plane past a circle of radius a whose center is positioned at $\tilde{z} = c$ rather than at the origin. If

$$c = \varepsilon + i\mu \tag{3-84}$$

the circle, on which ψ can be shown from equation 3-83 to vanish, has the equations

$$(\tilde{x} - \varepsilon)^2 + (\tilde{y} - \mu)^2 = a^2$$

$$\tilde{z} = c + ae^{i\theta} \tag{3-85}$$

The velocity field can be found by substituting equations 3-69 and 3-83 into 3-78, with the result

$$u - iv = V_\infty \frac{1 - [a^2/(\tilde{z} - c)^2]}{1 - (b^2/\tilde{z}^2)} \tag{3-86}$$

As is also the case with equation 3-79, 3-86 blows up when

$$\tilde{z} = \pm b \tag{3-87}$$

The parameters a, b, and $c = \varepsilon + i\mu$ must be adjusted so that such singularities do not map into points in the physical flow field and so cannot be outside the circle in the $\tilde{z}$ plane. Interesting things happen when they lie *on* that circle, and so on the body

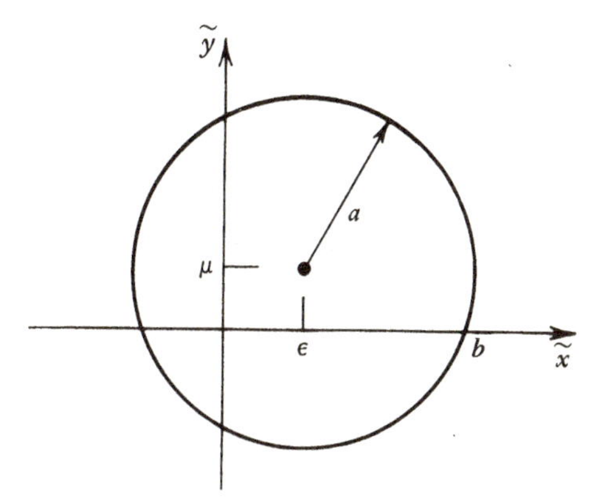

FIG. 3.20. The circle in the $\tilde{z}$ plane.

surface. Thus, suppose $\tilde{z} = +b$ lies on the circle, as indicated in Fig. 3.20, so

$$a^2 = (b - \varepsilon)^2 + \mu^2 \tag{3-88}$$

To keep the singularity at $\tilde{z} = -b$ on or within the circle, we require

$$\varepsilon \leq 0 \tag{3-89}$$

(unlike the case shown in Fig. 3.20).

Let us examine the body shape near the image of the singular point $\tilde{z} = b$. If

$$\tilde{z} = b + \delta e^{i\theta}$$

with $\delta \ll b$, application of the geometric series expansion of $(1 + x)^{-1}$ to equation 3-69 yields

$$\begin{aligned} z &= b + \delta e^{i\theta} + \frac{b}{1 + (\delta/b)e^{i\theta}} \\ &= b + \delta e^{i\theta} + b\left[1 - \frac{\delta}{b}e^{i\theta} + \left(\frac{\delta}{b}\right)^2 e^{2i\theta} - \cdots\right] \\ &\approx 2b + \frac{\delta^2}{b}e^{2i\theta} \end{aligned}$$

Thus, as we travel around the singular point from one point on the streamline $\psi = 0$ to another, as shown in Fig. 3.21, the argument of the vector from the singular point to our position changes by π in the $\tilde{z}$ plane, but by 2π in the physical z plane. Therefore, as indicated on the right side of Fig. 3.21, the streamline $\psi = 0$ is cusped at the singular point $z = 2b$ in the physical plane. Positioning the singularity $\tilde{z} = +b$ on the streamline gives it a sharp trailing edge.

To get some idea as to the shape overall of the stagnation streamline in the physical plane, consider the special case $\mu = 0$, in which equation 3-88 becomes $a = b - \varepsilon$. In the $\tilde{z}$ plane, the streamline $\psi = 0$ is then a circle centered at $\tilde{z} = \varepsilon$, a point on the negative $\tilde{x}$ axis, and intersects the $\tilde{x}$ axis at $\tilde{z} = b$, the singular point, and

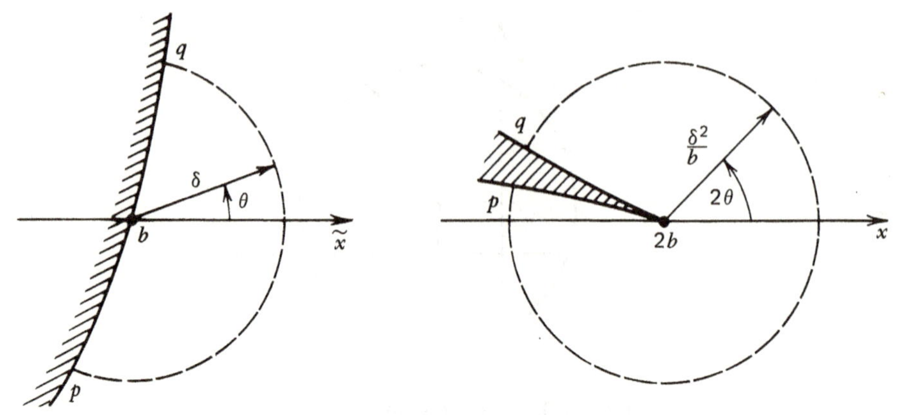

FIG. 3.21. Details of the stagnation streamline in the $\tilde{z}$ and z planes near the rear stagnation point.

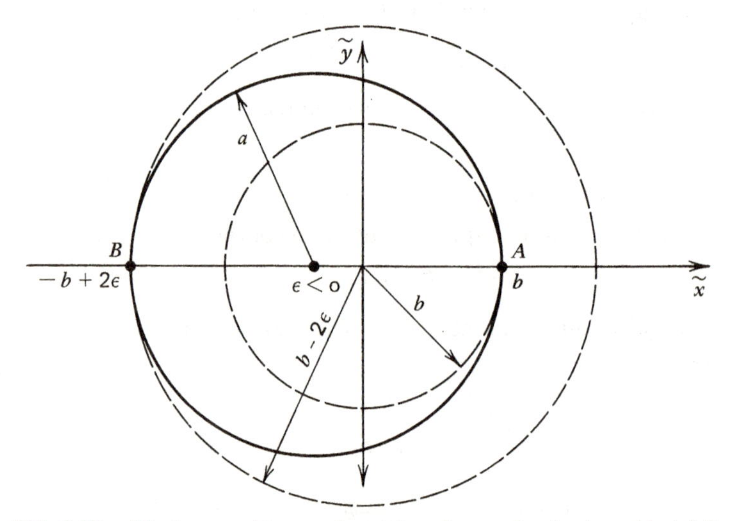

FIG. 3.22. Circles used to rough out the shape of a Joukowski airfoil.

at $\tilde{z} = -b + 2\varepsilon$. As indicated in Fig. 3.22, we label these intersections A and B, respectively.

Near A, the streamline $\psi = 0$ is closely approximated by a circle centered at the origin and of radius b, whereas, near B, it is close to the larger circle $\tilde{z} = (b - 2\varepsilon)e^{i\theta}$, as shown in Fig. 3.22. Therefore, we can get an idea of the shape of the streamline $\psi = 0$ in the physical plane by examining the images of two circles centered at $\tilde{z} = 0$, which is easy. If $\tilde{z} = Re^{i\theta}$, equation 3-69 shows

$$z = x + iy = Re^{i\theta} + \frac{b^2}{R}e^{-i\theta}$$

$$= \left(R + \frac{b^2}{R}\right)\cos\theta + i\left(R - \frac{b^2}{R}\right)\sin\theta$$

Thus, as we found earlier, the circle $\tilde{z} = Re^{i\theta}$ is mapped onto an ellipse in the z plane:

$$\left[\frac{x}{R + (b^2/R)}\right]^2 + \left[\frac{y}{R - (b^2/R)}\right]^2 = 1$$

The thickness ratio of the ellipse is

$$\tau = \frac{R - (b^2/R)}{R + (b^2/R)} \tag{3-90}$$

Thus, near the singular point A, the streamline $\psi = 0$ is mapped onto the ellipse that is the image of the circle of radius $R = b$ and so has a thickness ratio $\tau = 0$. This is just a straight line, in agreement with our expectation that the streamline is cusped near the singular point in the physical plane. Near the point B, the streamline maps onto an ellipse whose thickness ratio is found by setting $R = b - 2\varepsilon$ in equation 3-90:

$$\tau = \frac{(b - 2\varepsilon)^2 - b^2}{(b - 2\varepsilon)^2 + b^2}$$

$$\approx -2\frac{\varepsilon}{b} \qquad \text{if } |\varepsilon| \ll b$$

Since the streamline $\psi = 0$ lies between the circles $\tilde{z} = be^{i\theta}$ and $\tilde{z} = (b - 2\varepsilon)e^{i\theta}$, we expect its image in the physical plane, that is, the body about which equation 3-86 describes the flow field, to lie between the images of those two circles. Therefore, we have "solved" the problem illustrated in Fig. 3.23.

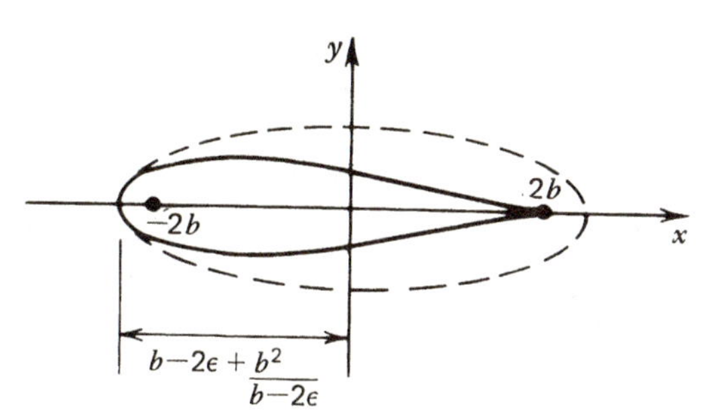

FIG. 3.23. Streamlines in the physical plane corresponding to the circles of Fig. 3.23.

The details of actually finding the body shape and the pressure distribution on it are somewhat messy. The results cannot be expressed in simple formulas like (3-82); you have to be content with numerical results for specific values of ε/b, as you will see in problem 21. However, all the numbers you get are substantially exact; you don't have to worry about increasing the number of nodes or panels, as in program DUBLET. Thus, the results can be used to test the convergence rate of numerical methods like DUBLET.

The shapes generated by applying the Joukowsky transformation (3-69) to the streamline $\psi = 0$ of the complex potential (3-83) (i.e., to the displaced circle of equation 3-85) are called *Joukowsky airfoils.* As may or may not be obvious, they are symmetric only when $\mu = 0$, so that the case $\mu \neq 0$ is properly deferred to Chapter 4.

3.11. PROBLEMS

1. Discuss the possibility of using the stream function

$$\psi = V_\infty y + \frac{Q}{2\pi}\theta$$

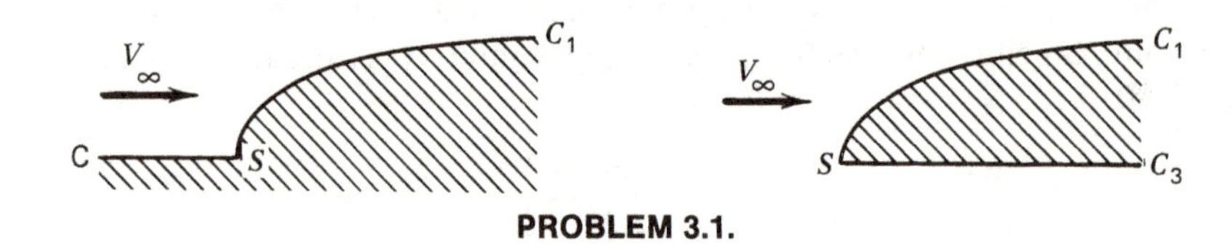

PROBLEM 3.1.

to describe the uniform flows about the contours pictured above which are constructed by picking out streamlines CSC_1 and C_1SC_3, respectively, from the streamline map of Fig. 3.5. That is, check whether this stream function meets all the conditions it should (there are three such conditions). For the case on the right side, also discuss the alternative possibility

$$\psi = V_\infty y + \frac{Q}{2\pi}\theta \qquad \text{for } y > 0$$

$$= V_\infty y + \frac{Q}{2} \qquad \text{for } y < 0$$

which would imply that you can "patch together" incompressible inviscid flows.

2. Find the asymptotic width of the semi-infinite board created by positioning a source of strength Q in a uniform flow of speed V_∞. Plot the pressure distribution on the board and its shape (C_p and Y/a versus x/a).

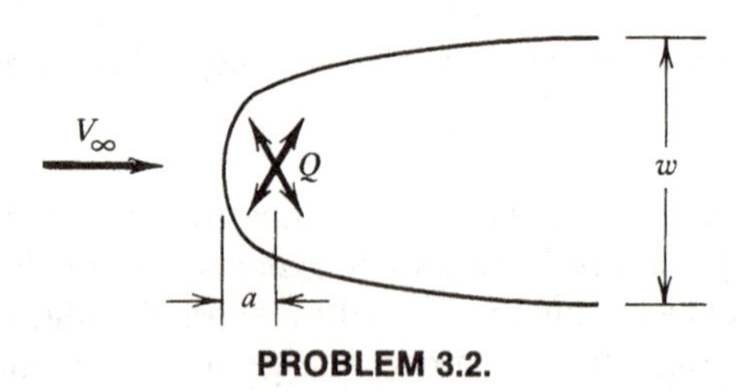

PROBLEM 3.2.

3. Find the force on the body described in problem 2. *Hint*: Use the momentum theorem, equation 2-12, with S_C a large circle centered at the source.

4. Plot the shapes of and pressure distributions on the Rankine ovals for which

$$\frac{Q}{2\pi V_\infty a} = 0.01, 0.1, 1.0$$

A computer program named RANKIN is listed in Section 3.12 for your use. In the program, dimensionless variables are used:

$$\frac{Q}{2\pi a V_\infty} \to \mathrm{Q}$$

$$\frac{x}{a}, \frac{y}{a}, \frac{x_0}{a} \to \mathrm{X, Y, X0}$$

$$\frac{u}{V_\infty}, \frac{v}{V_\infty} \to \mathrm{U, V}$$

PSI is the stream function, and CP, the pressure coefficient. Be sure to check the values output for ψ (under the heading "PSI") to make sure your results make sense.

5. For at least two of the following flows, plus any others you wish to invent, locate the stagnation streamline, plot some typical streamlines, and plot the velocity distribution along the stagnation streamline.
   (a) Flow past Rankine oval. Adjust the parameters (the strengths and locations of the source and sink) so as to emulate one of the bodies you studied in problem 4.
   (b) Flow past a circular cylinder with circulation (see problem 5 of Chapter 2).
   (c) Flow past three or more sources at different points on the x axis. Make the source furthest upstream have positive strength. Suit yourself as to whether or not the source strengths add up to zero.
   (d) Flow past a Rankine oval with circulation. Place vortices of equal strength on top of the two sources of equal but opposite strength.
   (e) Here's a really hard one. Put a vortex at the source furthest upstream of a source–sink pair, and adjust the angle of attack until the two stagnation points lie on the same streamline.

   A program called STRACE that should help substantially with these problems has been written for a microcomputer. Instructions on its use are available separately from the author.

6. Consider the symmetric airfoil shape

$$Y(x) = \sqrt{\frac{27}{16}}\tau\sqrt{\frac{x}{c}}(c - x)$$

   (a) Plot the shape for a thickness ratio $\tau = 0.1$.
   (b) Plot the pressure distribution (C_p versus x) predicted by thin-airfoil theory.
   (c) Calculate the pressure distribution numerically, using program DUBLET, which is based on line doublet distributions and listed under Section 3.12. Plot your results on the same graph you use to answer part(b).

   Your results will, of course, depend on the number of intervals N into which you divide the doublet distribution. How large does N have to be before further increases in N no longer change your results (so far as you can see on the scale of your graph)?

7. Use DUBLET to find the "exact" pressure distribution on an ellipse of thickness ratio $\tau = 0.1$. Compare your results with those of thin-airfoil theory.

8. The NACA four- and five-digit airfoils have a thickness distribution described by equation 1-2. Find the pressure distribution on a symmetric NACA four- or five-digit airfoil of thickness ratio 0.12 according to thin-airfoil theory.

9. Repeat problem 8, using program DUBLET. Compare results.

10. Since the doublet distribution method was derived by integrating by parts the stream function of a line source distribution, the piecewise-constant doublet distribution of program DUBLET must have an equivalent source distribution. What is it?

11. In view of your answer to problem 10, you ought to be able to use STRACE to plot the streamlines actually generated by the doublet distribution that DUBLET calculates for the body shape you analyzed in problem 6, at least when N is not too large. Do so. In particular, compare the stagnation streamline generated by the doublet distribution calculated by DUBLET with the body shape.

12. Show that a distribution of vortices of strength $\gamma(x)$, $x_s < x < x_f$, is equivalent to a distribution of doublets whose axes are aligned in the y direction and whose strength $\mu(x)$ satisfies

$$\mu(x) = \int_{x_s}^{x} \gamma(t)\,dt, \qquad x_s < x < \infty$$

That is, show that the two distributions have the same velocity potential. Note that the potential of a point y doublet at the origin is (cf. your solution to problem 4 of Chapter 2)

$$-\frac{M}{2\pi}\frac{y}{x^2 + y^2}$$

M being its strength.

13. Revise DUBLET to use a piecewise linear doublet distribution. That is, take

$$m(t) = m_i + (m_{i+1} - m_i)\frac{t - t_i}{t_{i+1} - t_i}$$

$$= \frac{t_{i+1} - t}{t_{i+1} - t_i} m_i + \frac{t - t_i}{t_{i+1} - t_i} m_{i+1} \qquad \text{for } t_i < t < t_{i+1} \text{ and } i = 1, \ldots, N$$

Then the integral appearing in the integral equation 3-52 can be approximated by

$$\int_{x_s}^{x_f} \frac{m(t)\,dt}{(x - t)^2 + y^2} = \sum_{i=1}^{N} \int_{t_i}^{t_{i+1}} \frac{m(t)\,dt}{(x - t)^2 + y^2}$$

$$= \sum_{i=2}^{N} m_i \left(\int_{t_{i-1}}^{t_i} \frac{t - t_{i-1}}{t_i - t_{i-1}} \frac{dt}{(x - t)^2 + y^2} + \int_{t_i}^{t_{i+1}} \frac{t_{i+1} - t}{t_{i+1} - t_i} \frac{dt}{(x - t)^2 + y^2} \right)$$

where we note that $m_1 = m(x_s)$ and $m_{N+1} = m(x_f)$ both $= 0$. The integrals required to satisfy the flow tangency condition and to compute the velocity distribution on the body can be worked out with the help of the table below:

$$X \equiv (t - x)^2 + y^2$$

$$\int \frac{dt}{X} = \frac{1}{y} \tan^{-1} \frac{t - x}{y} \equiv f_t$$

$$\int \frac{(t - x)\, dt}{X} = \ln[(x - t)^2 + y^2]^{1/2} \equiv f_l$$

$$\int \frac{dt}{X^2} = \frac{t - x}{2y^2 X} + \frac{1}{2y^2} f_t$$

$$\int \frac{(t - x)\, dt}{X^2} = -\frac{1}{2X}$$

$$\int \frac{(t - x)^2\, dt}{X^2} = -\frac{t - x}{2X} + \frac{1}{2} f_t$$

$$\int \frac{(t - x)^3\, dt}{X^2} = \frac{y^2}{2X} + f_l$$

14. The potential of a distribution of point (three-dimensional) sources of strength $q(x)$ per unit length along the x axis is (see problem 10 of Chapter 2)

$$\phi_s = \frac{-1}{4\pi} \int_{x_s}^{x_f} q(t)[(x - t)^2 + r^2]^{-1/2}\, dt$$

in which x and r are cylindrical coordinates.

(a) Show that

$$\lim_{r \to 0} r \frac{\partial \phi_s}{\partial r} = \frac{q(x)}{2\pi}$$

(b) Show that the source strength required for

$$\phi = V_\infty x + \phi_s$$

to solve the potential-flow problem for axisymmetric flow past a slender body of revolution whose meridional curve is described by $r = R(x)$ is

$$q(x) \approx V_\infty S'(x)$$

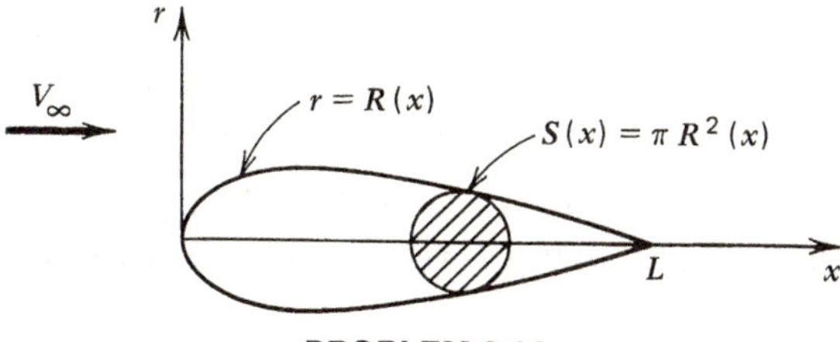

PROBLEM 3.14.

where $S(x) = \pi R^2$ is the cross-sectional area distribution and $R(x) \ll L$, the body length.

15. Find the velocity distribution on the slender ellipsoid of revolution of thickness ratio τ, for which

$$R^2(x) = \tau^2(1 - x^2) \qquad -1 \le x \le 1$$

Warning: In computing the x component of velocity, do *not* try to simplify the integrand beyond using the result of problem 14 for the source strength.

16. From problem 15, if not 14, demonstrate that on the body surface, the x component of velocity induced by the sources is proportional to τ^2, where τ is the thickness ratio of the body, and the r component is proportional to τ. Because this behavior differs from the two-dimensional case, equation 3-35 is not the appropriate approximation to Bernoulli's equation in computing the pressure distribution on a slender body of revolution. How much equation 3-35 be modified?

17. Show that

$$W = \frac{1}{2\pi}(Q + i\Gamma)\ln z$$

is the complex potential of a source of strength Q and vortex of strength Γ at the origin.

18. Plot the exact pressure distribution on an ellipse of thickness ratio $\tau = 0.1$. Compare with thin-airfoil theory and the results of program DUBLET.

19. Show that our result for the complex potential of uniform flow past an ellipse reduces to its proper limit when $\tau = 1$ (so that the ellipse becomes a circle).

20. Use equations 3-30 and 3-82 to find the exact strength of the line source distribution associated with flow past an ellipse.

21. Find the shape of and pressure distribution on the Joukowsky airfoil defined by $\varepsilon/b = -0.1$. Program JOUKOW, listed under Section 3.12, is at your disposal. Note that the program's EPS is the text's ε/b. Use the output as input to program DUBLET, and check out the latter's performance. The results of JOUKOW are, of course, exact.

22. Revise DUBLET to provide a numerical solution to the problem of axisymmetric flow past a body of revolution. According to problem 11 of Chapter 2, the Stokes stream function of such a flow should be constant on the body, whereas from problems 12 and 13 of Chapter 2, the Stokes stream function of an axial distribution of point sources in a uniform flow is

$$\psi = \tfrac{1}{2}V_\infty r^2 - \int_{x_s}^{x} q(t)\frac{x - t}{[(x - t)^2 + r^2]^{1/2}}\,dt$$

Here $q(t)$ is the strength of the source distribution, per unit length, which should therefore be determined so that ψ is constant (zero for a closed body) at $r = R(x)$, the body contour.

3.12. COMPUTER PROGRAMS

```
C     PROGRAM RANKIN
C
C                     LOCATE POINT ON RANKINE OVAL
C
      PRINT,        ' DIMENSIONLESS SOURCE STRENGTH =',
      READ,         Q
      PRINT 1000
      XO            =  - Q + SQRT(Q*Q + 1.)
  100 PRINT,         ' Y/A =',
      READ,         Y
      X2            =  XO*XO - Y*Y + 2.*Y*XO/TAN(Y/Q)
      IF (X2 .GE. 0.0)        GO TO 110
      PRINT,        ' Y IS TOO LARGE'
      GO TO 100
  110 X             =  SQRT(X2)
      PSI           =  Y + Q*ATAN2(Y,X+XO) - Q*ATAN2(Y,X-XO)
C
C                      COMPUTE PSI, VELOCITY, AND PRESSURE AT POINT
C
      RPLUS         =  (X+XO)**2 + Y*Y
      RMINUS        =  (X-XO)**2 + Y*Y
      U             =  1. + Q*(X+XO)/RPLUS - Q*(X-XO)/RMINUS
      V             =  Q*Y/RPLUS - Q*Y/RMINUS
      CP            =  1.0 - U*U - V*V
      PRINT 1010, X,PSI,U,V,CP
      GO TO 100
 1000 FORMAT(/,20X,'X/A     PSI      U       V       CP')
 1010 FORMAT(15X,5F8.3)
      END

C     PROGRAM DUBLET
C
C                     INCOMPRESSIBLE AERODYNAMICS OF SYMMETRIC AIRFOIL
C                     AT ZERO ANGLE OF ATTACK BY LINE DOUBLET DISTRIBUTION
C
      COMMON        T(100),M(100),N,XS,XF
      REAL          M,MPLOT
C
C                     INPUT NUMBER OF INTERVALS N
C
      PRINT,        ' N =',
      READ,         N
C
C                     DETERMINE ENDPOINTS OF DISTRIBUTION XS, XF
C
  100 PRINT,        ' XS,XF =',
      READ,         XS,XF
      CALL FINDM
      CALL PRESS(0.0,U0,CP0)
      CALL PRESS(1.0,U1,CP1)
      PRINT,        ' U AT X = 0,1 =',U0,U1
      PRINT,        ' DO YOU ACCEPT THESE RESULTS (Y/N)',
      READ 1000, IANS
      IF (IANS .NE. 1HY)      GO TO 100
```

```
C
C                      OUTPUT RESULTS
C
      PRINT 1010
      M(N+1)      =  0.0
      DO 200      I = 1,N+1
      MPLOT       =  M(I)*3.1415926585
  200 CALL PLOTXY(T(I),MPLOT,100.)
      PRINT 1020
      DO 210      I = 1,N
      XX          =  .5*(T(I) + T(I+1))
      YY          =  Y(XX)
  210 CALL PLOTXY(XX,YY,100.)
      PRINT 1030
      READ,       NPRINT
      DO 220      I = 1,NPRINT
      XX          =  (I - 1)/FLOAT(NPRINT - 1)
      CALL PRESS(XX,U,CP)
  220 CALL PLOTXY(XX,CP,40.)
 1000 FORMAT(A1)
 1010 FORMAT(/,' DOUBLET STRENGTH DISTRIBUTION',/
     +       ' M = M(I) FOR T(I) < T < T(I+1)',//
     +       4X,'T(I)',5X,'M(I)/2',/)
 1020 FORMAT(//,' BODY SHAPE',//,4X,'X',9X,'Y',/)
 1030 FORMAT(//,' BODY SURFACE PRESSURE DISTRIBUTION',//,
     +       4X,'X',8X,'CP',//,' INPUT NUMBER OF OUTPUT POINTS',)
      STOP
      END
C=============================================================
      SUBROUTINE FINDM
C
C                      FIND DOUBLET STRENGTH TO MEET
C                      FLOW TANGENCY CONDITION
C
      COMMON      T(100),M(100),N,XS,XF

      COMMON /COF/ A(101,111),NEQNS
      REAL M
      PI          =  3.1415926585
      NP          =  N + 1
      DO 100      I = 1,NP
      FRACT       =  .5*(1. - COS(PI*(I-1)/FLOAT(N)))
  100 T(I)        =  XS + (XF - XS)*FRACT
C
C                     SET UP LINEAR SYSTEM OF EQUATIONS
C
      DO 210      I = 1,N
      XI          =  .5*(T(I) + T(I+1))
      YI          =  Y(XI)
      FAC1        =  ATAN2(T(1) - XI,YI)
      DO 200      J = 1,N
      FAC2        =  ATAN2(T(J+1) - XI,YI)
      A(I,J)      =  (FAC2 - FAC1)/YI
  200 FAC1        =  FAC2
  210 A(I,NP)     =  1.0
```

```
C
C                   SOLVE FOR DOUBLET STRENGTH
C
      NEQNS       =  N
      CALL GAUSS(1)
      DO 300      I = 1,N
  300 M(I)        =  A(I,NP)
      RETURN
      END
C=================================================================
      SUBROUTINE PRESS(X,U,CP)
C
C                   FIND PRESSURE COEFFICIENT CP AT (X,Y(X))
C
      COMMON      T(100),M(100),N,XS,XF
      REAL        M
      YB          =  Y(X)
      U           =  1.0
      V           =  0.0
      VF1         =  1./((T(1) - X)**2 + YB*YB)
      UF1         =  (T(1) - X)*VF1
      DO 100      J = 1,N
      VF2         =  1./((T(J+1) - X)**2 + YB*YB)
      UF2         =  (T(J+1) - X)*VF2
      U           =  U + M(J)*(UF2 - UF1)
      V           =  V - M(J)*YB*(VF2 - VF1)
      VF1         =  VF2
  100 UF1         =  UF2
      CP          =  1.0 - U*U - V*V
      RETURN
      END
C=================================================================
      FUNCTION Y(X)
C
C                   ORDINATE OF BODY CONTOUR
C
C                   EXAMPLE GIVEN IS ELLIPSE OF THICKNESS RATIO 0.1
C
      Y           =  0.1*SQRT(X*(1. - X))
      RETURN
      END
C=================================================================
      SUBROUTINE PLOTXY(X,Y,YMULT)
C
C                   PLOT Y ON SAME LINE AS X AND Y ARE PRINTED
C
      COMMON /SKAL/ NZERO,YMULT
      YSLOP         =  .5/YMULT
      NPLOT       =  (Y + YSLOP)*YMULT
      IF (Y + YSLOP .LT. 0.0)    NPLOT = NPLOT - 1
      IF (NPLOT) 10,20,30
C
C                   --  NEGATIVE Y
C
   10 NTOX        =  NZERO + NPLOT
      IF (NTOX .LT. 1)        GO TO 40
      NTODOT      =  - NPLOT - 1
      IF (NTODOT .EQ. 0)      GO TO 15
      PRINT 1010,             X,Y,NTOX,NTODOT
      RETURN
```

```
   15 PRINT 1015,              X,Y,NTOX
      RETURN
C
C                   --  ZERO Y
C
   20 PRINT 1020,              X,Y,NZERO
      RETURN
C                   --  POSITIVE Y
C
   30 NTOX        =  NPLOT - 1
      IF (NTOX + NZERO .GT. 60)           GO TO 40
      IF (NTOX .EQ. 0)                    GO TO 35
      PRINT 1030,              X,Y,NZERO,NTOX
      RETURN
   35 PRINT 1035,              X,Y,NZERO
      RETURN
C
C                   --  Y OUT OF RANGE OF PLOT
C
   40 PRINT 1040,              X,Y
      RETURN
 1010 FORMAT(F8.4,F10.4,=X,1HX,=X,1H.)
 1015 FORMAT(F8.4,F10.4,=X,2HX.)
 1020 FORMAT(F8.4,F10.4,=X,1HX)
 1030 FORMAT(F8.4,F10.4,=X,1H.,=X,1HX)
 1035 FORMAT(F8.4,F10.4,=X,2H.X)
 1040 FORMAT(F8.4,F10.4)
      END
C=====================================================================
      SUBROUTINE GAUSS(NRHS)
C
C        SOLUTION OF LINEAR ALGEBRAIC SYSTEM BY
C        GAUSS ELIMINATION WITH PARTIAL PIVOTING
C
C                   [A]        =  COEFFICIENT MATRIX
C                   NEQNS      =  NUMBER OF EQUATIONS
C                   NRHS       =  NUMBER OF RIGHT-HAND SIDES
C
C                   RIGHT-HAND SIDES AND SOLUTIONS STORED IN
C                   COLUMNS NEQNS+1 THRU NEQNS+NRHS OF [A]
C
      COMMON /COF/ A(101,111),NEQNS
      NP          =  NEQNS + 1
      NTOT        =  NEQNS + NRHS
C
C                    GAUSS REDUCTION
C
      DO 150      I = 2,NEQNS
C
C                    --  SEARCH FOR LARGEST ENTRY IN (I-1)TH COLUMN
C                        ON OR BELOW MAIN DIAGONAL
C
      IM          =  I - 1
      IMAX        =  IM
      AMAX        =  ABS(A(IM,IM))
      DO 110      J = I,NEQNS
      IF (AMAX .GE. ABS(A(J,IM)))          GO TO 110
      IMAX        =  J
      AMAX        =  ABS(A(J,IM))
  110 CONTINUE
```

```
C
C                    --  SWITCH (I-1)TH AND IMAXTH EQUATIONS
C
      IF (IMAX .NE. IM)      GO TO 140
      DO 130      J = IM,NTOT
      TEMP        =  A(IM,J)
      A(IM,J)     =  A(IMAX,J)
      A(IMAX,J)   =  TEMP
  130 CONTINUE
C
C                    ELIMINATE (I-1)TH UNKNOWN FROM
C                    ITH THRU (NEQNS)TH EQUATIONS
C
  140 DO 150      J = I,NEQNS
      R           =  A(J,IM)/A(IM,IM)
      DO 150      K = I,NTOT
  150 A(J,K)      =  A(J,K) - R*A(IM,K)
C
C                    BACK SUBSTITUTION
C
      DO 220      K = NP,NTOT
      A(NEQNS,K)  =  A(NEQNS,K)/A(NEQNS,NEQNS)
      DO 210      L = 2,NEQNS
      I           =  NEQNS + 1 - L
      IP          =  I + 1
      DO 200      J = IP,NEQNS
  200 A(I,K)      =  A(I,K) - A(I,J)*A(J,K)
  210 A(I,K)      =  A(I,K)/A(I,I)
  220 CONTINUE
      RETURN
      END
C     PROGRAM JOUKOW
C
C        FIND SHAPE OF AND PRESSURE DISTRIBUTION ON
C        SYMMETRIC JOUKOWSKI AIRFOIL AT ZERO ANGLE OF ATTACK
C
      COMPLEX     Z,ZTILDE,ZPRIME,WPRIME,I
      I           =  (0.0,1.0)
      PI          =  3.1415926585
      PRINT 1000
      READ,       EPS
      EPS         =  - ABS(EPS)
      A           =  1.0 - EPS
      PRINT 1010
  100 READ,       THETA
      ZTILDE      =  A*CEXP(I*THETA*PI/180.) + EPS
      Z           =  ZTILDE + 1./ZTILDE
      ZPRIME      =  1. - 1./ZTILDE**2
      IF (CABS(ZPRIME) .LT. 1.E-10)      GO TO 200
      WPRIME      =  (1. - (A/(ZTILDE - EPS))**2)/ZPRIME
      CP          =  1. - CABS(WPRIME)**2
      PRINT 1020, Z,WPRIME,CP
      GO TO 100
  200 PRINT 1030
      GO TO 100
 1000 FORMAT(' INPUT THICKNESS PARAMETER EPS',)
 1010 FORMAT(' RESPOND TO QUESTION MARKS WITH THETA, IN DEGREES',
     +            //,7X,'X',9X,'Y',9X,'U',9X,'V',8X,'CP')
 1020 FORMAT(5F10.5)
 1030 FORMAT(' SINGULAR THETA')
      END
```

CHAPTER 4

LIFTING AIRFOILS IN INCOMPRESSIBLE IRROTATIONAL FLOW

We now remove the restrictions imposed on the methods developed in the last chapter, that the airfoil be symmetric about its chord line and at zero angle of attack. Consideration of airfoils with camber and/or at angle of attack is necessary, of course, if we are to get to the central problem of aerodynamics: the prediction and management of the lift force. To begin with, we shall work within the approximations of thin-airfoil theory but then develop a simple version of the widely used panel method and also an exact inverse method based on complex variables.

4.1. THE THIN-AIRFOIL THICKNESS AND CAMBER PROBLEMS

Consider a uniform flow past an airfoil whose upper surface is located at $y = Y_u(x)$ and lower surface at $y = Y_l(x)$, as shown in Fig. 4.1. By convention, the coordinates are arranged so that the *leading edge* is at the origin, and the *trailing edge* is on the x axis, at $x = c$ (the chord). The inclination to the x axis of the velocity vector of the uniform flow at infinity is then the *angle of attack* α.

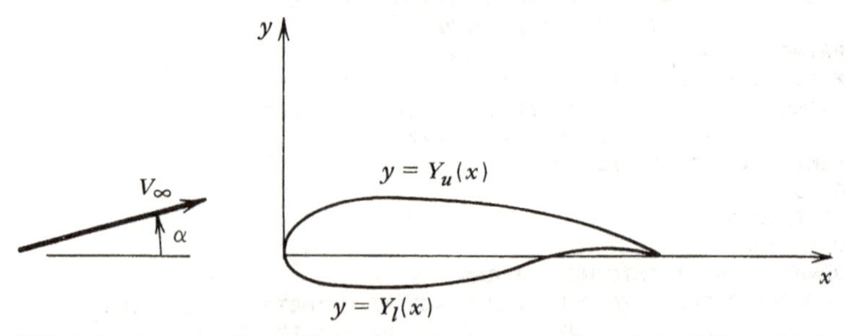

FIG. 4.1. Nomenclature for study of aerodynamics of the lifting airfoil.

Let $T(x)$ and $\bar{Y}(x)$ be the *thickness* and *camber* functions, respectively,

$$T(x) \equiv Y_u(x) - Y_l(x)$$
$$\bar{Y}(x) \equiv \tfrac{1}{2}[Y_u(x) + Y_l(x)] \qquad (4\text{-}1)$$

$\bar{Y}(x)$ then locates the mean line of the airfoil, and

$$Y_u = \bar{Y} + \tfrac{1}{2}T$$
$$Y_l = \bar{Y} - \tfrac{1}{2}T \qquad (4\text{-}2)$$

On the surface of the airfoil, we need to satisfy the flow tangency condition

$$\begin{aligned} v &= u\frac{dY_u}{dx} \quad \text{at } y = Y_u(x) \\ &\qquad\qquad\qquad\qquad \text{for } 0 < x < c \\ &= u\frac{dY_l}{dx} \quad \text{at } y = Y_l(x) \end{aligned} \qquad (4\text{-}3)$$

As in Chapter 3, we will approximate these equations for the case of a thin airfoil. We suppose first of all that it is thin enough that $u \approx V_\infty$ and further that we may apply the boundary conditions on the x axis. More precisely, we approximate the values of v on the upper and lower surfaces of the airfoil by its values on the upper and lower sides of the x axis, respectively,

$$v[x, Y_u(x)] \approx v(x, 0+)$$
$$v[x, Y_l(x)] \approx v(x, 0-) \qquad (4\text{-}4)$$

where

$$0+ \equiv \lim_{z \downarrow 0} z, \qquad 0- \equiv \lim_{z \uparrow 0} z$$

Thus, in the thin-airfoil approximation, the flow tangency conditions (4-3) become

$$\begin{aligned} v(x, 0+) &\approx V_\infty\left(\frac{d\bar{Y}}{dx} + \frac{1}{2}\frac{dT}{dx}\right) \\ &\qquad\qquad\qquad\qquad \text{for } 0 < x < c \\ v(x, 0-) &\approx V_\infty\left(\frac{d\bar{Y}}{dx} - \frac{1}{2}\frac{dT}{dx}\right) \end{aligned} \qquad (4\text{-}5)$$

The values of u and C_p on the body surface will also be approximated by their values at $y = 0\pm$. Maintaining a distinction between the upper and lower sides of the x axis allows u, v, and C_p to be discontinuous across the x axis, as they are across the airfoil. Note that even in those frequent cases for which the lower surface of the airfoil pops above the x axis (see Fig. 4.1), we still approximate the values of u, v, and C_p on the lower surface by their values at $y = 0-$.

It is convenient now to break up the velocity potential (and stream function) into three parts:

1. ϕ_∞, which will be used to satisfy the boundary condition at infinity.

2. ϕ_T, which will take care of the thickness terms in the body boundary condition.
3. ϕ_C, which will be associated with camber and/or the angle of attack.

We then write

$$\phi = \phi_\infty + \phi_T + \phi_C \tag{4-6}$$

We will require each term in equation 4-6 to satisfy Laplace's equation, so that ϕ will, also.

Specifically, ϕ_∞ is defined as the potential of a uniform flow whose velocity is V_∞:

$$\phi_\infty = V_\infty x \cos\alpha + V_\infty y \sin\alpha \tag{4-7}$$

If we then require the gradients of the thickness term ϕ_T and the camber term ϕ_C to vanish far from the body,

$$\nabla\phi \to \mathbf{V}_\infty$$

and the boundary condition at infinity is satisfied. In keeping with the thin-airfoil approximation, we suppose that the angle of attack α is small enough that equation 4-7 can be approximated by

$$\phi_\infty \approx V_\infty(x + \alpha y) \tag{4-8}$$

The associated velocity components are then V_∞ and $V_\infty \alpha$, respectively.

The thickness term is defined so as to satisfy the antisymmetric part of the flow tangency condition (4-5)

$$v_T = \frac{\partial \phi_T}{\partial y} = \pm\frac{1}{2}V_\infty\frac{dT}{dx} \qquad \text{at } y = 0\pm \text{ for } 0 < x < c \tag{4-9}$$

whereas the camber term must then be such that the total y velocity satisfies the boundary condition (4-5):

$$V_\infty\alpha + v_T + v_C = V_\infty\left(\frac{d\bar{Y}}{dx} \pm \frac{1}{2}\frac{dT}{dx}\right) \qquad \text{at } y = 0\pm \text{ for } 0 < x < c$$

Thus, on subtracting equation 4-9, we get the requirement

$$v_C = \frac{\partial \phi_c}{\partial y} = V_\infty\left(\frac{d\bar{Y}}{dx} - \alpha\right) \qquad \text{at } y = 0\pm \text{ for } 0 < x < c \tag{4-10}$$

We shall show that the thickness problem is solved by a distribution of sources along the x axis, whereas the camber problem requires a distribution of vortices. Any such distributions will, we know, meet the requirements of continuity and irrotationality and die out far from the body. All that has to be shown, then, is that source and vortex distributions can be used to satisfy the respective boundary conditions on the body (rather, on the x axis) of the thickness and camber problems.

To do so, we first look at the velocity field due to a source distribution of strength $q(t)$ per unit length. From equations 3-26 and 3-27,

$$v_s = \int_0^c \frac{q(t)}{2\pi} \frac{y}{(x-t)^2 + y^2}\, dt \tag{4-11}$$

$$u_s = \int_0^c \frac{q(t)}{2\pi} \frac{x-t}{(x-t)^2 + y^2}\, dt \tag{4-12}$$

Redoing the process that led to equation 3-30 with more attention to the sign of y yields

$$\lim_{y \to 0} v_s(x, y) = \frac{q(x)}{2\pi} \begin{cases} \dfrac{\pi}{2} - \left(-\dfrac{\pi}{2}\right) & \text{if } y > 0 \\ -\dfrac{\pi}{2} - \left(+\dfrac{\pi}{2}\right) & \text{if } y < 0 \end{cases}$$

Thus

$$v_s(x, 0\pm) = \pm\tfrac{1}{2} q(x) \tag{4-13}$$

Comparing this with the boundary condition on the thickness part of the problem, equation 4-9, we see that a source distribution along the chord line of the airfoil meets the requirements of the thickness contribution to the solution if its strength is

$$q(x) = V_\infty T'(x) \tag{4-14}$$

After accounting for the change in nomenclature, you should see that this is exactly the same as the result derived in Chapter 3 (equation 3-31, to be specific) for the symmetric airfoil at zero angle of attack.

Thus, we set $\phi_T = \phi_s$, $v_T = v_s$, and $u_T = u_s$. From equation 3-33, we obtain the following approximation for u_T on the body surface:

$$u_T(x, 0\pm) = \frac{V_\infty}{2\pi} \mathrm{P}\!\!\int_0^c T'(t) \frac{dt}{x-t} \qquad \text{for } 0 < x < c \tag{4-15}$$

Note that u_T is continuous across the x axis, whereas v_T is discontinuous,

$$\begin{aligned} u_T(x, 0+) &= u_T(x, 0-) \\ v_T(x, 0+) &= -v_T(x, 0-) \end{aligned} \tag{4-16}$$

The boundary condition on the camber part of the solution, equation 4-10, shows v_C to be continuous across the x axis. However, u_C must be allowed to be discontinuous:

$$u_C(x, 0+) \neq u_C(x, 0-)$$

Otherwise $u = V_\infty + u_C + u_T$ would be continuous across the x axis, which would imply u was the same on the upper and lower surfaces of the airfoil.

You may recall from equations 2-49 and 2-52 that the potential due to a point vortex has the same functional form as the stream function due to a point source, so that the x component of the velocity field due to one type of singularity looks like the

y velocity component due to the other. Given the behavior we expect near the x axis of u_C and v_C, along with what we know about u_T and v_T and the fact that the thickness problem is solved by a distribution of sources along the x axis, we therefore "try" representing ϕ_C by a distribution of vortices along the x axis of strength per unit length γ, say,

$$\phi_C = -\int_0^c \frac{\gamma(t)}{2\pi} \tan^{-1}\left(\frac{y}{x-t}\right) dt \tag{4-17}$$

The derivatives of ϕ_C in the limit as $y \to 0\pm$ have, in effect, already been evaluated in our study of source distributions (see the derivations of equations 3-30 and 3-33):

$$\begin{aligned} u_C(x, 0\pm) &= \lim_{y \to 0\pm} \frac{\partial \phi_C}{\partial x} \\ &= +\lim_{y \to 0\pm} \int_0^c \frac{\gamma(t)}{2\pi} \frac{y}{(x-t)^2 + y^2} dt \\ &= \pm\tfrac{1}{2}\gamma(x) \qquad \text{for } 0 < x < c \end{aligned} \tag{4-18}$$

$$\begin{aligned} v_C(x, 0\pm) &= -\lim_{y \to 0\pm} \int_0^c \frac{\gamma(t)}{2\pi} \frac{x-t}{(x-t)^2 + y^2} dt \\ &= -⨍_0^c \frac{\gamma(t)}{2\pi} \frac{dt}{x-t} \qquad \text{for } 0 < x < c \end{aligned} \tag{4-19}$$

Thus we see that u_C is, as desired, discontinuous across the x axis, whereas v_C is continuous. Substituting v_C from equation 4-19 into the boundary condition (4-10), we get

$$+\frac{1}{2\pi} ⨍_0^c \gamma(t) \frac{dt}{x-t} = V_\infty\left(\alpha - \frac{d\bar{Y}}{dx}\right) \qquad \text{for } 0 < x < c \tag{4-20}$$

Thus, to find the vortex strength, we must solve an *integral equation*. In a sense, this is what we did to find the source strength in order to satisfy equation 4-9, but the solution of equation 4-20 is much more difficult. On the other hand, once we find γ, the camber contribution to the x component of velocity on the airfoil is easily found from equation 4-18, whereas the calculation of the thickness contribution from equation 4-15 is often hard.

To summarize, corresponding to the splitting of the velocity potential into three parts, equation 4-6, the velocity components of the uniform flow past a thin airfoil can be written

$$\begin{aligned} u &= V_\infty + u_T + u_C \\ v &= V_\infty \alpha + v_T + v_C \end{aligned} \tag{4-21}$$

the first terms of which were found by taking the gradient of the approximation for ϕ_∞ given in equation 4-8. On the body surface, u_T, u_C, v_T, and v_C are given by equations 4-15, 4-18, 4-9, and 4-10, respectively, with the γ needed in equation 4-18 being defined as the solution of the integral equation 4-20.

Once we determine the velocity on the body surface, the pressure distribution can be calculated from Bernoulli's equation, equation 3-2,

$$p + \tfrac{1}{2}\rho(u^2 + v^2) = p_\infty + \tfrac{1}{2}\rho V_\infty^2 \tag{4-22}$$

This, too, can be simplified by thin-airfoil type approximations. Since, from equations 4-15, 4-18, and 4-20, u_T is proportional to the airfoil thickness, and u_C to the angle of attack and camber, we generally have

$$u_T + u_C \ll V_\infty$$

From equations 4-9 and 4-10, the same is true of v_T and v_C. Thus, from equations 4-21 and 4-22,

$$\begin{aligned} p &= p_\infty + \tfrac{1}{2}\rho V_\infty^2 - \tfrac{1}{2}\rho(V_\infty + u_T + u_C)^2 - \tfrac{1}{2}\rho(V_\infty\alpha + v_T + v_C)^2 \\ &\approx p_\infty - \rho V_\infty(u_T + u_C) \end{aligned} \tag{4-23}$$

in which we have discarded terms quadratic in small quantities. Compare equation 3-35.

4.2. FORCES AND MOMENTS ON A THIN AIRFOIL

We now want to relate the net force and moment on a thin airfoil to its pressure distribution and so to the strengths of the associated source and vortex distributions. The x and y components of the force per unit width on an airfoil were shown in equations 2-4 to be[1]

$$F'_x = \int_0^c \left(p_u \frac{dY_u}{dx} - p_l \frac{dY_l}{dx} \right) dx \tag{4-24}$$

$$F'_y = \int_0^c (p_l - p_u)\, dx \tag{4-25}$$

In the thin-airfoil approximation, quantities referring to the upper surface of the airfoil are approximated by evaluation at $y = 0+$ and lower-surface quantities are evaluated at $y = 0-$. Thus, from the thin-airfoil approximation to Bernoulli's equation, equation 4-23,

$$\begin{aligned} p_u &\simeq p_\infty - \rho V_\infty[u_T(x, 0+) + u_c(x, 0+)] \\ p_l &\simeq p_\infty - \rho V_\infty[u_T(x, 0-) + u_c(x, 0-)] \end{aligned} \tag{4-26}$$

Substituting for u_T and u_C from equations 4-15 and 4-18, we obtain

$$p_u - p_l \approx -\rho V_\infty \gamma$$

[1] Here and elsewhere, the primes on forces and moments indicate they are two-dimensional values, or per unit width in the spanwise direction.

and so, from equation 4-25,

$$F'_y \approx \rho V_\infty \int_0^c \gamma(x)\,dx = \rho V_\infty \Gamma \tag{4-27}$$

where Γ is the net *circulation* around the airfoil:

$$\Gamma \equiv \int_0^c \gamma(x)\,dx \tag{4-28}$$

The x component of the force is a bit more complicated. We first substitute equations 4-26 into 4-24 and then call on 4-15 and 4-18 to get

$$\begin{aligned} F'_x &\approx p_\infty \int_0^c \left(\frac{dY_u}{dx} - \frac{dY_l}{dx}\right) dx - \rho V_\infty \int_0^c u_T(x,0)\left(\frac{dY_u}{dx} - \frac{dY_l}{dx}\right) dx \\ &\quad - \rho V_\infty \int_0^c u_C(x,0+)\left(\frac{dY_u}{dx} + \frac{dY_l}{dx}\right) dx \\ &\approx p_\infty (Y_u - Y_l)\Big|_0^c - \rho V_\infty \int_0^c \left[\frac{V_\infty}{2\pi} \fint_0^c \frac{T'(t)\,dt}{x-t}\right] T'(x)\,dx \\ &\quad - \rho V_\infty \int_0^\infty \frac{\gamma(x)}{2} \cdot 2\frac{d\bar{Y}}{dx}(x)\,dx \end{aligned} \tag{4-29}$$

Note that the thickness and camber functions have been introduced from equations 4-2.

The term integrated out vanishes, since $Y_u = Y_l = 0$ at both $x = 0$ and $x = c$. The first integral is of the form

$$I \equiv \int_0^c dx \fint_0^c dt \frac{f(t)f(x)}{x-t}$$

which can be regarded as an integral over a square in the x–t plane, as shown in Fig. 4.2. The integrand is antisymmetric with respect to the line $x = t$; for example,

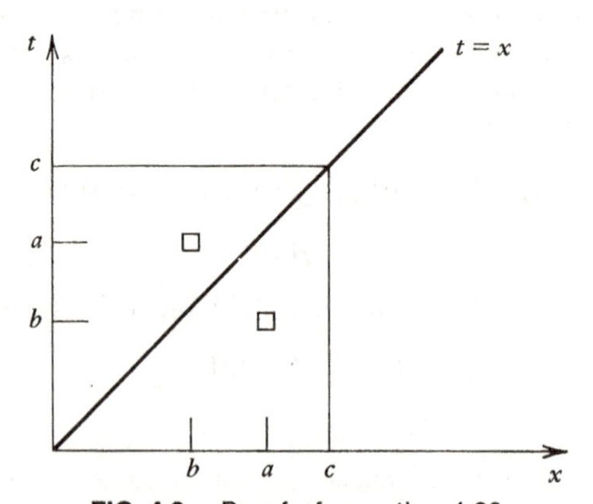

FIG. 4.2. Proof of equation 4-30.

its value at $x = a, t = b$ is

$$\frac{f(b)f(a)}{b - a}$$

whereas at $x = b, t = a$ it is

$$\frac{f(a)f(b)}{a - b}$$

Thus the value of the integral is zero:

$$\int_0^c dx \, ⨍_0^c dt \frac{f(t)f(x)}{x - t} = 0 \tag{4-30}$$

and the first integral in equation 4-29 vanishes. Then, using the integral equation 4-20 to eliminate $d\bar{Y}/dx$, we are left with

$$F'_x \approx \rho V_\infty \int_0^c \gamma(x)\left[-\alpha + \frac{1}{2\pi V_\infty} ⨍_0^c \frac{\gamma(t)\,dt}{x - t}\right] dx$$

The double integral is of the form shown in equation 4-30 to vanish, leaving

$$F'_x \approx -\rho V_\infty \alpha \int_0^c \gamma(x)\,dx = -\rho V_\infty \alpha \Gamma \tag{4-31}$$

in which the net circulation defined by equation 4-28 has been introduced.

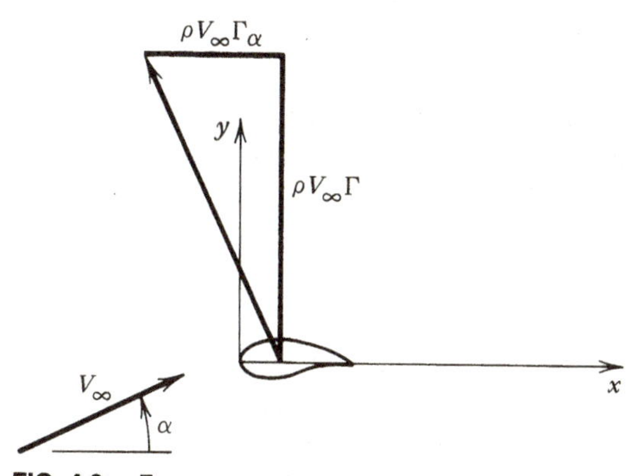

FIG. 4.3. Force on a thin airfoil with circulation Γ.

From the display of the results of equations 4-27 and 4-31 in Fig. 4.3, we see that, in the thin-airfoil approximation, the net force on the airfoil is directed at an angle of $\tan^{-1}\alpha$ to the y axis. But, for small α, $\tan^{-1}\alpha \approx \alpha$, so the net force is approximately perpendicular to the free stream. In fact, this is true even for "fat" airfoils at large angles of attack; recall the results for flow past a circular cylinder with circulation

[2] A general proof is outlined in problem 6 of Chapter 8.

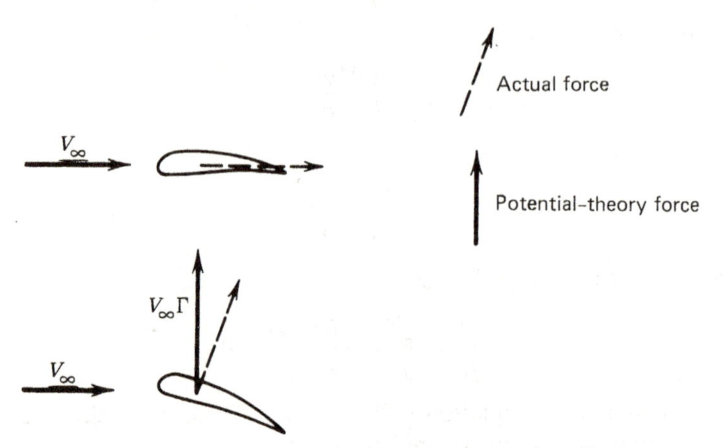

FIG. 4.4. Comparison of results of Kutta–Joukowski theorem with force actually felt by an airfoil.

that you got in problem 7 of Chapter 2.[2] The general result is called the

Kutta–Joukowski Theorem *An isolated two-dimensional airfoil in an incompressible inviscid flow feels a force per unit width of*

$$\mathbf{F}' = \rho_\infty \mathbf{V}_\infty \times \mathbf{\Gamma} \tag{4-32}$$

where Γ is the net circulation around the airfoil, and the direction of $\mathbf{\Gamma}$ is along the generator of the airfoil, the sense being determined by the right-hand rule.

Since the *lift* and *drag* forces are defined as the components of the aerodynamic force on a body in directions perpendicular and parallel to the onset flow, respectively, we thus have, according to the Kutta–Joukowski theorem,

$$L' = \rho V_\infty \Gamma$$
$$D' = 0 \tag{4-33}$$

in which L' and D' are the lift and drag per unit width. For an airfoil in a "real" (i.e., slightly viscous) flow, the second line of equation 4-33 is infinitely wrong, percentagewise. But, on a well-designed lifting airfoil, the true force is in fact close to that predicted by "potential theory," as is illustrated in Fig. 4.4.

That the net force on an airfoil is always perpendicular to the onset flow may sometimes seem peculiar. For example, suppose the airfoil is just a flat plate, with no thickness at all, as shown in Fig. 4.5. Wouldn't you expect the net force to be perpendicular to the plate? Well, it isn't. There is a component of force along the plate even in this case. To understand this, regard the plate as the limit of some airfoil that has finite thickness. The flow stagnates on the lower side of the leading edge, and must accelerate mightily to get around the leading edge, as sketched in Fig. 4.6. High velocity at the leading edge means low pressure there, which results in the limit in what is called the *leading edge suction force.*

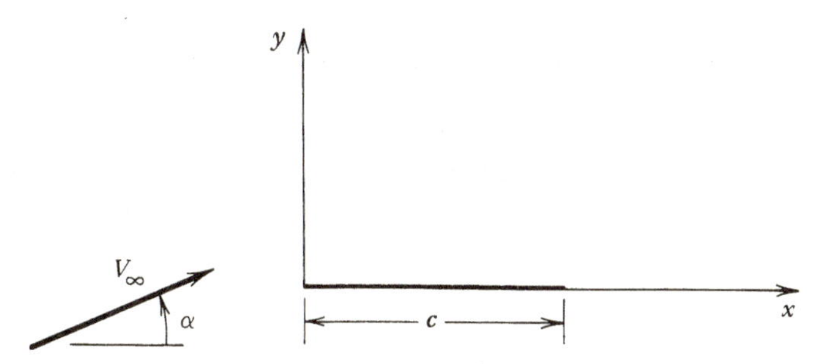

FIG. 4.5. A flat-plate airfoil at angle of attack.

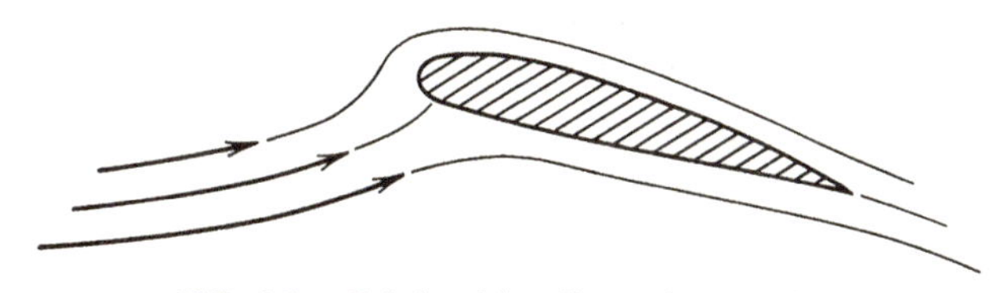

FIG. 4.6. Origin of leading-edge suction.

The moment of the forces per unit width on a thin airfoil was shown in equation 2-6 to be

$$\circlearrowleft\!\!+ M'_{\text{l.e.}} \approx -\int_0^c x(p_l - p_u)\, dx$$

where the subscript "l.e." indicates we are taking moments around the "leading edge" of the airfoil ($x = 0$). Substituting for the pressure difference from equation 4-27 gives

$$\circlearrowleft\!\!+ M'_{\text{l.e.}} \approx -\rho V_\infty \int x\gamma(x)\, dx$$

The ratio of $M'_{\text{i.e.}}$ to the force locates the *center of pressure* on the airfoil:

$$x_{\text{c.p.}} \equiv -\frac{M'_{\text{l.e}}}{L'} \approx \frac{\int_0^c x\gamma(x)\, dx}{\int_0^c \gamma(x)\, dx} \tag{4-34}$$

This can be thought of as the point where the lift acts, so far as $M'_{\text{l.e}}$ is concerned; see Fig. 4.7.

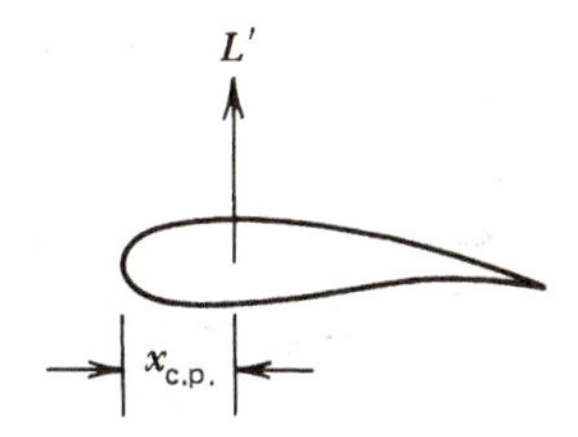

FIG. 4.7. The center of pressure.

Example 1: Symmetric Airfoil at Angle of Attack Suppose $Y_u(x) = -Y_l(x)$, so, from equation 4-1, $\bar{Y}(x) = 0$, as shown in Fig. 4.8. Let the airfoil be at an angle of attack α. We want to find the lift and moment on the airfoil, which, as equations 4-33 and 4-34 show, depend on the strength of the vortex distribution. Thus we must now turn to the solution of the integral equation 4-20, which reduces in this case to

$$\mathrm{P}\int_0^c \frac{\gamma(t)\,dt}{x-t} = 2\pi V_\infty \alpha \qquad \text{for } 0 < x < c \tag{4-35}$$

Notice that the integrand depends on x, but, according to the right side, the integral does not. This may remind you of the behavior of a similar integral we encountered in computing the velocity distribution on a thin ellipse; see equations 3-33 and 3-41. Let us therefore reintroduce the same changes of variables that were helpful then:

$$x = \frac{c}{2}(1 - \cos\theta_0)$$

$$t = \frac{c}{2}(1 - \cos\theta) \tag{4-36}$$

Also, let

$$\gamma(t) = g(\theta) \tag{4-37}$$

Then the integral equation 4-35 becomes

$$\int_0^\pi \frac{g(\theta)\sin\theta\,d\theta}{\cos\theta - \cos\theta_0} = 2\pi V_\infty \alpha \tag{4-38}$$

Again (and not for the last time) we pull from Appendix A the result

$$\int_0^\pi \frac{\cos n\theta}{\cos\theta - \cos\theta_0}\,d\theta = \pi\frac{\sin n\theta_0}{\sin\theta_0} \tag{4-39}$$

Since the right side of this integral reduces to π when $n = 1$, it looks like the integrand of equation 4-38 should be

$$\sin\theta\, g(\theta) = 2V_\infty \alpha \cos\theta$$

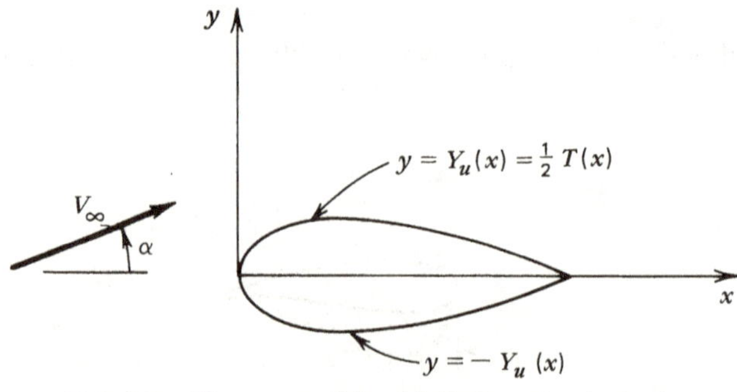

FIG. 4.8. The symmetric airfoil at angle of attack.

But equation 4-39 also shows that when $n = 0$ the integral vanishes, though the numerator of the integrand reduces to 1. Therefore, we can add to $\sin\theta\, g(\theta)$ any multiple of $\cos 0\theta = 1$ (i.e., any constant) *without changing the result of the integral on the left side of equation 4-38*:

$$\sin\theta\, g(\theta) = 2V_\infty \alpha \cos\theta + k$$

In terms of the undetermined constant k, the solution is

$$g(\theta) = \frac{k}{\sin\theta} + 2V_\infty \alpha \frac{\cos\theta}{\sin\theta} \tag{4-40}$$

It may be slightly disconcerting that this satisfies the integral equation 4-38 regardless of the value of the constant k. Let's see if that constant makes any difference in computing such quantities of interest as the lift on the airfoil. From the Kutta–Joukowski theorem (4-33), the lift per unit span on the airfoil is proportional to Γ, the circulation about the airfoil, which is related to γ by equation 4-28. Substituting equations 4-36 and 4-37 into equation 4-28 gives

$$\Gamma = \int_0^\pi g(\theta_0) \frac{c}{2} \sin\theta_0\, d\theta_0 \tag{4-41}$$

With $g(\theta_0)$ given by equation 4-40, the result is

$$\Gamma = \frac{kc\pi}{2} \tag{4-42}$$

Thus our as yet undetermined constant is not at all unimportant; the lift is directly proportional to it!

4.3. THE KUTTA CONDITION

We thus need a new physical principle. We set out to find a potential in the form of equation 4-6 and have now fixed all three of its components: ϕ_∞, by equation 4-8; ϕ_T, by taking it to be the potential of a source distribution whose strength we found in equation 4-14; and ϕ_C, by equation 4-17, with γ supposedly being determined by equation 4-20. For any γ that satisfies equation 4-20, the potential of equation 4-6 satisfies the conditions of continuity, irrotationality, and flow tangency on the airfoil surface and behaves as it should far from the airfoil. Yet, from the symmetric-airfoil example, these conditions are apparently insufficient even to determine the lift on the airfoil.

The missing principle is supplied by examining the flow near the trailing edge. For most airfoils the trailing edge is relatively sharp. A local analysis of the potential flow near a sharp convex edge (Appendix C) shows that there are two alternatives:

1. The velocity at the trailing edge is infinite.
2. The flow leaves the trailing edge smoothly, along the extension of the bisector of the trailing edge angle (the line $\theta = 0$ in Fig. 4.9).

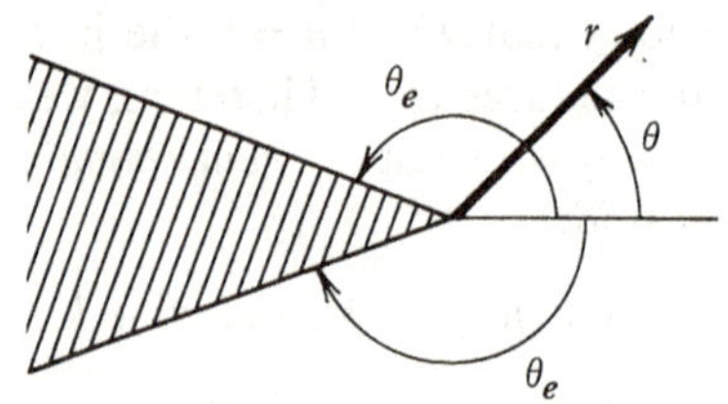

FIG. 4.9. Nomenclature for the flow near a sharp trailing edge.

Although the second alternative may seem more plausible, remember that we have idealized the flow as nonviscous. Singularities are acceptable, on the grounds that they may be relieved in reality by viscous effects.

In fact, however, if the flow is considered to be viscous, the first alternative is forbidden. This will be shown in detail in Chapter 7, since it is truly an effect of boundary-layer separation. For the moment, we assert the

> ***Kutta Condition*** *The flow leaves the trailing edge of a sharp-tailed airfoil smoothly; that is, the velocity is finite there.*

From the local analysis of Appendix C, we can derive these corollaries of the Kutta condition, which are often even more useful:

1. The streamline that leaves a sharp trailing edge is an extension of the bisector of the trailing edge angle.
2. Near the trailing edge, the flow speeds on the upper and lower surfaces of the airfoil are equal at equal distances from the trailing edge.
3. Unless the trailing edge is cusped ($\theta_e = \pi$ in Fig. 4.9), the flow stagnates there.

Returning now to the thin-airfoil approximation, we must be selective in how we state the Kutta condition. The last corollary, for example, is useless here; thin-airfoil theory always breaks down at a stagnation point and so cannot be expected to describe a stagnation point. The second corollary, however, may be translated into a requirement that the jump in u across the x axis vanishes at the trailing edge:

$$u(x, 0+) - u(x, 0-) \to 0 \text{ as } x \to c$$

Then equation 4-18 may be invoked to give this fourth corollary:

4. For a thin airfoil, the Kutta condition is that the vortex strength vanishes at the trailing edge.

Example 1: Symmetric Airfoil at Angle of Attack (Concluded) Let us now see how the Kutta condition fixes the circulation of a thin symmetric airfoil. First, translate equation 4-40 into Cartesian coordinates. From equation 4-36,

$$\sin\theta_0 = (1 - \cos^2\theta_0)^{1/2}$$

$$= 2\left[\frac{x}{c}\left(1 - \frac{x}{c}\right)\right]^{1/2}$$

and so, from equations 4-37 and 4-40,

$$\gamma(x) = g(\theta_0) = \frac{k + 2V_\infty\alpha\left(1 - 2\dfrac{x}{c}\right)}{2\sqrt{\dfrac{x}{c}\left(1 - \dfrac{x}{c}\right)}}$$

To satisfy the Kutta condition, pick k so that $\gamma(x)$ vanishes at $x = c$, namely,

$$k = 2V_\infty\alpha$$

Then

$$\gamma(x) = 2V_\infty\alpha\frac{\sqrt{1 - (x/c)}}{\sqrt{x/c}} \tag{4-43a}$$

and

$$g(\theta) = 2V_\infty\alpha\frac{1 + \cos\theta}{\sin\theta} \tag{4-43b}$$

There is still a singularity at the leading edge, but this is an acceptable approximation to the true situation. The flow stagnates on the lower surface near the leading edge, as sketched in Fig. 4.6. Although this edge is usually much less sharp than the trailing edge, the flow does accelerate to quite high speeds in rounding the leading edge and crossing to the upper surface.

In problem 2, you will show that the section lift and moment for the symmetric airfoil can now be found to be

$$L' = \pi\rho V_\infty^2\alpha c$$

$$\circlearrowleft M'_{\text{l.e.}} = -\frac{\pi}{4}\rho V_\infty^2\alpha c^2 \tag{4-44}$$

so that, from equation 4-34, the center of pressure is located at

$$x_{\text{c.p}} = -\frac{M'_{\text{l.e.}}}{L'} = \frac{c}{4} \tag{4-45}$$

4.4. CIRCULATION SPECIFICATION

You may be somewhat disturbed by the discovery that the conditions we set forth in Chapter 3—continuity, irrotationality, and boundary conditions on the airfoil surface and at infinity—are not sufficient to determine even the lift of an isolated airfoil. By introducing the Kutta condition, we were able to complete the solution, at least in this particular case. But the question then arises, does this device always work? Exactly what conditions *are* sufficient to specify the flow about a given airfoil uniquely? The answer, as shown in Appendix D, is that two solutions of the same flow problem are identical if they meet the four requirements stated above, and

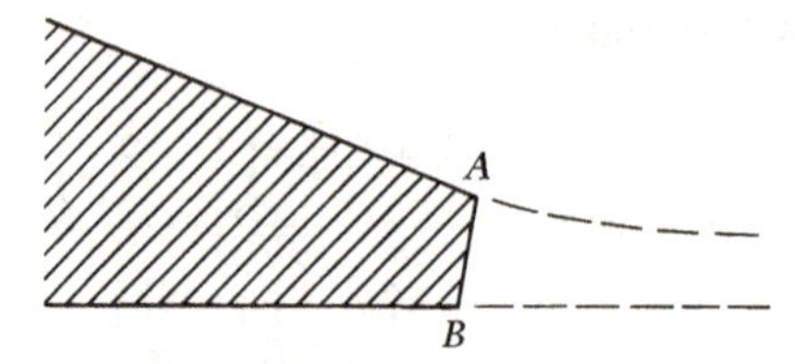

FIG. 4.10. Model for the flow near a thick trailing edge.

further, they produce the same circulation about the airfoil. The Kutta condition works, then, because it fixes the circulation about the airfoil.

However, the Kutta condition is restricted to sharp-tailed airfoils, which, strictly speaking, do not exist. And, as will be discussed in Chapter 7, we sometimes approximate the effects of viscosity on the pressure distribution by adding to the airfoil contour a measure of the boundary layer thickness. Although the sharp-edge approximation is often appropriate, how does one deal with cases in which it is not?

The usual procedure is to assume that there emanates from the trailing edge a thin wake, so thin and straight that it cannot support a pressure difference. The circulation-fixing condition is then that the pressure is continuous across the trailing edge, so that the velocities at points A and B in Fig. 4.10 are, from Bernoulli's equation, equal.

In the limit of a sharp trailing edge, this condition agrees with what we called "corollary 2" of the Kutta condition. For unsteady flows past thin airfoils, the circulation is also fixed by requiring continuity of pressure across the trailing edge. Because the Bernoulli equation must be modified if the flow is unsteady, this does not imply continuity of velocity. Another exceptional case is the jet-flapped airfoil, in which high lift is achieved by shooting a jet out the trailing edge. In this case there is a pressure jump across the jet, proportional to the curvature of the jet. Some recent analyses of the flows near ordinary trailing edges also admit a pressure jump at the trailing edge, due to wake curvature, but are too complicated to be generally useful, at least as of this writing (1983).

4.5. THE CAMBERED THIN AIRFOIL

We return now to the thin airfoil to consider the case where the camber $\bar{Y}(x)$ is nonzero. The problem is to find $\gamma(x)$ so as to satisfy the integral equation 4-20 and the Kutta condition

$$\gamma(c) = 0 \tag{4-46}$$

In terms of the trigonometric substitutions introduced in equations 4-36 and 4-37, equation 4-20 becomes

$$\frac{1}{2\pi V_\infty} \mathrm{P}\!\!\int_0^\pi \frac{g(\theta) \sin \theta \, d\theta}{\cos \theta - \cos \theta_0} = \alpha - s(\theta_0) \qquad \text{for } 0 < \theta_0 < \pi \tag{4-47}$$

where

$$\bar{Y}'(x) = s(\theta_0) \tag{4-48}$$

whereas the Kutta condition (4-46) is

$$g(\pi) = 0 \tag{4-49}$$

Since we know how to evaluate the integral in equation 4-47 when the integrand is a multiple of $\cos n\theta$, one approach to its solution would be to try to represent $g(\theta)\sin\theta$ by a *Fourier cosine series* (see Appendix E)

$$g(\theta)\sin\theta = \sum_{n=0} b_n \cos n\theta$$

However, especially since $\sin\pi = 0$, it would then be difficult to determine the b_n's so as also to satisfy the Kutta condition (4-49). A better approach is to start out with a parametric form of g that automatically satisfies the Kutta condition, and then to worry about equation 4-47. Such a form is

$$g(\theta) = 2V_\infty\left(A_0\frac{1+\cos\theta}{\sin\theta} + \sum_{n=1} A_n \sin n\theta\right) \tag{4-50}$$

You can verify that every term of this representation vanishes at $\theta = \pi$. You would need l'Hospital's rule to prove this for the A_0 term, except for the fact that the solution of the symmetric-airfoil problem, equation 4-43b, is of exactly this form, which of course explains its presence in equation 4-50.

Thus, using equation 4-50 for $g(\theta)$,

$$g(\theta)\sin\theta = 2V_\infty\left[A_0(1+\cos\theta) + \sum_{n=1} A_n \sin n\theta \sin\theta\right]$$

The contribution of the A_0 term to the integral in equation 4-47 is easily evaluated with the help of equation 4-39. So can the rest of the terms, since

$$\sin n\theta \sin\theta = \tfrac{1}{2}\cos(n-1)\theta - \tfrac{1}{2}\cos(n+1)\theta$$

Thus we get

$$\begin{aligned}\alpha - s(\theta_0) &= \frac{1}{\pi} \mathrm{P}\!\!\int_0^\pi \frac{d\theta}{\cos\theta - \cos\theta_0} \\ &\quad\times\left\{A_0(1+\cos\theta) + \sum_{n=1} \tfrac{1}{2}A_n[\cos(n-1)\theta - \cos(n+1)\theta]\right\} \\ &= \frac{1}{\sin\theta_0}\left\{A_0 \sin\theta_0 + \sum_{n=1} \tfrac{1}{2}A_n[\sin(n-1)\theta_0 - \sin(n+1)\theta_0]\right\} \\ &= A_0 - \sum_{n=1} A_n \cos n\theta_0 \end{aligned} \tag{4-51}$$

Now we must determine the coefficients $A_0, A_1, \ldots$ in terms of α and $s(\theta_0)$. As when seeking the coefficients of a Fourier series (see Appendix E), multiply equation 4-51 by $\cos m\theta_0$, $m = 0, 1, 2, \ldots$, and integrate over θ_0 from 0 to π:

$$\int_0^\pi [\alpha - s(\theta_0)]\cos m\theta_0\, d\theta_0 = A_0\int_0^\pi \cos m\theta_0\, d\theta_0 - \sum_{n=1} A_n \int_0^\pi \cos m\theta_0 \cos n\theta_0\, d\theta_0$$

Because of the orthogonality of the cosine functions over the interval $(0, \pi)$, (see equation E-10), all but one of the integrals on the right side are zero, the exceptional integral being the coefficient of A_m. Thus we get

$$\int_0^\pi [\alpha - s(\theta_0)]\cos 0\theta_0 \, d\theta_0 = \pi\alpha - \int_0^\pi s(\theta_0)\, d\theta_0 = \pi A_0$$

so

$$A_0 = \alpha - \frac{1}{\pi}\int_0^\pi s(\theta_0)\, d\theta_0 \qquad \text{(4-52)}$$

and, for $m > 0$,

$$\int_0^\pi [\alpha - s(\theta_0)]\cos m\theta_0 \, d\theta_0 = -\int_0^\pi s(\theta_0)\cos m\theta_0 \, d\theta_0 = -\frac{\pi}{2}A_m$$

so

$$A_m = \frac{2}{\pi}\int_0^\pi s(\theta_0)\cos m\theta_0 \, d\theta_0 \qquad \text{for } m = 1, 2, \ldots \qquad \text{(4-53)}$$

The idea, then, is as follows. Given a formula for the camber line shape $\bar{Y}(x)$, differentiate it to get $\bar{Y}'(x)$, and then substitute for x from the first of equations 4-36 to get, according to equation 4-48, $s(\theta_0)$. Then use equations 4-52 and 4-53 to get the coefficients of the modified sine series expansion of $g(\theta)$, equation 4-50, from which you can determine $\gamma(x)$ by reversing the x to θ transformation.

Equation 4-51 amounts to a *Fourier cosine series* expansion of $s(\theta_0) = \bar{Y}'(x)$. As noted in Appendix E, such a series converges even if the function being expanded has jump discontinuities. Thus the camber line can have abrupt changes in slope, or kinks, and the series will still be useful.

Example 2: Parabolic Camber For a simple example, consider the parabolic camber line illustrated in Fig. 4.11.

$$\bar{Y}(x) = 4\varepsilon\frac{x}{c}(c - x) \qquad \text{(4-54)}$$

Then

$$\bar{Y}'(x) = 4\varepsilon\left(1 - 2\frac{x}{c}\right)$$

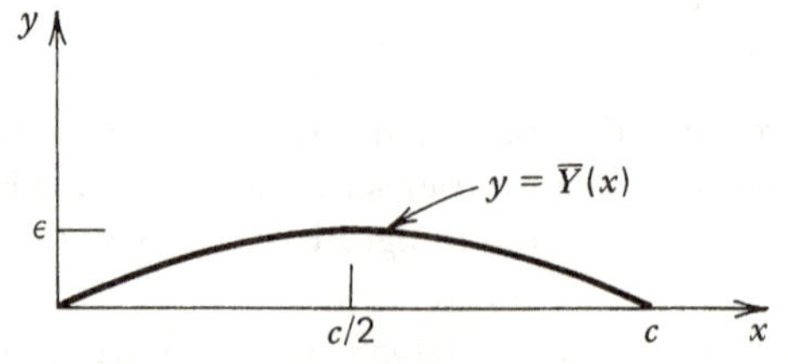

FIG. 4.11. The parabolic-arc airfoil.

so, from equations 4-36 and 4-48,

$$s(\theta_0) = 4\varepsilon[1 - (1 - \cos\theta_0)]$$
$$= 4\varepsilon \cos\theta_0 \quad (4\text{-}55)$$

Thus, in this case (and in any other where the camber line is described by a single polynomial), the Fourier coefficients of equation 4-51 can be identified without the integrations called for in equations 4-52 and 4-53. Just plugging equation 4-55 into 4-51, we see that

$$A_0 = \alpha$$
$$A_1 = 4\varepsilon$$
$$A_n = 0 \qquad \text{for } n \geq 2 \quad (4\text{-}56)$$

Of course, you get the same results by substituting equation 4-55 into equations 4-52 and 4-53.

4.6. AERODYNAMICS OF THE THIN AIRFOIL

Once the vortex strength is known, the pressure distribution could be found from the linearized Bernoulli equation 4-23. However, thin-airfoil theory is not accurate enough to give good data on the details of the pressue distribution. It is more useful for the lift, moment, and center of pressure, which we know in terms of $\gamma(x)$ from equations 4-28, 4-33, and 4-34.

The integrals in those equations are easier to evaluate if we make the trigonometric substitutions (4-36) and (4-37). Then, using equation 4-50 for $g(\theta)$, we find the section lift and moment to depend on only the first few A_n's. All the others get wiped out because of the orthogonality of the trigonometric functions, as you will show in problem 3:

$$L' = \pi\rho V_\infty^2 c(A_0 + \tfrac{1}{2}A_1)$$

$$M'_{\text{l.e.}} = \frac{-\pi}{4}\rho V_\infty^2 c^2(A_0 + A_1 - \tfrac{1}{2}A_2)$$

$$x_{\text{c.p.}} = \frac{c}{4}\frac{A_0 + A_1 - \tfrac{1}{2}A_2}{A_0 + \tfrac{1}{2}A_1} \quad (4\text{-}57)$$

Example 3 For the airfoil with parabolic camber, for which the A_n's are given by equation 4-56,

$$L' = \pi\rho V_\infty^2 c(\alpha + 2\varepsilon)$$

$$M' = -\frac{\pi}{4}\rho V_\infty^2 c^2(\alpha + 4\varepsilon)$$

$$x_{\text{c.p.}} = \frac{c}{4}\frac{\alpha + 4\varepsilon}{\alpha + 2\varepsilon} \quad (4\text{-}58)$$

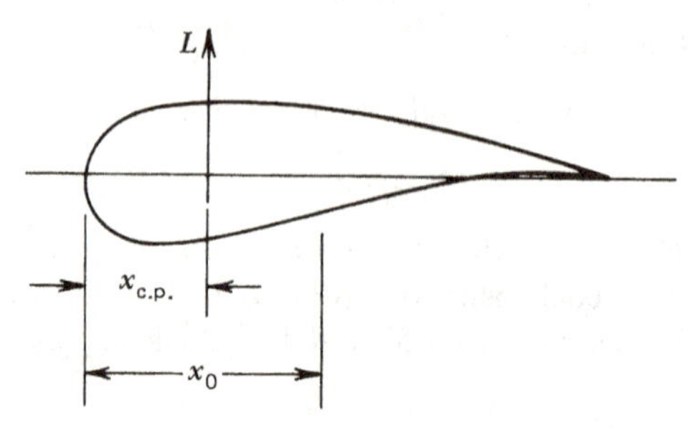

FIG. 4.12. The moment of the aerodynamic force about an arbitrary point on the airfoil.

As can be seen from the last of equations 4-58, the center of pressure on a cambered airfoil shifts with the angle of attack. This is a nuisance in doing dynamic analyses of aircraft motion. Thus it is of interest to note that there is one point along the chord about which the moment is, according to thin-airfoil theory, independent of angle of attack. This point is called the *aerodynamic center* of the airfoil. To locate it, note from Fig. 4.12 that since the lift acts at the center of pressure, the moment about any other point x_0, say, is given by

$$\begin{aligned}\circlearrowleft M'_{x0} &= L'(x_0 - x_{\text{c.p.}}) \\ &= \pi\rho V_\infty^2 c(A_0 + \tfrac{1}{2}A_1)\left(x_0 - \frac{c}{4}\frac{A_0 + A_1 - \frac{1}{2}A_2}{A_0 + \frac{1}{2}A_1}\right) \\ &= \pi\rho V_\infty^2 c\left[A_0\left(x_0 - \frac{c}{4}\right) + A_1\left(\frac{x_0}{2} - \frac{c}{4}\right) + A_2\frac{c}{8}\right]\end{aligned}$$

in which we have used equation 4-57 for L' and $x_{\text{c.p.}}$.

According to equations 4-52 and 4-53, of all the coefficients A_n, only A_0 depends on the angle of attack. Thus, M'_{x_0} becomes independent of the angle of attack if it is independent of A_0, which happens if

$$x_0 = \frac{c}{4}$$

Thus, *the aerodynamic center of a thin two-dimensional airfoil in incompressible potential flow is located at the quarter-chord point.*[3] The moment about the aerodynamic center is found from the equation above to be

$$M'_{\text{a.c.}} = \frac{-\pi}{8}\rho V_\infty^2 c^2(A_1 - A_2) \tag{4-59}$$

In terms of $M'_{\text{a.c.}}$, the moment about any other point (e.g., the mass center of the

[3] Do not confuse this with our earlier result that the center of pressure for a thin symmetric airfoil is also at the quarter-chord point. The center of pressure is the point where the lift acts, whereas the aerodynamic center is the point about which the moment of the lift force is independent of angle of attack. They do not generally coincide.

aircraft) x_0 can be calculated from

$$\circlearrowleft M'_{x0} = L'(x_0 - x_{\text{c.p.}}) = L'(x_0 - x_{\text{a.c.}}) + L'(x_{\text{a.c.}} - x_{\text{c.p.}}) = L'(x_0 - x_{\text{a.c.}}) + M'_{\text{a.c.}}$$

The dimensionless forms of the lift force and moment on a two-dimensional airfoil are the *section lift and moment coefficients*, c_l and c_m, respectively,

$$c_l \equiv \frac{L'}{\frac{1}{2}\rho_\infty V_\infty^2 c} \tag{4-60}$$

$$c_m \equiv \frac{M'}{\frac{1}{2}\rho_\infty V_\infty^2 c^2} \tag{4-61}$$

Substituting for L' and $M'_{\text{a.c.}}$ from equations 4-57 and 4-59 and then for the A_n's from equations 4-52 and 4-53, we get the thin-airfoil results

$$c_l = 2\pi(A_0 + \tfrac{1}{2}A_1) = 2\pi\left[\alpha - \frac{1}{\pi}\int_0^\pi s(\theta_0)(1 - \cos\theta_0)\,d\theta_0\right] \tag{4-62}$$

$$c_{\text{mac}} = -\frac{\pi}{4}(A_1 - A_2) = -\tfrac{1}{2}\int_0^\pi s(\theta_0)(\cos\theta_0 - \cos 2\theta_0)\,d\theta_0 \tag{4-63}$$

Thin-airfoil theory is seen from equation 4-62 to predict that the lift coefficient is linear in the angle of attack:

$$c_l = m_0(\alpha - \alpha_{L0}) \tag{4-64}$$

Here m_0 is the *lift-curve slope*

$$m_0 = 2\pi \tag{4-65}$$

and α_{L0} the *angle of zero lift*:

$$\alpha_{L0} = \frac{1}{\pi}\int_0^\pi s(\theta_0)(1 - \cos\theta_0)\,d\theta_0 \tag{4-66}$$

As will be seen in Chapter 7, equation 4-64 is in excellent agreement with the performance of real airfoils almost up to stall, although the values given by thin-airfoil theory for m_0 and α_{L0} in equations 4-65 and 4-66 are a bit off. The quantity

$$\alpha_a \equiv \alpha - \alpha_{L0} \tag{4-67}$$

is called the *absolute angle of attack* of the airfoil.

Example 4: Slatted, Flapped Symmetric Airfoil The range of angles of attack for which a fixed-geometry wing is useful is limited by stall, a viscous phenomenon to be discussed in Chapter 7. Outside this range its lift effectively vanishes, thus limiting its range of life coefficients. But, since the lift force required is roughly independent of speed, a high-speed aircraft needs a much higher lift coefficient to take off and

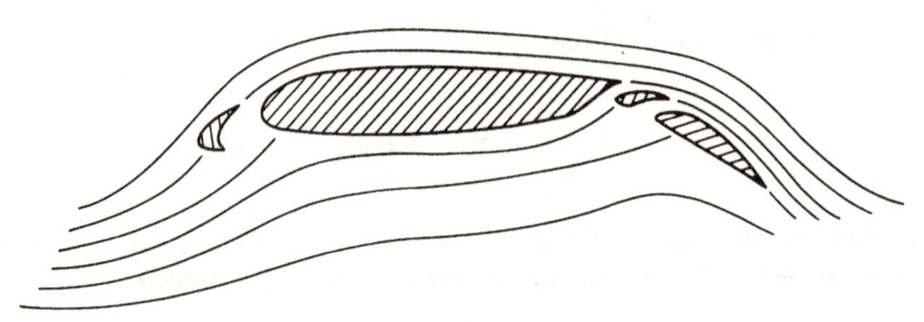

FIG. 4.13. Streamlines of the flow past a slatted, flapped airfoil. Copyright AIAA; from "High Lift Aerodynamics" by A.M.O. Smith. *Journal of Aircraft*, Vol. 12, 1975.

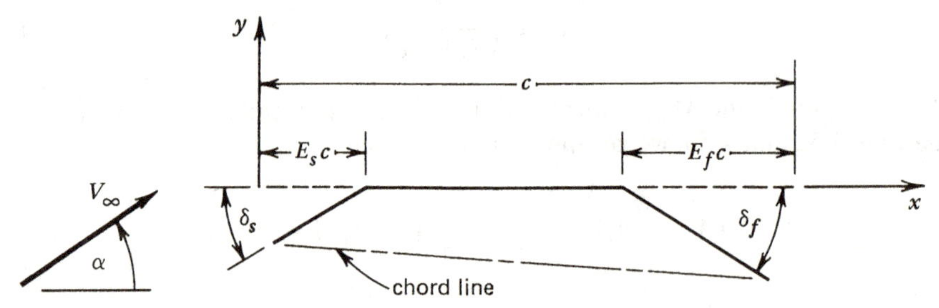

FIG. 4.14. Nomenclature for thin-airfoil theory of a slatted, flapped airfoil without gaps or camber.

land than it does to cruise. The traditional solution to this dilemma is to vary the camber (and so change the angle of zero lift) during flight. As noted in Chapter 1, even the Wright brothers warped their wings with a system of wires, although for purposes of control rather than to increase the maximum lift. For thicker (and stronger) wings, *flaps* and *slats* are used. That is, as is illustrated in Fig. 4.13, the trailing and leading edges are rotated to angles of attack different from those of the wing proper. Although it is advantageous to use slots between the different pieces of the airfoil,[4] we shall consider only "plain" slats and flaps, hinged to the main airfoil with no "leakage" at the hinges. We shall also suppose that the airfoil is symmetric when slat and flap are undeflected.

As shown in Fig. 4.14, let E_s and E_f be the lengths of the slat and flap, respectively, in units of the chord length, and δ_s and δ_f be their respective deflection angles. It is convenient to measure the angle of attack relative to the main section of the airfoil (i.e., the x axis), rather than to the chord line. Then the slope of the camber line is given by

$$\begin{aligned} \bar{Y}'(x) &\approx \delta_s && \text{for } 0 < x < E_s c \\ &\approx -\delta_f && \text{for } (1 - E_f)c < x < c \\ &= 0 && \text{otherwise} \end{aligned} \tag{4-68}$$

[4] For a good review of the various possibilities, see Ref. [1], especially p. 518.

To ease the introduction of the cosine variables, equations 4-36 and 4-48, let

$$E_s c = \frac{c}{2}(1 - \cos\theta_s)$$

$$(1 - E_f)c = \frac{c}{2}(1 - \cos\theta_f) \tag{4-69}$$

Then, in terms of the cosine variables, the camber-line slope is

$$\begin{aligned} s(\theta_0) &= \delta_s & 0 < \theta_0 < \theta_s \\ &= -\delta_f & \theta_f < \theta_0 < \pi \\ &= 0 & \theta_s < \theta_0 < \theta_f \end{aligned} \tag{4-70}$$

The coefficients A_n of the expansion of the vortex strength are easily found from equations 4-52 and 4-53 to be

$$\begin{aligned} A_0 &= \alpha - \frac{\delta_s}{\pi}\int_0^{\theta_s} d\theta_0 + \frac{\delta_f}{\pi}\int_{\theta_f}^{\pi} d\theta_0 \\ &= \alpha + \delta_f - \frac{1}{\pi}(\delta_s\theta_s + \delta_f\theta_f) \end{aligned}$$

$$\begin{aligned} A_n &= \frac{2}{\pi}\delta_s\int_0^{\theta_s}\cos n\theta_0\, d\theta_0 - \frac{2}{\pi}\delta_f\int_{\theta_f}^{\pi}\cos n\theta_0\, d\theta_0 \\ &= \frac{2}{\pi n}(\delta_s \sin n\theta_s + \delta_f \sin n\theta_f) \qquad n = 1, 2, 3, \ldots \end{aligned} \tag{4-71}$$

Substituting these results into equations 4-62 and 4-63, we find the angle of zero lift of equation 4-64 to be

$$\alpha_{L0} = \frac{\delta_s}{\pi}(\theta_s - \sin\theta_s) + \frac{\delta_f}{\pi}(\theta_f - \sin\theta_f - \pi)$$

whereas the moment about the aerodynamic center is

$$c_{\text{mac}} = -\frac{\delta_s}{2}\sin\theta_s(1 - \cos\theta_s) - \frac{\delta_f}{2}\sin\theta_f(1 - \cos\theta_f)$$

You will work out several example calculations in problem 6.

It should be emphasized that the foregoing procedure is much too crude to be useful for the design of real multielement airfoils. The gaps at the hinge lines are of crucial importance. The effects of gaps are primarily viscous in nature. See Ref. [1] for a discussion of these effects and for references to various design procedures and Ref. [2] for a good collection of practical data on flaps and slats.

4.7. THE LUMPED-VORTEX METHOD

An approximate method for solving the camber part of thin-airfoil problems can be set up as follows. Replace the distributed vortices of equation 4-17 by a single

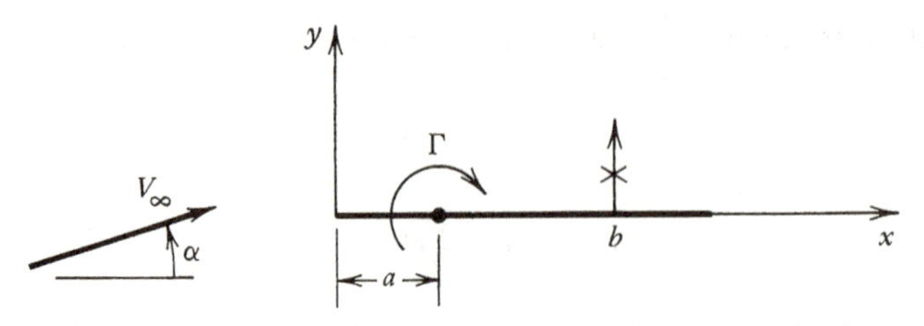

FIG. 4.15. Nomenclature for the lumped-vortex method.

concentrated vortex at the point $x = a$, say, as shown in Fig. 4.15. Since the vortex strength is now just a single number (Γ, say) rather than a function of x, we can no longer satisfy the flow tangency condition (4-10) all along the airfoil. Instead, to fix Γ, we will satisfy it at just one point, $x = b$, say. Since the velocity induced by the vortex at $x = b$, $y = 0$ has magnitude $\Gamma/2\pi(b - a)$ and is directed downward, equation 4-10 becomes

$$-\frac{\Gamma}{2\pi(b-a)} = V_\infty\left(\frac{d\bar{Y}}{dx} - \alpha\right) \qquad \text{at } x = b \tag{4-72}$$

To use the method, we must specify the points at which we locate the vortex ("a") and at which we satisfy flow tangency ("b"). We will choose them so that equation 4-72 agrees with the corresponding results of thin-airfoil theory for the case of an airfoil with parabolic camber. For that case, from equations 4-33 and 4-58,

$$\Gamma = \pi V_\infty c(\alpha + 2\varepsilon)$$

whereas from equation 4-54

$$\frac{d\bar{Y}}{dx}(b) = 4\varepsilon\left(1 - 2\frac{b}{c}\right)$$

Substituting these relations into equation 4-72, we get

$$-\frac{1}{2}V_\infty\frac{c}{b-a}(\alpha + 2\varepsilon) = V_\infty\left[4\varepsilon\left(1 - \frac{2b}{c}\right) - \alpha\right]$$

For this to be true for any α and ε, the coefficients of those variables must match, so

$$-\frac{1}{2}\frac{c}{b-a} = -1$$

$$-\frac{c}{b-a} = 4\left(1 - \frac{2b}{c}\right)$$

which can be solved for a and b:

$$a = \frac{c}{4}, \qquad b = \frac{3c}{4} \tag{4-73}$$

Thus, the idea is this: lump the distributed vortices of equation 4-17 into a concentrated vortex of strength Γ at the quarter-chord point $x = c/4$. To find Γ, satisfy the flow tangency condition at the three-quarter-chord point $x = 3c/4$. That is, from equation 4-72, set

$$\Gamma = \pi c V_\infty \left[\alpha - \frac{d\bar{Y}}{dx}\left(\frac{3c}{4}\right) \right] \tag{4-72a}$$

Naturally, this method works best for an airfoil whose camber line is not too far different from the parabolic case that is its basis. It may also be noted that the method as described above gives no data on the moment of the airfoil. However, it does play an important role as the foundation of the vortex-lattice method for wings of finite span, which will be discussed in Chapter 5.

Example 5 For a case in which the lumped-vortex method simply does not work, we need look no further than the flapped, slatted airfoil discussed above. If the three-quarter-chord point is between the flap and the slat, the camber-line slope is zero there, and from equation 4-72a, the results are completely insensitive to the flap and slat deflection. See problem 8 for a modification of the method, in which the distributed vortices are lumped into three concentrated vortices, at the quarter chords of the slat, the main airfoil, and the flap.

4.8. PANEL METHODS

Although we have been noting its deficiencies for multielement airfoils, in fact, thin-airfoil theory is not much used these days even for the analysis or design of single-element airfoils. It does give fairly good results for lift and moment coefficients but ignores the effect on those coefficients of the thickness distribution. This effect is small but must be accounted for in practice. Moreover, thin-airfoil theory gives good results on the pressure distribution only away from stagnation points. As will be discussed in Chapter 7, the pressure distribution has a marked effect on the behavior of the fluid in the boundary layer near the surface, and that behavior controls both the drag of the airfoil and its maximum lift. Therefore, the proper design of an airfoil requires an accurate prediction of its pressure distribution.

In the 1930s, Theodorsen developed a semianalytic method [3] that yielded quite accurate solutions of single-element airfoil problems. The method was based on complex variables, a direct version of the inverse methods described in Chapter 3 and later on in this chapter.

Much more powerful approaches can be based on distributions of sources and vortices or doublets. Of course, to avoid the inaccuracies of thin-airfoil theory, the flow-tangency condition must be satisfied on the body surface, and without the approximations of equations 4-5. Also, the singularities should be distributed on the body surface rather than on the chord line or any other line within or without the body. As is shown in Chapter 8, the method can then be guaranteed to treat any reasonable problem—including multielement airfoils—with any desired degree of accuracy. Also, unlike the complex-variable method, it can be extended to three-dimensional flows.

Such methods are generally called "panel methods," since the body surface is, as we shall see, approximated by a collection of "panels." There are a number of ways to set up a panel method. To begin with, there are choices even as to the type of singularity used. Most recent panel methods are based on a combination of sources and doublets oriented normal to the surface. With the source strength appropriately specified, the doublet strength (which is then the only unknown) turns out to be related to the velocity potential on the surface. The surface velocity can then be calculated simply by differentiating the doublet strength.

Methods based on doublets will be described in Chapter 8. Here we shall discuss the pioneering panel method due to Hess and Smith [4], which employs sources and vortices. Thus the potential may be decomposed in a manner reminiscent of equation 4-6,

$$\phi = \phi_\infty + \phi_S + \phi_V \tag{4-74a}$$

with ϕ_∞ being the potential of the uniform onset flow, and so given by equation 4-7, whereas

$$\phi_S \equiv \int \frac{q(s)}{2\pi} \ln r \, ds \tag{4-74b}$$

$$\phi_V \equiv -\int \frac{\gamma(s)}{2\pi} \theta \, ds \tag{4-74c}$$

in which the integrations are over the body surface. As shown in Fig. 4.16, s is distance measured along the surface, and (r, θ) are polar coordinates of the "field point" (x, y) relative to the point on the surface whose location is indicated by s. Thus, ϕ_S is the potential of a source distribution of strength $q(s)$ per unit length, and ϕ_V, the potential of a vortex distribution of strength $\gamma(s)$ per unit length.

Because of the superposition principle, this ϕ automatically satisfies the Laplace equation and the boundary condition at infinity and will be the solution we seek if $q(s)$ and $\gamma(s)$ are determined so as to meet the boundary condition of flow tangency and the Kutta condition. It may help you to think of the source strength as being governed by the flow tangency condition, and the vortex strength by the Kutta condition, although actually both singularity distributions are important in

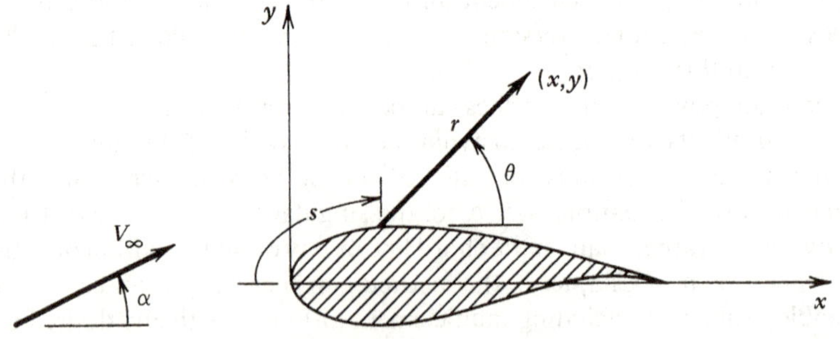

FIG. 4.16. Nomenclature for the analysis by surface singularity distribution (the panel method).

satisfying either condition. Such a viewpoint makes reasonable the simplification of equations 4-74 employed by Hess and Smith [4] namely, that the vortex strength is taken constant over the whole airfoil. That is, since flow tangency must be satisfied at all points on the body surface, the source strength must vary over the surface, but since the Kutta condition involves only the trailing edge, the vortex strength can be represented by a single number. Alternatively, recall from Section 4.4 that a potential flow is made unique by applying flow tangency (continuously) on the body surface and specifying the circulation about the airfoil, which is just the net vortex strength. Thus, if one distributes on or within the body surface vortices whose net strength is the correct circulation, the problem is solved if sources can be distributed over the body surface so as to make the total velocity field (comprised of the onset flow and the velocity fields due to the sources and vortices) tangent to the body surface, regardless of how the vortices are distributed. See also Chapter 8.

In any case, the integrals of equation 4-74 are hard to evaluate, even for simple forms of the source and vortex strengths, unless the surface on which the sources and vortices are distributed is a straight line. Thus we select a certain number N of points on the body contour, called *nodes*, and connect the nodes with straight lines, which become the *panels* of the method, as shown in Fig. 4.17. We then distribute the sources and vortices on the straight-line panels, so that the potential given by equations 4-7 and 4-74 may be written

$$\phi = V_\infty(x \cos \alpha + y \sin \alpha) + \sum_{j=1}^{N} \int_{\substack{\text{panel} \\ j}} \left[\frac{q(s)}{2\pi} \ln r - \frac{\gamma}{2\pi} \theta \right] ds \tag{4-75}$$

In most cases, equation 4-75 still allows an exact solution of the flow problem. The exceptional cases are those in which the sources and vortices must be distributed exactly on the body surface; to be mathematically precise, in which the potential cannot be "continued analytically" across the body surface. Even those cases can be well approximated by equation 4-75 simply by increasing the panel density so that the polygon formed by the panels better approximates the body shape.

As in the doublet-distribution method discussed in Chapter 3, we must now assume some parameterized form for the variation of the source strength over the panels (the vortex strength was already stipulated to be constant), so as to be able to evaluate the integrals of equation 4-75 in terms of the parameters. This is the only major approximation of the panel method, one that becomes more accurate (if the method works) as the number of panels increases and/or the reality of the assumptions as to the variations of the singularity strengths increases.

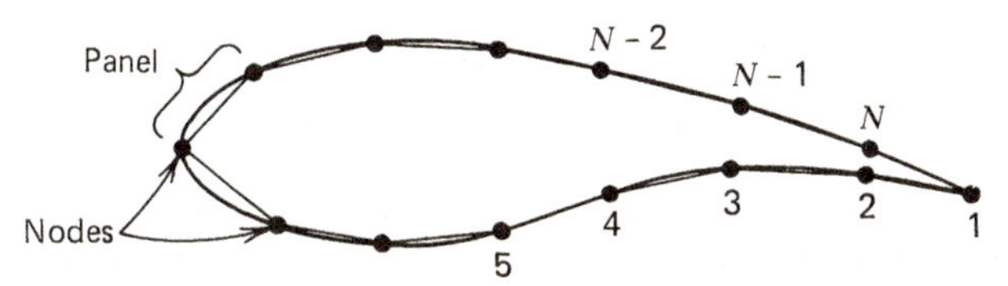

FIG. 4.17. Definition of nodes and panels.

FIG. 4.18. Approximation of the flow tangency condition in the panel method.

Following Hess and Smith, we take the source strength to be constant on each panel, but variable from one panel to the next:

$$q(s) = q_i \text{ on panel } i, \qquad i = 1, \ldots, N \tag{4-76}$$

The parameters to be determined are then the N source strengths q_i and the vortex strength γ. Since, whatever their values, equation 4-75 meets every condition of the problem except the flow-tangency condition on the body surface and the Kutta condition, we find the q_i's and γ by imposing the flow-tangency condition at N *control points* and some version of the Kutta condition.

As we shall see, for constant source and vortex strengths, the velocity is infinite at the end of each panel. This excludes the nodes from consideration as control points. The next most reasonable choice would be the points on the body midway between each adjacent pair of nodes, as indicated in Fig. 4.18*a*. However, for constant-strength source and vortex panels, it is just as accurate, and more convenient, to set the velocity component normal to each panel equal to zero at its midpoint, as in Fig. 4.18*b*. Similarly, for a Kutta condition, we equate the velocity components tangential to the panels adjacent to the trailing edge, again evaluating the components at the midpoints of the panels. If the lengths of the two trailing-edge panels are kept nearly equal as the number of panels is increased, this amounts to a requirement that the velocities at equal distances from the trailing edge approach one another as those distances are decreased, what we called the "second corollary" to the Kutta condition in Section 4.3.

To implement this method, we need some nomenclature. Let the ith panel be defined as the one between the ith and $(i + 1)$th nodes, and its inclination to the x axis be θ_i, as shown in Fig. 4.19. Specifically, let

$$\sin \theta_i = \frac{y_{i+1} - y_i}{l_i}$$

$$\cos \theta_i = \frac{x_{i+1} - x_i}{l_i} \tag{4-77}$$

where l_i is the length of the ith panel. Then

$$\hat{\mathbf{n}}_i \equiv -\sin \theta_i \hat{\mathbf{i}} + \cos \theta_i \hat{\mathbf{j}} \tag{4-78}$$

is a unit vector normal to the ith panel. Specifically, if the nodes are numbered so that as one goes from one node to the next, the body is on the right (as indicated in Fig. 4.17), $\hat{\mathbf{n}}_i$ is the unit vector directed outward from the body into the flow.

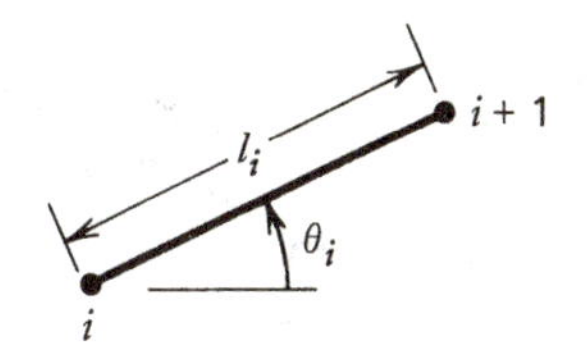

FIG. 4.19. The ith panel.

Similarly,

$$\hat{\mathbf{t}}_i \equiv \cos\theta_i \hat{\mathbf{i}} + \sin\theta_i \hat{\mathbf{j}} \tag{4-79}$$

is the unit vector tangential to the ith panel and directed from the ith node to the $(i + 1)$th node.

Now let the coordinates of the midpoint of the ith panel be

$$\bar{x}_i \equiv \frac{x_i + x_{i+1}}{2}$$

$$\bar{y}_i \equiv \frac{y_i + y_{i+1}}{2} \tag{4-80}$$

and write the velocity components at that point as

$$u_i \equiv u(\bar{x}_i, \bar{y}_i)$$

$$v_i \equiv v(\bar{x}_i, \bar{y}_i)$$

Then the flow tangency condition can be written

$$0 = -u_i \sin\theta_i + v_i \cos\theta_i \qquad \text{for } i = 1, \ldots, N \tag{4-81}$$

and the Kutta condition as

$$u_1 \cos\theta_1 + v_1 \sin\theta_1 = -u_N \cos\theta_N - v_N \sin\theta_N \tag{4-82}$$

The minus signs in equation 4-82 are due to the definition of the tangential direction; $\hat{\mathbf{t}}_1$ goes away from the trailing edge, and $\hat{\mathbf{t}}_N$ toward it.

The velocity components at the middle of the ith panel u_i, v_i are made up of contributions from the onset flow, the sources on each panel, and the vortices on each panel. Because the velocities induced by the sources and vortices on a panel are proportional to the source or vortex strength on that panel, we can write

$$u_i = V_\infty \cos\alpha + \sum_{j=1}^{N} q_j u_{sij} + \gamma \sum_{j=1}^{N} u_{vij}$$

$$v_i = V_\infty \sin\alpha + \sum_{j=1}^{N} q_j v_{sij} + \gamma \sum_{j=1}^{N} v_{vij} \tag{4-83}$$

where u_{sij}, for example, is the x component of the velocity at the midpoint of the ith panel due to a unit-strength source distribution on the jth panel.

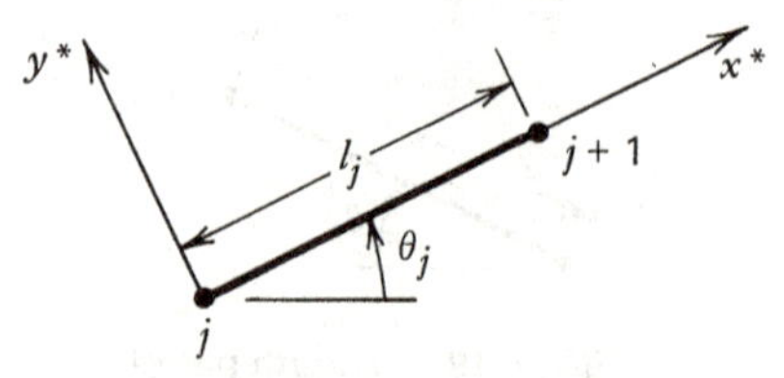

FIG. 4.20. Coordinate system fixed to the *j*th panel.

To evaluate u_{sij}, v_{sij}, u_{vij}, and v_{vij}, it is convenient to work in coordinates (x^*, y^*) oriented with the jth panel, as shown in Fig. 4.20. The global velocity components can be evaluated from

$$u = u^* \cos\theta_j - v^* \sin\theta_j$$

$$v = u^* \sin\theta_j + v^* \cos\theta_j \tag{4-84}$$

once we find the "local" components (u^*, v^*). The velocity components at (x_i, y_i) due to a unit-strength source distribution on the jth panel can then be written (see equations 4-11 and 4-12):

$$u_{sij}^* = \frac{1}{2\pi}\int_0^{l_j} \frac{x^* - t}{(x^* - t)^2 + y^{*2}}\,dt$$

$$v_{sij}^* = \frac{1}{2\pi}\int_0^{l_j} \frac{y^*}{(x^* - t)^2 + y^{*2}}\,dt \tag{4-85}$$

in which (x^*, y^*) are the local coordinates corresponding to (x_i, y_i). These integrals should not be strangers by this time:

$$u_{sij}^* = \frac{-1}{2\pi}\ln[(x^* - t)^2 + y^{*2}]^{\frac{1}{2}}\Big|_{t=0}^{t=l_j}$$

$$v_{sij}^* = \frac{1}{2\pi}\tan^{-1}\frac{y^*}{x^* - t}\Big|_{t=0}^{t=l_j}$$

Such results have a geometric interpretation, which obviates their translation back into the global coordinates. Referring to Fig. 4.21 for nomenclature, we write

$$u_{sij}^* = \frac{-1}{2\pi}\ln\frac{r_{ij+1}}{r_{ij}}$$

$$v_{sij}^* = \frac{\nu_l - \nu_0}{2\pi}$$

$$= \frac{\beta_{ij}}{2\pi} \tag{4-86}$$

As shown in Fig. 4.21, r_{ij} is the distance from the jth node to the middle of the ith panel, whereas β_{ij} is the angle subtended at the middle of the ith panel by the jth panel.

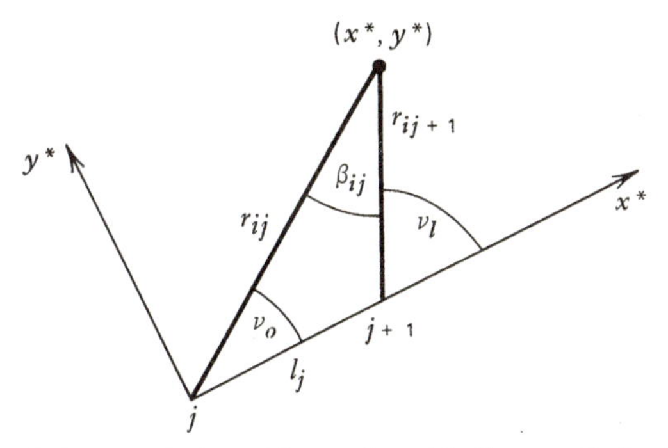

FIG. 4.21. Geometric interpretation of terms found in the results for the velocity field of constant-strength source and vortex panels.

It is easy to see from equation 4-86 that $u^*_{sii} = 0$, but the value of the y^* component of velocity induced by the source panel at its own midpoint is not so obvious. If the point (x^*, y^*) approaches the panel from outside the panels (i.e., if $y^* \downarrow 0$ for $0 < x^* < l_j$), $\beta_{ii} \rightarrow \pi$. However, if it approached the panel from the other side, β_{ii} would be negative and $\rightarrow -\pi$. Since the flow is outside the body, we are interested in working on the outside of the panels and set $\beta_{ii} = \pi$. A convenient FORTRAN expression for β_{ij} is then

$$\begin{aligned} \beta_{ij} &= \pi \qquad \text{if} \quad i = j \\ &= \text{ATAN } 2[(\bar{y}_i - y_{j+1})(\bar{x}_i - x_j) - (\bar{x}_i - x_{j+1})(\bar{y}_i - y_j), \\ &\quad (\bar{x}_i - x_{j+1})(\bar{x}_i - x_j) + (\bar{y}_i - y_{j+1})(\bar{y}_i - y_j)] \qquad \text{if } i \neq j \end{aligned} \tag{4-87}$$

The velocity induced at (x^*, y^*) by the vortices on the jth panel is now easy to evaluate. The basic formulas can be obtained from equations 4-18 and 4-19 and identified from equations 4-85 with quantities evaluated in equations 4-86:

$$\begin{aligned} u^*_{vij} &= -\frac{1}{2\pi}\int_0^{l_i} \frac{y^*}{(x^* - t)^2 + y^{*2}}\,dt = \frac{\beta_{ij}}{2\pi} \\ v^*_{vij} &= \frac{-1}{2\pi}\int_0^{l_i} \frac{x^* - t}{(x^* - t)^2 + y^{*2}}\,dt = \frac{1}{2\pi}\ln\frac{r_{ij+1}}{r_{ij}} \end{aligned} \tag{4-88}$$

The flow tangency conditions (4-81) may now be put into the form

$$\sum_{j=1}^{N} A_{ij}q_j + A_{iN+1}\gamma = b_i \tag{4-89}$$

where, with the help of equations 4-83 to 4-89,

$$\begin{aligned} A_{ij} &= -u_{sij}\sin\theta_i + v_{sij}\cos\theta_i \\ &= -u^*_{sij}(\cos\theta_j \sin\theta_i - \sin\theta_j\cos\theta_i) + v^*_{sij}(\sin\theta_j\sin\theta_i + \cos\theta_j\cos\theta_i) \end{aligned}$$

so

$$2\pi A_{ij} = \sin(\theta_i - \theta_j)\ln\frac{r_{ij+1}}{r_{ij}} + \cos(\theta_i - \theta_j)\beta_{ij} \tag{4-90}$$

and, similarly,

$$2\pi A_{iN+1} = \sum_{j=1}^{N} \cos(\theta_i - \theta_j)\ln\frac{r_{ij+1}}{r_{ij}} - \sin(\theta_i - \theta_j)\beta_{ij} \tag{4-91}$$

whereas

$$b_i = V_\infty \sin(\theta_i - \alpha) \tag{4-92}$$

The Kutta condition (4-82) can be put in similar form:

$$\sum_{j=1}^{N} A_{N+1,\,j}q_j + A_{N+1,\,N+1}\gamma = b_{N+1} \tag{4-93}$$

and, after the same sort of manipulation, we find

$$2\pi A_{N+1,\,j} = \sum_{k=1,\,N} \sin(\theta_k - \theta_j)\beta_{kj} - \cos(\theta_k - \theta_j)\ln\frac{r_{kj+1}}{r_{kj}}$$

$$2\pi A_{N+1,N+1} = \sum_{k=1,\,N}\sum_{j=1}^{N} \sin(\theta_k - \theta_j)\ln\frac{r_{kj+1}}{r_{kj}} + \cos(\theta_k - \theta_j)\beta_{kj}$$

$$b_{N+1} = -V_\infty \cos(\theta_1 - \alpha) - V_\infty \cos(\theta_N - \alpha) \tag{4-94}$$

Equations 4-89 and 4-93 comprise a set of $N + 1$ equations in the unknowns q_i, $i = 1, \ldots, N$, and γ. Once they are solved, we can compute the tangential velocity at the midpoint of each panel from

$$V_{ti} = V_\infty \cos(\theta_i - \alpha) + \sum_{j=1}^{N} \frac{q_j}{2\pi}\left[\sin(\theta_i - \theta_j)\beta_{ij} - \cos(\theta_i - \theta_j)\ln\frac{r_{ij+1}}{r_{ij}}\right]$$

$$+ \frac{\gamma}{2\pi}\sum_{j=1}^{N}\left[\sin(\theta_i - \theta_j)\ln\frac{r_{ij+1}}{r_{ij}} + \cos(\theta_i - \theta_j)\beta_{ij}\right] \tag{4-95}$$

as can be proved by manipulating equations 4-79, 4-83, 4-84, 4-86, and 4-88. Since $V_{ni} = 0$, the pressure coefficient at (x_i, y_i) can then be calculated from

$$C_p(\bar{x}_i, \bar{y}_i) = 1 - V_{ti}^2/V_\infty^2 \tag{4-96}$$

and the lift and moment estimated by assuming C_p constant over each panel (see problem 9).

4.8.1. Program PANEL

Two versions of the panel method described above have been prepared. One is designed to take advantage of a microcomputer's interactive graphics and allows you to use a cursor to position the nodes, and so to modify the airfoil shape, as well as to obtain graphic results for the pressure distribution. Instructions for its operation can be obtained separately. The remainder of this section will describe the other version of PANEL, which can be used to get hard-copy data from a mainframe computer, and is listed under Section 4.12.

The mainframe version of PANEL is set up to analyze the aerodynamics of NACA four-digit airfoils, whose thickness and camber distributions are given by in Section 1.2, which should be consulted for the coding system used to define these airfoils. It also handles NACA five-digit airfoils of the 230XX family, which have the same thickness distribution as do the four-digit airfoils but a different camber line.

The input required to run PANEL is as follows: NLOWER, the number of nodes on the lower surface; NUPPER, the number of nodes on the upper surface; the NACA number; and the angle of attack (in degrees). Subroutine SETUP distributes the nodes along the upper and lower surfaces according to a cosine formula, which concentrates the nodes near the leading and trailing edges. Although the program allows you to use different numbers of nodes on the upper and lower surfaces, it is not recommended that you do so. An unequal number of nodes yields trailing-edge panels of unequal length, which lowers the accuracy of the approximation to the Kutta condition. Also, the singularities of the velocity at the nodes are less serious if the panel midpoints on the opposite surface are kept as far away as possible.

Subroutine COFISH calculates the coefficients of equations 4-89 and 4-93, which are then solved by the subroutine GAUSS. Then VELDIS calculates and prints out the pressure distribution, after which the coefficients of lift, drag[5] and moment about the leading edge are computed in subroutine FANDM.

The input to and output from program PANEL are completely dimensionless. Because the equations that govern the potential are linear and (except for the boundary condition at infinity) homogeneous, it is easy to see that ϕ/V_∞ is independent of V_∞. So, then, is its gradient, $\mathbf{V}/V_\infty$. Furthermore, if one considers flows past bodies whose shapes differ only by a scale factor, so that (see Fig. 4.22)

$$Y_1(x) = L_1 f\left(\frac{x}{L_1}\right), \qquad Y_2(x) = L_2 f\left(\frac{x}{L_2}\right)$$

their slopes are the same at corresponding points,

$$\frac{dY_1}{dx} = f'\left(\frac{x}{L_1}\right), \qquad \frac{dY_2}{dx} = f'\left(\frac{x}{L_2}\right)$$

Thus, if the angles of attack of the two bodies are the same, the scaled velocity fields that satisfy the governing equations (equations 3-1a, 3-3a, 3-4a, and 3-5a, plus the Kutta condition) depend only on the scaled variables $(x/L, y/L)$. That is, for any

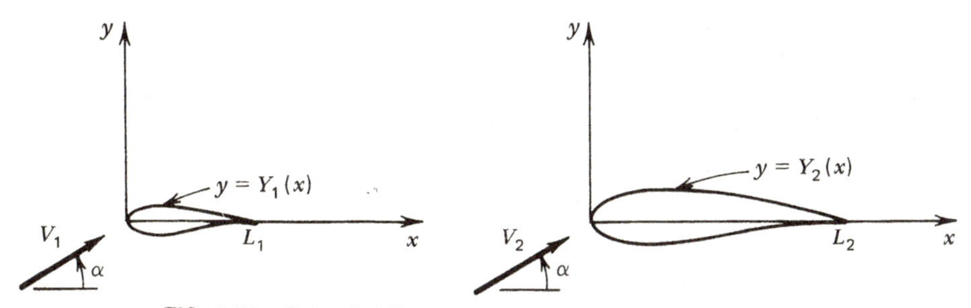

FIG. 4.22. Potential flows past geometrically similar bodies.

[5] Which should be zero, according to the Kutta–Joukowski theorem; the magnitude of the computed drag is a measure of the numerical inaccuracy of the method.

inviscid flow,

$$\mathbf{V}/V_\infty = F\left(\frac{x}{L}, \frac{y}{L}\right)$$

Program PANEL, therefore, scales all velocities with the onset flow speed V_∞ and all locations with the airfoil chord c. That is, all velocities calculated are actually velocities in units of V_∞, and all locations and coordinates are in units of c. Results for particular values of these parameters can be obtained by multiplying the output of PANEL by V_∞ or c, as is appropriate.

4.9. COMPLEX-VARIABLES METHODS

As noted in the preceding section, before panel methods came into being, analyses of incompressible irrotational flow past airfoils were often based on the theory of complex variables. Such methods are still used, in modified form, to devise grids for the numerical solutions of transonic and viscous flows by finite-difference methods.[6] Also, some classical results of this theory remain useful as a source of exact solutions against which numerical solutions can be tested. It is this latter aspect of the complex-variable methods that we consider here. That is, as in Chapter 3, we shall use an inverse approach to construct a few flows past bodies that look like airfoils. Now, however, we will allow the bodies to be cambered and/or at an angle of attack to the onset flow and so to generate lift.

It was shown in Chapter 3 that any analytic (i.e., differentiable) function of the complex variable $z = x + iy$ can be regarded as the "complex potential" of an incompressible irrotational flow, with its real and imaginary parts being the potential ϕ and stream function ψ of the flow, respectively,

$$W(z) = \phi + i\psi \tag{4-97}$$

We constructed symmetric solutions by letting W be the complex potential of uniform flow past a circle in the $\tilde{z} = \tilde{x} + i\tilde{y}$ plane, and relating z to $\tilde{z}$ by the so-called Joukowski transformation:

$$z = \tilde{z} + \frac{b^2}{\tilde{z}}, \qquad b > 0 \tag{4-98}$$

We shall follow a similar procedure here, but we now need a complex potential for the flow in the $\tilde{z}$ plane that allows both for an angle of attack of the flow at infinity and for circulation. As our model, we take the flow past a circular cylinder with circulation, for which (see problems 4 and 5 of Chapter 2)

$$\phi = V_\infty\left(x + \frac{a^2 x}{x^2 + y^2}\right) - \frac{\Gamma}{2\pi}\theta$$

$$\psi = V_\infty\left(y - \frac{a^2 y}{x^2 + y^2}\right) + \frac{\Gamma}{2\pi}\ln r$$

[6] A recent review of this work was given by Moretti [5].

where a is the radius of the cylinder, whose center is at the origin. Thus we write (see also problem 17 and Example 5 of Chapter 3)

$$W = V_\infty\left(\tilde{z} + \frac{a^2}{\tilde{z}}\right) + \frac{i\Gamma}{2\pi}\ln\frac{\tilde{z}}{\text{a}}$$

The flow described by this complex potential is directed along the $\tilde{x}$ axis far from the origin. To put it at the angle of attack α, just let $\tilde{z} \to \tilde{z}e^{-i\alpha}$ (as is suggested by Example 4 of Chapter 3):

$$W = V_\infty\left(\tilde{z}e^{-i\alpha} + \frac{a^2}{\tilde{z}e^{-i\alpha}}\right) + \frac{i\Gamma}{2\pi}\ln\frac{\tilde{z}e^{-i\alpha}}{a}$$

Finally, we shift the center of the circle to $\tilde{z} = c$ and so obtain

$$W = V_\infty\left[(\tilde{z} - c)e^{-i\alpha} + \frac{a^2e^{+i\alpha}}{\tilde{z} - c}\right] + \frac{i\Gamma}{\pi}\ln\frac{(\tilde{z} - c)e^{-i\alpha}}{a} \tag{4-99}$$

which is the complex potential of the flow about a circle of radius a in the $\tilde{z}$ plane, the circle being centered at $\tilde{z} = c$, the angle of attack at infinity being α, and the circulation being Γ.

From equation 3-78, we get the velocity in the physical z plane:

$$u - iv = \frac{dW}{dz} = \frac{V_\infty[e^{-i\alpha} - a^2e^{+i\alpha}/(\tilde{z} - c)^2] + \dfrac{i\Gamma}{2\pi}\dfrac{1}{\tilde{z} - c}}{1 - (b^2/\tilde{z}^2)} \tag{4-100}$$

As was the case with equation 3-86, this blows up when

$$\tilde{z} = \pm b$$

We position the singular point $\tilde{z} = b$ on the circle, and $\tilde{z} = -b$ within it, so as to produce an airfoil with cusped trailing edge when the circle is mapped into the z plane by the Joukowski transformation (4-98). Thus we require (compare equations 3-88 and 3-89)

$$a^2 = (b - \varepsilon)^2 + \mu^2 \tag{4-101}$$

$$\varepsilon \leq 0 \tag{4-102}$$

where

$$c = \varepsilon + i\mu \tag{4-103}$$

The Kutta condition that the velocity be finite at the trailing edge is then enforced by requiring the numerator of equation 4-100 to vanish (along with its denominator) at $\tilde{z} = b$:

$$0 = V_\infty\left[e^{-i\alpha} - \frac{a^2e^{+i\alpha}}{(b - c)^2}\right] + \frac{i\Gamma}{2\pi}\frac{1}{b - c}$$

This fixes the circulation:

$$\Gamma = \frac{-2\pi V_\infty}{i}\left[e^{-i\alpha}(b-c) - \frac{a^2 e^{+i\alpha}}{(b-c)}\right]$$

With the aid of equations 4-101 and 4-103, this may be simplified to

$$\Gamma = 4\pi V_\infty[(b-\varepsilon)\sin\alpha + \mu\cos\alpha] \tag{4-104}$$

To find the airfoil shape to which these results refer, set

$$\tilde{z} = c + ae^{i\theta} \tag{4-105}$$

which, as you can check, makes the imaginary part of the complex potential given by equation 4-99 (the stream function of the flow) vanish. The image of this streamline in the z plane is then found from the Joukowski transformation (4-98), using equations 4-101, 4-103, and 4-105.

Example 6 Consider the case

$$\varepsilon = \mu = 0$$

From equations 4-101 and 4-103, these imply

$$a = b, \qquad c = 0$$

so equation 4-105 becomes

$$\tilde{z} = ae^{i\theta}$$

Then, from equation 4-98, the streamline $\psi = 0$ maps onto

$$\begin{aligned} z &= ae^{i\theta} + ae^{-i\theta} \\ &= 2a\cos\theta \end{aligned}$$

whose real part (x) varies between $-2a$ and $+2a$ but whose imaginary part (y) is zero. Thus the airfoil to which the preceding results refer is a flat plate of chord $4a$. The circulation given by equation 4-104 is

$$\Gamma = 4a\pi V_\infty \sin\alpha$$

which differs from the result given by thin-airfoil theory (the first of equations 4-44 divided by ρV_∞) only in the difference between α and $\sin\alpha$.

Example 7 Now suppose

$$\varepsilon = 0$$

but allow μ to be nonzero. The center of the streamline circle in the $\tilde{z}$ plane is now on the imaginary axis, at $\tilde{z} = i\mu$; see Fig. 4.23. As in the previous example, both singular points $\tilde{z} = \pm b$ are on the circle, so that the image of the circle in the z plane (the airfoil) will have cusps at both its leading edge and its trailing edge.

Now look at the images of the points at the top and bottom of the circle

$$\tilde{z} = i(\mu + a),\ i(\mu - a)$$

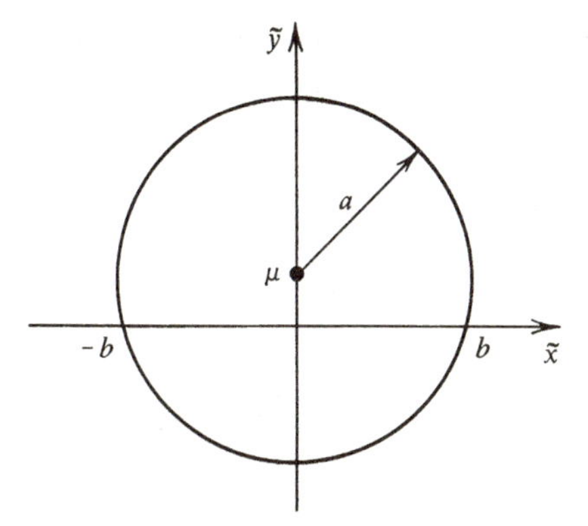

FIG. 4.23. The circle in the $\tilde{z}$ plane corresponding to a circular-arc airfoil.

From equation 4-98, if $\tilde{z} = i(\mu \pm a)$,

$$z = i(\mu \pm a) - i\frac{b^2}{\mu \pm a}$$

But, from Fig. 4.23,

$$b^2 = a^2 - \mu^2$$

so

$$\begin{aligned} z &= i(\mu \pm a) + i(\mu \mp a) \\ &= 2i\mu \end{aligned}$$

Thus the points $\tilde{z} = i(\mu \pm a)$ both map onto a point on the imaginary axis in the z plane. Putting this information together with the cusps at $z = \pm 2b$ (the images of $\tilde{z} = \pm b$), you should be ready to believe that the circle is mapped onto a circular arc of zero thickness, as shown in Fig. 4.24, which is in fact the case [6]. Thus the circulation for this case, which equation 4-104 predicts to be

$$\Gamma = 4\pi V_\infty(b \sin \alpha + \mu \cos \alpha)$$

may be compared with the result of thin-airfoil theory for the airfoil with parabolic camber, the first of equations 4-58, divided by ρV_∞. Note that the chord is now $4b$, whereas the camber ratio, what is called ε in equations 4-58, is $2\mu/4b$.

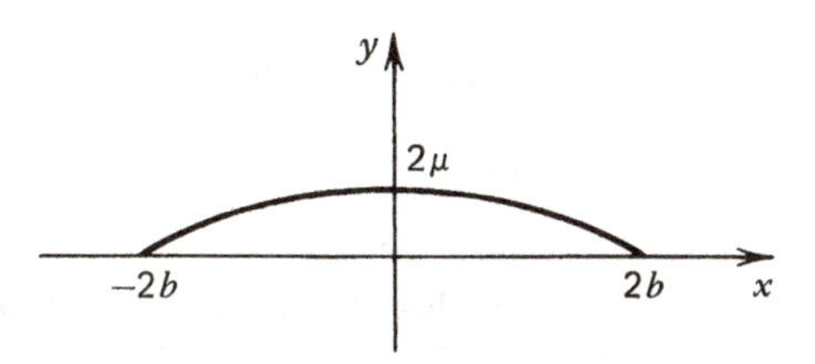

FIG. 4.24. The circular-arc airfoil.

It is often useful to write the Joukowski transformation (4-98) in somewhat different form, as follows. Add and subtract $2b$ from both sides of equation 4-98 to get

$$z + 2b = \tilde{z} + 2b + \frac{b^2}{\tilde{z}} = \frac{1}{\tilde{z}}(\tilde{z} + b)^2$$

$$z - 2b = \tilde{z} - 2b + \frac{b^2}{\tilde{z}} = \frac{1}{\tilde{z}}(\tilde{z} - b)^2$$

Then take the ratio of the two resultant equations, and find

$$\frac{z + 2b}{z - 2b} = \left(\frac{\tilde{z} + b}{\tilde{z} - b}\right)^2$$

Now the stage is set for a variation on the Joukowski transformation, namely, the *von Kármán–Trefftz transformation*

$$\left(\frac{z + 2b}{z - 2b}\right) = \left(\frac{\tilde{z} + b}{\tilde{z} - b}\right)^n$$

As with the Joukowski transformation, this one yields $\tilde{z} \to z$ far from the origin, and so keeps the onset flow the same in the z plane as in the $\tilde{z}$ plane. The main difference is in the shape of the transformed circle near $z = 2b$, which continues to be the image point of the singularity at $\tilde{z} = b$. Letting

$$\tilde{z} = b + \delta e^{i\theta}$$

and following the procedure we used in Chapter 3, we find that as we go around the singularity from one point to another on this circle, the path we trace in the physical plane is described by (if $\delta \ll b$)

$$z \approx 2b\left[1 + 2\left(\frac{\delta}{2b}\right)e^{in\theta}\right]$$

Since θ changes by π in this process, the trailing edge angle is not cusped but includes an angle equal to $(2 - n)\pi$, as shown in Fig. 4.25. Thus, using the von Kármán–Trefftz transformation, one can construct exact solutions of the equations of incompressible potential flow for airfoils with finite trailing-edge angles.

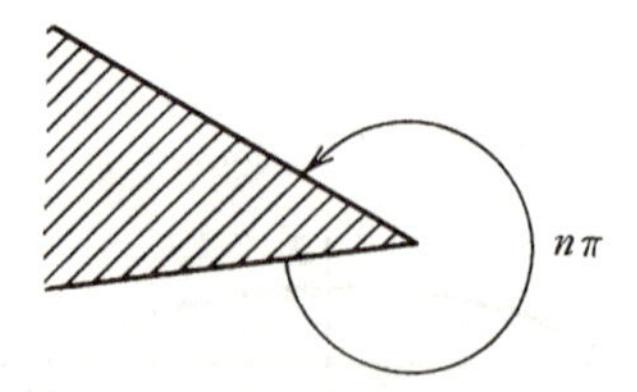

FIG. 4.25. The trailing edge of a Karman–Trefftz airfoil.

4.10. REFERENCES

1. Smith, A. M. O., "High-Lift Aerodynamics," *J. Aircr.* **12**, 501–529 (1975).
2. McCormick, B. W., *Aerodynamics, Aeronautics, and Flight Mechanics.* Wiley, New York (1979).
3. Theodorsen, T. and Garrick, I. E., *General Potential Theory of Arbitrary Wing Sections*, NACA TR 452 (1933).
4. Hess, J. L. and Smith, A. M. O., "Calculation of Potential Flow About Arbitrary Bodies," *Prog. Aeronaut. Sci.* **8**, 1–138 (1966).
5. Moretti, G., "Grid Generation Using Classical Techniques," NASA/ICASE Workshop on Numerical Grid Generation Techniques for Partial Differential Equations (1980).
6. Milne-Thompson, L. M., *Theoretical Aerodynamics*, 4th ed. Dover, New York (1973), Sect. 6.3.

4.11. PROBLEMS

1. For the NACA "1-series" of airfoils, the camber line was designed on the basis of thin-airfoil theory to yield a constant chordwise loading:

$$C_p(x, 0+) - C_p(x, 0-) = \text{constant}$$

Find $\bar{Y}(x)$ such that this is theoretically the case.

2. Verify equations 4-44 and 4-45 for the lift, moment, and center of pressure on a thin symmetric airfoil at angle of attack α.

3. Verify equations 4-57 for the lift and moment on a thin airfoil in terms of the coefficients A_n of the modified sine-series expansion of the vortex strength, equation 4-50.

4. Find the section lift and moment coefficients c_l and c_{mac}, the center of pressure, and the angle of zero lift for a thin airfoil with parabolic camber (see equation 4-54), whose lift and moment are given by equations 4-58.

5. The camber line of the NACA four-digit series of airfoils is given by a pair of parabolas; see equations 1-3. Find $c_l(\alpha)$ and c_{mac} for some member of this series of airfoils. Use thin-airfoil theory.

6. Verify the formulas derived in the text for the angle of zero lift α_{L0} and the pitching moment coefficient of a slatted, flapped symmetric airfoil, and evaluate them for the following cases:

E_s	0	0	0.15	0.15
E_f	0.15	0.30	0.15	0.30

Take $\delta_s = 5°$ and/or $10°$, and $\delta_f = 5°$ and/or $10°$.

7. Use the lumped-vortex method to recalculate the lift of the NACA four-digit airfoil studied in problem 5.

8. Find the lift and moment on a slatted, flapped symmetric airfoil by representing the vortex system with three concentrated vortices, at the quarter chords of the three sections of the airfoil, and determining their strengths so as to satisfy the approximate flow-tangency condition (4-10) at the three-quarter-chord points on the three sections of the airfoil. Compare your results with those you obtained in problem 6.

9. Derive formulas for the lift, drag, and moment coefficients usable in the panel method described in the text. Specifically, take the pressure coefficients to be constant on each of the N panels, find the contribution of each panel to the force and moment coefficients in terms of the coordinates of its endpoints, and sum the contributions.

10. Compare the results of program PANEL for an NACA four- or five-digit airfoil with those you obtained in problems 5 and 7 for the same airfoil.

11. Use program PANEL to compute the pressure distribution on the symmetric airfoil

$$Y_u(x) = -Y_l(x) = \sqrt{\frac{27}{16}}\tau\sqrt{\frac{x}{c}}(c-x)$$

for $\tau = 0.1$. Compare your results with those obtained by program DUBLET for the case of zero angle of attack.

12. Use PANEL to study the flow past one of the Rankine ovals you created in problem 4 of Chapter 3.

13. The moment coefficient printed out by PANEL is the moment about the leading edge. To get c_{mac}, you first need to locate the aerodynamic center. How can you do so, just using results obtained from PANEL? *Hint*: Fit formulas linear in the angle of attack to the data output by PANEL.

14. We shall see in Chapter 7 that the displacement effect of the boundary layer on an airfoil's pressure distribution can be calculated by solving an irrotational-flow problem that differs from the one studied in this chapter only in that V_n is not zero but is a prescribed function of the boundary-layer solution. How would you revise PANEL if the values of V_n at the middle of each panel were not zero, as is implied in equation 4-81, but were prescribed?

15. Modify program JOUKOW to describe the flow past an unsymmetric Joukowski airfoil at angle of attack. Find the exact solution for the case $\varepsilon/b = -0.1$, $\mu/b = 0.1$, and $\alpha = 5°$. Plot the body shape and the pressure distribution on its upper and lower surfaces.

4.12. COMPUTER PROGRAMS

```
C     PROGRAM PANEL
C
C                 SMITH-HESS (DOUGLAS) PANEL METHOD
C                 FOR SINGLE-ELEMENT LIFTING AIRFOIL
C                 IN TWO-DIMENSIONAL INCOMPRESSIBLE FLOW
C
      DIMENSION   Z(100)
      PI          =  3.1415926585
      CALL INDATA
      CALL SETUP
```

```
  100 PRINT 1000
      READ, ALPHA
      IF (ALPHA .GT. 90.)    GO TO 200
      COSALF      =  COS(ALPHA*PI/180.)
      SINALF      =  SIN(ALPHA*PI/180.)
      CALL COFISH(SINALF,COSALF)
      CALL GAUSS(1)
      CALL VELDIS(SINALF,COSALF)
      CALL FANDM(SINALF,COSALF)
      GO TO 100
  200 STOP
 1000 FORMAT(/////,' INPUT ALPHA IN DEGREES',)
      END
C===================================================
      SUBROUTINE SETUP
      COMMON /BOD/ NLOWER,NUPPER,NODTOT,X(100),Y(100),
     +    COSTHE(100),SINTHE(100)
      COMMON /NUM/ PI,PI2INV
      COMMON /SKAL/ NZERO,YMULT
      PI          = 3.1415926585
      PI2INV      = .5/PI
      NZERO       =  31
      YMULT       =  200.
C
C                 SET COORDINATES OF NODES ON BODY SURFACE
C
      PRINT 1000
      NPOINTS     = NLOWER
      SIGN        = -1.0
      NSTART      = 0
      DO 110  NSURF = 1,2
      DO 100  N = 1,NPOINTS
      FRACT       = FLOAT(N-1)/FLOAT(NPOINTS)
      Z           = .5*(1. - COS(PI*FRACT))
      I           = NSTART + N
      CALL BODY(Z,SIGN,X(I),Y(I))
      CALL PLOTXY(X(I),Y(I))
  100 CONTINUE
      NPOINTS     = NUPPER
      SIGN        = 1.0
      NSTART      = NLOWER
  110 CONTINUE
      NODTOT      = NLOWER + NUPPER
      X(NODTOT+1)= X(1)
      Y(NODTOT+1)= Y(1)
C
C                 SET SLOPES OF PANELS
C
      DO 200  I = 1,NODTOT
      DX          = X(I+1) - X(I)
      DY          = Y(I+1) - Y(I)
      DIST        = SQRT(DX*DX + DY*DY)
      SINTHE(I)   = DY/DIST
      COSTHE(I)   = DX/DIST
  200 CONTINUE
 1000 FORMAT(/////,' BODY SHAPE',//,4X,'X',9X,'Y',/)
      RETURN
      END
C==========================================================
```

```
      SUBROUTINE BODY(Z,SIGN,X,Y)
C
C                 RETURN COORDINATES OF POINT ON BODY SURFACE
C
C                    Z = NODE-SPACING PARAMETER
C                    X,Y = CARTESIAN COORDINATES
C                    SIGN = +1. FOR UPPER SURFACE
C                         = -1. FOR LOWER SURFACE
C
      COMMON/PAR/NACA,TAU,EPSMAX,PTMAX
      IF (SIGN .LT. 0.0)     Z = 1. - Z
      CALL NACA45(Z,THICK,CAMBER,BETA)
      X           = Z - SIGN*THICK*SIN(BETA)
      Y           = CAMBER + SIGN*THICK*COS(BETA)
      RETURN
      END
C=============================================================
      SUBROUTINE COFISH(SINALF,COSALF)
C
C                 SET COEFFICIENTS OF LINEAR SYSTEM
C
      COMMON /BOD/ NLOWER,NUPPER,NODTOT,X(100),Y(100),
     +    COSTHE(100),SINTHE(100)
      COMMON /COF/ A(101,111),KUTTA
      COMMON /NUM/ PI,PI2INV
      KUTTA       = NODTOT + 1
C
C           INITIALIZE COEFFICIENTS
C
      DO 90        J = 1,KUTTA
   90 A(KUTTA,J) = 0.0
C
C           SET VN = 0 AT MIDPOINT OF ITH PANEL
C
      DO 120  I = 1,NODTOT
      XMID        = .5*(X(I) + X(I+1))
      YMID        = .5*(Y(I) + Y(I+1))
      A(I,KUTTA) = 0.0
C
C           --  FIND CONTRIBUTION OF JTH PANEL
C
      DO 110  J = 1,NODTOT
      FLOG        = 0.0
      FTAN        = PI
      IF (J .EQ. I)  GO TO 100
      DXJ         = XMID - X(J)
      DXJP        = XMID - X(J+1)
      DYJ         = YMID - Y(J)
      DYJP        = YMID - Y(J+1)
      FLOG        = .5*ALOG((DXJP*DXJP+DYJP*DYJP)/(DXJ*DXJ+DYJ*DYJ))
      FTAN        = ATAN2(DYJP*DXJ-DXJP*DYJ,DXJP*DXJ+DYJP*DYJ)
  100 CTIMTJ      = COSTHE(I)*COSTHE(J) + SINTHE(I)*SINTHE(J)
      STIMTJ      = SINTHE(I)*COSTHE(J) - SINTHE(J)*COSTHE(I)
      A(I,J)      = PI2INV*(FTAN*CTIMTJ + FLOG*STIMTJ)
      B           = PI2INV*(FLOG*CTIMTJ - FTAN*STIMTJ)
      A(I,KUTTA) = A(I,KUTTA) + B
      IF ((I .GT. 1) .AND. (I .LT. NODTOT))  GO TO 110
C
C           --  IF ITH PANEL TOUCHES TRAILING EDGE,
C               ADD CONTRIBUTION TO KUTTA CONDITION
```

```
C
      A(KUTTA,J) = A(KUTTA,J) - B
      A(KUTTA,KUTTA)          = A(KUTTA,KUTTA) + A(I,J)
  110 CONTINUE
C
C                 FILL IN KNOWN SIDES
C
      A(I,KUTTA+1)            =  SINTHE(I)*COSALF -COSTHE(I)*SINALF
  120 CONTINUE
      A(KUTTA,KUTTA+1)        =  - (COSTHE(1) + COSTHE(NODTOT))*COSALF
     +                           - (SINTHE(1) + SINTHE(NODTOT))*SINALF
      RETURN
      END
C=============================================================
      SUBROUTINE VELDIS(SINALF,COSALF)
C
C                 COMPUTE AND PRINT OUT PRESSURE DISTRIBUTION
C
      COMMON /BOD/ NLOWER,NUPPER,NODTOT,X(100),Y(100),
     +    COSTHE(100),SINTHE(100)
      COMMON /COF/ A(101,111),KUTTA
      COMMON /CPD/ CP(100)
      COMMON /NUM/ PI,PI2INV
      COMMON /SKAL/ NZERO,YMULT
      DIMENSION  Q(150)
      YMULT       =  20.0
      PRINT 1000
C
C                RETRIEVE SOLUTION FROM A-MATRIX
C
      DO 50  I = 1,NODTOT
   50 Q(I)        = A(I,KUTTA+1)
      GAMMA       = A(KUTTA,KUTTA+1)
C
C                FIND VTAND CP AT MIDPOINT OF ITH PANEL
C
      DO 130  I = 1,NODTOT
      XMID        = .5*(X(I) + X(I+1))
      YMID        = .5*(Y(I) + Y(I+1))
      VTANG       = COSALF*COSTHE(I) + SINALF*SINTHE(I)
C
C                --  ADD CONTRIBUTIONS OF JTH PANEL
C
      DO 120  J = 1,NODTOT
      FLOG        = 0.0
      FTAN        = PI
      IF (J .EQ. I)  GO TO 100
      DXJ         = XMID - X(J)
      DXJP        = XMID - X(J+1)
      DYJ         = YMID - Y(J)
      DYJP        = YMID - Y(J+1)
      FLOG        = .5*ALOG((DXJP*DXJP+DYJP*DYJP)/(DXJ*DXJ+DYJ*DYJ))
      FTAN        = ATAN2(DYJP*DXJ-DXJP*DYJ,DXJP*DXJ+DYJP*DYJ)
  100 CTIMTJ      = COSTHE(I)*COSTHE(J) + SINTHE(I)*SINTHE(J)
      STIMTJ      = SINTHE(I)*COSTHE(J) - SINTHE(J)*COSTHE(I)
      AA          = PI2INV*(FTAN*CTIMTJ + FLOG*STIMTJ)
      B           = PI2INV*(FLOG*CTIMTJ - FTAN*STIMTJ)
      VTANG       = VTANG - B*Q(J) + GAMMA*AA
```

```
  120 CONTINUE
      CP(I)        = 1. - VTANG*VTANG
      CALL PLOTXY(XMID,CP(I))
  130 CONTINUE
 1000 FORMAT(/////,' PRESSURE DISTRIBUTION',//,4X,'X',8X,'CP',/)
      RETURN
      END
C=========================================================
      SUBROUTINE FANDM(SINALF,COSALF)
C
C                    COMPUTE AND PRINT OUT CD,CL,CM
C
      COMMON /BOD/ NLOWER,NUPPER,NODTOT,X(100),Y(100),
     +    COSTHE(100),SINTHE(100)
      COMMON /CPD/ CP(100)
      CFX          = 0.0
      CFY          = 0.0
      CM           = 0.0
      DO 100  I = 1,NODTOT
      XMID         = .5*(X(I) + X(I+1))
      YMID         = .5*(Y(I) + Y(I+1))
      DX           = X(I+1) - X(I)
      DY           = Y(I+1) - Y(I)
      CFX          = CFX + CP(I)*DY
      CFY          = CFY - CP(I)*DX
      CM           = CM + CP(I)*(DX*XMID + DY*YMID)
  100 CONTINUE
      CD           = CFX*COSALF + CFY*SINALF
      CL           = CFY*COSALF - CFX*SINALF
      PRINT 1000, CD,CL,CM
 1000 FORMAT(5(/),*    CD =*,F8.5,*    CL =*,F8.5,*    CM =*,F8.5)
      RETURN
      END
C=========================================================
      SUBROUTINE PLOTXY(X,Y)

      (SEE PROGRAM DUBLET)

C=========================================================
      SUBROUTINE GAUSS(NRHS)

      (SEE PROGRAM DUBLET)

C=========================================================
      SUBROUTINE INDATA

C
C                    SET PARAMETERS OF BODY SHAPE,
C                    FLOW SITUATION, AND NODE DISTRIBUTION
C
C                    USER MUST INPUT
C                       NLOWER = NUMBER OF NODES ON LOWER SURFACE
C                        NUPPER = NUMBER OF NODES ON UPPER SURFACE
C                     PLUS DATA ON BODY AND SUBROUTINE BODY
C
      COMMON /BOD/ NLOWER,NUPPER,NODTOT,X(100),Y(100),
     +    COSTHE(100),SINTHE(100)
      COMMON /PAR/ NACA,TAU,EPSMAX,PTMAX
      PRINT, ' INPUT NLOWER,NUPPER',
      READ,  NLOWER,NUPPER
```

```
      PRINT, ' INPUT NACA NUMBER',
      READ,NACA
      IEPS         = NACA/1000
      IPTMAX       = NACA/100 - 10*IEPS
      ITAU         = NACA - 1000*IEPS - 100*IPTMAX
      EPSMAX       = IEPS*0.01
      PTMAX        = IPTMAX*0.1
      TAU          = ITAU*0.01
      IF (IEPS .LT. 10)       RETURN
      PTMAX        = 0.2025
      EPSMAX       = 2.6595*PTMAX**3
      RETURN
      END
C=================================================================
      SUBROUTINE NACA45(Z,THICK,CAMBER,BETA)
      COMMON /PAR/ NACA,TAU,EPSMAX,PTMAX
C
C                   EVALUATE THICKNESS AND CAMBER
C                   FOR NACA 4- OR 5-DIGIT AIRFOIL
C
      THICK        = 0.0
      IF (Z .LT. 1.E-10)      GO TO 100
      THICK        = 5.*TAU*(.2969*SQRT(Z) - Z*(.126 + Z*(.3537
     +                - Z*(.2843 - Z*.1015))))
  100 IF (EPSMAX .EQ .0.0)    GO TO 130
      IF (NACA .GT. 9999)     GO TO 140
      IF (Z .GT. PTMAX)       GO TO 110
      CAMBER       = EPSMAX/PTMAX/PTMAX*(2.*PTMAX - Z)*Z
      DCAMDX       = 2.*EPSMAX/PTMAX/PTMAX*(PTMAX - Z)
      GO TO 120
  110 CAMBER       = EPSMAX/(1.-PTMAX)**2*(1. + Z - 2.*PTMAX)*(1. - Z)
      DCAMDX       = 2.*EPSMAX/(1.-PTMAX)**2*(PTMAX - Z)
  120 BETA         = ATAN(DCAMDX)
      RETURN
  130 CAMBER       = 0.0
      BETA         = 0.0
      RETURN
  140 IF (Z .GT. PTMAX)       GO TO 150
      W            = Z/PTMAX
      CAMBER       = EPSMAX*W*((W - 3.)*W + 3. - PTMAX)
      DCAMDX       = EPSMAX*3.*W*(1. - W)/PTMAX
      GO TO 120
  150 CAMBER       = EPSMAX*(1. - Z)
      DCAMDX       = - EPSMAX
      GO TO 120
      END
```

CHAPTER 5

WINGS OF FINITE SPAN

It is time we break out of our two-dimensional world and consider at least briefly the effect on a wing's performance of its three-dimensional character. At the same time, you will see how our results for airfoils can be used in a three-dimensional analysis, at least in the so-called lifting-line theory, and also how the analytical approach we have been taking can be extended to treat finite-span wings.

5.1. THE VORTEX SYSTEM FOR A THIN PLANAR WING OF FINITE SPAN

Consider the flow induced by the motion of a thin wing of finite span through an otherwise undisturbed fluid, as shown in Fig. 5.1. To begin with, we will aim simply for a qualitative description of the flow and so will make a number of simplifying assumptions that will be retracted later on.

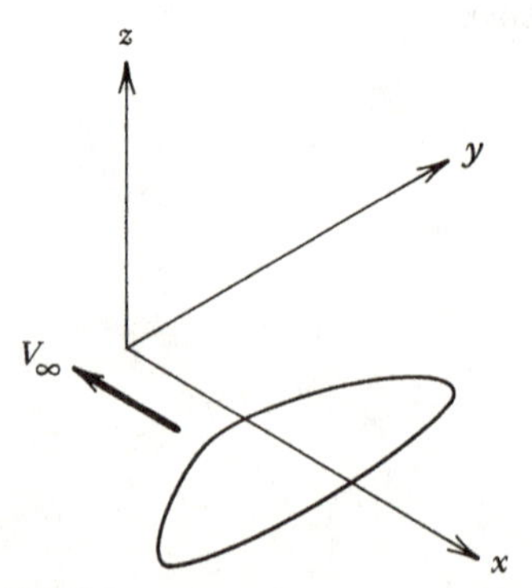

FIG. 5.1. Analysis of force exerted by a moving wing on the air.

It is convenient to adopt the viewpoint of an observer fixed in the undisturbed fluid, rather than our usual body-fixed frame of reference. This shift in coordinates means that all three velocity components are small, not just those perpendicular to the onset flow direction, as is the case when we work in coordinates fixed to a thin airfoil. In another departure from our usual *modus operandi*, instead of setting up a boundary-value problem for the velocity potential, we will look at the wing as a source of force on the fluid and try to determine the fluid flow that results from that force. Thus we write the equation of motion as

$$\rho \frac{D\mathbf{V}}{Dt} = -\nabla p + \mathbf{f} \tag{5-1}$$

Here **f** is the force per unit volume exerted by the wing on the fluid.

Since the fluid velocity induced by the motion of the wing is small, we approximate the convective acceleration

$$\frac{D\mathbf{V}}{Dt} \equiv \frac{\partial \mathbf{V}}{\partial t} + (\mathbf{V} \cdot \nabla)\mathbf{V}$$

by $\partial \mathbf{V}/\partial t$. Also, we suppose that the wing lies essentially in the x–y plane, while moving with velocity V_∞ in the negative x direction, and that the force per unit volume is mainly in the z direction. Then the equation of motion (5-1) can be approximated by

$$\rho \frac{\partial \mathbf{V}}{\partial t} \approx -\nabla p - f\hat{\mathbf{k}} \tag{5-2}$$

We can eliminate the pressure from equation 5-2 by taking its curl:

$$\rho \frac{\partial}{\partial t} \operatorname{curl} \mathbf{V} = 0 - \frac{\partial f}{\partial y}\hat{\mathbf{i}} + \frac{\partial f}{\partial x}\hat{\mathbf{j}} \tag{5-3}$$

Introducing the angular velocity $\boldsymbol{\omega}$ (see equation 2-34),

$$\boldsymbol{\omega} = \tfrac{1}{2} \operatorname{curl} \mathbf{V}$$

we thus obtain

$$2\rho \frac{\partial \omega_x}{\partial t} = -\frac{\partial f}{\partial y} \tag{5-4a}$$

$$2\rho \frac{\partial \omega_y}{\partial t} = \frac{\partial f}{\partial x} \tag{5-4b}$$

$$2\rho \frac{\partial \omega_z}{\partial t} = 0 \tag{5-4c}$$

Let us now specialize our study to a *uniformly loaded rectangular wing*; that is, one for which f is constant, except at the edges of the rectangular planform, where it changes abruptly to or from zero. We will see later how any more general case can be treated by superposing the solution that results. In the present case, as the leading

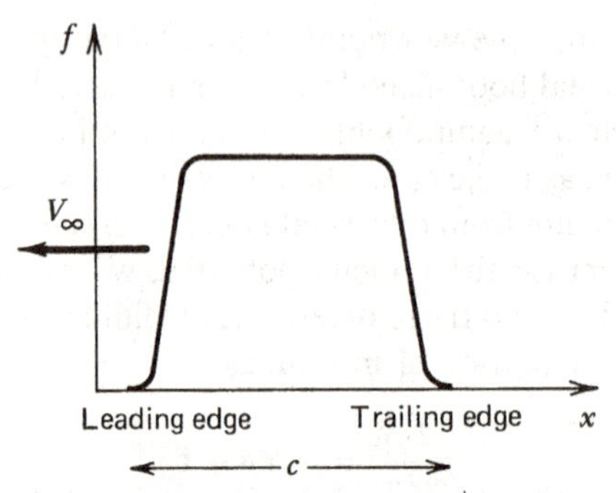

FIG. 5.2. Distribution of force on a "uniformly loaded" wing.

edge passes a given point, we see from Fig. 5.2 that $\partial f/\partial x$ becomes large and positive, and then returns to zero. Thus, from equation 5-4b, the y component of $\boldsymbol{\omega}$ will increase abruptly as the wing's leading edge goes by. The same reasoning shows ω_y to decrease with the arrival and passage of the trailing edge. In fact, it goes back to zero. For, away from the tips, the distribution of f is of constant shape but traveling with speed V_∞ in the x direction and so can be written

$$f(x, y, t) = F(x + V_\infty t)$$

Then

$$\frac{\partial f}{\partial x} = F' = \frac{1}{V_\infty}\frac{\partial f}{\partial t}$$

so equation 4-4b becomes

$$2\rho\frac{\partial \omega_y}{\partial t} = \frac{1}{V_\infty}\frac{\partial f}{\partial t}$$

which integrates to

$$\omega_y = \frac{1}{2\rho V_\infty} f$$

the x dependent constant of integration being evaluated at a time before the point in question is reached by the wing. Thus ω_y returns to zero when f does. However, while in contact with the wing, the fluid spins about a spanwise-oriented axis.

Similarly, we see from equation 5-4a that, as the right wing tip passes a point, ω_x increases, since $\partial f/\partial y < 0$ there. This rotation accumulates as the wing tip goes by and remains constant after the wing completes its passage. At the left tip, a negative rotation about the x direction builds up for the same reason.

As shown in Fig. 5.3, the resultant flow pattern thus consists of a set of vortices, some *bound* to the wing (they haven't yet been canceled out by the trailing edge's passing by), some trailing behind the wing tips (the *tip vortices*), and some left far behind in the wake where the wing started into motion (they were behind the leading edge to begin with, and so influenced only by the trailing edge). These last vortices, whose rotational speed $\omega_y < 0$, are called the *starting vortices*.

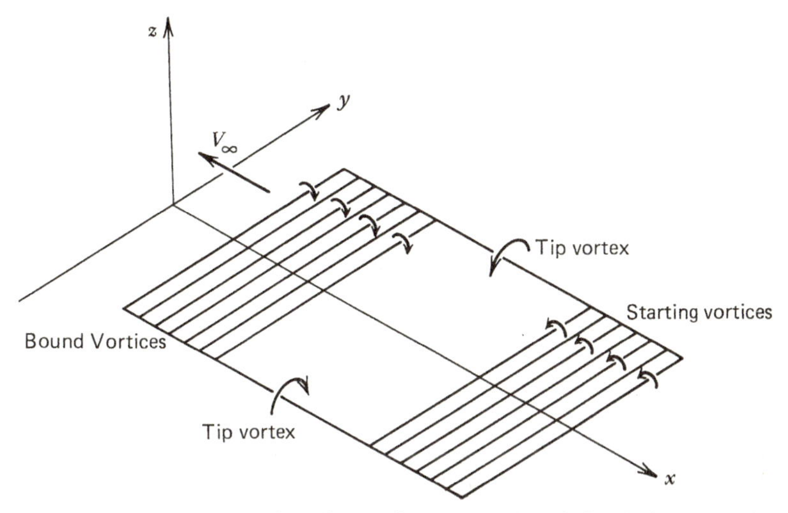

FIG. 5.3. Vortex system created by the motion of a uniformly loaded rectangular wing.

There is an important relation between the strength of the tip vortices and that of the bound vortices. Consider an elbow-shaped tubular sleeve, one of whose ends encompasses the tip vortex and the other, the bound vortices. Slit the tube as shown in Fig. 5.4, and then construct the closed contour $C_1 + C_2 + C_3 + C_4$. On this contour, and on the surface S of the sleeve it encloses,

$$\mathbf{V} = \nabla\phi$$

Since the curl of any gradient is zero,

$$\int_S \text{curl}\,\mathbf{V} \cdot \hat{\mathbf{n}}\, ds = 0$$

and so, from Stokes's theorem (see equation 2-25),

$$0 = \oint_{C_1+C_2+C_3+C_4} \mathbf{V} \cdot \mathbf{dl}$$

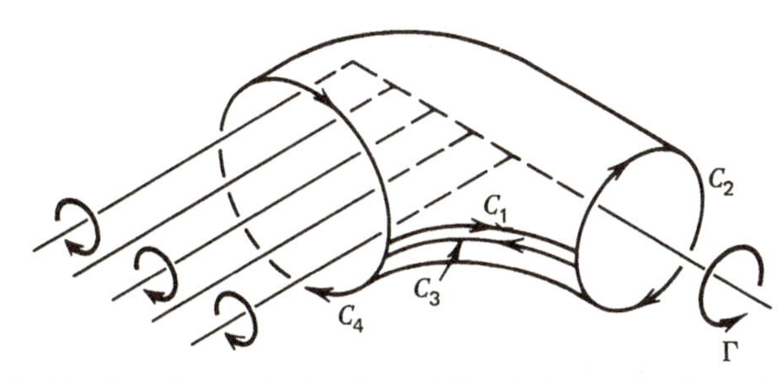

FIG. 5.4. Control surface for analyzing the relation between the bound vortices and the tip vortices of the uniformly loaded rectangular wing.

The line integrals on C_1 and C_3 cancel one another (**V** is the same at every corresponding point, but **dl** is oppositely directed), and so

$$\oint_{C_4} \mathbf{V} \cdot \mathbf{dl} = -\oint_{C_2} \mathbf{V} \cdot \mathbf{dl} \equiv \Gamma$$

where Γ is the net strength of the bound vortices, which is now seen to be equal to the strength of the right tip vortex. That is, letting the strength per unit length of the bound vortices aligned with the y axis be $\gamma_y(x)$,

$$\Gamma \equiv \int_0^c \gamma_y(x)\, dx$$

Similarly, we can show that the left tip vortex is of equal but opposite strength, as is the net strength of the starting vortices. The equality of the strengths of the two tip vortices to those of the bound and starting vortices can be related to the one of *Helmholtz's vortex theorems*, which states that a vortex cannot end in the fluid [1]. That is, if we have a vortex bound to the wing, it must continue back into the wake as a part of the tip vortex.

We now can build up a picture of the vortex system associated with a wing of arbitrary loading and planform. We simply regard the wing and its loading as being the sum of an infinite set of uniformly loaded rectangular wings. As indicated in Fig. 5.5, each rectangle contributes to the flow picture a distributed set of bound vortices aligned in the y direction plus "tip" vortices parallel to the x axis. The upshot is a set of vortices in both x and y directions on the wing (the *bound vortices*), plus a system of vortices parallel to the direction of flight that trail behind the wing's trailing edge (the *trailing vortices*).

Let $\gamma_y(x, y)$ be the strength per unit length of the bound vortices aligned in the y direction, and

$$\Gamma(y) \equiv \int_{\text{chord}} \gamma_y(x, y)\, dx \tag{5-5}$$

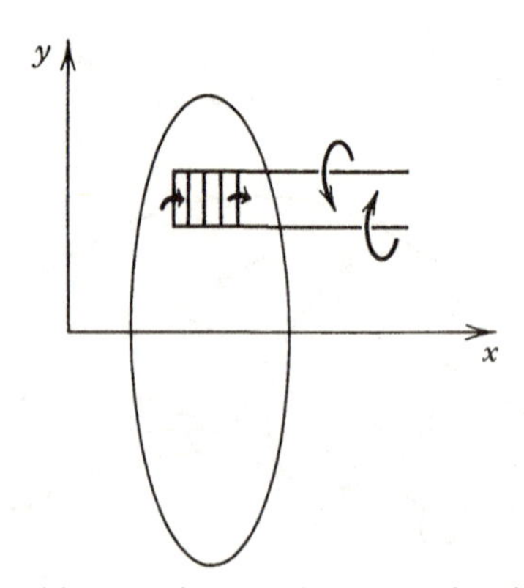

FIG. 5.5. Analysis of an arbitrary wing as the sum of uniformly loaded rectangular wings.

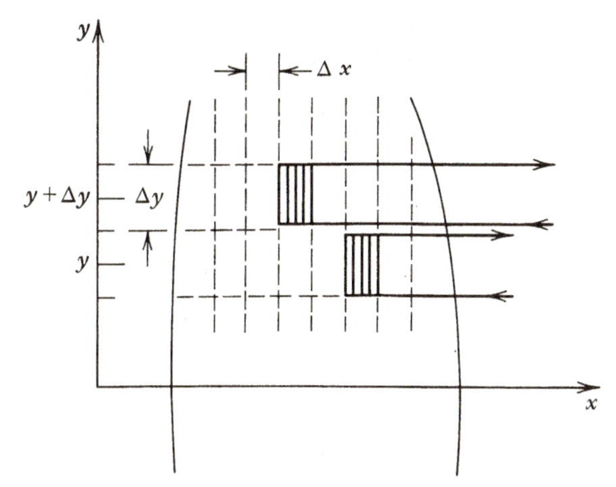

FIG. 5.6. Analysis of the trailing vortex system of an arbitrary wing.

be the circulation about the wing in a constant y plane. Also, let $\gamma_x(y)$ be the strength per unit width of the trailing vortex system. That is, $\gamma_x(y)\,\Delta y$ is the strength of the vortices aligned in the x direction that trail behind the trailing edge between y and $y + \Delta y$. If we cover the wing with rectangles $\Delta x \times \Delta y$, as shown in Fig. 5.6, the contributions to $\gamma_x(y)\Delta y$ come from the tip vortices of those rectangles just inboard of $y + \Delta y/2$ and also of those just outboard of the same station. The contributions from the inboard rectangles equal the net circulation around the bound vortices on those rectangles, whereas the outboard rectangles contribute the negative of the same quantity. For example, the rectangle that lies in the interval

$$(x, x + \Delta x) \times \left(y + \frac{\Delta y}{2}, y + 3\frac{\Delta y}{2}\right)$$

contributes a tip vortex of strength

$$-\int_x^{x+\Delta x} \gamma_y(x', y + \Delta y)\, dx' \tag{5-6}$$

whereas the rectangle described by

$$(x, x + \Delta x) \times \left(y - \frac{\Delta y}{2}, y + \frac{\Delta y}{2}\right)$$

contributes a tip vortex of strength

$$+\int_x^{x+\Delta x} \gamma_y(x', y)\, dx'$$

The result is

$$\begin{aligned}\gamma_x(y)\Delta y &= -\int_{\text{chord}} \gamma_y(x', y + \Delta y)\, dx' + \int_{\text{chord}} \gamma_y(x', y)\, dx' \\ &= -\Gamma(y + \Delta y) + \Gamma(y)\end{aligned}$$

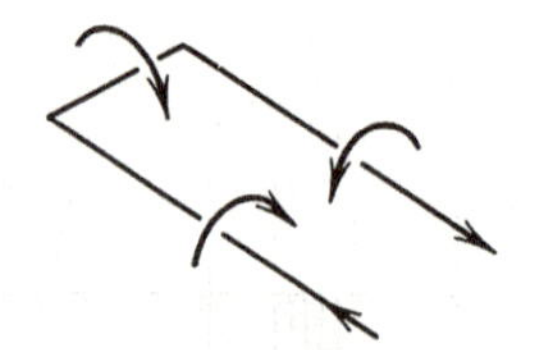

FIG. 5.7. A horseshoe vortex.

Here equation 5-5 was used to evaluate the integrals. Then

$$\gamma_x(y) = -\frac{\Gamma(y+\Delta y) - \Gamma(y)}{\Delta y}$$

$$\rightarrow -\frac{d\Gamma}{dy} \tag{5-7}$$

That is, *the strength per unit width of the trailing vortex system is equal in magnitude to the slope of the circulation distribution.*

As with the panel method described in Section 4.8, this model may be made exact by covering the surfaces of the wing (indeed, of the whole airplane!) with *horseshoe vortices*, that is, vortices that trail back from the body surface with undiminished strength, as shown in Fig. 5.7. Since vortices (or part thereof) that are not bound to the surface must drift with the flow,[1] the trailing vortex system is a stream surface. It is not easy to determine its shape exactly; it is usually approximated as a surface parallel to the undisturbed onset flow. The strengths of the distributed bound and trailing vortices must be determined so as to satisfy the boundary conditions on the body surface, including the Kutta condition. Usually, as in the panel method described in Chapter 4, this task requires a computer for implementation.

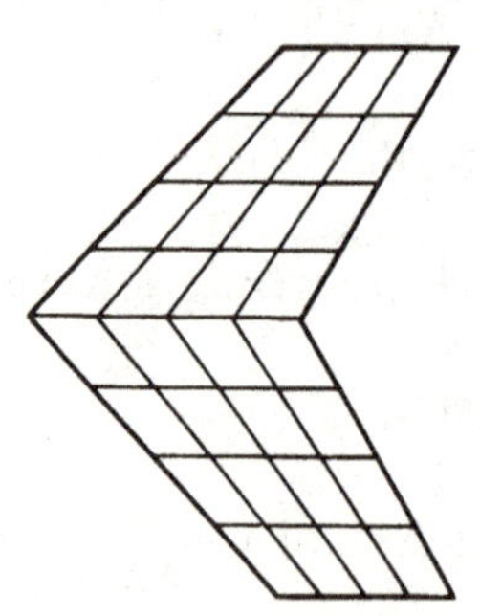

FIG. 5.8. Paneling for the vortex–lattice method.

5.2. THE VORTEX-LATTICE METHOD

A particularly interesting version of surface singularity distribution is the so-called *vortex-lattice method*, which is based on discrete horseshoe vortices, as follows.

The surface to be covered with vortices is broken into quadrilateral panels, as shown in Fig. 5.8. Although it is not necessary to do so, it simplifies matters to take

[1] See Section 7.13.

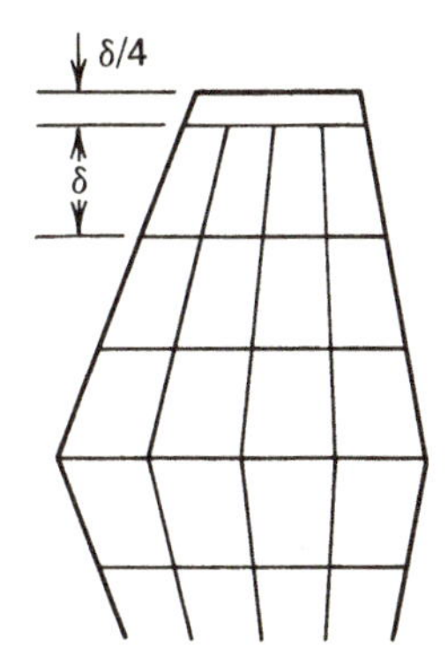

FIG. 5.9. Improving the accuracy of a vortex–lattice method by removing the panels from the wing tips.

the surface as plane; wing camber can be accounted for in a thin-airfoil-like approximation, by requiring flow tangency to the mean surface. Also, we will require two sides of every panel to be parallel to the free-stream direction.

Taking a cue from the results of Section 4.7, we lump the spanwise vorticity on each panel into a discrete vortex along its one-quarter-chord line. In accordance with the picture we now have of trailing vortices, we continue this vortex as a horseshoe into the wake with two semi-infinite filaments parallel to the free-stream direction.[2] The result is a system of horseshoe vortices, one for every panel on the surface. With justification residing in the result derived in Section 4.7 for a thin airfoil of parabolic camber, their strengths are determined by imposing the flow-tangency condition at the midpoint of the three-quarter chord of each panel.

The accuracy of this so-called *vortex-lattice method* seems to be enhanced for a given number of panels if the panels are of equal width δ in the spanwise direction, and if the panels do not extend to the wing tips, but only to a distance of $\delta/4$ from the tips, as shown in Fig. 5.9. The panels should also have equal lengths at any spanwise station.[3] Flapped and cranked wings are best treated by using separate systems of panels for the flaps and for geometrically distinct parts of the planform.

[2] Another possibility is to terminate the tip vortices at the bound vortex of the panel immediately downstream, so that the wing is covered by a system of rectangular vortices, with the trailing vortices emanating only from the panels on the wing's trailing edge. Such a viewpoint has advantages when the vortex-lattice method is used to describe the flow past wing-body combinations. It is equivalent to the use of piecewise-constant doublet distributions [2].

[3] Alternatively, one can use a "cosine spacing" in both the x and y directions for the location of both the vortices and the control points. For example, let the x coordinates of the vortices be located relative to the leading edge of the section at the spanwise station y by

$$x_{V_i} = \frac{c(y)}{2}\left(1 - \cos\frac{i\pi}{N+1}\right)$$

and the control points at

$$x_{C_i} = \frac{c(y)}{2}\left[1 - \cos\frac{(i+\frac{1}{2})}{N+1}\right]$$

for $i = 1, \ldots, N$, as shown in Fig. 5.10. See Ref. [3].

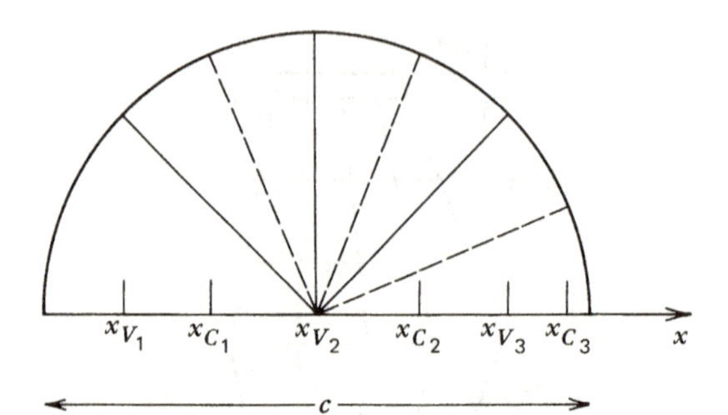

FIG. 5.10. Cosine spacing of the vortices and control points.

To be specific, consider the flat rectangular wing of zero thickness shown in Fig. 5.11. Let N_x and N_y be the numbers of panels to be used in the x and y directions, respectively, on each semispan (naturally, we will take advantage of the symmetry about the midspan), and Δx and Δy be the dimensions of each panel. Then

$$N_x \Delta x = c, \qquad 2\left(N_y \Delta y + \frac{\Delta y}{4}\right) = b \tag{5-8}$$

where b and c are the wing span and chord, respectively. We will use a double-subscript notation for each panel and its associated vortex. That is, let the ijth panel be the one centered at $y = (i - 1/2)\Delta y$, $x = (j - 1/2)\Delta x$. For example, the panel shaded in Fig. 5.11 is the "(3-2)th."

The boundary condition to be satisfied at the three-quarter-chord point of each panel is simply

$$\mathbf{V} \cdot \hat{\mathbf{k}} = 0$$

Contributions to the z component of velocity come from the onset flow (specifically, $V_\infty \sin \alpha$) and from the vortices attached to each panel. Since the vortex-induced velocity is generally downward, it is convenient to define it as being positive if in the negative z direction and to refer to it as a *downwash.*

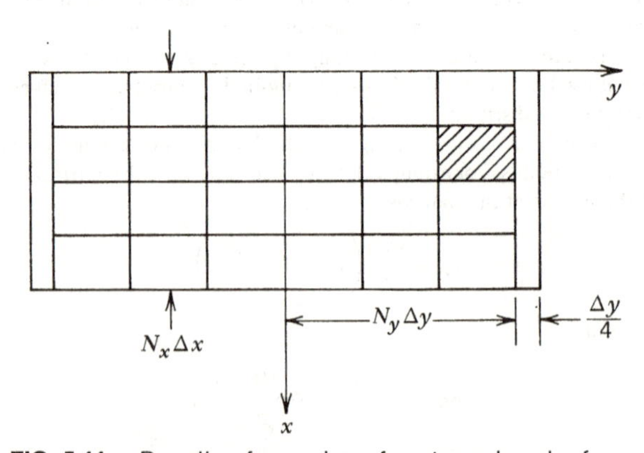

FIG. 5.11. Paneling for a wing of rectangular planform.

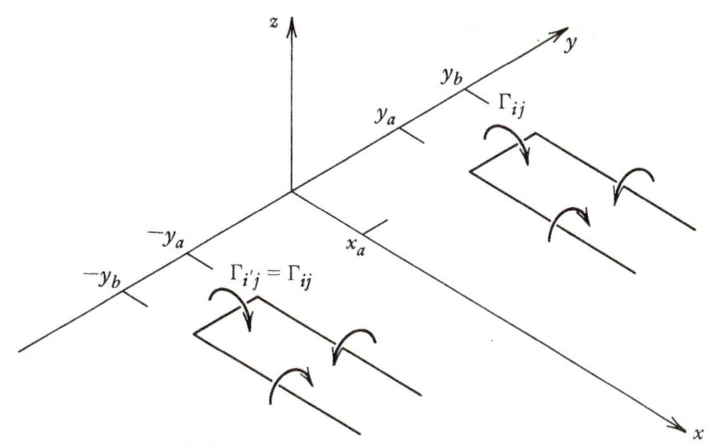

FIG. 5.12. Nomenclature for the downwash induced by a horseshoe vortex and its image in the plane of symmetry.

Let the downwash induced at $(x, y, 0)$ by the ijth vortex be $\Gamma_{ij}w_{ij}(x, y)$; that is, w_{ij} is the downwash this vortex would induce were its strength unity, since we know that the velocity field due to a vortex is directly proportional to its strength. It is shown in Appendix F that if the corners of this vortex are at (x_a, y_a) and (x_a, y_b) (see Fig. 5.12), then its downwash at $(x, y, 0)$ is

$$
\begin{aligned}
w_{ij}(x, y) &= \frac{1}{4\pi}\frac{1}{y - y_a}\left\{1 + \frac{\sqrt{(x - x_a)^2 + (y - y_a)^2}}{x - x_a}\right\} \\
&\quad - \frac{1}{4\pi}\frac{1}{y - y_b}\left\{1 + \frac{\sqrt{(x - x_a)^2 + (y - y_b)^2}}{x - x_a}\right\} \qquad \text{if } x \neq x_a \\
&= \frac{1}{4\pi}\left(\frac{1}{y - y_a} - \frac{1}{y - y_b}\right) \qquad \text{if } x = x_a \qquad (5\text{-}9)
\end{aligned}
$$

We can then write the flow tangency condition at the control point (x_p, y_p) in the form

$$V_\infty \sin\alpha - \sum_{i,j} \Gamma_{ij} w_{ij}(x_p, y_p) = 0 \qquad (5\text{-}10)$$

Since there are as many such equations of this form (i.e., as many control points) as there are panels, we thus obtain a system of linear equations for the vortex strengths Γ_{ij}.

To take advantage of the symmetry of the problem, we should combine the two terms involving vortices on opposite sides of the x axis, since the strengths of those vortices are equal. As shown in Fig. 5.12, if the ijth vortex has corners at (x_a, y_a) and (x_a, y_b), its mate has corners at $(x_{a'} - y_a)$ and $(x_{a'} - y_b)$. Calling the mate the $i'j$th vortex, we can rewrite equation 5-10 as

$$\sum_{ij} \Gamma_{ij}[w_{ij}(x_p, y_p) + w_{i'j}(x_p, y_p)] = V_\infty \sin\alpha \qquad (5\text{-}11)$$

where the sum now extends only over the vortices on the positive side of the x–z plane.

Finding the strengths of the horseshoe vortices is seen to involve the same processes that were required in the line doublet distribution method of Chapter 3 and the panel method of Chapter 4. First, you must set up a system of linear algebraic equations for the singularity (in this case, vortex) strengths by requiring flow tangency at an appropriate number of control points. Since each singularity contributes to the velocity everywhere in the flow field, the matrix of these equations is *full*; at every control point, every singularity is involved, and so all the coefficients of the unknown singularity strengths are nonzero.

5.3. INDUCED DRAG

A unique aspect of the vortex-lattice method is the way in which you compute the force and moment on a wing, once you know the vortex strengths. The force on each panel is calculated by applying the Kutta–Joukowski theorem (equation 4-32) to the vortex bound to that panel; that is, the force per unit length of the vortex bound to the ijth panel is taken to be

$$\mathbf{F}'_{ij} = \rho \mathbf{V}_{\text{eff}} \times \mathbf{\Gamma}_{ij} \tag{5-12}$$

in which the vector $\mathbf{\Gamma}_{ij}$ is directed along the axis of the vortex, whereas $\mathbf{V}_{\text{eff}}$ is the sum of the velocity far upstream $\mathbf{V}_\infty$ and the downwash induced by all the other vortices and so is the local "effective" onset flow velocity.

The force on the panel is perpendicular to the local effective onset velocity rather than to the onset flow far upstream. Thus, as shown in Fig. 5.13, the force has a component in the direction of flight; in other words, the panel experiences a drag force. This is called *induced drag*; it is due to the downwash induced by the trailing vortex system, without which $\mathbf{V}_{\text{eff}} = \mathbf{V}_\infty$ and the force would be purely in the lift direction. As can be seen from Fig. 5.13, if the angle of attack of the onset flow is α, and w is the downwash induced by the trailing vortices,

$$\mathbf{V}_{\text{eff}} = V_\infty \cos\alpha\,\hat{\mathbf{i}} + (V_\infty \sin\alpha - w)\hat{\mathbf{k}}$$

For the case of the rectangular wing depicted in Fig. 5.11, the bound vortices are in the y direction, so

$$\mathbf{\Gamma}_{ij} = \Gamma_{ij}\hat{\mathbf{j}}$$

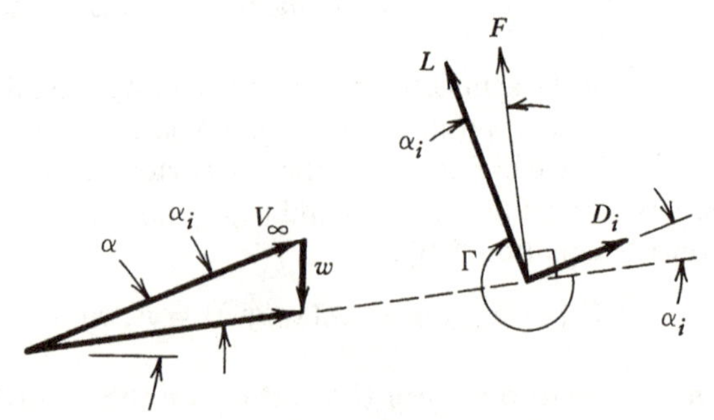

FIG. 5.13. Induced drag and induced angle of attack.

and the force on the ijth panel is found from equation 5-12 to be

$$F_{ij} = \rho \Delta y \Gamma_{ij}[V_\infty \cos\alpha\,\hat{\mathbf{k}} - (V_\infty \sin\alpha - w)\hat{\mathbf{i}}]$$

Resolving this into components along and perpendicular to the overall onset flow, we find, if α and w/V_∞ are small, the drag and lift on the ijth panel to be

$$D_{ij} \approx \rho\, \Delta y w \Gamma_{ij}$$

$$L_{ij} \approx \rho\, \Delta y V_\infty \Gamma_{ij} \tag{5-13}$$

In equation 5-13, w is calculated by summing over all the vortices. As in equation 5-11, we take advantage of the symmetry of the vortex system with respect to midspan and write

$$w = \sum_{kl} \Gamma_{kl}[w_{kl}(x_p, y_p) + w_{k'l}(x_p, y_p)] \tag{5-14}$$

in which (x_p, y_p) are now the coordinates of the quarter-chord point of the ijth panel.

The angle between $\mathbf{V}_{\text{eff}}$ and $\mathbf{V}_\infty$ is called the *induced angle of attack* α_i; it is the reduction in the effective angle of attack due to the trailing vortices. From Fig. 5.13,

$$\alpha_i \approx \frac{w}{V_\infty} \tag{5-15}$$

A computer program called VORLAT has been written to implement the vortex-lattice method for the case described here, a flat, untwisted, rectangular wing with equal-sized panels. It is listed in Section 5.10. Problem 2 gives you some exercises in its use. In running it, be sure to explore the effects of varying both N_x and N_y on your results. You should find they converge rather well.

You are invited in problem 4 to develop a vortex-lattice method that can treat tapered, swept wings. While the reprogramming required to extend VORLAT's capabilities to these cases is not extremely formidable, the results are not very satisfactory. In particular, the induced drag converges much too slowly for swept wings. The problem is a singularity connected with the kink in the bound vortices at the plane $y = 0$ (see Fig. 5.8). Various remedies are available [4], but to be confident of obtaining accurate results for swept wings, you must go to the more complex panel methods based on *continuous* distributions of sources and vortices or doublets. There are a variety of such methods, which differ from the vortex-lattice method most noticeably in the calculation of the force and moment. In the absence of discrete vortices, that is done by integration of the pressure distribution rather than by applying the Kutta–Joukowski theorem. Two-dimensional versions of these panel methods are described in Chapter 8.

5.4. LIFTING LINE THEORY

An alternative approximation for lift and induced drag can be found, sometimes analytically, by the *lifting-line theory* due to Prandtl [5]. It is convenient for us to describe this theory as a variation on the more modern vortex-lattice method, which was developed by Falkner [6].

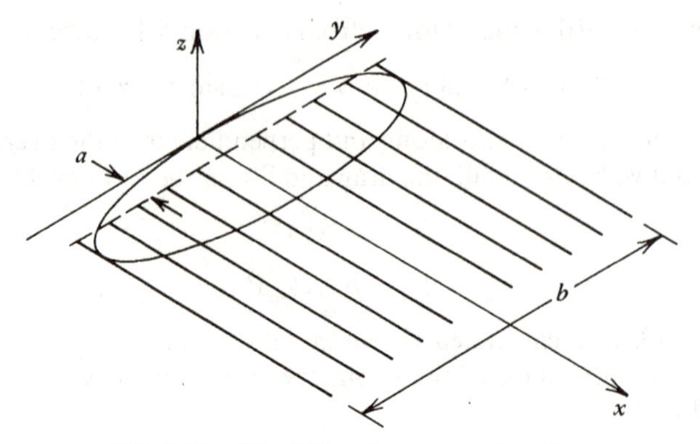

FIG. 5.14. The lifting line approximation.

From this point of view, lifting-line theory may be regarded as a vortex-lattice method that uses only one vortex panel in the chordwise direction but an infinite number of panels along the span. That is, it uses a single bound vortex, which is straight and parallel to the y axis, but whose strength is a continuous function $\Gamma(y)$. The trailing vortex system is then the semi-infinite plane

$$x > a, \qquad \text{say}$$

$$|y| < \frac{b}{2}$$

$$z = 0$$

as shown in Fig. 5.14. With this model, the theory is not adequate for highly swept wings, but it does very well otherwise.

As in the vortex-lattice theory, the lift per unit width L' and induced drag per unit width D' are calculated from the Kutta–Joukowski formula.[4] For the usual case of small angle of attack α and small downwash w (relative to V_∞), equations 5-13 yield

$$L'(y) \approx \rho V_\infty \Gamma(y) \tag{5-16}$$

$$D_i'(y) \approx \rho w(y)\Gamma(y)$$

$$\approx \rho V_\infty \Gamma(y)\alpha_i(y) \tag{5-17}$$

where $\alpha_i(y)$ is the induced angle of attack defined in equation 5-15 and Fig. 5.13.

The major difference between the lifting-line and vortex-lattice methods is the way in which the bound vortex strength $\Gamma(y)$ is calculated. Rather than satisfying flow tangency explicitly at any point along the wing, lifting-line theory simply assumes that, at any station along the wingspan, we can use two-dimensional results

[4] Note that L' and D' are not simply the total lift and drag divided by the span but are functions of y that give the distribution of the aerodynamic force along the span. Thus, for example, $L'(y)\,\Delta y$ is the lift on the part of the wing between y and $y + \Delta y$.

for the section lift coefficient c_l; that is,

$$c_l = m_0(\alpha - \alpha_{L0} - \alpha_i) \tag{5-18}$$

Here m_0 and α_{L0} are two-dimensional (theoretical or experimental) results for the lift-curve slope and angle of zero lift, respectively, and the induced angle of attack α_i accounts for the reduction in the effective angle of attack by the trailing vortices. Since

$$c_l \equiv \frac{L'}{\frac{1}{2}\rho_\infty V_\infty^2 c(y)} \tag{5-19}$$

where $c(y)$ is the local chord, the so-called *load distribution* $\Gamma(y)$ is found from equations 5-16, 5-18, and 5-19 to be

$$\begin{aligned} \Gamma(y) &= \frac{L'}{\rho V_\infty} = \tfrac{1}{2} c_l V_\infty c \\ &= \tfrac{1}{2} m_0 V_\infty c(\alpha - \alpha_{L0} - \alpha_i) \end{aligned} \tag{5-20}$$

The assumption of a locally two-dimensional flow in every constant y plane is clearly best if the wing's span b is large compared to its mean chord; that is, when its *aspect ratio*

$$AR \equiv b^2/S \tag{5-21}$$

S being the planform area (so S/b is the "average chord"), is large.

The calculation of α_i at the lifting line (i.e., the bound vortex) is greatly simplified by the rectangular conformation of the trailing vortex sheet. By symmetry, the downwash at the leading edge of the sheet (which, as shown in Fig. 5.15, we now put at $x = 0$) is exactly half what it would be were the vortex sheet doubly infinite (and so

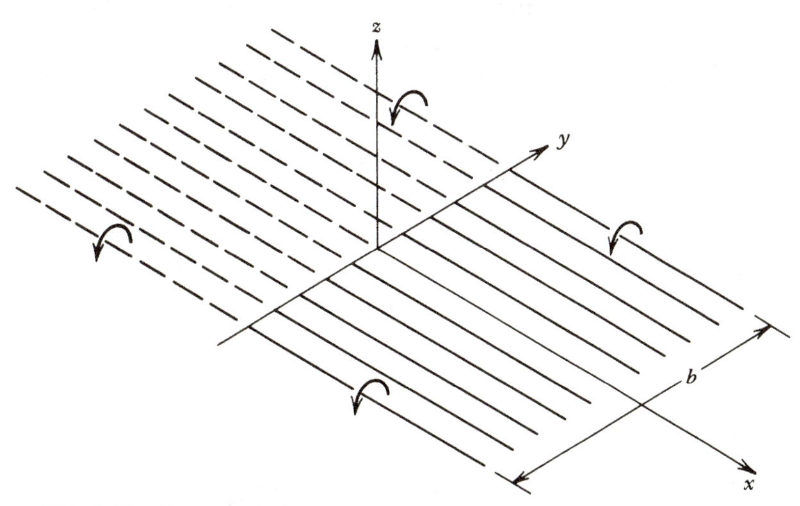

FIG. 5.15. Downwash due to the trailing vortex system of a lifting line.

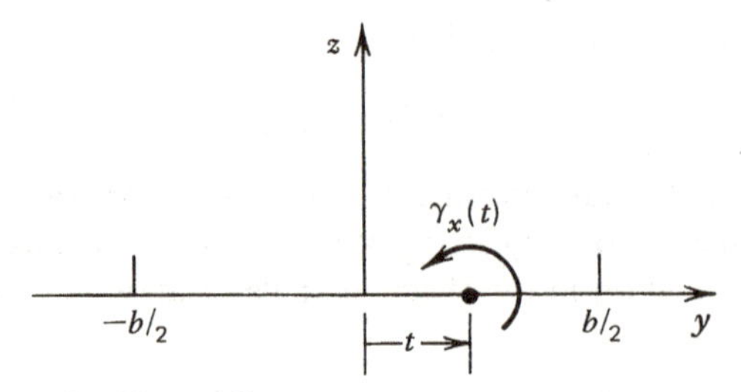

FIG. 5.16. The trailing vortex viewed from downstream.

two dimensional). Thus, we adopt the viewpoint shown in Fig. 5.16, and draw on two-dimensional results (equation 4-19) to get the downwash distribution on the lifting line:

$$2w(y) = \frac{1}{2\pi} \mathrm{P}\!\!\int_{-b/2}^{b/2} \frac{\gamma_x(t)\, dt}{t - y} \tag{5-22}$$

Eliminating γ_x and w with equations 5-7 and 5-15, we obtain

$$\alpha_i(y) = \frac{1}{4\pi V_\infty} \mathrm{P}\!\!\int_{-b/2}^{b/2} \frac{\Gamma'(t)\, dt}{y - t} \tag{5-23}$$

Plugging this into equation 5-20, we get an *integro-differential equation* for $\Gamma(y)$ (if α_{L0}, c, and m_0 are given functions of y):

$$\Gamma(y) = \tfrac{1}{2} V_\infty c m_0 \left[\alpha - \alpha_{L0} - \frac{1}{4\pi V_\infty} \mathrm{P}\!\!\int_{-b/2}^{b/2} \frac{\Gamma'(t)\, dt}{y - t} \right] \tag{5-24}$$

5.5. THE ELLIPTIC LIFT DISTRIBUTION

Rather than try to deal with the "direct" problem of finding $\Gamma(y)$ for a given wing planform and section distribution [i.e., rather than specify $\alpha_{L0}(y)$, $c(y)$, and $m_0(y)$], it is useful to begin with an "inverse" approach. We take the load distribution Γ to be elliptic:

$$\Gamma(y) = \Gamma_s \left[1 - \left(\frac{y}{b/2} \right)^2 \right]^{1/2} \tag{5-25}$$

As we shall see shortly, this distribution gives the least induced drag for a given lift and given span, regardless of how it is achieved, that is, regardless of $\alpha_{L0}(y)$ and $c(y)$.

Let us first calculate the variation of the induced angle of attack. Since

$$\Gamma'(y) = \Gamma_s \frac{1}{2} \left[1 - \left(\frac{2y}{b} \right)^2 \right]^{-1/2} \left(-2 \cdot \frac{4}{b^2} y \right)$$

we have, from equation 5-23,

$$\alpha_i(y) = -\frac{\Gamma_s}{\pi V_\infty b^2} \mathrm{P}\!\!\int_{-b/2}^{b/2} \frac{t\, dt}{(1 - 4t^2/b^2)^{1/2}(y - t)}$$

Because of the square-root term, it is convenient to introduce a trigonometric substitution:

$$t = \frac{b}{2}\cos\theta$$

$$y = \frac{b}{2}\cos\theta_0 \tag{5-26}$$

Then

$$\alpha_i(y) = -\frac{\Gamma_s}{\pi V_\infty b^2} ⨍_\pi^0 \frac{\frac{b}{2}\cos\theta\cdot\left(-\frac{b}{2}\sin\theta\right)d\theta}{(1-\cos^2\theta)^{1/2}\frac{b}{2}(\cos\theta_0-\cos\theta)}$$

$$= -\frac{\Gamma_s}{2\pi V_\infty b} ⨍_0^\pi \frac{\cos\theta\, d\theta}{\cos\theta_0 - \cos\theta}$$

But this is just a special case of that familiar formula from Appendix A

$$⨍_0^\pi \frac{\cos n\theta\, d\theta}{\cos\theta - \cos\theta_0} = \frac{\pi \sin n\theta_0}{\sin\theta_0} \tag{5-27}$$

Therefore

$$\alpha_i = \frac{\Gamma_s}{2V_\infty b} \tag{5-28}$$

and, from equation 5-15,

$$w = \frac{\Gamma_s}{2b} \tag{5-29}$$

Thus, *for an elliptic load distribution, the downwash induced by the trailing vortices is constant over the span.*

Before we go further, let us prove that this constant-downwash arrangement does, in fact, minimize the induced drag for a given lift and given span. In general, the total lift and drag are found by integrating equations 5-16 and 5-17:

$$L = \int_{-b/2}^{b/2} L'(y)\,dy = \rho V_\infty \int_{-b/2}^{b/2} \Gamma(y)\,dy \tag{5-30}$$

$$D_i = \int_{-b/2}^{b/2} D'(y)\,dy = \rho \int_{-b/2}^{b/2} w(y)\Gamma(y)\,dy \tag{5-31}$$

Let $\Gamma^*(y)$ be the optimum load distribution. Then, if $\Gamma(y)$ were changed slightly from $\Gamma^*(y)$ without changing the lift, the change in the induced drag ought to be of the order of the *square* of the change in Γ. This can be seen from Fig. 5.17, which is a conceptual plot of induced drag versus the load distribution function. Similarly, we see that if a nonoptimum Γ is changed slightly, the change in induced drag is of the order of the first power of the change in Γ.

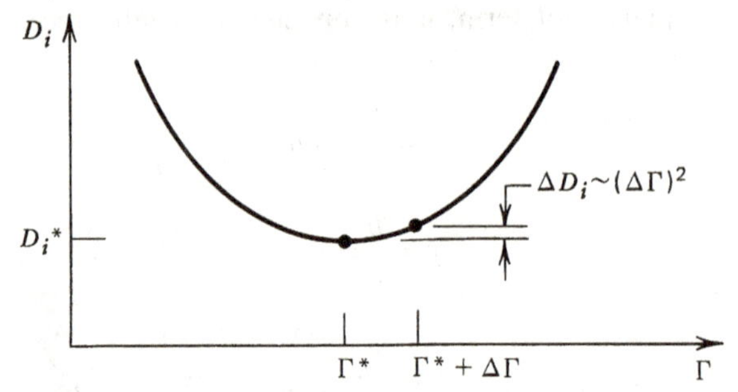

FIG. 5.17. Conceptual view of the optimum (minimum-drag) load distribution.

Therefore, we shall show that the elliptic load distribution $\Gamma_e(y)$, say, is optimum by proving that, if Γ is changed from Γ_e without changing the total lift, the change in induced drag is of the order of the square of the change in Γ.[5] Let the change in the load distribution be $\delta\Gamma(y)$, so that

$$\Gamma(y) = \Gamma_e(y) + \delta\Gamma(y) \tag{5-32}$$

Because the altered load distribution $\Gamma(y)$ is to produce the same lift as does the elliptic distribution $\Gamma_e(y)$, we see from equations 5-30 and 5-32 that

$$\int_{-b/2}^{b/2} \delta\Gamma(y)\,dy = 0 \tag{5-33}$$

The downwash distribution associated with $\Gamma(y)$ is, from equations 5-15 and 5-23,

$$\begin{aligned} w(y) &= \frac{1}{4\pi} ⨍_{-b/2}^{b/2} \frac{\Gamma_e'(t)\,dt}{y-t} + \frac{1}{4\pi} ⨍_{-b/2}^{b/2} \frac{\delta\Gamma'(t)\,dt}{y-t} \\ &= w_e + \frac{1}{4\pi} ⨍_{-b/2}^{b/2} \frac{\delta\Gamma'(t)\,dt}{y-t} \end{aligned} \tag{5-34}$$

in which w_e is the downwash due to the elliptic distribution. Substituting equations 5-32 and 5-34 into 5-31, we get

$$\begin{aligned} D_i &= \rho \int_{-b/2}^{b/2} w_e \Gamma_e(y)\,dy + \rho \int_{-b/2}^{b/2} w_e\,\delta\Gamma(y)\,dy \\ &\quad + \frac{\rho}{4\pi} \int_{-b/2}^{b/2} ⨍_{-b/2}^{b/2} \frac{\Gamma_e(y)\delta\Gamma'(t)}{y-t}\,dt\,dy + \frac{\rho}{4\pi} \int_{-b/2}^{b/2} ⨍_{-b/2}^{b/2} \frac{\delta\Gamma(y)\delta\Gamma'(t)}{y-t}\,dt\,dy \end{aligned} \tag{5-35}$$

The first integral in equation 5-35 is, from equation 5-31, just the induced drag of the elliptic distribution, D_e, say. Recalling from equation 5-29 that w_e is constant, we see from equation 5-33 that the second integral vanishes. As for the third, let

$$f(t) \equiv ⨍_{-b/2}^{b/2} \frac{\Gamma_e(y)\,dy}{y-t}$$

[5] Although all this really shows is that the drag induced by Γ_e for a given lift is a local extremum, it is, in fact, a true minimum.

Then the third integral in equation 5-35 can be written

$$\frac{\rho}{4\pi}\int_{-b/2}^{b/2} \delta\Gamma'(t)f(t)\,dt$$

Integrating by parts yields

$$\frac{\rho}{4\pi}\delta\Gamma(t)f(t)\Big|_{-b/2}^{b/2} - \frac{\rho}{4\pi}\int_{-b/2}^{b/2} \delta\Gamma(t)f'(t)\,dt$$

But $\Gamma = \Gamma_e = 0$ at the wing tips, so $\delta\Gamma(\pm b/2) = 0$ also, and the part integrated out vanishes. Also, from the definition of $f(t)$,

$$\begin{aligned} f'(t) &= \frac{d}{dt}\,\mathrm{P}\!\!\int_{-b/2}^{b/2} \frac{\Gamma_e(y)\,dy}{y-t} \\ &= \mathrm{P}\!\!\int_{-b/2}^{b/2} \Gamma_e(y)\frac{\partial}{\partial t}\left(\frac{1}{y-t}\right) dy \\ &= -\mathrm{P}\!\!\int_{-b/2}^{b/2} \Gamma_e(y)\frac{\partial}{\partial y}\left(\frac{1}{y-t}\right) dy \end{aligned}$$

Now integrating by parts with respect to y, we get

$$f'(t) = -\frac{\Gamma_e}{y-t}\Bigg|_{y=-b/2}^{y=b/2} + \mathrm{P}\!\!\int_{-b/2}^{b/2} \frac{\Gamma_e'(y)\,dy}{y-t}$$

Again the part integrated out vanishes, whereas what remains is, from equations 5-15 and 5-23, proportional to the constant w_e:

$$f'(t) = 4\pi\,w_e$$

Thus the third integral in equation 5-35 is

$$-\frac{\rho}{4\pi}\int_{-b/2}^{b/2} \delta\Gamma(t)4\pi w_e\,dt$$

which, from the constancy of w_e and equation 5-33, is just zero. Equation 5-35 therefore has been reduced to

$$D_i = D_e + O(\Delta\Gamma)^2$$

which is what we wanted to show: that any change in Γ from Γ_e that does not change the lift increases the induced drag by something quadratic in $\Delta\Gamma$. Thus the elliptic loading is, as claimed, optimal, in the sense that it minimizes the induced drag for a given lift and span.

5.6. THE OPTIMAL WING

Let us now examine the performance and geometry of the optimal wing, the wing whose induced drag is minimum for a given lift and span, because of its elliptic load distribution. To compute its lift, we must substitute for $\Gamma(y)$ in equation 5-30 from equation 5-25. Again it is convenient to use the cosine transformations (5-26):

$$L = \rho V_\infty \Gamma_s \int_{-b/2}^{b/2} \left[1 - \left(\frac{2y}{b}\right)^2\right]^{1/2} dy$$

$$= \rho V_\infty \Gamma_s \int_{\pi}^{0} (1 - \cos^2\theta)^{1/2}\left(-\frac{b}{2}\sin\theta\right) d\theta$$

$$= \rho V_\infty \Gamma_s \frac{\pi b}{4} \tag{5-36}$$

Since α_i is constant when the loading is elliptic, we have, from equations 5-28, 5-30, and 5-31,

$$D_i = \alpha_i L$$

$$= \frac{\Gamma_s}{2V_\infty b} L$$

But, from equation 5-36,

$$\Gamma_s = \frac{4}{\pi b}\frac{L}{\rho V_\infty}$$

and so

$$D_i = \frac{1}{\pi}\frac{L^2}{\frac{1}{2}\rho V_\infty^2 b^2} \tag{5-37}$$

In coefficient form, the induced drag is

$$C_{D_i} \equiv \frac{D_i}{\frac{1}{2}\rho V_\infty^2 S} = \frac{1}{\pi}\frac{S}{b^2}\left(\frac{L}{\frac{1}{2}\rho V_\infty^2 S}\right)^2$$

$$= \frac{1}{\pi \mathcal{R}} C_L^2 \tag{5-38}$$

where the lift coefficient is defined as

$$C_L \equiv \frac{L}{\frac{1}{2}\rho V_\infty^2 S} \tag{5-39}$$

and the aspect ratio is defined in equation 5-21.

Note that these results are consequences of an elliptic *load* (Γ) distribution and say nothing about the planform shape $[c(y)]$. To bring in the planform, we return to the basic equation 5-20, substitute for Γ and α_i the elliptic-loading results (5-25) and (5-28),

$$\Gamma_s\left[1 - \left(\frac{2y}{b}\right)^2\right]^{1/2} = \tfrac{1}{2} m_0 V_\infty c\left(\alpha - \alpha_{L0} - \frac{\Gamma_s}{2V_\infty b}\right)$$

and solve for $c(y)$ in terms of $m_0(y)$ and $\alpha_{L0}(y)$:

$$c(y) = \frac{2\Gamma_s}{V_\infty m_0} \frac{[1 - (2y/b)^2]^{1/2}}{[\alpha - \alpha_{L0} - (\Gamma_s/2V_\infty b)]} \tag{5-40}$$

If m_0 and α_{L0} are constant along the span, the chord distribution required to produce the optimum (elliptic) load distribution is seen to be elliptic, also; that is,

$$c = c_s\left[1 - \left(\frac{2y}{b}\right)^2\right]^{1/2} \tag{5-41}$$

where c_s is the root chord. As shown in problem 5, the total lift coefficient for the flat elliptic wing is

$$C_L = \frac{m_0(\alpha - \alpha_{L0})}{1 + m_0/\pi \mathit{AR}} \tag{5-42}$$

However, we can also use equation 5-40 to determine the optimal distribution of α_{L0} (and hence of camber and/or twist) corresponding to a given planform [and hence $c(y)$]:

$$\alpha_{L0}(y) = \alpha - \frac{\Gamma_s}{2V_\infty b}\left[1 + \frac{4b}{m_0 c(y)}\left[1 - \left(\frac{2y}{b}\right)^2\right]^{1/2}\right] \tag{5-43}$$

Remember, to minimize drag for a given lift and span, it is the *loading* $\Gamma(y)$ that must be elliptic, not necessarily the planform shape.

One of the beauties of an optimum design is that even when, for one reason or another, you cannot meet exactly the conditions that make the design optimum, your performance is likely to be fairly close to that of the optimum configuration. Returning to the conceptual sketch Fig. 5.17, we see that induced drag changes very little with small departures from the optimum load distribution. Thus the dependence of induced drag on lift and aspect ratio is generally well predicted by the famous formula derived for the optimum wing, equation 5-38. Prandtl showed this in his 1921 paper [5] by plotting experimental data for C_L and C_D for cambered rectangular wings of aspect ratios 1 to 7. As can be seen from Fig. 5.18*a*, these data can, for values of C_L not too large, be well approximated by a parabola, and the focal length of the parabola increases with aspect ratio, as predicted by equation 5-38. That the minimum drag is not at zero lift reflects the camber of the wings; that it is not zero is, of course, due to viscous effects.

The agreement of the data with equation 5-38 is better than qualitative, however. Prandtl subtracted $C_L^2/\pi \mathit{AR}$ from his measured drags and found that the induced drag was almost independent of C_L and AR. This can be seen from Fig. 5.18*b*, in which the induced drag has been further modified by adding $C_L^2/5\pi$ (Prandtl did it, not me). Comparing the parabola $C_L^2/5\pi$ with the data, you can see that the difference between the total drag of the rectangular wings and the induced drag of an optimum wing of the same aspect ratio increases somewhat with C_L. As we shall see in Chapter 7, this, too, is a viscous effect, which can be fairly well predicted by two-dimensional data.

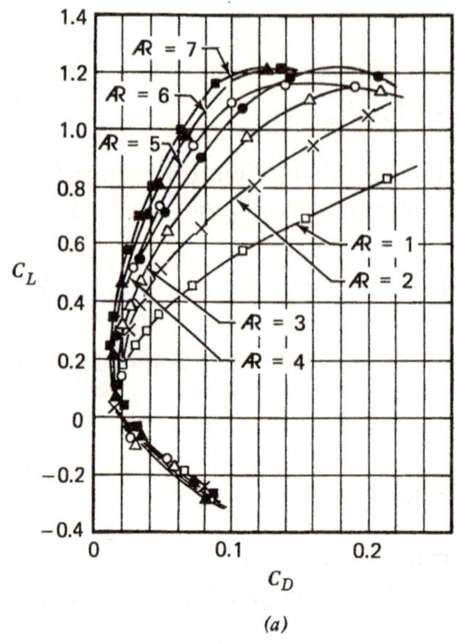

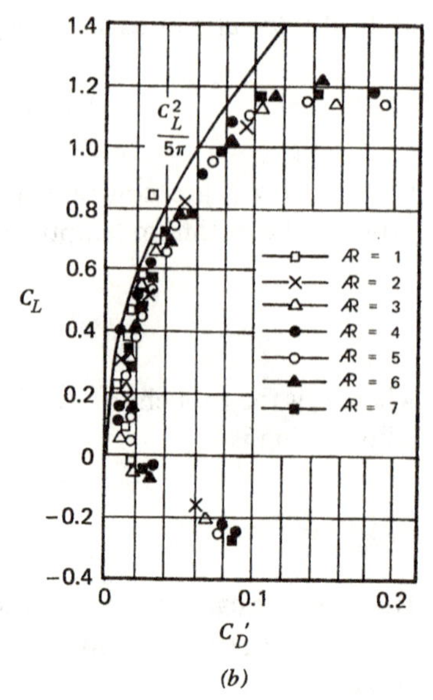

FIG. 5.18. Drag polars for cambered rectangular wings of various aspect ratios [5]. (*a*) Basic data. (*b*) Collapsing of data by removing theoretical induced drag (according to lifting-line theory for optimal wing) from measured drags.

Lift and drag data are often presented in the form illustrated by Fig. 5.18; graphs of lift versus drag are called *drag polars*.

The dependence of lift on angle of attack for an arbitrary wing is also often quite close to that predicted for the optimum flat wing, equation 5-42. Rewriting that formula as

$$C_L = m_0\left(\alpha - \frac{C_L}{\pi \mathcal{R}} - \alpha_{L0}\right) \tag{5-44}$$

you can understand Prandtl's definition of a reduced angle of attack

$$\alpha' = \alpha - \frac{C_L}{\pi \mathcal{R}} + \frac{C_L}{5\pi}$$

and appreciate the remarkable independence of $\mathcal{R}$ of the experimental data on rectangular wings when plotted as in Fig. 5.19. As you can see from Figs. 5.18 and 5.19 both, lifting-line theory, ostensibly accurate only for "large" aspect ratios, is in fact quite good down to $\mathcal{R} = 1.0$. You certainly wouldn't suspect that from the raw data of Fig. 5.19*a*, in which the dependence of C_L on α is decidedly nonlinear. Of course, equation 5-42 itself does not predict that nonlinearity. It is only when it is rearranged as equation 5-44 that it gains validity.

5.7. NONELLIPTIC LIFT DISTRIBUTIONS

Now let us turn to the direct problem: given $\alpha_{L0}(y)$, $c(y)$, and $m_0(y)$—that is, given a specific wing and its sectional aerodynamic properties—find $\Gamma(y)$ so as to

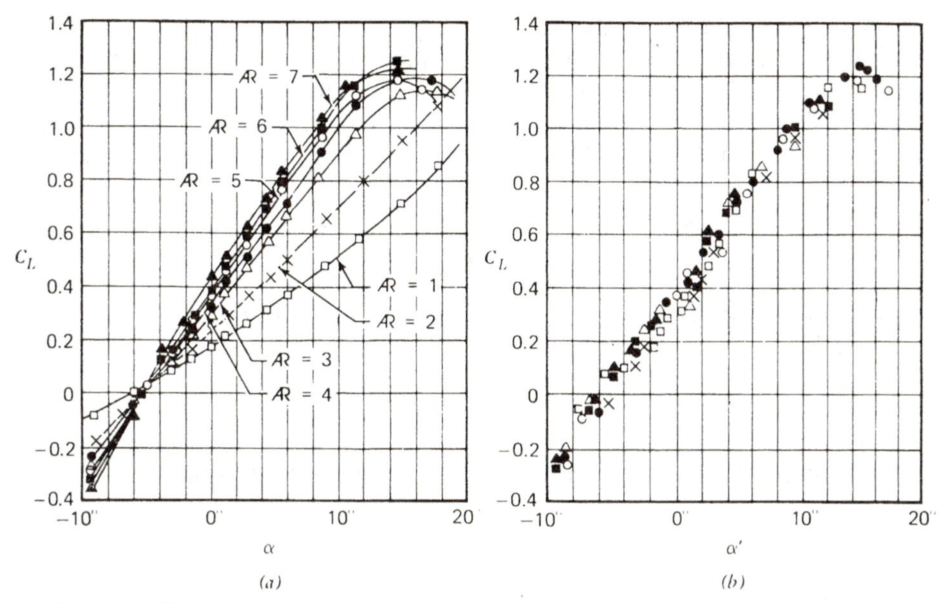

FIG. 5.19. Lift versus angle of attack for cambered rectangular wings of various aspect ratios [5]. (*a*) Basic data. (*b*) Collapsing of data by defining reduced angle of attack according to lifting-line theory for optimal wing.

satisfy the integrodifferential equation 5-24. As in the camber problem of thin-airfoil theory, it is convenient to introduce trigonometric variables by equation 5-26 and a Fourier series for $\Gamma(y)$:

$$\Gamma(y) = g(\theta_0) = \tfrac{1}{2}m_{0_s}c_s V_\infty \sum_{n=1} A_n \sin n\theta_0 \tag{5-45}$$

in which the constant outside the summation is included for later convenience (the subscript "s" denotes values of m_0 and c at the plane of symmetry, $y = 0$). Then

$$\Gamma'(t)\,dt = \frac{d\Gamma}{d\theta}\frac{d\theta}{dt}\,dt = g'(\theta)\,d\theta$$

$$= \tfrac{1}{2}m_{0_s}c_s V_\infty \sum_{n=1} nA_n \cos n\theta\,d\theta \tag{5-46}$$

and

$$⨍_{-b/2}^{b/2} \frac{\Gamma'(t)\,dt}{y-t} = \tfrac{1}{2}m_{0_s}c_s V_\infty \sum_{n=1} nA_n ⨍_{\pi}^{0} \frac{\cos n\theta\,d\theta}{\frac{b}{2}(\cos\theta_0 - \cos\theta)}$$

Using equation 5-27 to evaluate the integral, and substituting the result and equation 5-45 into equation 5-24, we get

$$\sum_{n=1} A_n \sin n\theta_0 = \frac{cm_0}{c_s m_{0_s}}(\alpha - \alpha_{L0}) - \frac{cm_0}{4b}\sum_{n=1} nA_n \frac{\sin n\theta_0}{\sin\theta_0} \tag{5-47}$$

To solve equation 5-47 for the A_n's is not easy. In the *collocation method*, you first truncate the series (5-45) for $g(\theta_0)$ (set $A_n = 0$ above a certain n) so as to reduce the problem to one with a finite number of unknowns and then evaluate equation 5-47 enough different values of θ_0 to find the unknowns. To check whether the truncation was too crude, you must rework the problem, truncating the series after a different number of terms.

The job is made a little easier in the case of a *symmetric loading*

$$\Gamma(y) = \Gamma(-y)$$

since the even coefficients $A_2, A_4, \ldots$ are then all zero. To see this, note that equation 5-26 implies

$$-y = \frac{b}{2}\cos(\pi - \theta_0)$$

Thus, in terms of the function $g(\theta_0)$ defined by equation 5-44, symmetric loading means

$$g(\theta_0) = g(\pi - \theta_0)$$

or, on substituting for g its series expansion (5-45),

$$\sum_{n=1} A_n \sin n\theta_0 = \sum_{n=1} A_n \sin n(\pi - \theta_0)$$

$$= \sum_{n=1} A_n(\sin n\pi \cos n\theta_0 - \cos n\pi \sin n\theta_0)$$

$$= -\sum_{n=1} A_n \cos n\pi \sin n\theta_0$$

Because of the orthogonality of the sine functions, this can be true only if for every n and θ_0

$$A_n \sin n\theta_0 = -A_n \cos n\pi \sin n\theta_0$$

Since

$$\cos n\pi = +1 \qquad \text{if } n \text{ is even}$$
$$= -1 \qquad \text{if } n \text{ is odd}$$

this shows that

$$A_n = 0 \qquad \text{if } n \text{ is even}$$

for any case in which the loading is symmetric about the midspan.

Example 1: Flat Rectangular Wing Keuthe and Chow [7] work out the case of an untwisted uncambered rectangular wing with the series (5-45) truncated after A_8. We will check the accuracy of their results by truncating the series instead after A_6:

$$g(\theta_0) \approx \tfrac{1}{2} m_{0_s} c_s V_\infty (A_1 \sin\theta_0 + A_3 \sin 3\theta_0 + A_5 \sin 5\theta_0)$$

To find A_1, A_3, and A_5, we satisfy equation 5-47 for $\theta = \pi/6$, $\pi/3$, and $\pi/2$. For the special case of a flat rectangular wing,

$$m_0 = m_{0_s}$$

$$c(y) = c_s$$

$$\alpha_{LO} = 0$$

Kuethe and Chow take $m_{0_s} = 2\pi$ and consider the case $A\!\!\!R = 6$, in which

$$b = 6c_s$$

Then the equations derived from (5-47) for the stated values of θ_0 are

$$\alpha = A_1\left(1 + \frac{\pi}{12 \cdot \frac{1}{2}}\right)\left(\frac{1}{2}\right) + A_3\left(1 + \frac{3\pi}{12 \cdot \frac{1}{2}}\right)(1) + A_5\left(1 + \frac{5\pi}{12 \cdot \frac{1}{2}}\right)\left(\frac{1}{2}\right)$$

$$\alpha = A_1\left(1 + \frac{\pi}{12 \cdot \frac{\sqrt{3}}{2}}\right)\left(\frac{\sqrt{3}}{2}\right) + A_3\left(1 + \frac{3\pi}{12 \cdot \frac{\sqrt{3}}{2}}\right)(0) + A_5\left(1 + \frac{5\pi}{12 \cdot \frac{\sqrt{3}}{2}}\right)\left(\frac{-\sqrt{3}}{2}\right)$$

$$\alpha = A_1\left(1 + \frac{\pi}{12 \cdot 1}\right)(1) + A_3\left(1 + \frac{3\pi}{12 \cdot 1}\right)(-1) + A_5\left(1 + \frac{5\pi}{12 \cdot 1}\right)(1)$$

whose solution is

$$A_1 = 0.9160\alpha$$

$$A_3 = 0.1069\alpha$$

$$A_5 = 0.0152\alpha$$

These results are quite close to the results obtained by Kuethe and Chow for the same problem with truncation after A_8:

$$A_1 = 0.9174\alpha$$

$$A_3 = 0.1104\alpha$$

$$A_5 = 0.0218\alpha$$

$$A_7 = 0.0038\alpha$$

Once the coefficients A_n are obtained to satisfactory accuracy, one can calculate life and drag coefficients, as follows. From equations 5-39, 5-30, 5-26, and 5-45,

$$C_L = \frac{L}{\frac{1}{2}\rho V_\infty^2 S} = \frac{\rho V_\infty \int_{-b/2}^{b/2} \Gamma(y)\, dy}{\frac{1}{2}\rho V_\infty^2 S}$$

$$= \frac{2}{V_\infty S}\int_\pi^0 \tfrac{1}{2} m_{0_s} c_s V_\infty \sum_{n=1} A_n \sin n\theta\left(-\frac{b}{2}\sin\theta\right) d\theta$$

$$= \frac{m_{0_s} c_s b}{2S} \sum_{n=1} A_n \int_0^\pi \sin n\theta \sin\theta\, d\theta$$

The integral vanishes unless $n = 1$, in which case it equals $\pi/2$. Thus, for any wing whose loading is described by the Fourier series (5-45),

$$C_L = \frac{\pi}{4} m_{0_s} \frac{bc_s}{S} A_1 \tag{5-48}$$

Now for the induced drag coefficient. The induced angle of attack that appears in equation 5-31 can be calculated by substituting equation 5-46 into 5-23, as we did in deriving equation 5-47, with the result

$$\alpha_i(y) = \frac{m_{0_s} c_s}{4b} \sum_{k=1} kA_k \frac{\sin k\theta}{\sin \theta}$$

Using this and equation 5-45 in equation 5-31, we then obtain

$$C_{D_i} = \frac{2}{V_\infty S} \cdot \tfrac{1}{2} m_{0_s} c_s V_\infty \frac{m_{0_s} c_s}{4b} \sum_{n=1} A_n \sum_{k=1} kA_k \int_\pi^0 \sin n\theta \frac{\sin k\theta}{\sin \theta} \left(-\frac{b}{2} \sin \theta \right) d\theta$$

$$= \frac{m_{0_s}^2}{8} \cdot \frac{c_s^2}{S} \sum_{n=1} A_n \sum_{k=1} kA_k \int_0^\pi \sin n\theta \sin k\theta \, d\theta$$

The integral vanishes unless $k = n$, in which case it reduces to $\pi/2$. Thus

$$C_{D_i} = \frac{\pi}{16} m_{0_s}^2 \frac{c_s^2}{S} \sum_{n=1} nA_n^2 \tag{5-49}$$

You will evaluate these expressions for C_L and C_{D_i} for the case in which the A_n's are given above (flat rectangular wing of aspect ratio 6.0) in problem 11.

You can see that once you get away from the elliptic loading, lifting-line theory becomes rather cumbersome and hard to implement without a computer. As long as you're going to use a computer, you may as well use a method that gives you better results. Lifting-line theory is often quite good for lift and drag calculations, down to aspect ratios as low as 1.0, as we saw in Figs. 5.18 and 5.19. However, it gives no useful information on the chordwise load distribution and fails to account properly for effects of sweepback. Thus, except for the case of elliptic loading or for rough preliminary estimates of performance, vortex-lattice or (especially for swept wings) other panel methods are preferred to lifting-line theory.

5.8. REFERENCES

1. Milne-Thompson, L. M., *Theoretical Aerodynamics*, 4th ed. Dover, New York (1973), Sec. 6.3.
2. Maskew, B., "Prediction of Subsonic Aerodynamic Characteristics: A Case for Low-Order Panel Methods," *J. Airc.* **19**, 157–163 (1982).
3. DeJarnette, F. R., "Arrangement of Vortex Lattices on Subsonic Wings," in *Vortex-Lattice Methods*, Workshop held at Langley Research Center, May 17–18 (1976). NASA SP-405.
4. Hough, G. R., "Lattice Arrangements for Rapid Convergence," in *Vortex-Lattice*

Methods. Workshop held at Langley Research Center, May 17–18 (1976). NASA SP-405.

5. Prandtl, L., "Applications of Modern Hydrodynamics to Aeronautics," NACA Report, 116 (1921).
6. Falkner, V. M., *The Calculation of Aerodynamic Loading on Surfaces of Any Shape*, Great Britain, ARC R & M 1910 (1946).
7. Kuethe, A. M. and Chow, C.-Y., *Foundations of Aerodynamics*, 3rd ed. Wiley, New York (1976).

5.9. PROBLEMS

1. If the load distribution on a finite-span wing is as shown, sketch the trailing vortex system, indicating the magnitude and direction of the trailing vortices.

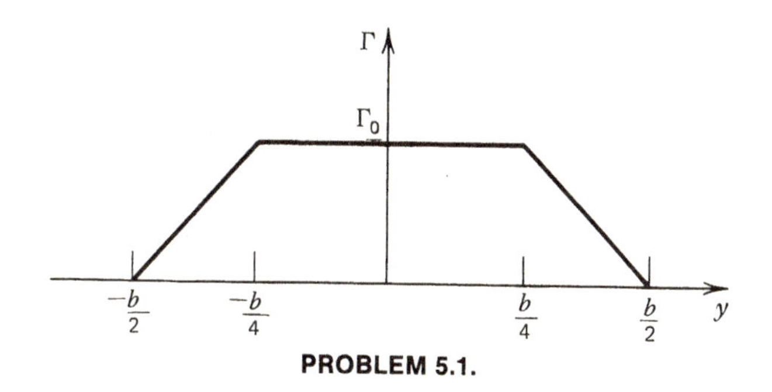

PROBLEM 5.1.

2. Use the vortex-lattice method to compute the lift, induced drag, moment, and center of pressure of flat rectangular wings of aspect ratio 2.0 and 6.0. How large must N_x and N_y be before the results become essentially independent of N_x and N_y?
3. Revise VORLAT to use the cosine spacing suggested in the footnote to the discussion of Fig. 5.9. Compare its rate of convergence with that of the evenly spaced method for the case $\mathcal{R} = 2.0$.
4. Devise a vortex-lattice method capable of treating wings with camber, taper, and sweep. You may restrict the analysis to wings with straight leading and trailing edges

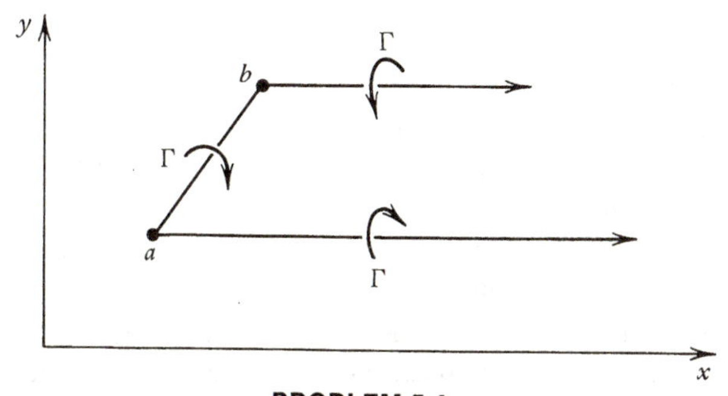

PROBLEM 5.4.

and tips parallel to the plane of symmetry $y = 0$. Satisfy boundary conditions in the $z = 0$ plane

$$\left.\frac{\partial \phi}{\partial z}\right|_{z=0} = V_\infty \left(\alpha + \frac{\partial z_1}{\partial x} \right)$$

where $z = z_1(x, y)$ locates the mean surface of the wing, and use the skewed horseshoe vortices in the $z = 0$ plane shown in the sketch. It can be shown by the methods of Appendix F that the downwash due to such a vortex is

$$w(x, y, 0) = \frac{\Gamma}{4\pi} \left\{ \frac{1}{y - y_a} \left[1 + \frac{(y_b - y_a)R_a}{(x - x_a)(y_b - y_a) - (x_b - x_a)(y - y_a)} \right] - \frac{1}{y - y_b} \left[1 + \frac{(y_b - y_a)R_b}{(x - x_b)(y_b - y_a) - (x_b - x_a)(y - y_b)} \right] \right\}$$

where

$$R_{\substack{a\\b}} = \left[\left(x - x_{\substack{a\\b}} \right)^2 + \left(y - y_{\substack{a\\b}} \right)^2 \right]^{1/2}$$

If

$$\frac{x - x_b}{y - y_b} = \frac{x - x_a}{y - y_a} = \frac{x_b - x_a}{y_b - y_a}$$

use instead

$$w(x, y, 0) = \frac{\Gamma}{4\pi} \left[\frac{1}{y - y_a} \left(1 + \frac{x - x_a}{R_a} \right) - \frac{1}{y - y_b} \left(1 + \frac{x - x_b}{R_b} \right) \right]$$

Compare the use of cosine spacing with evenly spaced vortices.

5. Show that, for an elliptic-planform wing with constant α_{L_0} and m_0, the lift coefficient is, according to lifting-line theory,

$$C_L = \frac{m_0(\alpha - \alpha_{L0})}{1 + m_0/\pi AR}$$

Note that the area of the wing is

$$S = \frac{\pi}{4} bc_s$$

where c_s is the maximum chord.

6. For an untwisted constant-section wing of elliptical planform and aspect ratio 5, find the sectional and total lift and drag coefficients, according to lifting-line theory. Take the lift-curve slope m_0 to be 5.7, $V = 45$ m/sec; the wing loading $W/S = 1000$ N/m^2 (consider the level flight case, for which $L = W$), the altitude to be slightly above sea level, and the span $b = 10$ m.

7. For the data of problem 6, plot the absolute and induced angles of attack as functions of spanwise position y, and calculate the power required to overcome the induced drag.

8. Use equations 5-48 and 5-49 to show that the elliptic loading is indeed optimal, in the sense that it gives minimum induced drag force (*not* drag coefficient) for a given lift force (*not* lift coefficient), given span, and given *dynamic pressure* $\frac{1}{2}\rho V_\infty^2$.
9. A wing is to be designed so as to produce 6000 lb of lift at 100 ft/sec at an altitude at which $\rho = 0.0025$ slug/ft^3. Consider three planforms: elliptical, rectangular, and tapered with a tip-to-root chord ratio of 0.5. For each case, find (according to lifting-time theory)
   (a) The load distribution that minimizes induced drag for a given span.
   (b) The associated required distribution of absolute angle of attack $\alpha - \alpha_{LO}$ (take $m_0 = 2\pi$), in terms of the span b, root chord c_s, and $\Gamma_{\max}$.

   If the maximum section lift coefficient is 0.5, what minimum areas are required for the three wings? Note that $S = \pi/4bc_{\max}$ for an ellipse. *Hints*:
   (a) Knowing L, ρ, and V_∞, determine SC_L, and hence $\Gamma_{\max}b$.
   (b) Find the value of y at which $c_l = c_{l_{\max}}$. Then, knowing $c_{l_{\max}}$, determine $(\Gamma/c)_{\max}$.
   (c) Thus you can calculate $S = \text{constant} \times bc_{\max}$.
10. For the wings of problem 2, plot the spanwise distribution of lift predicted by vortex-lattice theory. Compare with the optimum distribution according to lifting-line theory. Also compare results for $\partial C_L/\partial\alpha$ and C_{D_i}/C_L^2.
11. Use lifting-line theory to compute the lift and induced drag coefficients of a flat rectangular wing with aspect ratio 6. Use the A_n's obtained by Kuethe and Chow [7] for a four-term approximation to the load distribution and in this book for a three-term approximation. Compare with the results of the vortex-lattice method (see problem 2).

5.10. COMPUTER PROGRAM

```
C         PROGRAM VORLAT
C
C            VORTEX-LATTICE METHOD FOR FLAT RECTANGULAR WING
C
          DIMENSION     GAM(100)
          COMMON      DX,DY,AR,PI
          COMMON /COF/ A(100,101),NEQNS
          PI            =  3.1415926585
CC
C                   INPUT ASPECT RATIO (AR), NUMBERS OF VORTICES
C                   IN X- AND Y-DIRECTIONS (NX,NY) AND
C                   ANGLE OF ATTACK IN DEGREES (ALPHA)
C
          PRINT,        ' INPUT AR',
          READ,         AR
          PRINT,        ' INPUT NX,NY',
          READ,         NX,NY
          PRINT,        ' INPUT ALPHA',
          READ,         ALPHA
          DX            =  1./FLOAT(NX)
          DY            =  AR/(2.*NY + .5)
          NEQNS         =  NX*NY
```

```
      COSALF     =  COS(ALPHA*PI/180.)
      SINALF     =  SIN(ALPHA*PI/180.)
      PRINT 1000
C
C                SET COEFFICIENTS OF EQUATIONS FOR VORTEX STRENGTHS
C
      DO 100     I = 1,NY
      DO 100     J = 1,NX
      IJ         =  (I - 1)*NX + J
      A(IJ,NEQNS+1)         =  SINALF
      DO 100     K = 1,NY
      DO 100     L = 1,NX
      KL         =  (K - 1)*NX + L
      CALL DNWASH(I,J,K,L,A(KL,IJ),7HCONTROL)
  100 CONTINUE
C
C                SOLVE FOR VORTEX STRENGTHS
C
      CALL GAUSS(1)
      DO 200     I = 1,NY
      DO 200     J = 1,NX
      IJ         =  (I - 1)*NX + J
  200 GAM(IJ)    =  A(IJ,NEQNS+1)
C
C                COMPUTE FORCE AND MOMENT COEFFICIENTS
C
      CMT        =  0.0
      CDT        =  0.0
      CLT        =  0.0
      DO 320     I = 1,NY
      CX         =  0.0
      CZ         =  0.0
      CM         =  0.0
      DO 310     J = 1,NX
      IJ         =  (I - 1)*NX + J
      W          =  0.0
      DO 300     K = 1,NY
      DO 300     L = 1,NX
      KL         =  (K - 1)*NX + L
      CALL DNWASH(K,L,I,J,DELW,4HSELF)
      W          =  W + DELW*GAM(KL)
  300 CONTINUE
      CX         =  CX + GAM(IJ)*(W - SINALF)*2.
      CZ         =  CZ + GAM(IJ)*COSALF*2.
      CM         =  CM - GAM(IJ)*DX*(J - .75)*COSALF*2.
  310 CONTINUE
      CL         =  CZ*COSALF - CX*SINALF
      CD         =  CZ*SINALF + CX*COSALF
      CLT        =  CLT + CL*DY*2./AR
      CDT        =  CDT + CD*DY*2./AR
      CMT        =  CMT + CM*2.*DY/AR
      XCP        =  - CM/CL
      Y          =  (I - .5)*DY
      PRINT 1010, Y,CL,CD,XCP
  320 CONTINUE
      XCP        =  - CMT/CLT
      CDOCL2     =  CDT/CLT**2
      PRINT 1020, CLT,CDT,CDOCL2,CMT,XCP
```

```
 1000 FORMAT(/////,'   Y        CL(Y)      CD(Y)      XCP(Y)',/)
 1010 FORMAT(F6.3,3F10.5)
 1020 FORMAT(/////,' CL =',F12.5,/,' CD =',F14.7,/,' CD/CL2 =',F7.4,
     +           /,' CMLE =',F11.6,/,' XCP =',F11.5)
      STOP
      END
C=================================================================
      SUBROUTINE DNWASH(I,J,K,L,W,IND)
C
C                 COMPUTE DOWNWASH ON PANEL CENTERED AT (L-.5)DX,(K-.5)DY
C                 DUE TO VORTICES AT PANELS CENTERED AT (J-.5)DX,+-(I-.5)DY
C
      COMMON      DX,DY,AR,PI
      XA          =  DX*(J - .75)
      YA          =  DY*(I - 1)
      YB          =  DY*I
      IF (IND .EQ. 7HCONTROL)XP = DX*(L - .25)
      IF (IND .EQ. 4HSELF)    XP = DX*(L - .75)
      YP          =  DY*(K - .5)
      W           =  WHV(XP,YP,XA,YA) - WHV(XP,YP,XA,YB)
     +               - WHV(XP,YP,XA,-YA) + WHV(XP,YP,XA,-YB)
      W           =  W*.25/PI
      RETURN
      END
C=====================================================================
      FUNCTION WHV(X1,Y1,X2,Y2)
      IF (X1 .EQ. X2)          GO TO 100
      WHV         =  (1. + SQRT((X1-X2)**2 + (Y1-Y2)**2)/(X1 - X2))
     +               /(Y1 - Y2)
      RETURN
  100 WHV         =  1./(Y1 - Y2)
      RETURN
      END
C=====================================================================
      SUBROUTINE GAUSS(NRHS)

      (SEE PROGRAM DUBLET)
```

CHAPTER 6

THE NAVIER–STOKES EQUATIONS

The conservation laws reviewed in Chapter 2, on which everything we have done since is based, assume that the only stress in a fluid is the normal stress or pressure. This is not in fact the case; fluids in motion are also subject to tangential shear stresses. For air, water, and other engineering fluids, shear stresses can be related to the velocity field. The momentum conservation principle for such fluids is expressed by the Navier–Stokes equations, whose derivation is the sole purpose of this chapter. Although the equations are usually too difficult to solve in their exact form, they are the basis for the very useful approximation of boundary-layer theory to be discussed in Chapter 7. It helps to understand the nature of such approximations if you know the source of the basic equations.

6.1. THE STRESS AT A POINT

The Navier–Stokes equations are nothing more than the expression of Newton's law

$$\mathbf{F} = \frac{d}{dt} m\mathbf{V}$$

for a special class of fluids, which happen to include air and water under practical conditions. In applying this law to a moving fluid, we must pay attention to its fine print: Newton's laws apply to a "fixed" portion of matter, that is, to matter whose composition does not change with time. For this reason, in this chapter we are often concerned with what is called a *fluid particle*, by which we mean the fluid that is bounded by an imaginary surface that moves with the flow so that no fluid crosses any part of the surface. If $\mathscr{V}$ is the region of space occupied by the fluid particle, its momentum is

$$\int_{\mathscr{V}} \rho \mathbf{V} \, d\mathscr{V}$$

but we cannot equate the force **F** acting on the particle to

$$\frac{d}{dt} \int_{\mathscr{V}} \rho \mathbf{V} \, d\mathscr{V}$$

unless we allow $\mathscr{V}$ to move in space with the fluid. If, as is usually more convenient for analysis, $\mathscr{V}$ is fixed in space, we must also account for the momentum that enters and leaves $\mathscr{V}$ through its boundary S. In this case, the correct expression of Newton's law is

$$\frac{d}{dt} \int_{\mathscr{V}} \rho \mathbf{V} \, d\mathscr{V} = -\int_{S} \rho \mathbf{V}(\mathbf{V} \cdot \hat{\mathbf{n}}) \, dS + \mathbf{F} \tag{6-1}$$

where $\hat{\mathbf{n}}$ is the unit vector normal to S and directed out of $\mathscr{V}$, as is shown in Fig. 6.1.

The force **F** on the fluid in $\mathscr{V}$ may have a number of different sources. *Body forces* act on the fluid as a whole and contribute to **F** in proportion to the amount of fluid within $\mathscr{V}$. The most prominent example is gravity, another one being electromagnetic forces, but body forces are relatively unimportant in most of aerodynamics. *Surface forces* are internal to the fluid. They arise because to analyze the motion of the fluid within $\mathscr{V}$, we must isolate it from that outside $\mathscr{V}$. The idea is that, if we replace the fluid outside $\mathscr{V}$ by a proper system of forces on its boundary S, the fluid inside won't know the difference; its motion will be exactly the same as if it were surrounded by other fluid, which, of course, is the actual case.

In Chapter 2 we supposed that the surface force was directed normal to S. Letting the magnitude of the force per unit area be p, we then have

$$\mathbf{F} = -\int_{S} p\hat{\mathbf{n}} \, dS$$

and equation 6-1 reduces to equation 2-8.

More generally, the surface force will have components tangential to the surface, too. Consider, for example, a solid bar in tension. To find the internal force on a surface normal to the axis, "cut" the bar on such a surface; that is, isolate part of the bar from the rest of it, and replace the ignored part by a system of forces distributed

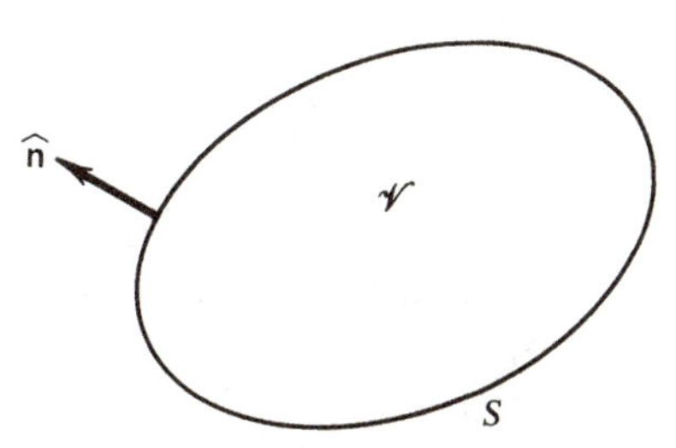

FIG. 6.1. General control volume for application of Newton's law.

over the surface of the cut, as shown in Fig. 6.2. Since the isolated part must be in equilibrium, you expect to find internal forces normal to the surface of the cut. Alternatively, the internal force may be ascribed to the displacement of the atoms and/or molecules that make up the bar from their positions of equilibrium with respect to the forces they exert on one another. Interparticle forces are generally repulsive if the particles get too close and attractive if the distance between them gets beyond a certain value.

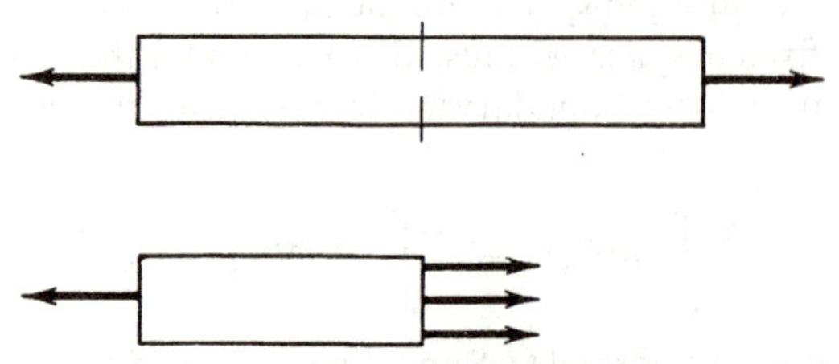

FIG. 6.2. Stresses in a bar in tension on a surface normal to the line of action of the applied force.

However, suppose you cut the bar on a bias, as in Fig. 6.3. For equilibrium, the internal forces should still be generally opposite to the tensile force applied on the end of the bar, and so cannot be purely normal to the surface of the cut. Thus, in general, the internal force has both normal and tangential (or *shear*) components, and the split between them depends on the orientation of the surface on which they act.

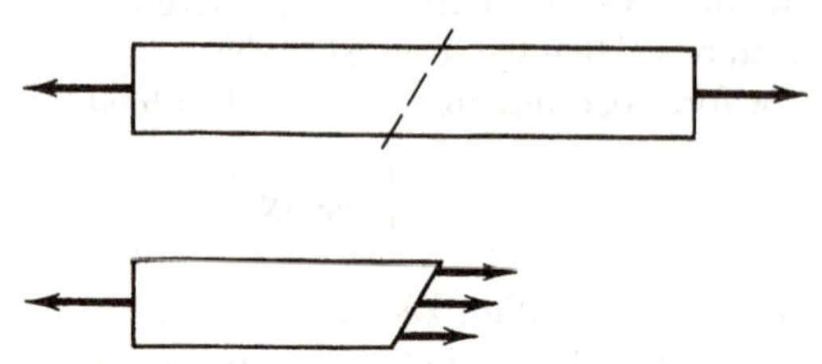

FIG. 6.3. Stresses in a bar in tension on a surface oriented at an angle to the line of action of the applied force.

The *stress at a point* in a material (fluid or solid) is the internal force per unit area that must be exerted on a test surface passing through that point in order to represent the interaction between the material on one side of the surface and that on the other side. Since the interaction force on the first side due to the second must be equal and opposite to the force exerted on the second side by the first, see Fig. 6.4,

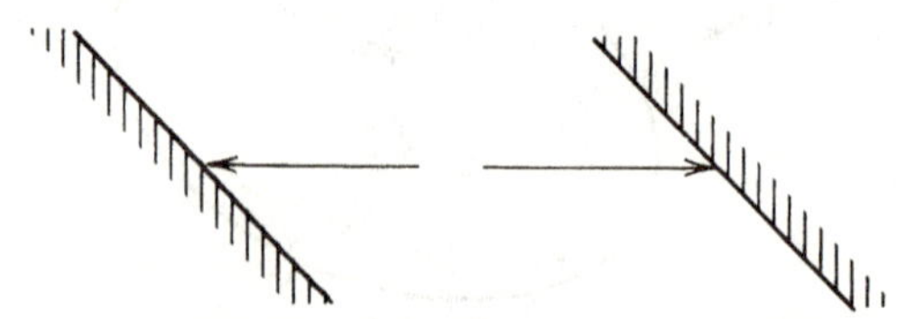

FIG. 6.4. Newton's third law, applied to internal forces.

you have to be careful, in specifying the stress at a point, as to which side you're talking about.

Also, to specify the state of stress at a point, it is not enough to pass just one test surface through the point. Consider the situations sketched in Fig. 6.5. If we pass surfaces through points A in each bar normal to the axis of the bar, we are likely to see only stresses normal to the surfaces. However, if we pass the test surfaces through A parallel to the axis of the bar, it seems possible to find no stress at all in the case marked (*a*), but there certainly would be a normal stress on the test surface in the case marked (*b*).

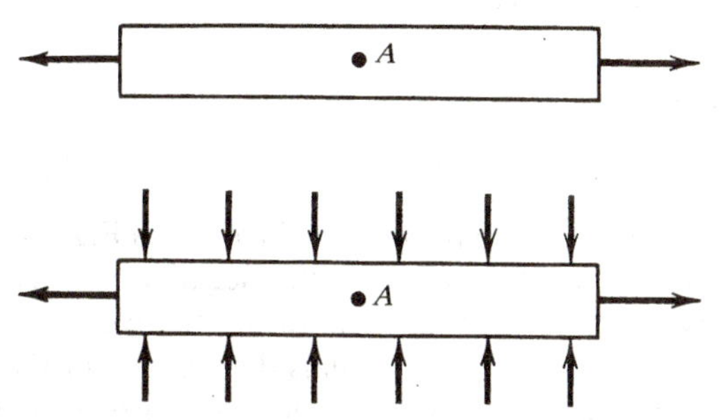

FIG. 6.5. Loadings on solid bars that may yield the same stress on some surfaces through corresponding points.

It can be shown that if you know the stresses on three mutually perpendicular planes passing through a point, you can determine the stress on any other plane passing through the same point. Thus we need some nomenclature for the stresses acting on surfaces perpendicular to the x, y, and z directions, say. Let us define

$(\sigma_{xx}, \sigma_{xy}, \sigma_{xz}) \equiv$ components of force per unit area on surface whose outward normal (i.e., normal directed outward from the material on which the force acts) is in $+x$ direction

$(\sigma_{yx}, \sigma_{yy}, \sigma_{yz}) \equiv$ components of force per unit area on surface whose outward normal is in $+y$ direction

$(\sigma_{zx}, \sigma_{zy}, \sigma_{zz}) \equiv$ components of force per unit area on surface whose outward normal is in $+z$ direction

This is rather cumbersome, isn't it? A much more compact nomenclature is possible if you label the axes x_1, x_2, and x_3 instead of x, y, and z, as shown in Fig. 6.6. Then we can condense the three lines of the preceding definition into

$\sigma_{ij} \equiv$ component in $+x_j$ direction of force per unit area on surface whose outward normal is in $+x_i$ direction

When the two subscripts on σ are the same, the stress is a *normal stress*:

σ_{ii} = normal stress on surface normal to x_i direction

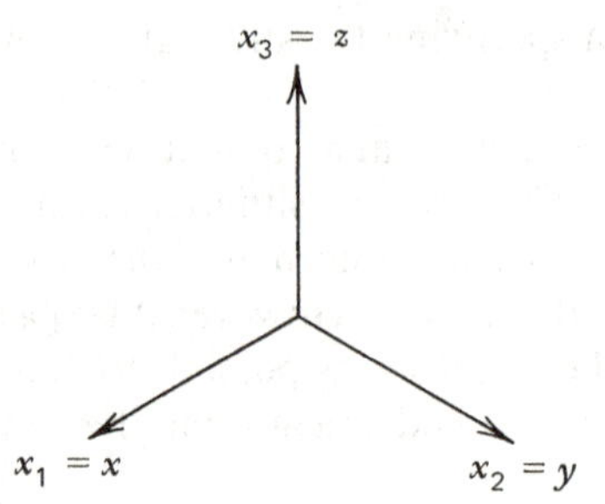

FIG. 6.6. Indicial notation for axes of Cartesian coordinate system.

whereas otherwise it is a *shear stress*:

$\sigma_{ij}(i \neq j) =$ shear stress in jth direction on surface normal to x_i direction

Because of the action–reaction principle referred to in Fig. 6.4, the internal force on a surface whose outward normal is in the negative x_i direction is equal and opposite to the internal force on the neighboring surface, whose normal is in the positive x_i direction. Therefore, σ_{ij} also equals the *negative* of the x_j component of the internal force per unit area on a surface whose outward normal is in the *negative* x_i direction.

6.2. NEWTON'S SECOND LAW FOR FLUIDS

With the definition of stress at a point now established, we will now specialize equation 6-1 to the case where the volume $\mathscr{V}$ is a differentially small rectangular parallelepiped with sides parallel to the coordinate directions, the side parallel to the x_i axis being of length Δx_i, as shown in Fig. 6.7. This ith component of equation 6-1 is

$$\frac{d}{dt}\int_{\mathscr{V}} \rho V_i \, d\mathscr{V} = -\int_S \rho V_i \mathbf{V} \cdot \hat{\mathbf{n}} \, dS + F_i \tag{6-2}$$

Suppose the center of $\mathscr{V}$ to be the point with coordinates (x_1, x_2, x_3); see Fig. 6.7. If

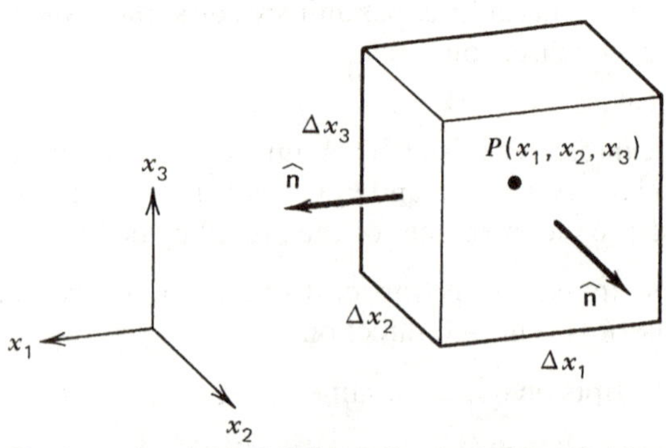

FIG. 6.7. Rectangular control volume for application of Newton's law.

$\mathscr{V}$ is sufficiently small, the left side of equation 6-2 is approximately

$$\left.\frac{\partial \rho V_i}{\partial t}\right|_{x_1 x_2 x_3} \Delta x_1 \, \Delta x_2 \, \Delta x_3$$

As for the integral on the right side of equation 6-2, consider first the contribution of the parts of S that are normal to the x_1 direction. One is centered at $(x_1 - \Delta x_1/2, x_2, x_3)$, the other at $(x_1 + \Delta x_1/2, x_2, x_3)$, and both have areas $\Delta x_2 \, \Delta x_3$. Recalling that $\hat{\mathbf{n}}$ is directed out of the volume, the contributions are, in the limit as $\Delta x_1 \to 0$,

$$\left.\rho V_i V_1\right|_{x_1 - (\Delta x_1/2), x_2 x_3} \Delta x_2 \, \Delta x_3 - \left.\rho V_i V_1\right|_{x_1 + (\Delta x_1/2), x_2 x_3} \Delta x_2 \, \Delta x_3$$

$$\approx -\frac{\partial}{\partial x_1}(\rho V_i V_1) \, \Delta x_1 \, \Delta x_2 \, \Delta x_3$$

Similar results can be obtained for the other parts of S, so that the integral on the right side of equation 6-2 reduces to[1]

$$-\sum_{j=1}^{3} \frac{\partial}{\partial x_j}(\rho V_i V_j) \, \Delta x_1 \, \Delta x_2 \, \Delta x_3$$

Now for the force F_i. It is not hard to account for the weight of the fluid, but, as is usually justifiable in aerodynamics, we will ignore it, and suppose that F_i is due only to the stresses exerted by the surrounding fluid. Because the part of S centered at $(x_1 - \Delta x_1/2, x_2, x_3)$ has an outward normal in the negative x_1 direction, the component in the x_i direction of the force on that surface is

$$-\left.\sigma_{1i}\right|_{x_1 - (\Delta x_1/2), x_2 x_3} \Delta x_2 \, \Delta x_3$$

while the x_i component of the force on the opposite surface is

$$+\left.\sigma_{1i}\right|_{x_1 + (\Delta x_1/2), x_2 x_3} \Delta x_2 \, \Delta x_3$$

These add up to

$$\left.\frac{\partial \sigma_{1i}}{\partial x_1}\right|_{x_1 x_2 x_3} \Delta x_1 \, \Delta x_2 \, \Delta x_3$$

as $\Delta x_1 \to 0$, and similar consideration of the other parts of S leads to

$$F_i = \sum_{j=1}^{3} \frac{\partial \sigma_{ji}}{\partial x_j} \Delta x_1 \, \Delta x_2 \, \Delta x_3$$

[1] From problem 1, this result may be written

$$-\operatorname{div}(\rho V_i \mathbf{V}) \, \Delta x_1 \, \Delta x_2 \, \Delta x_3$$

which follows directly by application of the divergence theorem, equation 2-17, to the integral on the right side of equation 6-2.

Substituting all this into equation 6-2, and dividing by the volume $\Delta x_1\, \Delta x_2\, \Delta x_3$, we get

$$\frac{\partial \rho V_i}{\partial t} = -\sum_j \frac{\partial \rho V_i V_j}{\partial x_j} + \sum_j \frac{\partial \sigma_{ji}}{\partial x_j} \qquad \text{for } i = 1, 2, 3 \tag{6-3}$$

in which, for brevity, we introduce the convention that the summation index j should be understood to range over 1 to 3. In conventional Cartesian coordinates, the part of equation 6-3 corresponding to $x_i = x$ would be

$$\frac{\partial(\rho V_x)}{\partial t} = -\frac{\partial(\rho V_x V_x)}{\partial x} - \frac{\partial(\rho V_x V_y)}{\partial y} - \frac{\partial(\rho V_x V_z)}{\partial z} + \frac{\partial \sigma_{xx}}{\partial x} + \frac{\partial \sigma_{yx}}{\partial y} + \frac{\partial \sigma_{zx}}{\partial z}$$

All this may be very impressive, but not too useful as it stands. Even when $\rho =$ constant, equations 6-3 and the continuity equation constitute but 4 equations in 12 unknowns. In the ideal case of no shear stresses that we've considered in previous chapters,

$$\begin{aligned} \sigma_{ij} &= -p \qquad \text{if} \quad i = j \\ &= 0 \qquad\ \ \text{otherwise} \end{aligned}$$

which cuts the number of unknowns down to the number of equations. However, a lot of work is required to make equation 6-3 useful outside the ideal case.

6.3. SYMMETRY OF STRESSES

An immediate reduction in the number of unknowns can be obtained by analyzing the rotational motion of a small fluid element. For simplicity, suppose that, at the time of observation, the element is a cube with side ε. A corollary of Newton's law is that the net moment of the external forces acting on the element, taken about its mass center, equals the time rate of change of its angular momentum:

$$\begin{aligned} \sum \mathbf{M} &= \frac{d}{dt}\mathbf{H} \\ = \sum \mathbf{r} \times \mathbf{F} &= \frac{d}{dt}\int_{\mathscr{V}(t)} \mathbf{r} \times \rho \mathbf{V}\, d\mathscr{V} \end{aligned} \tag{6-4}$$

where $\mathbf{r}$ is the vector distance from the center of the cube. To avoid having to account for the fluxes of angular momentum through the sides of the cube, we require that the integral on the right side be over a region $\mathscr{V}(t)$ that drifts with the fluid; that is, it is over a *fluid particle* that is in the shape of a cube at the time of observation but which moves so that its sides are impervious to the fluid.

Let us look at the x_3 component of this equation, for example. Assume the cube to be so small that the stresses are uniform over each of its sides, so that the resultant of the normal stresses passes through the center of the cube. As can be seen from Fig. 6.8, only the shear stresses σ_{12} and σ_{21} contribute to the x_3 component of the

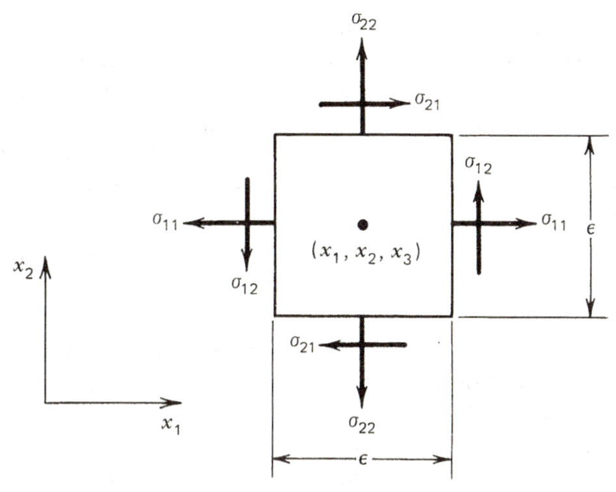

FIG. 6.8. Definition of stresses in a plane.

moment about the mass center. Thus, since the area of each face of the cube is ε^2 and the moment arms of the various shear forces about the mass center are each $\varepsilon/2$,

$$+\sum M_3 = \frac{\varepsilon}{2}\sigma_{12}\bigg|_{x_1+\frac{\varepsilon}{2},x_2}\varepsilon^2 + \frac{\varepsilon}{2}\sigma_{12}\bigg|_{x_1-\frac{\varepsilon}{2},x_2}\varepsilon^2 - \frac{\varepsilon}{2}\sigma_{21}\bigg|_{x_1,x_2+\frac{\varepsilon}{2}}\varepsilon^2 - \frac{\varepsilon}{2}\sigma_{21}\bigg|_{x_1,x_2-\frac{\varepsilon}{2}}\varepsilon^2$$

$$\approx \varepsilon^3(\sigma_{12}-\sigma_{21})\big|_{x_1x_2}$$

if ε is small.

But the x_3 component of the angular momentum, because it involves an integral of the distance from the center of the cube over its volume, must be proportional to ε^4. Dividing both sides of the x_3 component of equation 6-4 by ε^3 and letting $\varepsilon \to 0$, we thus obtain

$$\sigma_{12} - \sigma_{21}\big|_{x_1x_2} \propto \varepsilon \to 0$$

A similar analysis can be made for the other components of equation 6-4 with the general result

$$\sigma_{ij} = \sigma_{ji} \tag{6-5}$$

Thus the nine stresses σ_{ij} are not all independent; there are really only six independent σ_{ij}'s. That cuts the number of unknowns in equation 6-3 down to nine, and we now need only five more equations to balance the number of unknowns.

6.4. MOLECULAR VIEW OF STRESS IN A FLUID

Valuable information on the relation between the stresses at a point in a fluid and other properties of the flow can be obtained by adopting a molecular viewpoint. As you know, all matter is composed of discrete particles—atoms, molecules, and subatomic particles—separated from one another by nothingness. We can generally ignore this fact, because of the enormous number of particles in any volume of significant size. More precisely, the average distance between particles is extremely small compared to the distance over which the flow properties vary measurably, a distance we shall call the *characteristic length* of the flow.

All gross properties of a fluid flow can be interpreted from a molecular viewpoint. The density at a point is just the mass of the molecules in some test volume $\Delta\mathscr{V}$ surrounding the point, divided by $\Delta\mathscr{V}$. To get a definite answer, $\Delta\mathscr{V}$ must be small on the scale of the characteristic length of the flow but large enough to contain many molecules. The temperature T at a point is proportional to the average kinetic energy (relative to any mean motion) of the molecules in the test volume. This definition leads to a kinetic-theory explanation of changes of state. In a solid, the particles vibrate about positions of equilibrium with respect to the forces exerted by neighboring particles. Adding heat increases the energy of this vibratory motion, until the particles actually change neighbors. When the particles no longer retain the same neighbors for more than brief periods of time, the solid is said to be liquified. Adding still more heat further increases the motion of the particles until they are mostly in free flight, only occasionally interacting with any other particle. That is the kinetic-theory description of a gas.

We now will look at the stresses in a fluid from the viewpoint of kinetic theory. For convenience, we shall restrict our considerations to a gas. Although this is good enough for aerodynamicists, our major results will be applicable to other fluids, too, including water.

First recall the definition of stress in a continuum: it is the internal force (per unit area) that must be applied to the surface of a moving *fluid particle* in order to imitate the effect of the fluid's actual surroundings, so that the fluid within moves under the action of the stresses exactly as it does when actually immersed in the surrounding fluid.

From the continuum viewpoint, the surface of the fluid particle is impermeable. The surface moves normal to itself with the same speed as the local flow, so no fluid crosses the surface. In the molecular picture, molecules are continually crossing the surface of the fluid particle in both directions. However, through any section ΔS of the particles' surface, and in any sensible time interval Δt, as many molecules enter the particle as leave it.

This permits a molecular model of stress in which we imagine that a force is exerted on each molecule that is leaving the fluid particle in such a way as to turn it around and make it imitate some entering molecule, as illustrated in Fig. 6.9. It's highly artificial, to be sure, but no more so than the idea of isolating a fluid particle from its surroundings, an idea which it imitates quite exactly.

To implement this model, set up a local x_1, x_2, x_3 coordinate system with origin at the center of a certain section ΔS of the fluid particles' surface, and with the x_1 axis pointing out of the fluid, as shown in Fig. 6.10. Because the surface is moving with the fluid, the mean velocity of the fluid at the origin has the same component in the x_1 direction as does the coordinate system.

According to the impulse-momentum principle, the momentum of a molecule subjected to a force $\mathbf{F}$ changes in a time interval Δt by

$$\Delta m\mathbf{V} = \int_{\Delta t} \mathbf{F}\, dt$$

where m is the mass of the molecule. Thus the time average (over Δt) of the force exerted on a "leaving" molecule is related to the difference between its momentum and that of the "associated entering" molecule by

$$m\mathbf{V}\big|_{\text{entering}} - m\mathbf{V}\big|_{\text{leaving}} = \mathbf{F}_{\text{average}}\, \Delta t$$

The total force exerted on the section ΔS of interest can then be obtained from

$$\mathbf{F}_{\text{total}} = \frac{N_l}{\Delta t}(m\mathbf{V}|_e - m\mathbf{V}|_l) \tag{6-6}$$

where N_l is the number of molecules that want to leave the fluid particle through ΔS in time Δt.

To begin with, suppose that the gas is "static"; that is, that the average velocity (taken over all the molecules) is zero. To make it really simple, suppose that the

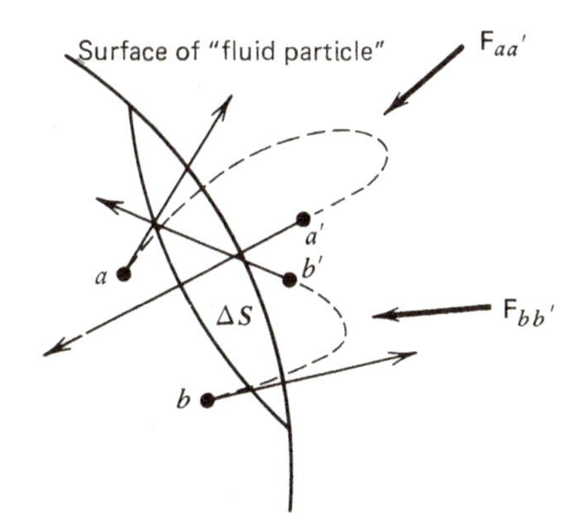

FIG. 6.9. Molecular view of stress in a gas.

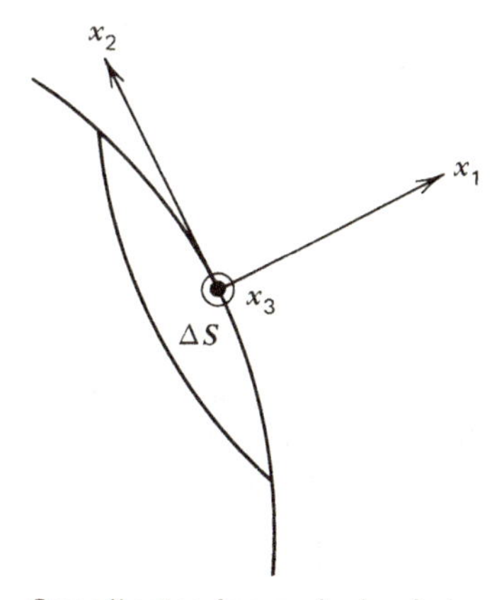

FIG. 6.10. Coordinates for analysis of stress in a gas.

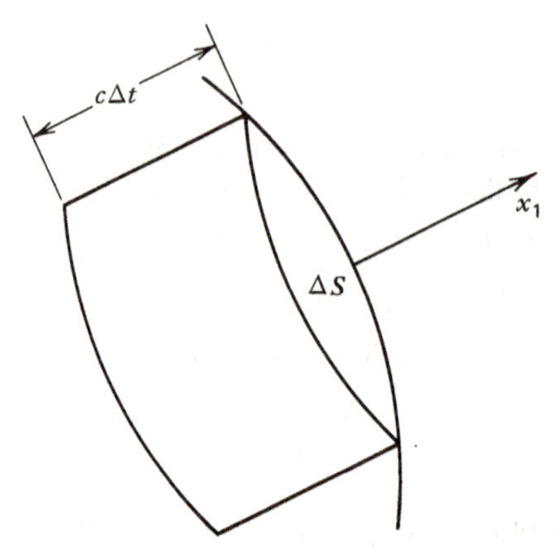

FIG. 6.11. Molecular transport through the surface of the control volume.

molecules all move with the same speed c and that they move only in one of six directions: $+x_1$, $-x_1$, $+x_2$, $-x_2$, $+x_3$, and $-x_3$. Assuming the fluid properties are the same in each of these directions—that is, that the fluid is *isotropic*—one sixth of the molecules must be moving in each direction. Then the number that will leave the fluid particle in time Δt through ΔS must be the number that lie in a cylinder of base ΔS and height $c\,\Delta t$ (see Fig. 6.11) and are headed in the $+x_1$ direction, or

$$N_l = \tfrac{1}{6} n\,\Delta S\, c\,\Delta t$$

where n is the number of molecules per unit volume. Their momentum on leaving is

$$m\mathbf{V}|_l = mc\hat{\mathbf{e}}_1$$

while, when "turned around" to simulate an entering molecule, it's

$$m\mathbf{V}|_e = -mc\hat{\mathbf{e}}_1$$

Substituting all this into equation 6-6, we get

$$\begin{aligned}\mathbf{F}_{\text{total}} &= \frac{1}{\Delta t}\tfrac{1}{6} nc\,\Delta t\,\Delta S(-mc\hat{\mathbf{e}}_1 - mc\hat{\mathbf{e}}_1)\\ &= -\tfrac{1}{3} nmc^2\,\Delta S\hat{\mathbf{e}}_1\end{aligned}$$

Dividing this force by ΔS gives us the stress being exerted on ΔS, whose magnitude is, in this case, the fluid pressure p:

$$p = \tfrac{1}{3} nmc^2 \tag{6-7}$$

As it turns out, this result is exactly right for a static fluid. Recalling that the temperature is proportional to the average kinetic energy of the molecules, $mc^2/2$, we have in fact derived the *perfect gas law*

$$p = n\mathscr{R}T$$

where $\mathscr{R}$ is the "universal gas constant."

Now suppose that the fluid is flowing, that is, that the average fluid velocity is *not* zero. Specifically, consider the special case in which the fluid velocity **V** varies with x_1, the distance normal to the surface on which we seek the stresses, but not with the transverse coordinates x_2 and x_3. The density will be taken to be constant.

As before, we suppose that the individual molecules move in one of only six directions— $\pm x_1$, $\pm x_2$, and $\pm x_3$—and let c be the speed of a molecule relative to the mean flow. Since the surface ΔS moves normal to itself (in the x_1 direction) with the fluid (with speed V_1), the number of molecules that leave the fluid particle (the region below the x_2–x_3 plane) through ΔS in time Δt is, just as in the static case,

$$N_l = \tfrac{1}{6} nc \, \Delta t \, \Delta S$$

Now, however, the magnitude of the momentum of the entering molecules differs from that of the leaving molecules. In a gas, a molecule spends most of its time in essentially unaccelerated motion, altering its velocity only when it comes sufficiently close to another molecule. The average distance a molecule travels at constant velocity (between "collisions") is called the *mean free path* λ. Thus the entering molecules bring with them, on the average, the properties of the average flow a distance on the order of λ above ΔS, whereas the leaving molecules have the properties of the average flow a distance proportional to λ below ΔS. We express this by writing the average momenta of entering and leaving molecules as

$$m\mathbf{V}\big|_e = m[\mathbf{V}(\alpha\lambda) - c\hat{\mathbf{e}}_1]$$

$$m\mathbf{V}\big|_l = m[\mathbf{V}(-\alpha\lambda) + c\hat{\mathbf{e}}_1]$$

where α is a number of the order of one. Thus, for this case, equation 6-6 yields

$$\mathbf{F}_{\text{total}} = \frac{1}{\Delta t} \tfrac{1}{6} nc \, \Delta t \, \Delta S \, m\{-2c\hat{\mathbf{e}}_1 + [\mathbf{V}(\alpha\lambda) - \mathbf{V}(-\alpha\lambda)]\}$$

for the force exerted on ΔS over the time interval Δt. The first term yields a normal force no different from that found in the static case, but we now have additional terms proportional to the change in the average fluid velocity over the distance $2\alpha\lambda$. If this change is small—if the distance $2\alpha\lambda$ is small compared to the "characteristic length" for variations of the mean velocity **V**—we can accurately approximate the change by the product of $2\alpha\lambda$ and its slope at $x_1 = 0$, as shown in Fig. 6.12, and so get

$$\mathbf{F}_{\text{total}} = -\tfrac{1}{3} nmc^2 \, \Delta S \, \hat{\mathbf{e}}_1 + \tfrac{1}{6} nmc \, \Delta S \cdot 2\alpha\lambda \frac{\partial \mathbf{V}}{\partial x_1}\bigg|_{x_1 = 0}$$

Dividing by ΔS and using equation 6-7 then yields the stresses on ΔS:

$$\begin{aligned} \sigma_{11} &= -p + \frac{1}{3}\rho c \alpha\lambda \frac{\partial V_1}{\partial x_1} \\ \sigma_{12} &= \frac{1}{3}\rho c \alpha\lambda \frac{\partial V_2}{\partial x_1} \\ \sigma_{13} &= \frac{1}{3}\rho c \alpha\lambda \frac{\partial V_3}{\partial x_1} \end{aligned} \tag{6-8}$$

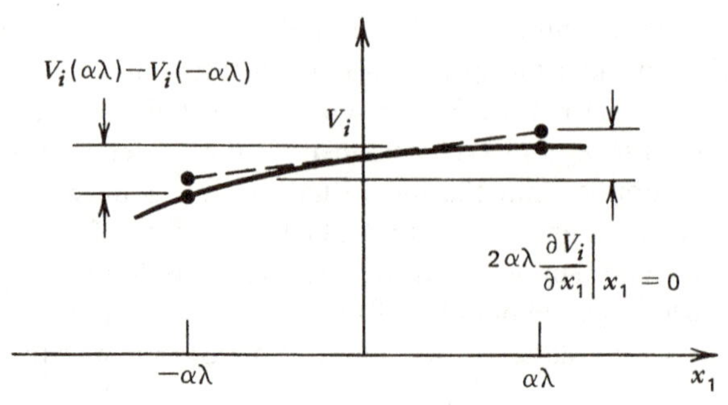

FIG. 6.12. Estimate of change in velocity of molecules crossing the control surface in a moving gas.

in which we observe that the product of the molecular mass m and the number of molecules per unit volume n is the fluid density ρ.

Unless $\partial V_1/\partial x_1 = 0$, our model does not actually balance the number of entering and leaving molecules. But, for an incompressible flow, the continuity equation 2-31 is

$$\rho \operatorname{div} \mathbf{V} = 0 = \rho \sum_i \frac{\partial V_i}{\partial x_i} \tag{6-9}$$

In the case under consideration, $\mathbf{V} = \mathbf{V}(x_1)$, and equation 6-9 reduces to

$$\frac{\partial V_1}{\partial x_1} = 0 \tag{6-10}$$

so that the model (and equation 6-8) are accurate after all.

We will consider shortly an example that does meet the conditions under which equation 6-8 is valid. For later reference, however, we will interpret equation 6-8 as indicating that, in general, each stress component is linear in the derivatives of the mean velocity components, whereas the normal stress components also contain a term independent of the velocity. The most general way to write this is

$$\sigma_{ij} = -p_{ii}\delta_{ij} + \sum_{kl} A_{ijkl} \frac{\partial V_l}{\partial x_k} \tag{6-11}$$

in which δ_{ij}, the *Kronecker delta*, turns on only for normal stresses:

$$\begin{aligned} \delta_{ij} &\equiv 1 \qquad \text{if } i = j \\ &= 0 \qquad \text{otherwise} \end{aligned} \tag{6-12}$$

6.5. THE NO-SLIP CONDITION

Let us now apply our molecular viewpoint to the flow adjacent to a fixed solid surface. Using arguments similar to those that led to equations 6-8, we shall discover an important fundamental fact concerning flows past solids.

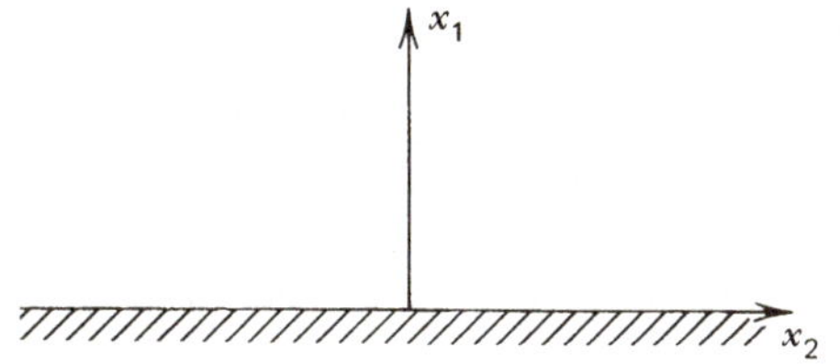

FIG. 6.13. Coordinates for analysis of velocity tangential to a solid surface.

We know that the velocity normal to a solid surface is zero. Otherwise the surface would not be solid. What about the tangential velocity component? Let the x_1 direction be away from the surface, as in Fig. 6.13, and $\mathbf{V}^+$, $\mathbf{V}^-$ be the average velocities of molecules that leave and arrive at the surface $x_1 = 0$, respectively. Since as many molecules leave the surface as arrive at it per unit time, the average fluid velocity at the surface is just the average of $\mathbf{V}^+$ and $\mathbf{V}^-$:

$$\mathbf{V}(0) = \tfrac{1}{2}\mathbf{V}^+(0) + \tfrac{1}{2}\mathbf{V}^-(0) \tag{6-13}$$

Because the atoms and molecules of the solid surface are closely packed, the typical gas molecule gets "trapped" in the structure (more precisely, by the force fields exerted by the surface molecules), colliding many times with the solid molecules before reentering the flow. Its direction on reemission is therefore essentially random, and so the average velocity of molecules leaving the surface has no component tangential to the surface:

$$V_2^+(0) = 0$$

The arriving molecules, however, have the velocity they acquired at their last collision, which, on the average, occurred at $x_1 = \alpha\lambda$, say, where λ is the mean free path and α is a number of order unity. Thus

$$V_2^-(0) = V_2(\alpha\lambda)$$

the average tangential component at $\alpha\lambda$ above the surface, and equation 6-13 yields

$$V_2(0) = \tfrac{1}{2}V_2(\alpha\lambda) \tag{6-14}$$

In most aerodynamic applications, $\alpha\lambda$ is extremely small compared to the characteristic length for $\mathbf{V}$, which means that $\mathbf{V}$ changes very little over the distance $\alpha\lambda$. Then

$$V_2(\alpha\lambda) \approx V_2(0)$$

and equation 6-14 shows that $V_2(0) = 0$, as is $V_3(0)$. Thus the tangential velocity adjacent to a solid surface is zero.

This is called the *no-slip condition*. Note that it is not satisfied by the inviscid solutions we discussed in Chapters 3 and 4. As you shall show in problem 10, those solutions do happen to satisfy the Navier–Stokes equations, so that their failure to meet the no-slip condition is their only error, a fact whose ramifications will be explored in Chapter 7.

6.6. UNIDIRECTIONAL FLOWS

While equation 6-8 is not sufficiently general to deal with arbitrary fluid flows, it does apply to an important special class of flows that happens to include almost all the cases for which the general equations to be derived below can be solved analytically. These are flows in which the streamlines are all parallel, so that the velocity field can be written

$$\mathbf{V} = V_1(x_1, x_2, x_3)\hat{\mathbf{e}}_1 \tag{6-15}$$

if we choose the x_1 axis parallel to the streamlines. One example of such a *unidirectional flow* will be considered shortly and another at the end of the chapter.

Under condition (6-15), the mass conservation equation for an incompressible flow, equation 6-9, reduces to

$$\frac{\partial V_1}{\partial x_1} = 0 \tag{6-16}$$

Then the x_1 component of Newton's law, equation 6-3, simplifies to

$$\rho \frac{\partial V_1}{\partial t} = \frac{\partial \sigma_{11}}{\partial x_1} + \frac{\partial \sigma_{21}}{\partial x_2} + \frac{\partial \sigma_{31}}{\partial x_3} \tag{6-17}$$

In view of equation 6-16, the first of equation 6-8 yields[2]

$$\sigma_{11} = -p \tag{6-18}$$

To use those equations for the other stress components needed in equation 6-17, first relabel the axes of Fig. 6.10 so that the one normal to ΔS is either x_2 or x_3. Then we get, in turn,

$$\sigma_{21} = \mu \frac{\partial V_1}{\partial x_2}$$

$$\sigma_{31} = \mu \frac{\partial V_1}{\partial x_3} \tag{6-19}$$

in which the coefficient of the velocity derivatives has been simplified to

$$\mu = \tfrac{1}{3}\rho c \alpha \lambda \tag{6-20}$$

which is called the *viscosity coefficient* or absolute *viscosity* of the fluid. As will be seen in the next section, μ is a fluid property that depends mainly on the temperature. Taking it to be constant, we combine equations 6-17 to 6-19 to get, for the unidirectional flow of an incompressible fluid.

$$\rho \frac{\partial V_1}{\partial t} = -\frac{\partial p}{\partial x_1} + \mu\left(\frac{\partial^2 V_1}{\partial x_2^2} + \frac{\partial^2 V_1}{\partial x_3^2}\right) \tag{6-21}$$

[2] If you look at the fine print of our derivation of equation 6-8, you will see that it is valid only if $\mathbf{V} = \mathbf{V}(x_1)$, which implies equation 6-10, in apparent agreement with equation 6-16. However, here we are allowing $\mathbf{V} = V_1\hat{\mathbf{e}}_1$ to depend on x_2 and x_3. Nevertheless, as can be verified from equation 6-56, equation 6-18 is valid under the condition (6-15).

Example 1: Couette Flow Consider a steady unidirectional flow between two infinite parallel plates a distance h apart, as shown in Fig. 6.14. Let the upper plate move with speed U, while the lower is fixed. This may be considered as an approximation to the flow between two concentric cylinders, one of which is fixed while the other rotates, if the gap between them is small compared to their radii; see Fig. 6.14.

With the axis system shown in Fig. 6.14, V_1 is independent of x_3. Also, as is especially clear from the concentric-cylinder picture,

$$\partial p/\partial x_1 = 0$$

Thus equation 6-21 reduces, for this case of steady flow, to

$$0 = \mu \frac{d^2 V_1}{dx_2^2}$$

which is easily integrated to

$$V_1 = a + bx_2$$

To find the constants of integration a and b, apply the no-slip condition at the two walls:

$$\begin{aligned} V_1 &= 0 \qquad \text{at } x_2 = 0 \\ &= U \qquad \text{at } x_2 = h \end{aligned}$$

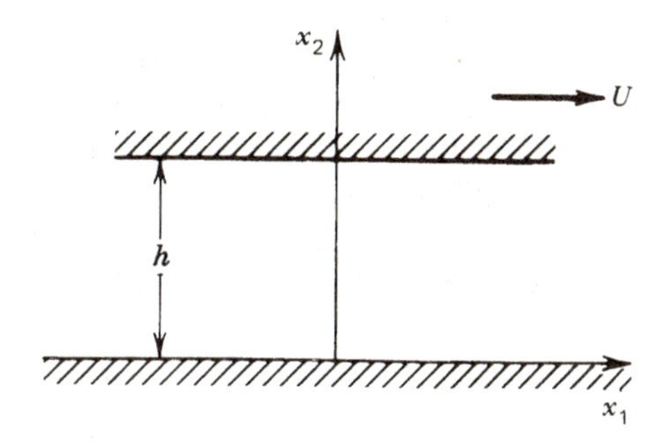

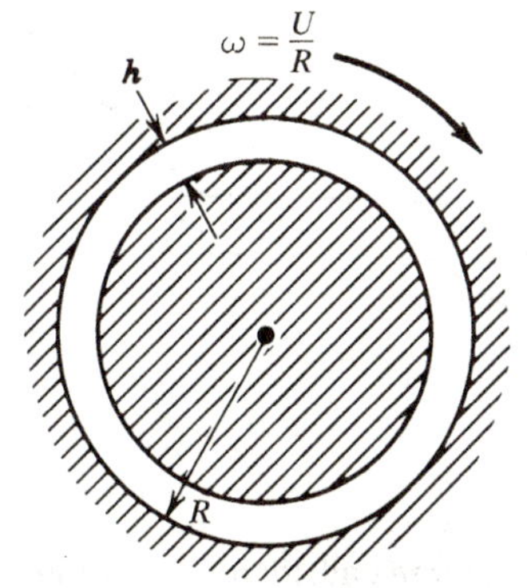

FIG. 6.14. Couette flow. (*a*) Rectilinear view. (*b*) View as an approximation to the flow between two cylinders.

Thus we get the linear velocity profile

$$V_1 = U\frac{x_2}{h}$$

Substituting this into equation 6-19, we find the shear stress to be

$$\sigma_{21} = \frac{\mu U}{h}$$

Thus the stress is uniform throughout the flow, and this formula gives directly the force per unit area required to drag the upper plate along, since the force exerted by the fluid on the plate is equal and opposite to the force exerted on the fluid by the plate.

6.7. THE VISCOSITY COEFFICIENT

The viscosity coefficient introduced in equation 6-19 is an important property of the fluid whenever shear stresses must be considered. According to equation 6-20, μ is proportional to the mean free path, the mean distance a molecule travels between collisions. A rough estimate for the mean free path can be obtained as follows. Suppose the molecules to behave like hard spheres of diameter d, which alter their velocity only when they come in contact with one another. We will compute the average distance between a molecule's collisions by dividing the distance it travels by the number of collisions it experiences while traveling that distance. To ease the estimate of the collision frequency, suppose all the other molecules in the gas to be fixed in space. Then, as can be seen from Fig. 6.15, a collision occurs when the center of the moving molecule comes within a distance equal to its diameter d of the center of one of the stationary "target" molecules. The number of collisions the molecule experiences as it moves a certain distance thus equals the number of target molecules whose centers lie in a kinky cylinder of diameter $2d$ whose axis coincides with the

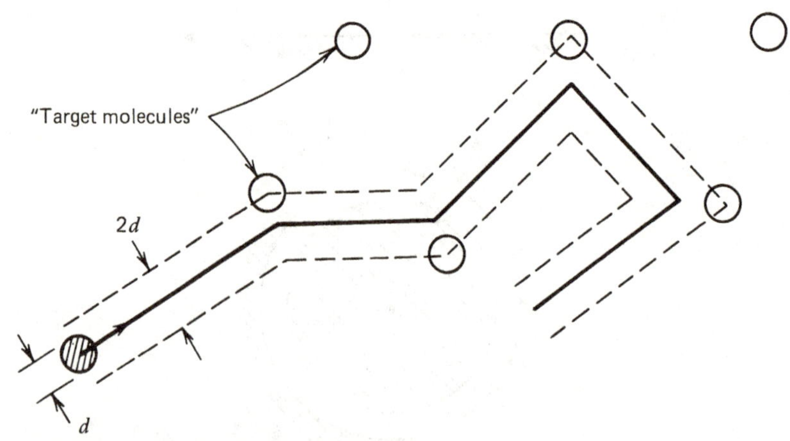

FIG. 6.15. Estimation of mean free path for a "hard-sphere" molecule moving through a gas of stationary hard spheres.

path of the moving molecule's center. This, in turn, is n (the number density) times the volume of that cylinder, which, if the distance between kinks (collisions) is large compared to the cylinder diameter, is approximately the distance s traveled by the molecule times πd^2. Thus the number of collisions the molecule has while traveling a distance s is

$$ns\pi d^2$$

and the mean free path is

$$\lambda = \frac{s}{ns\pi d^2} = \frac{1}{\pi n d^2} \tag{6-22}$$

From equation 6-20, the viscosity coefficient is, therefore,

$$\mu = \frac{\rho c \alpha}{\pi n d^2} = \frac{m c \alpha}{\pi d^2} \tag{6-23}$$

where we recall that the density $\rho = nm$, m being the mass of a molecule. Since α is a number of the order of unity, whereas m and d are properties of the gas under consideration, equation 6-23 implies that the viscosity of a given gas depends only on c, the mean speed of its molecules relative to the mean velocity. But the temperature T is proportional to the average kinetic energy of the motion of the molecules relative to the mean motion, $mc^2/2$, and so

$$\mu \sim \sqrt{T}$$

This is not quite correct; our hard-sphere model is much too crude. However, it is true that the viscosity of a gas is an increasing function of its temperature and that the viscosity of any fluid depends on its temperature but is relatively independent of pressure and density. Figure 6.16 shows how μ varies with temperature for some common liquids and gases. Note that it decreases with temperature for liquids.

6.8. PASCAL'S LAW

For flows more complex than the Couette flow example considered above, equation 6-8 is not adequate. We now turn to its generalization, equation 6-11, which we will simplify drastically by deducing certain relations among the coefficients p_{ii} and A_{ijkl}. First, we will show that the velocity-independent contributions to the normal stresses, $-p_{11}$, $-p_{22}$, and $-p_{33}$, all have the same magnitude. It suffices to consider a static fluid, in which case the forces on the fluid within the prism in Fig. 6.17 must balance. If the sides normal to the x_1' axis and x_1–x_2 plane are ε and δ, respectively, summing forces in the x_1 and x_2 directions yields

$$p_{11}' \varepsilon\delta \cos\theta = p_{11} \varepsilon\delta \cos\theta$$
$$p_{11}' \varepsilon\delta \sin\theta = p_{22} \varepsilon\delta \sin\theta$$

from which we get

$$p_{11}' = p_{11} = p_{22}$$

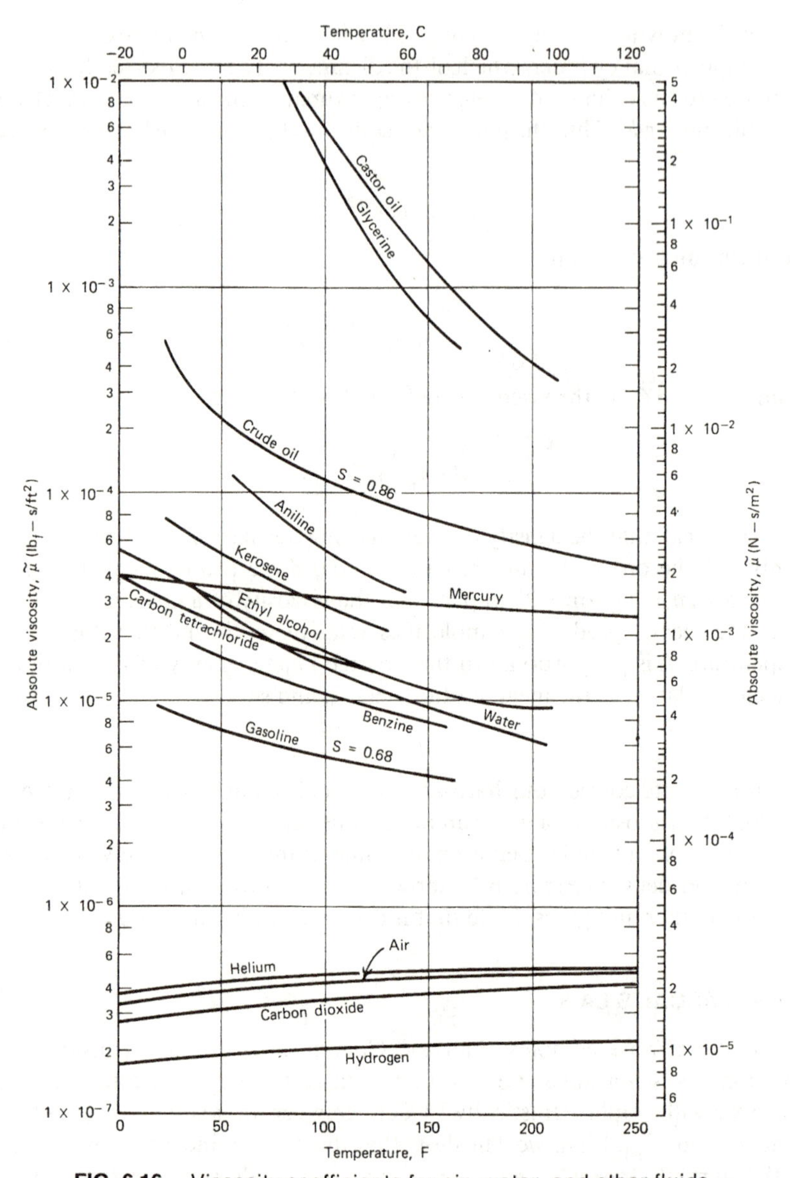

FIG. 6.16. Viscosity coefficients for air, water, and other fluids.

This is *Pascal's law*, in essence; the pressure in a static fluid (which, in our model, is the velocity-independent part of the normal stress) is the same in every direction. Thus we rewrite equation 6-11 as

$$\sigma_{ij} = -p\delta_{ij} + \sum_{kl} A_{ijkl} \frac{\partial V_l}{\partial x_k} \tag{6-24}$$

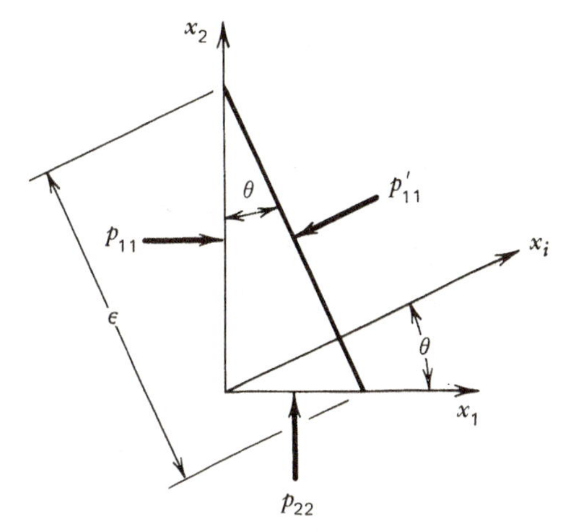

FIG. 6.17. Control volume for examining normal stress components in a static fluid.

6.9. STRAIN VERSUS ROTATION

The 3^4 A_{ijkl}'s in equation 6-24 are supposed to be independent of the velocity field but may depend on fluid properties like p and ρ. We can cut down the task of specifying these coefficients considerably by recalling the symmetry of the stresses; see equation 6-5, which implies

$$A_{ijkl} = A_{jikl} \tag{6-25}$$

and so reduces the number of independent A_{ijkl}'s to 54 (6 choices of i and j, 3 each for k and l).

A similar reduction in the number of independent coefficients can be obtained by examining what the existence of a derivative $\partial V_l/\partial x_k$ implies about the motion of the fluid. Consider a small fluid particle in the shape of a cylinder of length l lined up with the x_1 direction as shown in Fig. 6.18. If the left end has velocity **V**, the velocity of the right end is approximately

$$\mathbf{V} + \frac{\partial \mathbf{V}}{\partial x_1} l$$

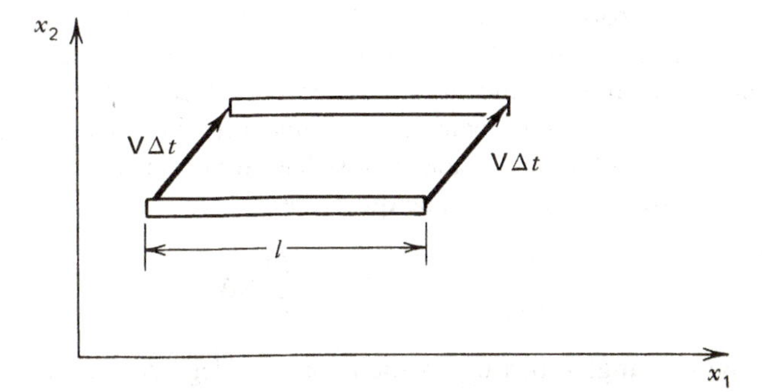

FIG. 6.18. Translation of a fluid particle.

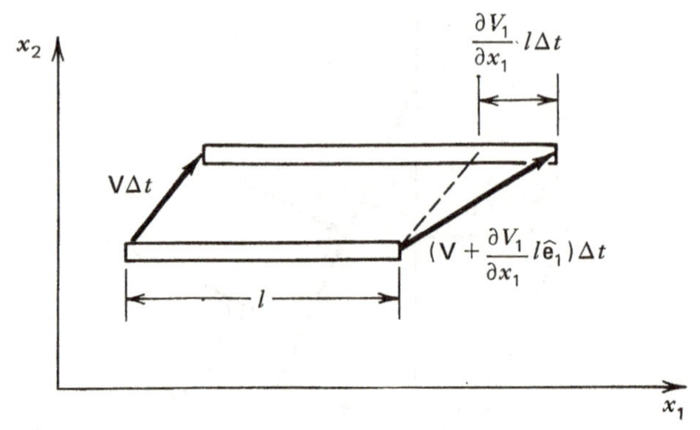

FIG. 6.19. Extension of a fluid particle.

If $\partial \mathbf{V}/\partial x_1$ is zero, the particle is simply translating with velocity $\mathbf{V}$, as shown in Fig. 6.18. According to equation 6-24, no stress is required to sustain such a motion, which makes sense: you expect a stress to be associated with a deformation of the fluid, not just a translation.

Suppose now that $\partial V_1/\partial x_1$ is the only nonzero derivative and that it is positive. Then the velocity of the right end is approximately (for small l)

$$\mathbf{V} + \frac{\partial V_1}{\partial x_1} l \hat{\mathbf{e}}_1$$

Following the motion of this fluid particle through time Δt, we see, from Fig. 6.19, that the element extends in length an amount

$$\frac{\partial V_1}{\partial x_1} l \, \Delta t$$

As in solid mechanics, the percentage increase in length of a fluid particle is called a *normal strain.* In this case, the relative increase in length is $\partial V_1/\partial x_1 \; \Delta t$, so that $\partial V_1/\partial x_1$ is the *rate of normal strain* (specifically, the rate of normal strain in the x_1 direction, because the original length was in that direction).

Now suppose $\partial V_2/\partial x_1$ is the only nonzero derivative. Then, as shown in Fig. 6.20, the particle is seen to rotate in the x_1–x_2 plane counterclockwise through an angle $\partial V_2/\partial x_1 \; \Delta t$ in the time interval Δt. A simple rotation without change in shape, however, is not a deformation and should not be associated with stress. On the other hand, it is hard to tell what is happening to this linear fluid particle in such a motion. Consider instead an *L*-shaped particle, whose legs are initially lined up with the x_1 and x_2 axes, as shown in Fig. 6.21. The quantities

$$\frac{\partial V_2}{\partial x_1} \Delta t \qquad \text{and} \qquad \frac{\partial V_1}{\partial x_2} \Delta t$$

are seen to be the angles through which the two legs rotate, in the first case counterclockwise and in the second clockwise. Thus the angle between the two legs

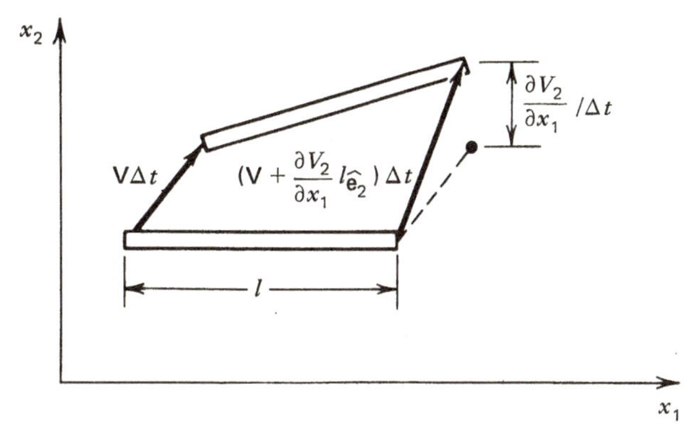

FIG. 6.20. Rotation of a fluid particle.

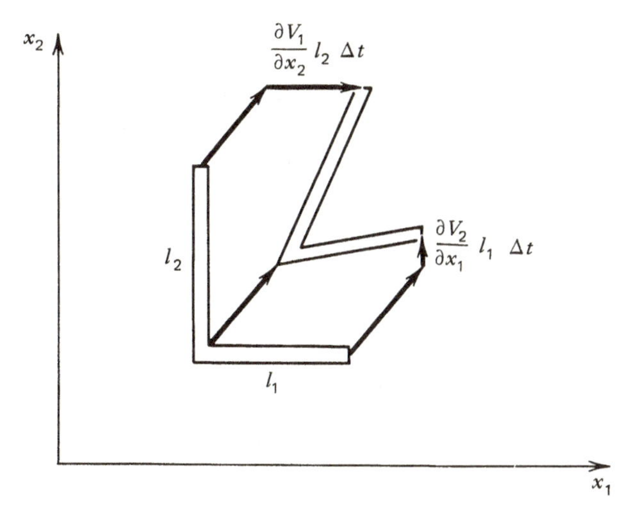

FIG. 6.21. Shear strain in a fluid.

is reduced by

$$\left(\frac{\partial V_2}{\partial x_1} + \frac{\partial V_1}{\partial x_2}\right) \Delta t$$

which, again as in solid mechanics, is the *shear strain* experienced by the element. Note that, in a simple rotation,

$$\frac{\partial V_2}{\partial x_1} = -\frac{\partial V_1}{\partial x_2}$$

In fact, you may recall (see equation 2-34) that the angular velocity of a fluid particle is

$$\boldsymbol{\omega} = \tfrac{1}{2} \text{curl } \mathbf{V} \tag{6-26}$$

whose x_3 component is

$$\omega_3 = \frac{1}{2}\left(\frac{\partial V_2}{\partial x_1} - \frac{\partial V_1}{\partial x_2}\right)$$

which, from Fig. 6.21, you can see is the average rate of counterclockwise rotation of the two legs of the L.

We will call

$$\dot{\varepsilon}_{ij} \equiv \frac{1}{2}\left(\frac{\partial V_j}{\partial x_i} + \frac{\partial V_i}{\partial x_j}\right) \tag{6-27}$$

a *rate of strain*. If $i = j$, it is a rate of normal strain, while otherwise it is associated with a shear strain. Also, though it is not a conventional notation, we will let

$$\omega_{ij} \equiv \frac{1}{2}\left(\frac{\partial V_j}{\partial x_i} - \frac{\partial V_i}{\partial x_j}\right) \tag{6-28}$$

which then represents the angular velocity in the i–j plane.

As noted above, it makes sense that stress is proportional to a rate of strain, but not to a rate of rotation. To put this information into equation 6-24, observe, from equations 6-27 and 6-28, that

$$\frac{\partial V_l}{\partial x_k} = \dot{\varepsilon}_{kl} + \omega_{kl}$$

and write equation 6-24 as

$$\begin{aligned}
\sigma_{ij} + p\delta_{ij} - \sum_{kl} A_{ijkl}\dot{\varepsilon}_{kl} &= \sum_{kl} A_{ijkl}\omega_{kl} \\
&= \sum_{lk} A_{ijlk}\omega_{lk} \\
&= \frac{1}{2}\sum_{lk} (A_{ijkl}\omega_{kl} + A_{ijlk}\omega_{lk})
\end{aligned}$$

in which the middle line stems from the simple observation that it makes no difference what we call the subscripts if we're just going to sum over them. But, from equation 6-28,

$$\omega_{kl} = -\omega_{lk}$$

Thus

$$\sigma_{ij} = -p\delta_{ij} + \sum_{kl} A_{ijkl}\dot{\varepsilon}_{kl} + \tfrac{1}{2}\sum_{kl} (A_{ijkl} - A_{ijlk})\omega_{kl}$$

Therefore, the stress will depend only on the rate of strain and not on the rate of rotation if the coefficients A_{ijkl} are symmetric with respect to the second pair of indexes

$$A_{ijkl} = A_{ijlk} \tag{6-29}$$

as well as with respect to the first pair (see equation 6-25), and the stress/rate-of-strain relation is

$$\sigma_{ij} = -p\delta_{ij} + \sum_{kl} A_{ijkl}\dot{\varepsilon}_{kl} \tag{6-30}$$

It should be noted that the rate of strain, like the stress, is symmetric; from the definition (6-27),

$$\dot{\varepsilon}_{ij} = \dot{\varepsilon}_{ji} \tag{6-31}$$

6.10. ISOTROPY

The double symmetry of the A_{ijkl}'s, equations 6-25 and 6-29, reduces to 36 the number of independent coefficients. We now will reduce that number to a mere three on the assumption (certainly true for gases, and also for many liquids, including water) that the fluid is *isotropic*, that its properties are the same in any direction. Then a normal rate of strain in the x_1 direction should induce the same stress in the x_1 direction as a similar value of $\dot{\varepsilon}_{22}$ would induce in the x_2 direction or that $\dot{\varepsilon}_{33}$ would in the x_3 direction. Therefore,

$$A_{1111} = A_{2222} = A_{3333} = A, \text{ say} \tag{6-32}$$

Also, a normal rate of strain in one direction may induce normal stresses in transverse directions, but the stresses must be the same in all transverse directions. Thus

$$A_{2211} = A_{3311} = B, \text{ say} \tag{6-33}$$

Again, the same rate of strain in the x_2 or x_3 direction would have the same effects in their transverse directions, so

$$A_{1122} = A_{3322} = A_{1133} = A_{2233} = B \tag{6-34}$$

Further, isotropy rules out any connection between any normal stress and a rate of shear strain. For example, suppose A_{1112} were positive, and the fluid were experiencing a rate of strain in the x_1–x_2 plane, as is shown in Fig. 6.22. If the x_1 and x_2 axes were going to the right and up, respectively, as shown in Fig. 6.22*a*, the rate of strain $\dot{\varepsilon}_{12}$ would be positive and so contribute positively to σ_{11}. However, if the x_1 axis were reversed, but the strain kept the same, as shown in Fig. 6.22*b*, the $\dot{\varepsilon}_{12}$ would be negative, and a positive A_{1112} would produce a negative contribution to σ_{11}. Since the value of the stress should be independent of the choice of coordinate system, this inconsistency is resolved only if $A_{1112} = 0$. Similarly,

$$A_{iijk} = 0 \qquad \text{unless} \quad j = k \tag{6-35}$$

whereas a parallel argument shows that no shear stress can be associated with a normal rate of strain:

$$A_{ijkk} = 0 \qquad \text{unless} \quad i = j \tag{6-36}$$

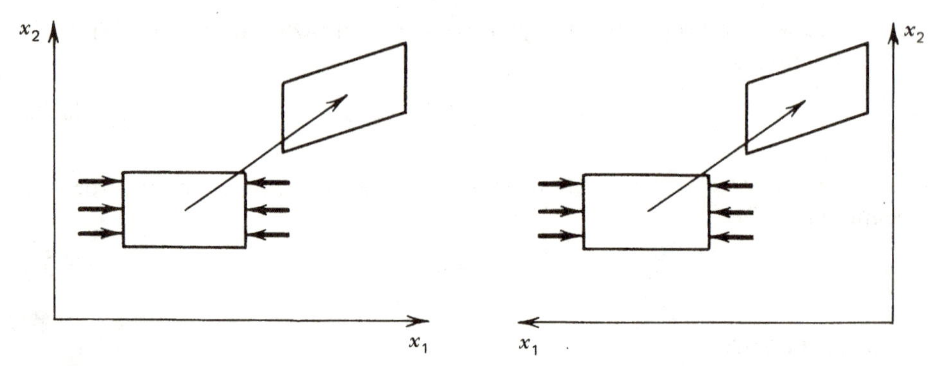

FIG. 6.22. Demonstration that a shear strain cannot induce a normal stress in an isotropic fluid.

At this stage, it is useful to take time out to keep score. If you write out equation 6-30 in matrix form, it should now look like this:

$$\begin{bmatrix} \sigma_{11} \\ \sigma_{22} \\ \sigma_{33} \\ \sigma_{12} \\ \sigma_{23} \\ \sigma_{31} \end{bmatrix} = -p \begin{bmatrix} 1 \\ 1 \\ 1 \\ 0 \\ 0 \\ 0 \end{bmatrix} + \begin{bmatrix} A & B & B & 0 & 0 & 0 \\ B & A & B & 0 & 0 & 0 \\ B & B & A & 0 & 0 & 0 \\ 0 & 0 & 0 & X & X & X \\ 0 & 0 & 0 & X & X & X \\ 0 & 0 & 0 & X & X & X \end{bmatrix} \begin{bmatrix} \dot{\varepsilon}_{11} \\ \dot{\varepsilon}_{22} \\ \dot{\varepsilon}_{33} \\ \dot{\varepsilon}_{12} \\ \dot{\varepsilon}_{23} \\ \dot{\varepsilon}_{31} \end{bmatrix} \tag{6-37}$$

The A's and B's were introduced in equations 6-32 and 6-33, respectively, whereas the X's represent A_{ijkl}'s we have yet to deal with. Specifically, because of the symmetry of the $\dot{\varepsilon}_{kl}$'s and the A_{ijkl}'s, equations 6-29 and 6-31, the first X in the fifth row is $A_{2312} + A_{2321} = 2A_{2312}$. However, note that we have already cut the 81 coefficients of equation 6-11 down to just 11 independent coefficients!

The final reductions come from the observations that

$$A_{1212} = A_{2323} = A_{3131} = C, \text{ say} \tag{6-38}$$

and that

$$A_{1223} = A_{1231} = \cdots = 0 \tag{6-39}$$

Equation 6-38 should be easy to believe; it says only that the connections between a shear stress component and its corresponding rate of shear strain should be the same for each component. Equation 6-39 implies that no shear stress component is associated with anything other than the corresponding rate of shear strain. To show this, consider the deforming fluid particle shown in Fig. 6.23. Observed from either of the two axes systems shown, ε_{12} is positive. If $A_{3112} > 0$, however, the stress component σ_{31} on the face of the particle showing (the face whose outward normal is in the $+x_3$ direction) would have to be to the right if you used the axis system in Fig. 6.23*a* and to the left if you used the one in Fig. 6.23*b*.

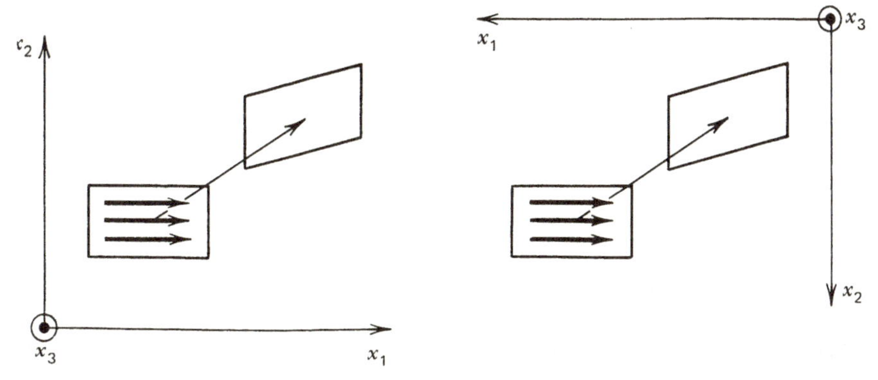

FIG. 6.23. Demonstration that a shear strain is connected only to the corresponding shear stress.

Thus, through equations 6-23–6-36, 6-38, and 6-39, isotropy has finally reduced equations 6-30 to

$$\begin{aligned}
\sigma_{11} &= -p + A\dot{\varepsilon}_{11} + B(\dot{\varepsilon}_{22} + \dot{\varepsilon}_{33}) \\
\sigma_{22} &= -p + A\dot{\varepsilon}_{22} + B(\dot{\varepsilon}_{11} + \dot{\varepsilon}_{33}) \\
\sigma_{33} &= -p + A\dot{\varepsilon}_{33} + B(\dot{\varepsilon}_{11} + \dot{\varepsilon}_{22}) \\
\sigma_{12} &= 2C\dot{\varepsilon}_{12} \\
\sigma_{23} &= 2C\dot{\varepsilon}_{23} \\
\sigma_{31} &= 2C\dot{\varepsilon}_{31}
\end{aligned} \tag{6-40}$$

6.11. VECTORS AND TENSORS

The remaining nonzero independent coefficients in equation 6-37 can be cut by another third (from three to two) because stress and rate of strain are *tensors.* To which the natural reply is, "What's a tensor?"

For one thing, a vector is a tensor. That statement will become more helpful when we recall what a vector is: "a quantity with magnitude and direction" is the definition most of us learn. The significance of this definition lies in the fact that, although a vector can be represented by a single symbol (like **M**, **H**, **V**, or **r**), we generally work with vectors in terms of their components along certain coordinate axes. The values of these components depend on the coordinate system being used. However, what the definition requires is that, *as you change the coordinate system, the components change so that they represent a quantity whose magnitude and direction are unchanged.*

Thus, let $\hat{\mathbf{e}}_1, \hat{\mathbf{e}}_2, \hat{\mathbf{e}}_3$ be unit vectors along the x_1, x_2, and x_3 axes, respectively, and (V_1, V_2, V_3) the components of **V** along the same axes, as shown in Fig. 6.24. Then we write

$$\mathbf{V} = V_1\hat{\mathbf{e}}_1 + V_2\hat{\mathbf{e}}_2 + V_3\hat{\mathbf{e}}_3$$

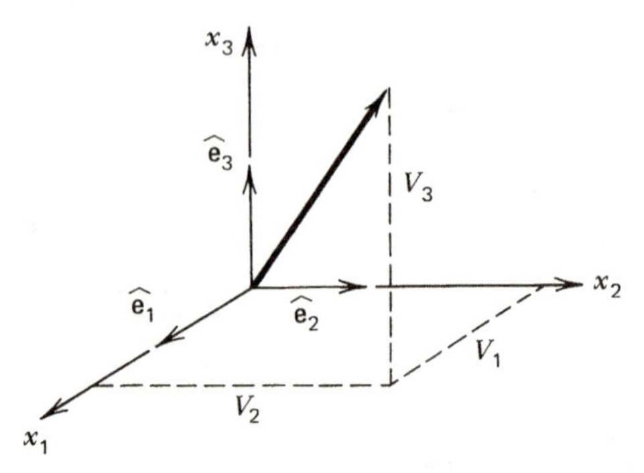

FIG. 6.24. Representation of a vector in terms of unit vectors and components along those unit vectors.

or, taking advantage of the indicial notation,

$$\mathbf{V} = \sum_i V_i \hat{\mathbf{e}}_i$$

If we use some other coordinate system, with unit vectors $(\hat{\mathbf{e}}'_1, \hat{\mathbf{e}}'_2, \hat{\mathbf{e}}'_3)$, say, the components of $\mathbf{V}$ might change to (V'_1, V'_2, V'_3), but they must add up to the same $\mathbf{V}$:

$$\mathbf{V} = \sum_j V'_j \hat{\mathbf{e}}'_j$$

Thus

$$\sum_j V'_j \hat{\mathbf{e}}'_j = \sum_i V_i \hat{\mathbf{e}}_i \tag{6-41}$$

To extract from equation 6-41 relations between the components of $\mathbf{V}$ along the two coordinate axes, just take its dot product with a typical one of the unit vectors, say, $\hat{\mathbf{e}}'_2$. Since the $\hat{\mathbf{e}}'_j$'s are mutually perpendicular,

$$\hat{\mathbf{e}}'_j \cdot \hat{\mathbf{e}}'_2 = 0 \qquad \text{if } j = 1 \text{ or } 3$$

and since the $\hat{\mathbf{e}}'_j$'s have unit length,

$$\hat{\mathbf{e}}'_2 \cdot \hat{\mathbf{e}}'_2 = |\hat{\mathbf{e}}'_2|^2 = 1$$

Thus

$$V'_2 = \sum_i V_i \hat{\mathbf{e}}_i \cdot \hat{\mathbf{e}}'_2$$

which may be generalized to

$$V'_k = \sum_i V_i \hat{\mathbf{e}}_i \cdot \hat{\mathbf{e}}'_k \tag{6-42}$$

Similarly,

$$V_i = \sum_j V'_j \hat{\mathbf{e}}'_j \cdot \hat{\mathbf{e}}_i \tag{6-43}$$

Equations 6-42 and 6-43 show how the components of a vector change from one coordinate system to another in order that the entity of the vector—its magnitude[3] and direction—be independent of the choice of coordinates. You have seen equations like them before, I'm sure, though perhaps not in this form; see problem 6 for a review of the same process in nomenclature that may be more familiar.

Equations 6-42 and 6-43 help us to recognize a vector, given information only on its components. If a quantity is defined such that its components in different coordinate systems satisfy equations 6-42 and 6-43, we can turn our derivation around and say that that quantity is a vector.

Example 2 Suppose we have sets of numbers (V_1, V_2, V_3) and (V'_1, V'_2, V'_3) that are defined in their respective coordinate systems by

$$V_i = \frac{\partial \phi}{\partial x_i}, \qquad V'_i = \frac{\partial \phi'}{\partial x'_i} \tag{6-44}$$

where $\phi'(\mathbf{x}') = \phi(\mathbf{x})$. Are these the components of a vector?

Solution Check it out by using the chain rule:

$$V'_i = \frac{\partial \phi'}{\partial x'_i} = \sum_j \frac{\partial \phi}{\partial x_j} \frac{\partial x_j}{\partial x'_i}$$

Since the x_j's and x'_i's are components of the position vector,

$$\mathbf{x} = \sum_j x_j \hat{\mathbf{e}}_j = \sum_j x'_i \hat{\mathbf{e}}'_i$$

they satisfy equation 6-43,

$$x_j = \sum_j x'_i \hat{\mathbf{e}}'_i \cdot \hat{\mathbf{e}}_j$$

and so

$$\frac{\partial x_j}{\partial x'_i} = \hat{\mathbf{e}}'_i \cdot \hat{\mathbf{e}}_j \tag{6-45}$$

Then the components of $\mathbf{V}$ are seen to satisfy equation 6-42, and the quantity whose components are defined by equation 6-44 is a vector. In fact, you should have recognized it as the gradient of ϕ, defined so that its magnitude and direction are those of the maximum rate of change of ϕ, and so a vector by that definition.

This leads to a definition of a tensor, or at least a tensor of rank 2:

A ***tensor of rank 2*** *is a quantity with* 3^2 components $A_{11}, A_{12}, A_{13}, A_{21}, \ldots$ *in a certain coordinate system. In changing from a Cartesian coordinate system whose unit vectors are* $\hat{\mathbf{e}}_i$ *to one whose unit vectors are* $\hat{\mathbf{e}}'_j$, *the components* A_{ik} *change to* A'_{jl}, *with*

$$A'_{jl} = \sum_{ik} A_{ik} (\hat{\mathbf{e}}_i \cdot \hat{\mathbf{e}}'_j)(\hat{\mathbf{e}}_k \cdot \hat{\mathbf{e}}'_l) \tag{6-46}$$

[3] See problem 5 for an explicit proof that the magnitude of $\mathbf{V}$ is the same in both coordinate systems.

Conversely

$$A_{ik} = \sum_{jl} A'_{jl}(\hat{\mathbf{e}}'_j \cdot \hat{\mathbf{e}}_i)(\hat{\mathbf{e}}'_l \cdot \hat{\mathbf{e}}_k) \tag{6-47}$$

Comparing equations 6-46 and 6-47 with 6-42 and 6-43, you should be able to understand that *a vector is a tensor of rank 1* and to guess how a tensor of rank higher than 2 might be defined, were we to need such a definition.

So far as we are concerned, the most important property of second-order tensors is that the *trace* of such a tensor—the sum of its diagonal elements—

$$\text{tr}(A') \equiv \sum_j A'_{jj} \tag{6-48}$$

is invariant with respect to rotation of the coordinate system. To show this, substitute in equation 6-48 for A'_{jj} from equation 6-46, the definition of a tensor in terms of the effect of coordinate rotation on its components:

$$\text{tr}(A') = \sum_{jik} A_{ik}(\hat{\mathbf{e}}_i \cdot \hat{\mathbf{e}}'_j)(\hat{\mathbf{e}}_k \cdot \hat{\mathbf{e}}'_j)$$

But, as you will show in problem 4,

$$\sum_j (\hat{\mathbf{e}}'_j \cdot \hat{\mathbf{e}}_i)(\hat{\mathbf{e}}'_j \cdot \hat{\mathbf{e}}_k) = \delta_{ik} \tag{6-49}$$

Thus

$$\text{tr}(A') = \sum_{ik} A_{ik}\delta_{ik} = \sum_i A_{ii}$$
$$= \text{tr}(A) \tag{6-50}$$

6.12. THE STRESS TENSOR

Of course, a vector $\mathbf{V}$ is defined by its magnitude and direction, and the way in which its components change with coordinate system, equations 6-42 and 6-43, simply reflects the lack of dependence on coordinate system of its definition. Similarly, the components of certain physical entities change with coordinate system according to equations 6-46 and 6-47 and so are the components of a tensor because of certain physical facts implied by their definitions. In particular, we shall now show that the stress components defined in Section 6.1 satisfy equations 6-46 and 6-47, so that stress is a tensor.

Consider, for simplicity, a rotation of the coordinates (x_1, x_2, x_3) about the x_3 axis through an angle θ, as shown in Fig. 6.25. To see how the stress components at a point change, construct a small control volume around the point. Let it be in the shape of a prism, with two sides normal to the $x_3 = x'_3$ axis, and the others normal to the x_1, x_2, and x'_1 axes. For further simplicity, suppose that the stresses on the faces perpendicular to the x_3 axis are equal and so, being oppositely directed, yield no net force on the prism.

As shown in Fig. 6.25, let the length of the side normal to the x'_1 axis be ε. Also, let the side normal to the x_1–x_2 plane have length δ. The names of the stresses acting on the faces that are parallel to the x_3 axis, and their directions if positive, are shown in Fig. 6.26.

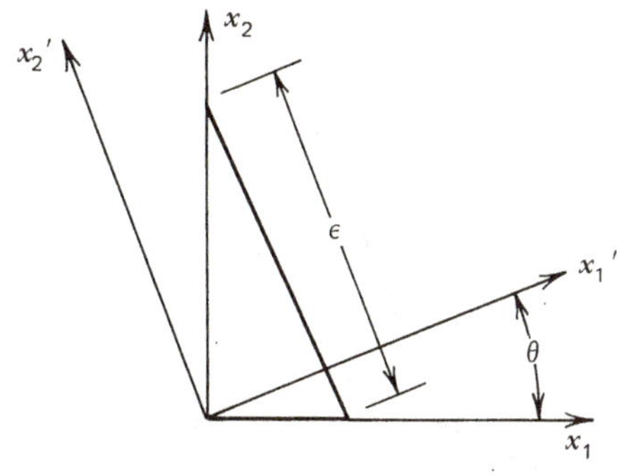

FIG. 6.25. Control volume for demonstration that stress is a tensor.

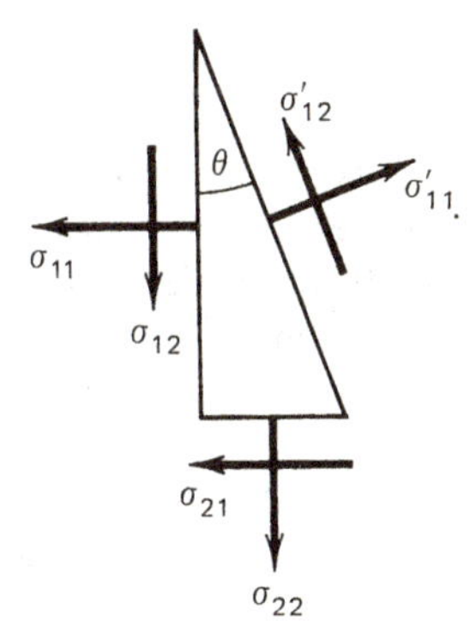

FIG. 6.26. Nomenclature for proof that stress is a tensor.

We now apply Newton's law of motion to the fluid in this prism. Its momentum is

$$\int_{\mathscr{V}} \rho \mathbf{V}\, d\mathscr{V} = \overline{\rho \mathbf{V}} \tfrac{1}{2} \varepsilon^2 \cos\theta \sin\theta \delta$$

in which $\overline{\rho \mathbf{V}}$ is the average momentum per unit volume in the prism. The time rate of change of this quantity (following the fluid particle whose shape is the prism under discussion at the time of observation) equals the net force on the prism, which we resolve into components along the primed axes:

$$\mathbf{F} = \sum_i F'_i \hat{\mathbf{e}}'_i$$

Noting that the areas of the faces of the prism normal to the x'_1, x_1, and x_2 axes are, respectively,

$$\varepsilon\,\delta$$

$$\varepsilon\,\delta \cos\theta = \varepsilon\,\delta \hat{\mathbf{e}}_1 \cdot \hat{\mathbf{e}}'_1$$

$$\varepsilon\,\delta \sin\theta = \varepsilon\,\delta \hat{\mathbf{e}}_2 \cdot \hat{\mathbf{e}}'_1$$

we can write the force on the prism as

$$\begin{aligned} \mathbf{F} = {} & \varepsilon\delta(\sigma'_{11}\hat{\mathbf{e}}'_1 + \sigma'_{12}\hat{\mathbf{e}}'_2) + \varepsilon\delta\hat{\mathbf{e}}_1 \cdot \hat{\mathbf{e}}'_1(-\sigma_{12}\hat{\mathbf{e}}_1 - \sigma_{12}\hat{\mathbf{e}}_2) \\ & + \varepsilon\delta\hat{\mathbf{e}}_2 \cdot \hat{\mathbf{e}}'_1(-\sigma_{21}\hat{\mathbf{e}}_1 - \sigma_{22}\hat{\mathbf{e}}_2) + \text{components in } \hat{\mathbf{e}}_3 \text{ direction} \qquad \text{(6-51)} \end{aligned}$$

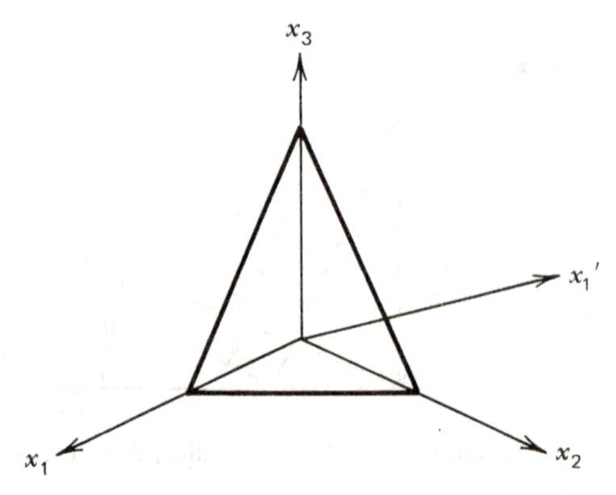

FIG. 6.27. Control volume for more general three-dimensional demonstration that stress is a tensor.

in which we assume that the prism is so small that the stresses are essentially uniform over the faces of the prism. Since the momentum of the fluid in the prism is proportional to $\varepsilon^2 \delta$, if we divide through Newton's law by $\varepsilon \delta$ and let $\varepsilon \to 0$, we see that, to the approximation represented by equation 6-51, $\mathbf{F} = 0$.[4] Resolving equation 6-51 into the x_1' and x_2' directions, we get

$$\begin{aligned}\sigma_{11}' = {} & \sigma_{11}(\hat{\mathbf{e}}_1 \cdot \hat{\mathbf{e}}_1')(\hat{\mathbf{e}}_1 \cdot \hat{\mathbf{e}}_1') + \sigma_{21}(\hat{\mathbf{e}}_2 \cdot \hat{\mathbf{e}}_1')(\hat{\mathbf{e}}_1 \cdot \hat{\mathbf{e}}_1') \\ & + \sigma_{12}(\hat{\mathbf{e}}_1 \cdot \hat{\mathbf{e}}_1')(\hat{\mathbf{e}}_2 \cdot \hat{\mathbf{e}}_1') + \sigma_{22}(\hat{\mathbf{e}}_2 \cdot \hat{\mathbf{e}}_1')(\hat{\mathbf{e}}_2 \cdot \hat{\mathbf{e}}_1') \\ \sigma_{12}' = {} & \sigma_{11}(\hat{\mathbf{e}}_1 \cdot \hat{\mathbf{e}}_1')(\hat{\mathbf{e}}_1 \cdot \hat{\mathbf{e}}_2') + \sigma_{21}(\hat{\mathbf{e}}_2 \cdot \hat{\mathbf{e}}_1')(\hat{\mathbf{e}}_1 \cdot \hat{\mathbf{e}}_2') \\ & + \sigma_{12}(\hat{\mathbf{e}}_1 \cdot \hat{\mathbf{e}}_1')(\hat{\mathbf{e}}_2 \cdot \hat{\mathbf{e}}_2') + \sigma_{22}(\hat{\mathbf{e}}_2 \cdot \hat{\mathbf{e}}_1')(\hat{\mathbf{e}}_2 \cdot \hat{\mathbf{e}}_2')\end{aligned}$$

At least for the special case under consideration, we have thus shown that the stress components in the rotated coordinate system satisfy the condition (6-46) on the components of a tensor. A generalization of our proof could be carried out along similar lines; we'd just have to look at the fluid in a tetrahedron bounded by planes normal to the x_1, x_2, and x_3 axes, and the x_1', x_2', or x_3' axis, as sketched in Fig. 6.27. The result would be a verification of the result we advertised above: stress is a tensor. That is, to satisfy Newton's laws, *the stress components change with coordinate system in the ways dictated by the definition of a tensor.*

6.13. THE RATE-OF-STRAIN TENSOR

Now we will show that the rate-of-strain components are defined by equation 6-27 in such a way that they also are the components of a second-order tensor. To do so, we must examine what happens when the coordinates x_1, x_2, and x_3 are rotated into their primed counterparts x_1', x_2', and x_3'. Since $\mathbf{x}$ and $\mathbf{V}$ are vectors, their

[4] Because the stresses can actually vary over the faces of the prism, there may be a net force on the fluid, proportional to $\varepsilon^2\delta$, and so able to change the momentum of the fluid.

components satisfy equations 6-42 and 6-43:

$$V'_m = \sum_l V_l \hat{\mathbf{e}}_l \cdot \hat{\mathbf{e}}'_m$$

$$x_k = \sum_n x'_n \hat{\mathbf{e}}'_n \cdot \hat{\mathbf{e}}_k$$

Thus, using the chain rule and equation 6-45,

$$\frac{\partial V'_m}{\partial x'_n} = \sum_k \frac{\partial x_k}{\partial x'_n} \frac{\partial}{\partial x_k} \sum_l V_l \hat{\mathbf{e}}_l \cdot \hat{\mathbf{e}}'_m$$

$$= \sum_{kl} \hat{\mathbf{e}}'_n \cdot \hat{\mathbf{e}}_k \frac{\partial V_l}{\partial x_k} \hat{\mathbf{e}}_l \cdot \hat{\mathbf{e}}'_m$$

Similarly

$$\frac{\partial V'_n}{\partial x'_m} = \sum_{kl} \frac{\partial V_k}{\partial x_l} (\hat{\mathbf{e}}_l \cdot \hat{\mathbf{e}}'_m)(\hat{\mathbf{e}}_k \cdot \hat{\mathbf{e}}'_n)$$

so

$$\dot{\varepsilon}'_{nm} \equiv \frac{1}{2}\left(\frac{\partial V'_m}{\partial x'_n} + \frac{\partial V'_n}{\partial x'_m}\right)$$

$$= \sum_{kl} \dot{\varepsilon}_{kl} (\hat{\mathbf{e}}_k \cdot \hat{\mathbf{e}}'_n)(\hat{\mathbf{e}}_l \cdot \hat{\mathbf{e}}'_m)$$

which is the condition that the $\dot{\varepsilon}_{kl}$'s be components of a second-order tensor, equation 6-46.

This proof is rather formal. See Appendix G for a demonstration that rate of strain is a tensor in order that the rate of deformation of a fluid particle be independent of the coordinate system in which it is observed.

6.14. THE TWO COEFFICIENTS OF VISCOSITY

Now, as was promised, we will use the fact that the stress and rate-of-strain components each satisfy equation 6-50 to get a relation among the coefficients A, B, and C of equation 6-40. First, rewrite the first three of equations 6-40 as

$$\sigma_{ii} = -p + (A - B)\dot{\varepsilon}_{ii} + B\,\mathrm{tr}(\dot{\varepsilon}) \tag{6-52}$$

Then consider a simple rotation of axes in the x_1–x_2 plane through an angle θ, so that, as you can see from Fig. 6.25,

$$\hat{\mathbf{e}}_1 \cdot \hat{\mathbf{e}}'_1 = \cos\theta$$

$$\hat{\mathbf{e}}_2 \cdot \hat{\mathbf{e}}'_1 = \sin\theta$$

$$\hat{\mathbf{e}}_3 \cdot \hat{\mathbf{e}}'_1 = 0$$

The normal stress in the x'_1 direction must satisfy equation 6-46, which reduces in this case to

$$\sigma'_{11} = \sigma_{11}\cos^2\theta + \sigma_{12}\cos\theta\sin\theta + \sigma_{21}\sin\theta\cos\theta + \sigma_{22}\sin^2\theta$$

Substituting for the σ_{ij}'s from equations 6-40 and 6-52 gives

$$\begin{aligned}\sigma'_{11} &= \cos^2\theta[-p + (A - B)\dot{\varepsilon}_{11} + B\,\mathrm{tr}(\varepsilon)] \\ &\quad + 2\sin\theta\cos\theta\, 2C\dot{\varepsilon}_{12} + \sin^2\theta[-p + (A - B)\dot{\varepsilon}_{22} + B\,\mathrm{tr}(\varepsilon)] \\ &= -p + B\,\mathrm{tr}(\dot{\varepsilon}) + (A - B)(\cos^2\theta\,\dot{\varepsilon}_{11} + \sin^2\theta\,\dot{\varepsilon}_{22}) + 4\sin\theta\cos\theta\, C\dot{\varepsilon}_{12} \qquad (6\text{-}53)\end{aligned}$$

in which we have recalled the symmetry of the stress tensor, equation 6-5. But σ'_{11} must also satisfy equation 6-52,

$$\sigma'_{11} = -p + B\,\mathrm{tr}(\dot{\varepsilon}') + (A - B)\dot{\varepsilon}'_{11}$$

which, since $\dot{\varepsilon}'_{11}$ must satisfy equation 6-46, may be written

$$\sigma'_{11} = -p + B\,\mathrm{tr}(\dot{\varepsilon}') + (A - B)(\dot{\varepsilon}_{11}\cos^2\theta + 2\dot{\varepsilon}_{12}\sin\theta\cos\theta + \dot{\varepsilon}_{22}\sin^2\theta) \qquad (6\text{-}54)$$

Comparing the two expressions 6-53 and 6-54 for σ'_{11}, and noting that the trace of the rate-of-strain tensor satisfies equation 6-50, we thus obtain

$$A - B = 2C$$

and so rewrite equation 6-52 as

$$\sigma_{ii} = -p + 2C\dot{\varepsilon}_{11} + B\,\mathrm{tr}(\dot{\varepsilon})$$

This may be combined with the last three of equations 6-40 to get the very compact stress/rate-of-strain relation,

$$\sigma_{ij} = 2C\dot{\varepsilon}_{ij} + [B\,\mathrm{tr}(\dot{\varepsilon}) - p]\,\delta_{ij} \qquad (6\text{-}55)$$

Equation 6-55 reduces to equation 6-19 in the special case of unidirectional flow, equation 6-15, if we identify C with the absolute viscosity μ. The coefficient B is called the *bulk viscosity*, which we shall write as μ_B. Since, from equations 6-27 and 6-48,

$$\mathrm{tr}(\dot{\varepsilon}) = \sum_i \frac{1}{2}\left(\frac{\partial V_i}{\partial x_i} + \frac{\partial V_i}{\partial x_i}\right) = \sum_i \frac{\partial V_i}{\partial x_i}$$

which you ought to recognize as the *divergence* of the velocity vector, we rewrite equation 6-55 as

$$\sigma_{ij} = 2\mu\dot{\varepsilon}_{ij} + (\mu_B\,\mathrm{div}\,\mathbf{V} - p)\,\delta_{ij} \qquad (6\text{-}56)$$

In problem 8, you will show that div $\mathbf{V}$ is the rate of change of the volume of a fluid particle, which accounts for the name "bulk" given to the viscosity coefficient μ_B; it is like the "bulk modulus" of elasticity.

A third kind of viscosity is the *kinematic viscosity*; this is simply the ratio of the absolute viscosity (almost always labeled μ) to the density, a ratio that is pervasive in viscous flows. It is usually symbolized by ν:

$$\nu = \mu/\rho$$

6.15. THE NAVIER–STOKES EQUATIONS

We've come a long way from our definition of the stress at a point. Let's review where we've been. Equation 6-56 is based on the linear relation (6-11) between the stress at a point in a fluid and the velocity derivatives at the same point. That relation was suggested by a kinetic-theory analysis, which led to equation 6-8 for the special case $\mathbf{V} = \mathbf{V}(x_1)$. Because of Pascal's law equation 6-11 was reduced to equation 6-24, which, in turn, was altered to equation 6-30 to agree with the fact that stress in a fluid should have to do with deformation, not mere rotation. On assuming that the fluid is isotropic, we reduced equation 6-30 to equation 6-40. The final formula (6-56) was then obtained by using the fact that the stress and rate-of-strain components change from one coordinate system to another according to the laws for tensors.

Substituting the linear isotropic stress/rate-of-strain relations (6-56) into Newton's law, equation 6-3, yields the *Navier–Stokes equations*, which govern the flow of air, water, and other interesting fluids. We will consider only incompressible flows, for which the continuity equation is equation 6-9. This wipes out the bulk viscosity contribution to equation 6-56, but, even so, the Navier–Stokes equations are extremely formidable. Don't expect to solve them without a computer, except in rather special cases (one of which we'll discuss shortly).

We can maneuver the Navier–Stokes equations into relatively compact form, however, as follows. Combining equations 6-3, 6-9, 6-27, and 6-56, we get

$$\frac{\partial \rho V_i}{\partial t} + \sum_j \frac{\partial}{\partial x_j} \rho V_i V_j = \sum_j \frac{\partial}{\partial x_j}\left[\mu\left(\frac{\partial V_j}{\partial x_i} + \frac{\partial V_i}{\partial x_j}\right) - p\,\delta_{ij}\right]$$

Take μ as well as ρ to be constant, and carry out the indicated differentiations as follows:

$$\rho\frac{\partial V_i}{\partial t} + \rho\sum_j V_i\frac{\partial V_j}{\partial x_j} + V_j\frac{\partial V_i}{\partial x_j} = -\frac{\partial p}{\partial x_i} + \mu\sum_j \frac{\partial^2 V_j}{\partial x_i \partial x_j} + \frac{\partial^2 V_i}{\partial x_j^2}$$

The continuity equation 6-9 can now be used to eliminate the first terms of the sums on either side of the equation, leaving

$$\rho\frac{DV_i}{Dt} = \mu\nabla^2 V_i - \frac{\partial p}{\partial x_i} \tag{6-57}$$

in which we have introduced the convective derivative from equation 2-30, and the Laplacian from equation 2-38. The vector form of this equation is

$$\rho\frac{D\mathbf{V}}{Dt} = \mu\nabla^2\mathbf{V} - \nabla p \tag{6-58}$$

I mentioned "special cases" for which you might be able to solve these equations.

Most of these are *unidirectional flows*, in which the velocity is in one particular direction, say the x_1 direction, so $V_2 = V_3 = 0$. The continuity equation 6-9 then reduces to equation 6-16, whereas the Navier–Stokes equation 6-57 becomes

$$\rho\frac{\partial V_1}{\partial t} = \mu\nabla^2 V_1 - \frac{\partial p}{\partial x_1} \tag{6-59}$$

$$\frac{\partial p}{\partial x_2} = \frac{\partial p}{\partial x_3} = 0 \tag{6-60}$$

Example 3: Plane Poiseuille Flow Consider the steady unidirectional flow through a rectangular duct of high aspect ratio, high enough that the flow may be approximated as two dimensional, so that in the coordinates defined in Fig. 6.28 everything is independent of x_3. Let h be the height of the duct. Then the boundary conditions supplied by the no-slip condition are

$$V_1\left(\pm\frac{h}{2}\right) = 0$$

In this case, equation 6-59 reduces to

$$0 = \mu\frac{d^2V_1}{dx_2^2} - \frac{dp}{dx_1}$$

(if you try $dp/dx_1 = 0$, you eventually find V_1 is identically zero). Since, from equation 6-60, p is independent of x_2, this can be integrated to

$$V_1 = a + bx_2 + \frac{1}{2\mu}x_2^2\frac{dp}{dx_1}$$

Applying the boundary conditions gives

$$V_1 = \frac{1}{2\mu}\left(x_2^2 - \frac{h^2}{4}\right)\frac{dp}{dx_1}$$

which, when integrated across the duct, can be used to find the pressure gradient along the duct required to drive a given rate of flow through it. Let the volume rate

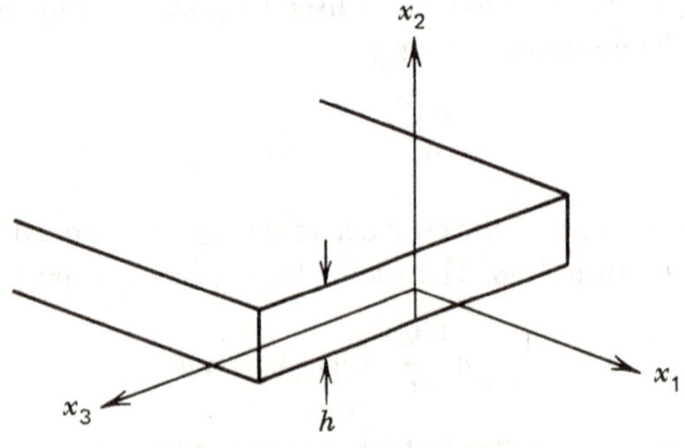

FIG. 6.28. Poiseuille flow.

of flow be Q:

$$Q \equiv \int_{-h/2}^{h/2} V_1 \, dx_2$$

$$= -\frac{h^3}{12\mu} \frac{dp}{dx_1}$$

Then

$$\frac{dp}{dx_1} = -\frac{12\mu}{h^3} Q$$

6.16. PROBLEMS

1. Express the divergence of a vector, the gradient of a scalar, and the dot product of two vectors in the indicial notation.
2. What are the differences between

$$\sum_i A_i \quad \text{and} \quad \sum_j A_j$$

$$\sum_j B_{ij} \quad \text{and} \quad \sum_k B_{ik}$$

$$\sum_{jk} C_{ijk} \quad \text{and} \quad \sum_{jk} C_{ikj}$$

3. What are the differences among

$$\sum_{ij} A_i B_j \, \delta_{ij}, \qquad \sum_i A_i B_j \, \delta_{ij}, \qquad \text{and } A_i B_j \, \delta_{ij}$$

and between

$$\sum_i A_i \, \delta_{ij} \quad \text{and} \quad A_i \, \delta_{ij}$$

Caution: In many books that use subscript notation, a repeated index implies summation over that index, in which case there are no differences among the members of either of these groups.

4. By substituting for V_i in equation 6-42 its expression in terms of the primed components of $\mathbf{V}$, equation 6-43, derive equation 6-49.
5. Show that if the components V_i and V'_j satisfy equations 6-42 and 6-43, then the sum of the squares of the components (the square of the length of the vector whose components they are) is the same in the primed and unprimed coordinate systems.
6. Refer to Fig. 6.25, and consider a rotation of coordinates about the x_3 axis through an angle θ. Use trigonometry to find the relations between the components of a vector $\mathbf{V}$ in the primed and unprimed systems. Show that these relations are consistent with equations 6-42 and 6-43.
7. Is the quantity whose components are defined in different coordinate systems by

$$V_i = 1/x_i, V'_i = 1/x'_i$$

a vector? Here x_i and x'_i are the components of the position vector in the unprimed and primed coordinate systems, respectively.

8. Show that the divergence of the velocity vector represents the rate of increase of the volume of a fluid particle, per unit volume per unit time. *Hint*: Take the fluid particle to be initially in the shape of a rectangular parallelepiped, and compute its change of volume in time Δt in terms of the normal strain components. Then divide by Δt and the original volume.

9. For the two-dimensional steady flow of an incompressible viscous Newtonian fluid, show that the rate of change of vorticity along a streamline is

$$\frac{D\omega}{Dt} = \frac{\mu}{\rho}\nabla^2\omega$$

Hint: See problem 3 of Chapter 2. Here you should take the curl of the Navier–Stokes equation 6-57.

10. Show that a solution of the equations of steady irrotational incompressible flow—equations 3-1b, 3-2, and 3-3b—also satisfies the steady-flow version of Navier–Stokes equation 6-57.

11. Show that, in plane Poisseuille flow, the maximum velocity in the duct is 1.5 times the average velocity.

CHAPTER 7

THE BOUNDARY LAYER

The Navier–Stokes equations derived in Chapter 6 are general enough to cover any aerodynamic situation. In fact, they are usually more than enough. They are hard to solve, even numerically, in any realistic case, and contain terms that complicate their solution but are never of any significant size. In this chapter, we will simplify the Navier–Stokes equations by introducing the very powerful boundary-layer approximation. We then develop a numerical method that solves the simplified equations in cases like flow past an airfoil. Also, we draw qualitative conclusions that provide criteria for airfoil design, which we test by examining experimental data on certain well-studied wing sections.

7.1. THE LAMINAR BOUNDARY LAYER

For two-dimensional motion of an incompressible fluid, the *Navier–Stokes equations*, equation 6-57, reduce to

$$\rho\left(\frac{\partial u}{\partial t} + u\frac{\partial u}{\partial x} + v\frac{\partial u}{\partial y}\right) = -\frac{\partial p}{\partial x} + \mu\left(\frac{\partial^2 u}{\partial x^2} + \frac{\partial^2 u}{\partial y^2}\right) \tag{7-1}$$

$$\rho\left(\frac{\partial v}{\partial t} + u\frac{\partial v}{\partial x} + v\frac{\partial v}{\partial y}\right) = -\frac{\partial p}{\partial y} + \mu\left(\frac{\partial^2 v}{\partial x^2} + \frac{\partial^2 v}{\partial y^2}\right) \tag{7-2}$$

in which we now revert to x–y and u–v notation. The continuity equation is, as for inviscid flows,

$$\text{div}\,\mathbf{V} = \frac{\partial u}{\partial x} + \frac{\partial v}{\partial y} = 0 \tag{7-3}$$

In the case of uniform flow past a solid body, the boundary conditions are that $\mathbf{V}$ and p take on prescribed values far upstream of the body and that

$$\mathbf{V} = 0 \tag{7-4}$$

on the body surface. The *no-slip* condition (7-4) differs crucially from the *flow-tangency* condition that is sufficient in inviscid flows. It may be regarded as an empirical observation or as due to an equilibration between the fluid molecules close to the boundary and the particles of the solid (see Section 6.5).

As usual, we will restrict our interests to steady flows, and so delete the time derivatives from equations 7-1 and 7-2. As we shall see later (under Section 7.7), this step is rather more drastic and less justified than it may appear, but some of our most important conclusions will turn out to be valid anyway.

Even in this special case, exact solutions of equations 7-1 through 7-4 can be found analytically only for very special geometries. High-speed computers help somewhat, but numerical solutions of the full Navier–Stokes equations are still very expensive.

Fortunately, for many practical situations, it is possible to make important simplifications of the Navier–Stokes equations. Some of the terms in equations 7-1 and 7-2 can be neglected in much of the flow. To see this, we need a means for estimating the various derivatives of u, v, and p. If the flow speed far from the body is V_∞, u and v themselves may, in general, be about as large as V_∞, which we express by

$$u \sim V_\infty, \qquad v \sim V_\infty \tag{7-5}$$

To estimate the derivatives of u and v, we first take note of the no-slip condition (7-4) and conclude that u and v may change by as much as V_∞ (roughly) from one part of the flow to another:

$$\Delta u \sim V_\infty, \qquad \Delta v \sim V_\infty \tag{7-6}$$

Now let L be the *characteristic length* of the flow, by which we mean the distance over which significant changes in the flow properties take place, as shown schematically in Fig. 7.1. Sometimes this distance can be estimated (to within a factor of 2 or 3, which is close enough for estimation purposes) from the boundary conditions, or from dimensional analysis. For example, if we are considering flow past an airfoil of chord c, we expect $L \sim c$.

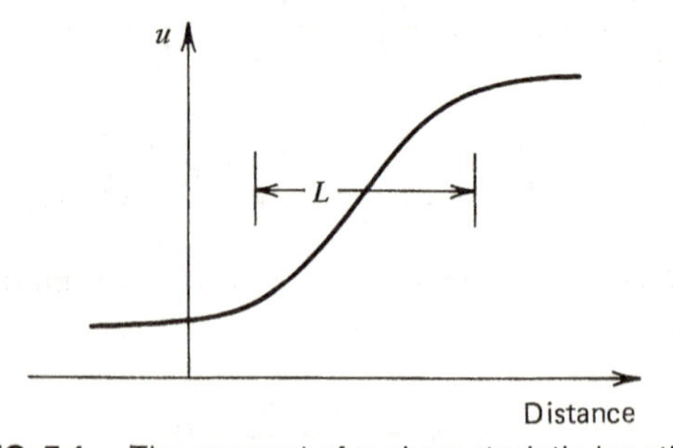

FIG. 7.1. The concept of a characteristic length.

In any case, the concept of a characteristic length illustrated in Fig. 7.1 suggests we write

$$\frac{\partial u}{\partial x} \sim \frac{\Delta u}{L} \sim \frac{V_\infty}{L} \tag{7-7}$$

with similar estimates for the y derivative of u and for the derivatives of v. Also,

$$\frac{\partial p}{\partial x}, \frac{\partial p}{\partial y} \sim \frac{\Delta p}{L} \tag{7-8}$$

However, it is not true that $\Delta p \sim p_\infty$; the pressure need not (indeed, *should* not) be zero anywhere in the flow. Thus we leave Δp as is for the moment.

On the basis of estimates like equation 7-7, both terms in the continuity equation 7-3 seem to be of the same size, as of course they must be if the equation is to be satisfied. However, the terms of the x-motion equation 7-1 have the following estimates:

$$\begin{aligned}
\rho u \frac{\partial u}{\partial x} &\sim \frac{\rho V_\infty^2}{L} \\
\rho v \frac{\partial u}{\partial y} &\sim \frac{\rho V_\infty^2}{L} \\
\frac{\partial p}{\partial x} &\sim \frac{\Delta p}{L} \\
\mu \frac{\partial^2 u}{\partial x^2} &\sim \mu \frac{V_\infty}{L^2} \\
\mu \frac{\partial^2 u}{\partial y^2} &\sim \mu \frac{V_\infty}{L^2}
\end{aligned} \tag{7-9}$$

Note that we have, in effect, assumed

$$\Delta\left(\frac{\partial u}{\partial x}\right) \sim \frac{\partial u}{\partial x} \sim \frac{V_\infty}{L}$$

so

$$\frac{\partial^2 u}{\partial x^2} \sim \frac{1}{L} \Delta\left(\frac{\partial u}{\partial x}\right) \sim \frac{V_\infty}{L^2}$$

Fluid dynamics is characterized by a balance between pressure-induced forces and the inertia terms $\rho\, D\mathbf{V}/Dt$. *Therefore*, we get an estimate for Δp: it must be such that

$$\frac{\partial p}{\partial x} \sim \rho u \frac{\partial u}{\partial x}, \rho v \frac{\partial u}{\partial y}$$

so

$$\Delta p \sim \rho V_\infty^2 \tag{7-10}$$

It is for this reason that pressures are made dimensionless with respect to the "dynamic pressure" $\frac{1}{2}\rho V_\infty^2$ rather than p_∞ in the definition of the *pressure coefficient*:

$$C_p \equiv \frac{p - p_\infty}{\frac{1}{2}\rho V_\infty^2} \tag{7-11}$$

The "$\frac{1}{2}$" is used to make $C_p = 1$ at stagnation points in an incompressible inviscid flow.

The ratio of the inertia terms to the viscous terms is then

$$\frac{\rho\, Du/Dt}{\mu\, \nabla^2 u} \sim \frac{\rho V_\infty^2}{L^2} \bigg/ \frac{\mu V_\infty}{L^2} = \frac{\rho V_\infty L}{\mu} \tag{7-12}$$

which is the *Reynolds number* Re:

$$\text{Re} \equiv \frac{\rho V_\infty L}{\mu} \tag{7-13}$$

If the characteristic length L is in fact the chord length, and $V_\infty = 100$ m/sec, say, the Reynolds number is of the order of 10^7, and we are perfectly justified in neglecting the viscous terms, as we did in Chapters 3 and 4. This looks even better when you observe (see problem 10 of Chapter 6) that an irrotational velocity field does in fact satisfy the Navier–Stokes equations exactly.

However, if we throw out the viscous terms, there is no way we can satisfy the no-slip boundary condition (7-4); all we can require is flow tangency. This may suggest that when the Reynolds number Re is large, there is a thin region near the body in which viscous effects are important, and in which an adjustment is made from no slip at the body to flow tangency at the edge of the region. For that is the case; the region is called the *boundary layer*.

The problem with our derivation of equation 7-12 is our assumption that the characteristic length is isotropic. The boundary-layer idea suggests that the fluid properties change much more rapidly in the direction away from the surface than along it, at least when one is close to the surface. Thus, let x be measured parallel to the surface and y normal to the surface, as shown in Fig. 7.2. Let δ be the thickness of the boundary layer and hence the characteristic length for variations in the y direction near the wall. We continue to use L for the characteristic length for variations in the flow direction (and for variations in any direction, outside the boundary layer). The ratio of δ to L should be such that, within the boundary layer,

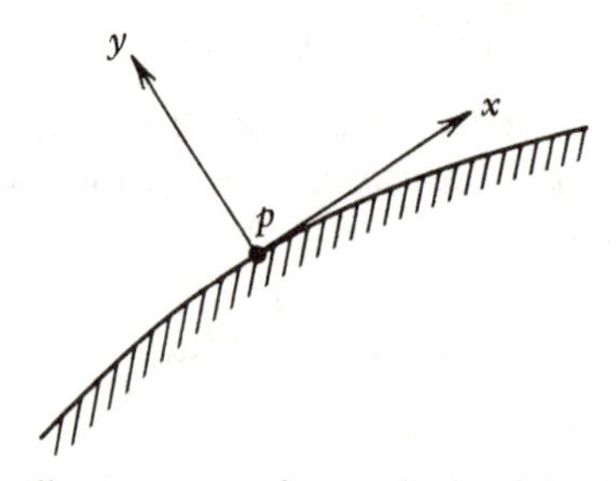

FIG. 7.2. Coordinate system for analysis of the boundary layer.

the viscous terms are as important as the inertia terms and so cannot be neglected, even though the Reynolds number Re is large. Outside the boundary layer, where $y \gg \delta$, the characteristic length for variations in the y direction will be as large as in any other (namely, L), and the preceding proof that the viscous terms can be dropped if Re $\gg 1$ is valid.

Let V_e be the flow speed just outside the boundary layer. Then $u \approx V_e$ for $y \approx \delta$, and so

$$u \sim V_e$$

is our estimate for u. Since V_e vanishes at stagnation points, the change in u in the x direction can be as big as V_e, and

$$\frac{\partial u}{\partial x} \sim \frac{V_e}{L}$$

From the no-slip condition (7-4), $u = 0$ at $y = 0$. Thus its change in the y direction can also be as big as V_e, so

$$\frac{\partial u}{\partial y} \sim \frac{V_e}{\delta}$$

within the boundary layer.

Now, from the continuity equation 7-3

$$\frac{\partial v}{\partial y} = -\frac{\partial u}{\partial x} \sim \frac{V_e}{L}$$

However, our order-of-magnitude estimate for $\partial v/\partial y$ is

$$\frac{\partial v}{\partial y} \sim \frac{\Delta v}{\delta}$$

Since $v = 0$ at $y = 0$, the two preceding equations yield the estimate

$$v \sim \Delta v \sim \frac{\delta}{L} V_e \qquad (7\text{-}14)$$

within the boundary layer. Thus, the velocity component along the surface u is much larger within the boundary layer than the one normal to the surface v.

Our estimates for the terms in equations 7-1 and 7-2 are then as follows:

$$\rho u \frac{\partial u}{\partial x} \sim \rho \frac{V_e^2}{L}$$

$$\rho v \frac{\partial u}{\partial y} \sim \rho \frac{\delta}{L} V_e \frac{V_e}{\delta} = \rho \frac{V_e^2}{L}$$

$$\mu \frac{\partial^2 u}{\partial x^2} \sim \mu \frac{V_e}{L^2}$$

$$\mu \frac{\partial^2 u}{\partial y^2} \sim \mu \frac{V_e}{\delta^2}$$

$$\rho u \frac{\partial v}{\partial x} \sim \rho V_e \frac{\delta}{L} \frac{V_e}{L} = \rho \frac{V_e^2 \delta}{L^2}$$

$$\rho v \frac{\partial v}{\partial y} \sim \rho \frac{\delta}{L} V_e \frac{\delta}{L} \frac{V_e}{\delta} = \rho \frac{V_e^2 \delta}{L^2}$$

$$\mu \frac{\partial^2 v}{\partial x^2} \sim \mu \frac{\delta}{L} \frac{V_e}{L^2}$$

$$\mu \frac{\partial^2 v}{\partial y^2} \sim \mu \frac{\delta}{L} \frac{V_e}{\delta^2} = \frac{\mu V_e}{L^2}$$

As in the analysis that led to equation 7-10, no estimates are given for the derivatives of p; rather, the fact that the equations must be balanced will be used to size those derivatives.

The estimates made above show that, in equations 7-1 and 7-2 both,

1. The two inertia terms are of the same order of magnitude within the boundary layer.
2. The viscous term involving x derivatives is much smaller than the one involving y derivatives, assuming $\delta \ll L$.
3. The larger of the viscous terms is comparable in importance with the inertia terms within the boundary layer if

$$\rho \frac{V_e^2}{L} \sim \frac{\mu}{\delta^2} V_e$$

This gives us an estimate for the thickness of the boundary layer required to achieve this balance,

$$\frac{\delta}{L} \sim \sqrt{\frac{\mu}{\rho V_e L}} = \frac{1}{\sqrt{\text{Re}}} \tag{7-15}$$

which is indeed small.

Further, assuming that the pressure gradient terms are as large as the inertia terms, we see

$$\frac{\partial p}{\partial x} \sim \rho \frac{V_e^2}{L}$$
$$\frac{\partial p}{\partial y} \sim \frac{\delta}{L} \rho \frac{V_e^2}{L} \tag{7-16}$$

The second of these estimates shows that the pressure is essentially constant through the boundary layer; its change in the y direction is

$$\sim \delta \frac{\partial p}{\partial y} \sim \left(\frac{\delta}{L}\right)^2 \rho V_e^2 \tag{7-17}$$

Thus, the flow within the boundary layer is governed by equation 7-3 and

$$\rho\left(u\frac{\partial u}{\partial x}+v\frac{\partial u}{\partial y}\right)=-\frac{\partial p}{\partial x}+\mu\frac{\partial^2 u}{\partial y^2} \tag{7-18}$$

$$0=-\frac{\partial p}{\partial y} \tag{7-19}$$

where x is measured along the body surface and y is normal to it.

Equations 7-14 and 7-15 show that within the boundary layer the velocity is nearly parallel to the surface. Hence the name *laminar* given to the boundary layer in which equations 7-18 and 7-19 are (to a very good approximation) valid; the fluid flows in layers ("lamina") past one another at speeds varying from zero at the surface to V_e far away.

We have implicitly assumed in our analysis that the body surface is essentially flat on the scale of the boundary layer; more precisely, that its radius of curvature, R, is large compared to the boundary layer thickness. Near a blunt leading edge, R is relatively small (recall from Chapter 3 that it was of the order of τ^2 for an ellipse of thickness ratio τ) and, as you will see shortly, δ is not zero. Therefore, we must be prepared to correct our analysis for the curvature of the surface. Cebeci and Bradshaw [1] show that the main change is to admit a pressure gradient in the normal direction, due to centrifugal effects:

$$\frac{\partial p}{\partial y}=\rho\frac{u^2}{R}$$

7.2. USE OF THE BOUNDARY LAYER EQUATIONS

The estimates that led to equations 7-18 and 7-19 as appropriate large-Reynolds-number approximations to the Navier–Stokes equations 7-1 and 7-2 make inviscid analyses practical. The flow field has two scales; L, the characteristic length for flow variations along the body surface, and δ, that for variations in the direction normal to the surface. If the Reynolds number is large, $\delta \ll L$, and the boundary-layer equations 7-18 and 7-19 apply within a distance of order δ near the body surface. If the distance from the body surface is $\gg \delta$, the length scale for variations in the direction away from the body is no longer δ, but L, and the Navier–Stokes equations may properly be approximated by the inviscid Euler equations.

Because the boundary layer is so thin, and the velocity component normal to the surface is so small within the boundary layer, we can calculate a first approximation to the velocity and pressure fields as if the flow were inviscid, using a flow-tangency condition at the body surface and the usual boundary condition at infinity. Then we expand the scale in the region close to the body surface to study the transition from flow tangency outside the boundary layer to the no-slip condition

$$u=v=0 \qquad \text{at } y=0 \tag{7-20}$$

The quantity V_e was introduced as the flow speed "outside" the boundary layer, so

$$u \to V_e \qquad \text{as } y/\delta \to \infty \tag{7-21}$$

But, as we go far from the surface on the scale of the boundary layer, we are still close to the wall on the scale of the inviscid flow. Thus, if the boundary-layer velocity field is to merge smoothly into the inviscid flow field, V_e is also the tangential velocity component on the body surface predicted by the inviscid analysis. That is, V_e is both

$$V_e = \lim_{y/\delta \to \infty} u$$

and

$$V_e = \lim_{y/L \to 0} u$$

Example 1 If L is 1 m, and Re $= 10^6$, then δ is about 1 mm. If $y = 10$ mm, $y/L \approx 0$, but $y/\delta = 10$ and little change takes place on the scale of the boundary layer as y/δ increases beyond 10.

Since, therefore, V_e is known from an inviscid-flow analysis at the time one starts to analyze the boundary layer, equations 7-20 and 7-21 are boundary conditions on the solution of the boundary-layer equations 7-3 and 7-18. The details of that solution will be discussed later, but two points are worth mentioning now:

1. As is suggested by the presence of the x derivative of u in equations 7-3 and 7-18, you need data on the boundary-layer "velocity profile" $u(x, y)$ at some station $x = x_0$, say. That makes it possible to calculate the velocity in the boundary layer downstream of x_0. The Navier–Stokes equations 7-1 and 7-2, in contrast, contain double x derivatives of both u and v. Thus, for their solution, you need to specify u and v both upstream and downstream of the region of interest. So doing is one of the things that makes solving the Navier–Stokes equations so much more difficult than dealing with the boundary-layer equations.
2. The pressure gradient term in equation 7-18 can be taken from the inviscid solution; pressure is not an unknown so far as the boundary layer is concerned. From equation 7-17 or 7-19, the pressure does not change through the boundary layer, and so has the value predicted by the inviscid analysis just outside the boundary layer (at the body surface on the scale of the inviscid flow), which is related to V_e by Bernoulli's equation:

$$p + \tfrac{1}{2}\rho V_e^2 = \text{constant} \tag{7-22}$$

7.2.1. Skin Friction

Once we calculate the velocity distribution through the boundary layer, we can determine the distribution of shear stress on the body surface. From equations 6-27 and 6-55,

$$\sigma_{xy} = \mu\left(\frac{\partial u}{\partial y} + \frac{\partial v}{\partial x}\right)$$

which we will call τ within the boundary layer, where $\partial v/\partial x \ll \partial u/\partial y$, so that it simplifies to

$$\tau = \mu \frac{\partial u}{\partial y} \tag{7-23}$$

We will call the value of the shear stress at the body surface ("wall") τ_w,

$$\tau_w = \mu \left.\frac{\partial u}{\partial y}\right|_{y=0} \tag{7-24}$$

The dimensionless form of the wall shear stress is the *skin-friction coefficient* c_f:

$$c_f \equiv \frac{\tau_w}{\frac{1}{2}\rho V_e^2} \tag{7-25}$$

By integrating the shear stress over the body surface, we obtain the *skin-friction drag* on the body, and so get our first major correction to the Kutta–Joukowski formula.

7.2.2. Displacement Thickness

After calculating the boundary-layer behavior on the basis of the inviscid pressure distribution, we can correct that pressure distribution for the presence of the boundary layer, as follows. From the solution of the boundary-layer equations, compute the velocity component normal to the wall $v(x, y)$ at a value of y large compared to δ but small compared to L. Then use this result as a boundary condition on an inviscid analysis, that is, set

$$V_n\big|_{y/L=0} = \lim_{y/\delta \to \infty} v(x, y) \tag{7-26}$$

Again, note that we are saying y is both large and small; the justification is, as before, that the scales on which y is measured within the boundary layer and outside differ so widely.

To calculate the right side of equation 7-26 from the boundary-layer solution, use the continuity equation 7-3 to get

$$\begin{aligned} v(x, y) &= \int_0^y \frac{\partial v}{\partial y}(x, y^*)\, dy^* \\ &= -\int_0^y \frac{\partial u}{\partial x}(x, y^*)\, dy^* \\ &= -\frac{\partial}{\partial x}\int_0^y u\, dy^* \end{aligned}$$

Because *I* know where we're headed, I'll write this as

$$v(x, y) = \frac{\partial}{\partial x}\int_0^y [V_e(x) - u(x, y^*)]\, dy^* - y\frac{dV_e}{dx}$$

Now, if y^* is large compared to δ, $u(x, y^*) \approx V_e(x)$ (recall equation 7-21), and so the

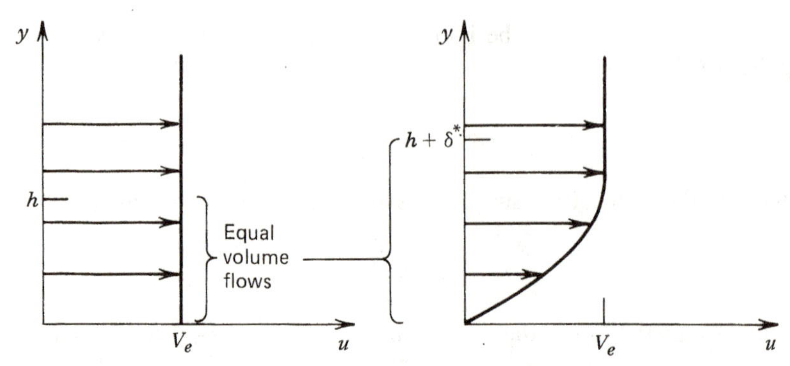

FIG. 7.3. Definition of the displacement thickness of the boundary layer.

integral is changed very little if we let its upper limit go to ∞. The integral is then independent of y, and we can write

$$v(x, y) = \frac{d}{dx} V_e \delta^* - y \frac{dV_e}{dx} \tag{7-27}$$

where

$$\delta^*(x) \equiv \int_0^\infty \left[1 - \frac{u(x, y)}{V_e(x)}\right] dy \tag{7-28}$$

is the so-called *displacement thickness* of the boundary layer. The source of this name is the fact that it is the distance the external flow streamlines are displaced by the boundary layer. That is, as shown in Fig. 7.3, were it not for the boundary layer, the volume rate of flow that would fit between the wall and some distance h small compared to L is $V_e h$. Because of the boundary layer, if h is large compared to δ, the flow actually occupies the distance $h + \delta^*$:

$$V_e h = \int_0^{h+\delta^*} u \, dy$$

Finally, by evaluating equation 7-27 at $y = 0$ (on the scale of the inviscid flow), we get the result required in equation 7-26, which then becomes

$$V_n \big|_{y/L=0} = \frac{d}{dx}(V_e \delta^*) \tag{7-29}$$

To recapitulate, the flow about a body is first analyzed as if inviscid, with $V_n = 0$ at solid surfaces. The resulting surface tangential velocity distribution is then taken to be the velocity $V_e(x)$ outside the boundary layer, which is analyzed with no-slip boundary conditions, and for which the displacement thickness distribution $\delta^*(x)$ is calculated. The inviscid flow is then reanalyzed using the boundary condition of equation 7-29. The result is a pressure distribution that accounts for the presence of the boundary layer. Figure 7.4, taken from Cebeci and Bradshaw [1], shows the improvement in predicted C_p distribution (and c_l) that results. I hasten to add that the inviscid results are not usually so bad as for the airfoil shown in Fig. 7.4.

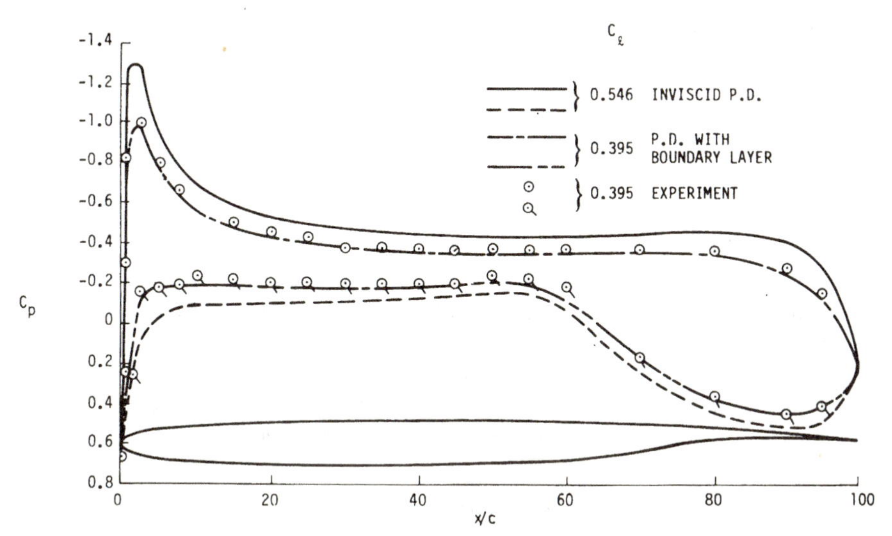

FIG. 7.4. Effect of boundary-layer displacement on the pressure distribution and lift of a modern airfoil [1].

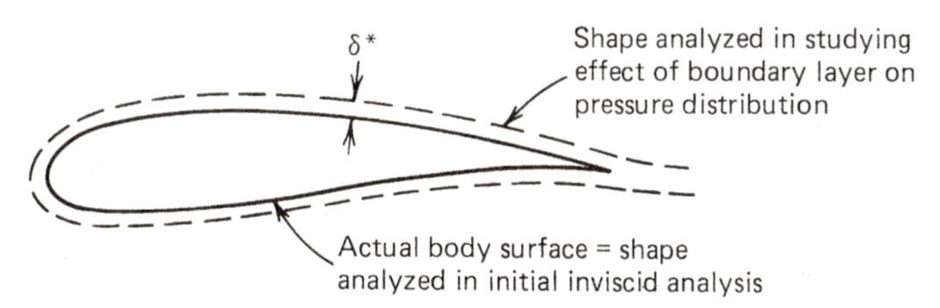

FIG. 7.5. Analysis of displacement effect by altering the body surface shape.

The analysis of the effect of the boundary layer on the pressure distribution is sometimes described as calculating the inviscid flow past a body whose shape differs from the physical one by the addition of the displacement thickness to its contour, as illustrated in Fig. 7.5. This is equivalent to the present description if $\delta^* \ll L$ but may be harder to carry out. In particular, if a panel method like the Smith–Hess method described in Chapter 4 is used for the original inviscid calculation, the coefficient matrix in the system of linear equations for the source and vortex strengths is exactly the same in both inviscid calculations if equation 7-29 is used; V_n comes in only on the known side. Although the trailing edge is effectively thickened by the displacement effect of the boundary layer, the condition used to control circulation is also the same, as was discussed in Chapter 4.[1] Thus, if one stores the matrix (or its inverse), the computation of the corrected source and vortex strengths is very cheap. The most time-consuming part of PANEL is the construction of the coefficient matrix.

[1] Strictly speaking, you ought to calculate the wake behind the body and its displacement effect, but that is often unnecessary.

7.2.3. Momentum Thickness

Another useful output of a boundary-layer analysis is the *momentum thickness*

$$\theta \equiv \int_0^\infty \frac{u}{V_e}\left(1 - \frac{u}{V_e}\right) dy \qquad (7\text{-}30)$$

This plays a role in some empirical formulas for drag estimation [2]. Its connection with drag can be seen from the following example.

Example 2 Compute the drag force of a flat plate aligned with an onset flow, as shown in Fig. 7.6.

We will start by applying Newton's law to the control volume pictured in Fig. 7.6, whose height h is supposed to be big enough that shear stresses are negligible on its upper surface. Then the x component of Newton's law, equation 6-1, yields

$$\begin{aligned} D' &= -\oint_{\substack{\text{control}\\\text{volume}}} \rho \mathbf{V} \cdot \hat{\mathbf{n}} u \, dS \\ &= \int_0^h \rho u^2 \Big|_{x=0} dy - \int_0^h \rho u^2 \Big|_{x=L} dy - \int_0^L \rho v u \Big|_{y=h} dx \end{aligned} \qquad (7\text{-}31)$$

Here D' is the drag force per unit width of the plate (for one side thereof), whose reaction on the fluid is in the negative x direction.

At $x = 0$, $u = V_e = V_\infty$ over the entire upstream face of the control surface. At $y = h$, $u \approx V_e$, and v is small but not exactly zero. There must be mass flow through the upper surface of the control volume in order to satisfy mass conservation. Specifically,

$$\begin{aligned} \dot{m} &\equiv \int_0^L \rho v \Big|_{y=h} dx \\ &= \int_0^h \rho u \Big|_{x=0} dy - \int_0^h \rho u \Big|_{x=L} dy \\ &= \rho V_e h - \int_0^h \rho u \Big|_{x=L} dy \end{aligned}$$

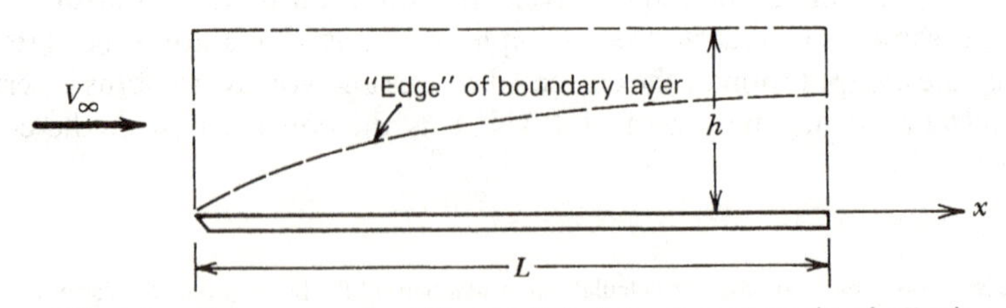

FIG. 7.6. Drag per unit width of a flat plate at zero angle of attack.

Then equation 7-31 yields

$$D' = \int_0^h \left(\rho V_e^2 - \rho u^2 \Big|_{x=L} \right) dy - V_e \left(\rho V_e h - \int_0^h \rho u \Big|_{x=L} dy \right)$$

$$= \rho V_e^2 \int_0^h \frac{u}{V_e}\left(1 - \frac{u}{V_e}\right)\Big|_{x=L} dy$$

Since the integrand vanishes for y large enough that $u \approx V_e$, we can now let $h \to \infty$ and so obtain, using the definition (7-30),

$$D' = \rho V_e^2 \,\theta \big|_{x=L} \tag{7-32}$$

The drag of a flat plate is directly proportional to the momentum thickness at its trailing edge.

7.3. THE MOMENTUM INTEGRAL EQUATION

The major outputs of a boundary-layer analysis—the distributions of δ^*, θ, and c_f—are connected by the *von Kármán momentum integral equation*, which is derived as follows. Adding u times the continuity equation 7-3 to the x-momentum equation 7-18 yields

$$\rho\left(u\frac{\partial u}{\partial x} + u\frac{\partial u}{\partial x} + v\frac{\partial u}{\partial y} + u\frac{\partial v}{\partial y}\right) = -\frac{dp}{dx} + \frac{\partial \tau}{\partial y} \tag{7-33}$$

in which we have taken advantage of equation 7-19 to write the x derivative of p as an ordinary derivative and introduced the shear stress τ from equation 7-23.

Using the formula for the derivative of a product, and eliminating the pressure with the Bernoulli equation 7-22, we can write equation 7-33 as

$$\rho\left(\frac{\partial u^2}{\partial x} + \frac{\partial uv}{\partial y}\right) = \rho V_e \frac{dV_e}{dx} + \frac{\partial \tau}{\partial y}$$

Integrate this equation from $y = 0$ to $y = h$, an as yet unspecified constant:

$$\rho \int_0^h \frac{\partial u^2}{\partial x} dy + \rho uv \Big|_0^h = \rho V_e \frac{dV_e}{dx} h + \tau \Big|_0^h$$

The change of v from $y = 0$ to $y = h$ can be obtained by integrating the continuity equation 7-3 from 0 to h:

$$v \Big|_0^h = \int_0^h \frac{\partial v}{\partial y} dy = -\int_0^h \frac{\partial u}{\partial x} dy$$

From the no-slip condition (7-4), u and v both vanish at $y = 0$, where $\tau = \tau_w$, the wall shear stress. If we suppose $y = h$ to be outside the boundary layer, $u \simeq V_e$ and $\tau \approx 0$ at $y = h$. Then the preceding two equations can be combined into

$$\rho \int_0^h \frac{\partial u^2}{\partial x} dy - \rho V_e \int_0^h \frac{\partial u}{\partial x} dy = \rho V_e \frac{dV_e}{dx} h - \tau_w$$

Interchanging the order of integration and differentiation, and adding and subtracting certain appropriate terms to the integrands, we may write this as

$$\rho \frac{d}{dx}\left[\int_0^h (u^2 - V_e^2)\, dy + V_e^2 h\right] - \rho V_e \frac{d}{dx}\left[\int_0^h (u - V_e)\, dy + V_e h\right]$$
$$= \rho V_e \frac{dV_e}{dx} h - \tau_w$$

The three terms proportional to h cancel. Since the integrands now vanish for $y > h$, we can let $h \to \infty$ in the limits of the integrals and so get

$$\tau_w = \rho \frac{d}{dx} V_e^2 \int_0^\infty \left(1 - \frac{u^2}{V_e^2}\right) dy - \rho V_e \frac{d}{dx} V_e \int_0^\infty \left(1 - \frac{u}{V_e}\right) dy$$
$$= \rho \frac{d}{dx} [V_e^2(\theta + \delta^*)] - \rho V_e \frac{d}{dx} (V_e \delta^*)$$

in which we have introduced the displacement and momentum thicknesses from the definitions (7-28) and (7-30). On expanding the derivatives and dividing through by ρV_e^2, we finally obtain the *momentum integral equation* (first derived and used by von Kármán in 1921 [3]),

$$\frac{d\theta}{dx} + \frac{\theta}{V_e}(2 + H)\frac{dV_e}{dx} = \tfrac{1}{2} c_f \tag{7-34}$$

in which

$$H \equiv \delta^*/\theta \tag{7-35}$$

is called the *shape factor* of the boundary-layer velocity profile, and c_f, the skin-friction coefficient, was defined back in equation 7-25.

Equation 7-34 contains far too many unknowns, θ, H, and c_f,[2] to be useful by itself, but it does have the virtue of focusing on the major outputs of interest. To yield those outputs, it must be supplemented with other equations. A variety of methods for so doing exist. Many involve assumptions as to the functional form of the "velocity profile" $u(x, y)$ and/or data fitting.

7.4. VELOCITY PROFILE FITTING: LAMINAR BOUNDARY LAYERS

The first applications of the momentum integral equation to the solution of the boundary layer were by von Kármán [3] and Pohlhausen [4]. They assumed that the velocity profile $u(x, y)$ can be described by a simple polynomial in y out to some distance $\delta(x)$ (the "boundary-layer thickness," in effect), beyond which $u = V_e(x)$. The x-dependent coefficients of the polynomial were determined so as to satisfy the momentum integral equation, the wall shear-stress formula (7-24), and certain continuity requirements.

[2] $V_e(x)$, it should be recalled, is known from a potential-flow calculation.

To illustrate the method, which generally requires the solution of an ordinary differential equation for $\delta(x)$, we will consider the special case of flow along a flat plate at zero angle of attack, for which V_e = constant. While Pohlhausen used a quartic velocity profile, we shall be content with the quadratic formula

$$\frac{u(x, y)}{V_e(x)} = A + B\frac{y}{\delta} + C\left(\frac{y}{\delta}\right)^2 \qquad \text{for } y < \delta(x)$$
$$= 1 \qquad \text{for } y > \delta(x)$$

From the no-slip condition (7-4), $A = 0$. To avoid discontinuities in either u or $\partial u/\partial y$ at $y = \delta$, we require

$$\left.\frac{u}{V_e}\right|_{y=\delta+} = 1 = B + C = \left.\frac{u}{V_e}\right|_{y=\delta-}$$

$$\left.\frac{1}{V_e}\frac{\partial u}{\partial y}\right|_{y=\delta+} = 0 = \frac{B}{\delta} + 2\frac{C}{\delta} = \left.\frac{1}{V_e}\frac{\partial u}{\partial y}\right|_{y=\delta-}$$

and so end up with the single-parameter velocity profile

$$\frac{u}{V_e} = 2\frac{y}{\delta} - \left(\frac{y}{\delta}\right)^2 \qquad \text{for } y < \delta(x)$$
$$= 1 \qquad \text{for } y > \delta(x) \tag{7-36}$$

To determine the parameter $\delta(x)$, we will use the momentum integral equation 7-34. Substituting equation 7-36 into the definitions (7-24), (7-25), and (7-30) yields

$$c_f = 4\frac{\nu}{V_e\delta}$$

$$\theta = \frac{2}{15}\delta$$

and substituting these results into (7-34) gives

$$\frac{d\delta^2}{dx} = 30\frac{\nu}{V_e}$$

whose solution is

$$\delta^2 = 30\frac{\nu x}{V_e}$$

Then

$$\theta = \frac{0.73x}{\sqrt{\text{Re}_x}}$$

$$c_f = \frac{0.73}{\sqrt{\text{Re}_x}} \tag{7-37}$$

where

$$\text{Re}_x \equiv \rho V_e x/\mu$$

These results are in excellent agreement with those obtained from the exact solution of the laminar boundary-layer growth along a flat plate (see Chapter 10), according to which their only error is that the constant 0.73 should be replaced by 0.664. Pohlhausen's quartic, which allows inflection points in the velocity profile, works quite well for laminar boundary layers in negative and not-too-positive pressure gradients.

7.5. THWAITES'S METHOD FOR LAMINAR BOUNDARY LAYERS

An alternative method, which also relies on the momentum integral equation but makes no specific assumption on the form of the velocity profile, was devised by Thwaites [5]. Since his method is somewhat more accurate than Pohlhausen's in positive pressure gradients, we will look at its utilization in some detail.

Our basic aim is to supplement the momentum integral equation 7-34 with algebraic relations among the unknowns, θ, H, and c_f. Such relations will certainly be simpler if in terms of dimensionless variables; thus we make θ and x dimensionless by forming Reynolds numbers

$$\text{Re}_\theta \equiv \frac{\rho V_e \theta}{\mu}$$

$$\text{Re}_x \equiv \frac{\rho V_e x}{\mu} \tag{7-38}$$

Although H and c_f are already dimensionless, the skin friction coefficient is a rather strong function of the Reynolds number, as can be seen, for example, from the flat-plate solution discussed in the preceding section, equation 7-37. Thus we introduce the parameter

$$l \equiv \tfrac{1}{2}\,\text{Re}_\theta\, c_f \tag{7-39}$$

which is independent of the Reynolds number, not only for the flat-plate problem but for broad classes of exact solutions of the boundary-layer equations; see Section 10.2 for a discussion of the class of solutions called "similar."

Thus, we multiply the momentum integral equation 7-34 through by Re_θ to get

$$\frac{\rho V_e \theta}{\mu}\frac{d\theta}{dx} + \frac{\rho\theta^2}{\mu}(2 + H)\frac{dV_e}{dx} = l$$

Thwaites defined a dimensionless pressure-gradient parameter

$$\lambda \equiv \frac{\rho\theta^2}{\mu}\frac{dV_e}{dx} \tag{7-40}$$

and rewrote the above equation as

$$\frac{\rho V_e}{\mu}\frac{d\theta^2}{dx} = 2[l - (2 + H)\lambda] \tag{7-41}$$

If we use Pohlhausen's quartic velocity profile, l and H can be shown to be functions only of λ. Thwaites found that, to an excellent approximation, the same is true of the available, exact solutions of the laminar boundary layer equations. That is, when the values of l and H for such solutions are plotted against λ, the data cluster around a single curve with very little scatter. In fact, the right side of equation 7-41 is very well approximated by the linear formula

$$2[l - (2 + H)\lambda] \approx 0.45 - 6\lambda$$

When this and the definition (7-40) of λ are substituted into equation 7-41, we obtain

$$\frac{\rho V_e}{\mu}\frac{d\theta^2}{dx} = 0.45 - \frac{6\rho\theta^2}{\mu}\frac{dV_e}{dx}$$

Bringing the dV_e/dx term over to the left side and multiplying through by V_e^5, we get

$$\frac{\rho}{\mu}\left(V_e^6\frac{d\theta^2}{dx} + \theta^2 6V_e^5\frac{dV_e}{dx}\right)$$
$$= \frac{\rho}{\mu}\frac{d}{dx}(\theta^2 V_e^6) = 0.45\, V_e^5 \tag{7-42}$$

Thus, for any given $V_e(x)$ and initial value of θ, $\theta(x)$ can be determined by integration of a single first-order ordinary differential equation, a really simple task even for a programmable calculator. Once θ is known, λ can be calculated from equation 7-40, and $l(\lambda)$ and $H(\lambda)$ from Thwaites's graphs or, better yet, from correlation formulas given by Cebeci and Bradshaw [1]:

$$l(\lambda) = 0.22 + 1.57\lambda - 1.8\lambda^2 \qquad \text{for } 0 < \lambda < 0.1$$
$$= 0.22 + 1.402\lambda + \frac{0.018\lambda}{\lambda + 0.107} \qquad \text{for } -0.1 < \lambda < 0$$
$$H(\lambda) = 2.61 - 3.75\lambda + 5.24\lambda^2 \qquad \text{for } 0 < \lambda < 0.1$$
$$= 2.088 + \frac{0.0731}{\lambda + 0.14} \qquad \text{for } -0.1 < \lambda < 0 \tag{7-43}$$

Since Thwaites's method requires solution of the first-order ordinary differential equation 7-42, it needs one initial condition to get it started. Two cases occur fairly frequently. Either the boundary layer starts with zero thickness, as at the leading edge of a flat plate, or, as when considering flow past an airfoil, the boundary layer starts at a stagnation point; see Fig. 7.7. In both cases, Thwaites's equation 7-42 can be integrated analytically.

Example 3: Boundary Layer Growth along a Flat Plate Suppose, as in the preceding section, that V_e is constant. Then equation 7-42 reduces to

$$\frac{\rho}{\mu}V_e\frac{d\theta^2}{dx} = 0.45$$

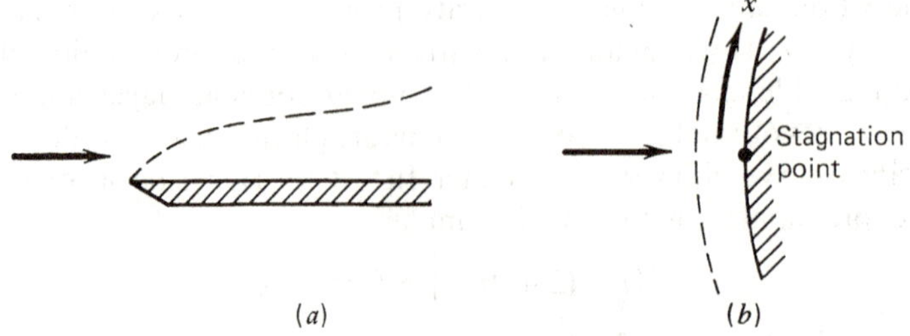

FIG. 7.7. Common starting conditions for analysis of boundary layer growth. (*a*) Flat plate. (*b*) Stagnation point.

which integrates to

$$\theta^2 = 0.45\frac{x\mu}{\rho V_e} + \text{constant}$$

Evaluating the constant of integration at $x = 0$, where the boundary-layer thickness is also zero, we get

$$\frac{\theta}{x} = \frac{0.671}{\sqrt{\text{Re}_x}}$$

which differs from the exact result mentioned above by only 1%.

Example 4: Boundary Layer at a Stagnation Point The velocity of a nonviscous flow at a stagnation point is generally analytic[3] and so can be expanded in a power series at that point. If x is measured along the surface from the stagnation point, we thus have, near the origin,

$$\begin{aligned} V_e(x) &= \frac{dV_e}{dx}(0)x + \frac{1}{2}\frac{d^2V_e}{dx^2}(0)x^2 + \cdots \\ &\approx \dot{V}_0 x \end{aligned}$$

in which $\dot{V}_0$ is the velocity gradient at the stagnation point. Substituting this into equation 7-42, we get

$$\frac{\rho}{\mu}\frac{d}{dx}(\dot{V}_0^6 x^6 \theta^2) = 0.45\dot{V}_0^5 x^5$$

which is easily integrated:

$$\frac{\rho}{\mu}\dot{V}_0^6 x^6 \theta^2 = 0.075\dot{V}_0^5 x^6 + \text{constant}$$

Assuming that the value of θ at $x = 0$ is finite, we see, after dividing through by x^6

[3] Except when the stagnation point is in a concave corner or at the apex of a wedge; see Appendix C, in which it is shown that the velocity near such a point follows a power law, $V_e \propto x^m$, with $0 < m < 1$.

and letting $x \to 0$, that the integration constant must be zero and so obtain

$$\theta(0) = \sqrt{\frac{0.075\,\mu}{\rho \dot{V}_0}} \tag{7-44}$$

Note that the boundary layer has a nonzero thickness from its start at a stagnation point. Is that what you expected?

7.6. FORM DRAG

In problem 4, you will use Thwaites's method to show that if $V_e \propto x^m$, the displacement thickness δ^* of a laminar boundary layer grows with x according to

$$\delta^* \propto \frac{x}{\sqrt{\dfrac{\rho V_e x}{\mu}}}$$

$$\propto \left(\frac{x}{V_e}\right)^{1/2}$$

$$\propto x^{(1-m)/2}$$

Thus, if V_e increases with $x(m > 0)$, the displacement thickness grows more slowly than if it decreases ($m < 0$). Now, near the trailing edge of an airfoil, the inviscid-flow streamlines diverge, so that V_e decreases (to zero if the trailing edge is sharp but not cusped, see Chapter 4), which leads to a relatively rapid growth of δ^*. This prevents the inviscid-flow streamlines from expanding to follow the airfoil contour exactly, so that V_e is higher than it would be were the flow perfectly inviscid. From the Bernouli equation, the pressure is therefore lower than its inviscid value near the trailing edge, as shown in Fig. 7.8.

The failure of the pressure to recover its inviscid stagnation value at the trailing edge leads to an imbalance in the pressure distribution on the body, which, from Fig. 7.8, clearly adds to the component in the direction of $\mathbf{V}_\infty$ of the force exerted on

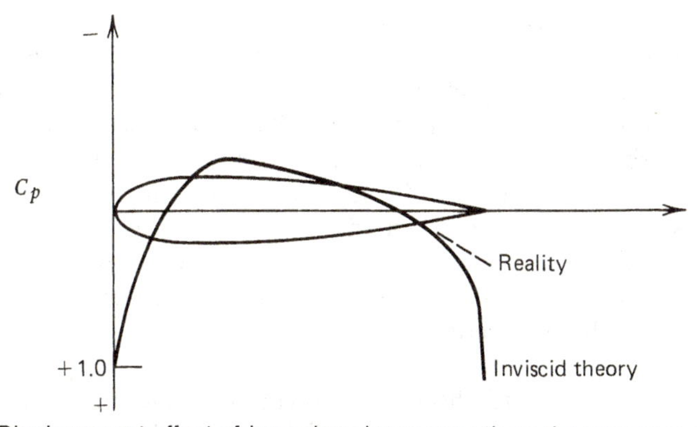

FIG. 7.8. Displacement effect of boundary-layer growth on the pressure distribution.

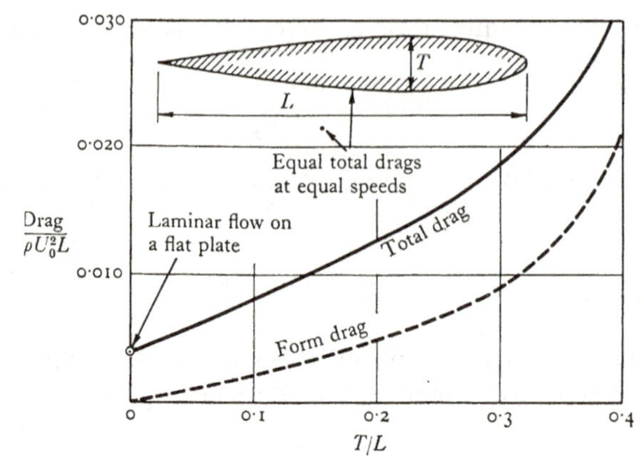

FIG. 7.9. Effect of thickness on form and total drag [6].

the airfoil; that is, to the drag component. Drag due to the imbalance of pressure induced by boundary-layer behavior is called *form drag* (or *profile drag*). It is distinct from the *skin-friction drag* mentioned earlier, which is the resultant of the shear forces exerted by the fluid on the body. Skin-friction drag depends only weakly on the shape of the body; it is mainly a function of the surface area and the Reynolds number. Form drag, however, is a strong function of the pressure gradient on the body, which, in turn, can be significantly altered by varying the body shape. In particular, it generally increases with the thickness of the body, as shown in Fig. 7.9 [6].

7.7. TURBULENT FLOWS

We are mainly interested in flows that are steady, in the sense that the boundary conditions do not change with time. For example, an airfoil of fixed geometry may be considered to be immersed in a flow that is uniform in space and constant in time far upstream. Nevertheless, it is not always permissible, even in such cases, simply to delete the time derivatives in equations 7-1 and 7-2. The flow may be *turbulent* and experience an unsteadiness that has nothing to do with the boundary conditions. The source of the unsteadiness is an *unstable response to the small disturbances* that can never be completely eliminated in practical situations. Flows that are stable to small disturbances are called *laminar*. Since the viscosity of a flow helps to damp out disturbances, and since the relative importance of the viscous terms is inversely proportional to the Reynolds number, the stability of a flow depends on the Reynolds number. If the Reynolds number is large enough, any small disturbance of the flow can be wildly amplified. The flow is unstable and exhibits a time dependence that is not strictly random—it is still governed by the Navier–Stokes equations—but is effectively random because it is so sensitive to the initial disturbance.

In such cases, there is little hope of following the fluid motion in detail. It is necessary and, for practical purposes, sufficient to seek some sort of average description of the flow. The situation is analogous to that which leads us to use the continuum viewpoint rather than regard a fluid as a collection of interacting molecules (which it is). The analogy goes deeper. To some extent a turbulent flow looks like the flow of a collection of fluid parcels of varying size (and, unfortunately, composition). That is, there is high correlation between the fluid velocities at points close together—closer than the "size" of the parcel—but essentially none between the velocities at points further apart. These parcels are called *eddies.*

For "steady" turbulent flows (again, ones in which the boundary conditions are independent of time), we can average the flow properties over time. For example, we define the time *average* of $u(x, y, t)$ as

$$\bar{u}(x, y) \equiv \lim_{T \to \infty} \frac{1}{T} \int_{t_0}^{t_0 + T} u(x, y, t)\, dt \tag{7-45}$$

and label $u - \bar{u}$, the *fluctuating part of* u, with the symbol u', so

$$u(x, y, t) = \bar{u}(x, y) + u'(x, y, t) \tag{7-46}$$

In an unsteady turbulent flow—one in which the boundary conditions change with time in a prescribed manner—we can average out the turbulence by using an *ensemble average.* Imagine the process to be executed a large number of times. The ensemble average of u, $\bar{u}(x, y, t)$, is the average value of u at the position (x, y) and at the time t after the start of the process, the average being taken over the different executions of the process.

Now the Navier–Stokes equations and the continuity equation are valid at every instant of every execution of a flow process. By averaging every term of equations 7-1 to 7-3, we can try to set up equations among the time-averaged flow properties. This process is simplified by using the so-called *Reynolds rules of averaging,* which are easily deduced from equations 7-45 and 7-46. If $a = a(x, y, t)$ and $b = b(x, y, t)$,

$$\overline{a + b} = \bar{a} + \bar{b}$$

$$\overline{a'} = 0$$

$$\overline{\frac{\partial a}{\partial x}} = \frac{\partial \bar{a}}{\partial x}, \qquad \overline{\frac{\partial a}{\partial y}} = \frac{\partial \bar{a}}{\partial y}$$

$$\overline{\bar{a} b} = \bar{a} \bar{b}$$

Also, for a steady flow,

$$\begin{aligned} \overline{\frac{\partial a}{\partial t}} &\equiv \lim_{T \to \infty} \frac{1}{T} \int_{t_0}^{t_0 + T} \frac{\partial a}{\partial t}\, dt \\ &= \lim_{T \to \infty} \frac{1}{T} [a(x, y, t_0 + T) - a(x, y, t_0)] \\ &= 0 \end{aligned}$$

but

$$\overline{ab} = \overline{(\bar{a} + a')(\bar{b} + b')}$$
$$= \overline{\bar{a}\bar{b}} + \overline{a'\bar{b}} + \overline{\bar{a}b'} + \overline{a'b'}$$
$$= \bar{a}\,\bar{b} + \overline{a'b'}$$

Note that, even though the averages of a' and b' are zero, the average of their product $a'b'$ need not be; see problem 6.

Averaging the continuity equation 7-3 thus yields an equation of the same form, but in terms of average velocity components:

$$\frac{\partial \bar{u}}{\partial x} + \frac{\partial \bar{v}}{\partial y} = 0 \tag{7-47}$$

Before averaging the Navier–Stokes equation 7-1, it is useful to add ρu div $\mathbf{V}$ to the left side (since div $\mathbf{V} = 0$, this does not change its validity), which yields, with the formula for the derivative of a product,

$$\rho\left(\frac{\partial u}{\partial t} + \frac{\partial u^2}{\partial x} + \frac{\partial uv}{\partial y}\right) = -\frac{\partial p}{\partial x} + \mu\left(\frac{\partial^2 u}{\partial x^2} + \frac{\partial^2 u}{\partial y^2}\right)$$

Averaging this equation according to the rules given above, we get

$$\rho\left(\frac{\partial \bar{u}^2}{\partial x} + \frac{\partial \overline{u'^2}}{\partial x} + \frac{\partial \bar{u}\,\bar{v}}{\partial y} + \frac{\partial \overline{u'v'}}{\partial y}\right) = -\frac{\partial \bar{p}}{\partial x} + \mu\left(\frac{\partial^2 \bar{u}}{\partial x^2} + \frac{\partial^2 \bar{u}}{\partial y^2}\right)$$

Now subtract $\bar{u}$ times the averaged continuity equation 7-47 and transpose the terms involving products of fluctuating components:

$$\rho\left(\bar{u}\frac{\partial \bar{u}}{\partial x} + \bar{v}\frac{\partial \bar{u}}{\partial y}\right) = -\frac{\partial \bar{p}}{\partial x} + \frac{\partial}{\partial x}\left(\mu\frac{\partial \bar{u}}{\partial x} - \rho\overline{u'^2}\right) + \frac{\partial}{\partial y}\left(\mu\frac{\partial \bar{u}}{\partial y} - \rho\overline{u'v'}\right) \tag{7-48}$$

Similarly treating equation 7-2 yields

$$\rho\left(\bar{u}\frac{\partial \bar{v}}{\partial x} + \bar{v}\frac{\partial \bar{v}}{\partial y}\right) = -\frac{\partial \bar{p}}{\partial y} + \frac{\partial}{\partial x}\left(\mu\frac{\partial \bar{v}}{\partial x} - \rho\overline{u'v'}\right) + \frac{\partial}{\partial y}\left(\mu\frac{\partial \bar{v}}{\partial y} - \rho\overline{v'^2}\right) \tag{7-49}$$

These are the *time-averaged Navier–Stokes equations* for an incompressible flow.

Our objective in averaging the continuity and Navier–Stokes equations was to obtain equations for the average flow properties $\bar{u}$, $\bar{v}$, and $\bar{p}$. Equations 7-48 and 7-49 show we were not completely successful; they contain derivatives of terms of the form

$$-\rho\overline{V_i'V_j'}$$

which because of their placement in equations 7-48 and 7-49 are called *Reynolds stresses* or *turbulent shear stresses.*[4] That is, the time-averaged continuity and

[4] It may help to recall that, in Chapter 6, the form of the laminar stresses, $\mu\,\partial V_i/\partial x_j$, was derived by applying mean-free path ideas to the average rate of transport of i momentum in the j direction, a quantity that looks very much like $\rho\overline{V_i'V_j'}$.

Navier–Stokes equations constitute only three equations in six unknowns. The problem of *closure*—of supplementing the time-averaged Navier–Stokes equations so as to form a closed set, with enough equations to determine all the unknowns—is not completely solved (as of 1983). A typical approximation is to assume a formula for the Reynolds stresses in terms of the average velocity and its first derivative, with constants adjusted to fit experimental data. Such approaches work very well for flows not too different in character from the ones that constituted the database.

If it assumed that the characteristic length for flow variations in the y direction is small compared with that for variations in the x direction, an analysis of the time-averaged Navier–Stokes equations 7-48 and 7-49 yields results essentially equivalent to those obtained in the laminar case. Since the continuity equation 7-47 satisfied by the mean velocity is of the same form as that which governs a steady laminar flow, equation 7-3, we find, as in equation 7-14, that the y component of the average velocity is small compared to the x component, so that the streamlines of the average flow are nearly parallel to the surface. However, the fluctuating components u' and v' are much closer in magnitude, and the instantaneous streamlines are not nearly so parallel to the surface as when viscous forces are strong enough to damp out disturbances, which is why the term "laminar" is reserved for stable flows.

Also, the Reynolds stresses, $-\rho\overline{u'v'}$, $-\rho\overline{u'^2}$, and $-\rho\overline{v'^2}$ are then of roughly the same size. Thus it is no longer true, as in the laminar case, that the arguments of the x derivatives on the right sides of the time-averaged Navier–Stokes equations are small compared to the arguments of the y derivatives. However, y derivatives still dominate x derivatives, and equation 7-18 is changed only by the inclusion of the Reynolds stress $-\rho\overline{u'v'}$:

$$\rho\bar{u}\frac{\partial\bar{u}}{\partial x}+\rho\bar{v}\frac{\partial\bar{u}}{\partial y}=-\frac{\partial\bar{p}}{\partial x}+\frac{\partial}{\partial y}\left(\mu\frac{\partial\bar{u}}{\partial y}-\rho\overline{u'v'}\right) \tag{7-50}$$

On the other hand, the shear-stress contribution to the y momentum equation 7-19 can no longer be safely neglected; the proper version is

$$\frac{\partial\bar{p}}{\partial y}=\frac{\partial}{\partial y}\left(\mu\frac{\partial\bar{v}}{\partial y}-\rho\overline{v'^2}\right) \tag{7-51}$$

But because the boundary layer is so thin, we can still take the pressure in equation 7-50 from the inviscid pressure distribution, and so still use the Bernoulli equation 7-22 to eliminate the pressure.

Thus, the boundary-layer idea is as valid in turbulent flow as in laminar flow. The main difference is in the order of magnitude of the terms neglected, which is $(\delta/L)^2$ in laminar flows but (δ/L) in turbulent cases.

The wildness of the flow within a turbulent boundary layer—it is called "turbulent" for good reason—leads to rates of momentum transport that are much higher than in the laminar case. Since the shear stress on the fluid is directly proportional to the rate of momentum transport, shear stresses are relatively high in turbulent boundary layers. This is true even at the body surface, where because the

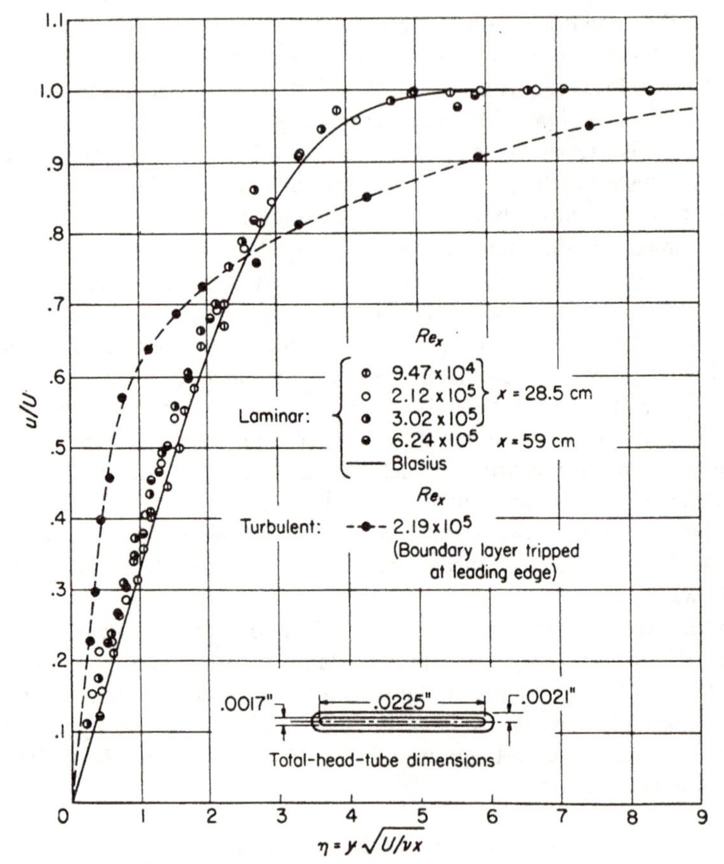

FIG. 7.10. Velocity profiles in laminar and turbulent boundary layers on flat plate [7].

no-slip condition applies to the fluctuating part of the velocity as well as to its mean value, $\rho\overline{u'v'}$ is actually zero. The higher momentum transport makes the (mean) velocity profile much "fuller" in turbulent flow than in the laminar case, as is illustrated in Fig. 7.10, so that $\partial u/\partial y$ is much larger at the surface. Thus skin-friction drag increases markedly when the Reynolds number becomes large enough that the flow becomes turbulent, as is demonstrated in Fig. 7.11 [7], which gives the variation of the wall shear stress on a flat plate at zero angle of attack with the Reynolds number based on distance from the leading edge. Note that above a Reynolds number based on plate length of about 3×10^5 the flow on the upstream end of the plate may be laminar, whereas it is turbulent downstream; see below for further discussion of the process of "transition" from laminar to turbulent flow.

Although the differential equations 7-47 and 7-50 are insufficient to predict the mean velocity profile, even with $\bar{p}$ taken as known from inviscid results, they have the virtue of being identical in form with their laminar counterparts, equations 7-3 and 7-18. Thus, if we add $\rho\bar{u}$ times equation 7-47 to equation 7-50, we get an equation of

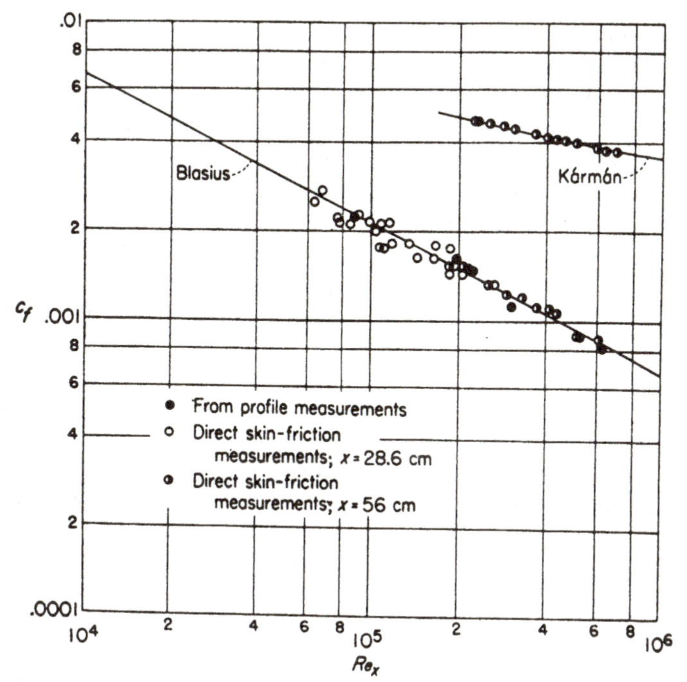

FIG. 7.11. Skin-friction coefficient of laminar and turbulent boundary layers on a flat plate [7].

the form of 7-33, except that u and v are replaced by their mean values, whereas τ is given by

$$\tau = \mu \frac{\partial u}{\partial y} - \rho \overline{u'v'} \tag{7-52}$$

instead of by equation 7-23. On tracing through the steps of the derivation that led from equation 7-33 to the *momentum integral equation* 7-34, we find that they are as valid in turbulent flow as in the laminar case. Of course, just as when the flow is laminar, equation 7-34 is insufficient to determine its unknowns θ, H, and c_f by itself. Let us see how it may be used as part of a solution procedure when the flow is turbulent.

7.8. VELOCITY PROFILE FITTING: TURBULENT BOUNDARY LAYERS

As in the Pohlhausen method for laminar flow, many methods for calculating turbulent boundary layers are based in part on an assumed parametric form for the velocity profile $u(y)$. At first glance, it may seem that the Pohlhausen method itself could be used in turbulent flows. The momentum integral equation 7-34, the no-slip condition (7-4), and the continuity conditions imposed at $y = \delta$ are the same whether the flow is laminar or turbulent. So is the wall shear-stress formula (7-24);

since the no-slip condition applies at all times, $u' = v' = 0$ at the wall, and the turbulence stresses therefore also vanish there.

However, the growth of a turbulent boundary layer is not accurately predicted by the Pohlhausen method. Why? Because of the form assumed for the velocity profile. The turbulent boundary layer, as we shall see, has layers within the layer, each having its own distinguishing characteristics. This obviates a global description of the velocity profile by a simple polynomial.

To see what formula is appropriate, let us look in some detail at the structure of a turbulent boundary layer. There are three more or less distinct regions within the layer: an *outer layer*, which is relatively sensitive to the external flow properties; an *inner layer*, in which turbulent mixing is the dominant influence; and a *laminar sublayer*, closest to the surface, in which, because the no-slip condition forces the fluctuating velocity components to zero at the wall, the turbulent stresses are negligible compared to $\mu\, \partial u/\partial y$. As shown in Fig. 7.12, there is a gradual transition between the laminar sublayer and the inner layer, sometimes called the *buffer layer*. Note the logarithmic scale of Fig. 7.12, which greatly exaggerates the inner layer and the laminar sublayer. The outer layer actually accounts for 80 to 90% of the boundary-layer thickness. The quantity u_τ that appears on the graph is the *friction velocity*

$$u_\tau \equiv \sqrt{\frac{\tau_w}{\rho}} \tag{7-53}$$

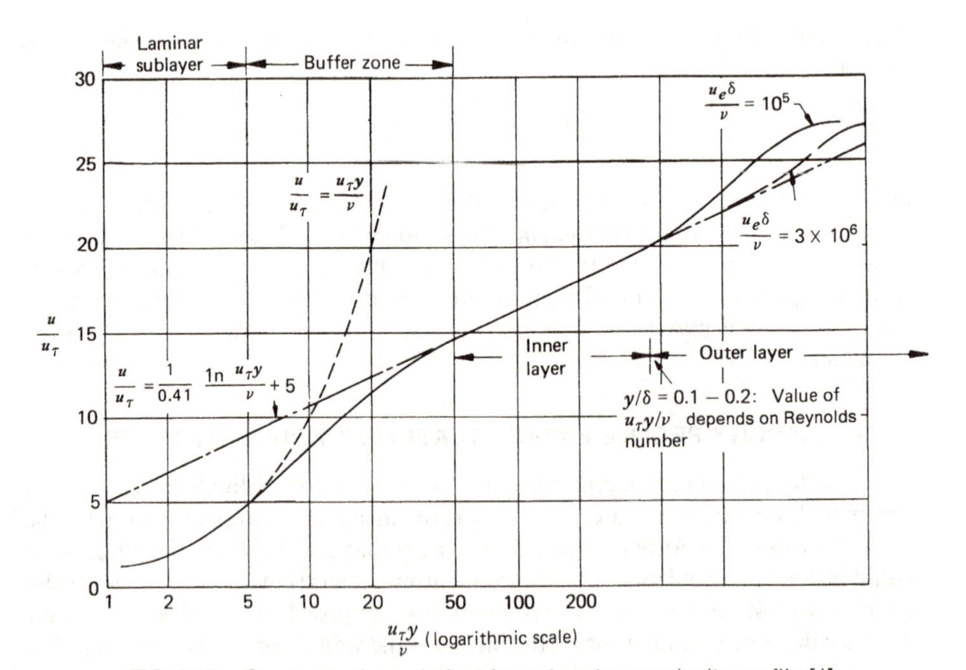

FIG. 7.12. Structure of a turbulent boundary-layer velocity profile [1].

that is, a quantity related to the wall shear stress and having the dimensions of velocity.

In the laminar sublayer, the shear stress is given by equation 7-23. Because the laminar sublayer is so thin, τ will not vary much from τ_w, its value at the wall, and we can integrate this equation to get

$$\tau_w y = \mu u$$

the constant of integration being evaluated at $y = 0$, where $u = 0$ also. Eliminating τ_w by introducing the friction velocity from equation 7-53, we then get

$$u^+ \equiv \frac{u}{u_\tau} \approx \frac{\rho u_\tau y}{\mu} \equiv y^+ \qquad \text{in laminar sublayer} \tag{7-54}$$

Note that y^+ is a Reynolds number based on u_τ and the distance from the wall.

Outside of the sublayer, we must include the turbulent stress term in equation 7-52, and equation 7-54 does not hold. In the inner layer, we assume the flow to be so dominated by turbulent mixing that it is independent of the external flow. The only scale available for the mean velocity in the inner layer is

$$\sqrt{\tau/\rho}$$

which we again take to be nearly constant and so equal to u_τ. The slope of the mean velocity profile is then supposed to satisfy

$$\frac{\partial u}{\partial y} \approx \frac{u_\tau}{l} \tag{7-55}$$

where l, a characteristic length for local flow variations in the y direction, is called the *mixing length*. Since the only length scale available is the distance from the wall y, we take

$$l = \kappa y \tag{7-56}$$

where κ is a dimensionless constant, so that equation 7-55 becomes

$$\frac{\partial u}{\partial y} = \frac{u_\tau}{\kappa y}$$

Integrating, we get

$$\frac{u}{u_\tau} = \frac{1}{\kappa} \ln y + C'$$

By letting

$$C' = \frac{1}{\kappa} \ln \frac{\rho u_\tau}{\mu} + C$$

and introducing the nomenclature of equation 7-54, we obtain the standard form of the famous *law of the wall*, or *log law*:

$$u^+ \equiv \frac{u}{u_\tau} = \frac{1}{\kappa} \ln y^+ + C \qquad \text{in the inner layer} \tag{7-57}$$

Equation 7-57 fits available experimental data remarkably well with $\kappa = 0.4$ and $C = 5.2$, or $\kappa = 0.41$ and $C = 5.0$. Figure 7.12 is typical in showing the fraction of the velocity profile that is well represented by the log law.

In the outer layer, dimensional analysis leads to the *velocity defect law*

$$\frac{V_e - u}{u_\tau} = f\left(\frac{y}{\delta(x)}; x\right)$$

Coles [8] found that the experimental data available (from a variety of sources) could be fit very well by letting f be the product of some function of x and another function of y/δ only. This was his *law of the wake*, which leads to the formula

$$u^+ = \frac{1}{\kappa} \ln y^+ + C + \frac{2}{\kappa} \Pi(x) \sin^2 \frac{\pi y}{2\delta(x)} \qquad \text{for } y \leq \delta(x) \tag{7-58}$$

which is very accurate outside the laminar sublayer—that is, in both the inner and outer layers—for properly chosen Π, δ, and u_τ.

If we ignore the laminar sublayer, we can calculate θ and δ^* in terms of Π, δ, and u_τ by plugging equation 7-58 into the definitions 7-28 and 7-30. Coles [9] gives the formulas

$$\frac{\delta^*}{\delta} = \frac{u_\tau}{\kappa V_e}(1 + \Pi)$$

$$\frac{\theta}{\delta} = \frac{\delta^*}{\delta} - \frac{2}{\kappa}\left(\frac{u_\tau}{V_e}\right)^2 \left\{1 + \Pi\left(1 + \frac{1}{\pi} S_i(\pi)\right) + \frac{3}{2}\Pi^2\right\}$$

where

$$S_i(\pi) = \int_0^\pi \frac{\sin x}{x}\, dx \doteq 1.8516$$

Evaluating equation 7-58 at $y = \delta$, where $u = V_e$, gives a third equation,

$$\frac{V_e}{u_\tau} = \frac{1}{\kappa} \ln \frac{\delta u_\tau}{\nu} + C + \frac{2}{\kappa} \Pi(x) \tag{7-59}$$

From equations 7-25 and 7-53,

$$c_f = 2\left(\frac{u_\tau}{V_e}\right)^2$$

and so we have, counting the momentum integral equation 7-34, five equations in the six unknowns θ, δ^*, c_f, Π, δ, and u_τ.

Honorable people differ as to how one should fill in the missing equation. However, all the better methods complete the set of governing equations with a differential equation. Usually this is derived by multiplying the differential momentum equation 7-33 through by some weighting factor and then integrating it across the boundary layer, as we did in deriving the integral momentum equation 7-34. For example, the integral *moment-of-momentum equation* is derived by multiplying equation 7-33 by y and integrating the result over y from 0 to δ, whereas the *mechanical energy equation*

$$\rho \frac{d}{dx}(V_e^3 \delta_3) = 2D$$

with

$$\delta_3 \equiv \int_0^\infty \frac{u}{V_e}\left(1 - \frac{u^2}{V_e^2}\right) dy, \qquad D \equiv \int_0^\infty \tau \frac{\partial u}{\partial y}\, dy$$

is obtained by using the x-velocity component u as a weighting factor. In 1968, Stanford University hosted what has come to be known as the "turbulent olympics," in which 29 different methods for computing turbulent boundary layers competed on their ability to match certain carefully analyzed experimental data. The better integral methods were based either on the moment-of-momentum equation or on the mechanical energy equation.[5] However, the details of these methods being rather messy, I prefer to discuss a simpler (though sometimes less accurate) method due to M.R. Head.[6]

7.9. HEAD'S METHOD FOR TURBULENT BOUNDARY LAYERS

Head's method is based on the concept of an entrainment velocity. If $\delta(x)$ is the boundary-layer thickness, the volume rate of flow within the boundary layer at x is

$$Q(x) = \int_0^{\delta(x)} u\, dy \tag{7-60}$$

The *entrainment velocity* E is the rate at which Q increases with x,

$$E = \frac{dQ}{dx}$$

Some idea as to the physical significance of E can be gained from Fig. 7.13.

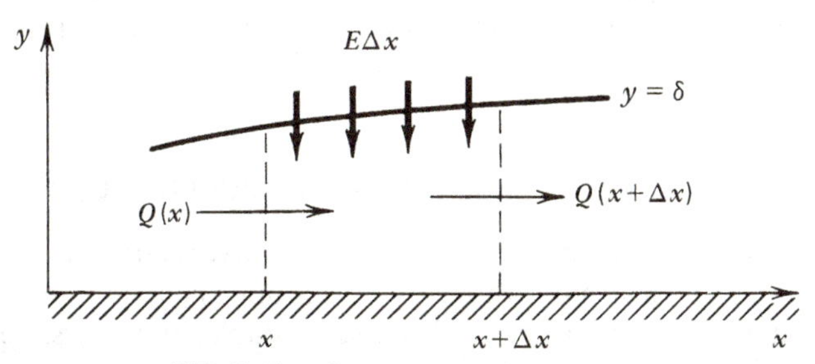

FIG. 7.13. Concept of entrainment.

[5] See White [10] for a good discussion of the basic differences in their formulations.
[6] See Cebeci and Bradshaw [1].

Combining equation 7-60 with the definition 7-28 of the displacement thickness, we find

$$\delta^* = \delta - \frac{Q}{V_e}$$

Then

$$E = \frac{d}{dx} V_e(\delta - \delta^*) \qquad (7\text{-}61)$$

which we write

$$E = \frac{d}{dx}(V_e \theta H_1)$$

where

$$H_1 \equiv \frac{\delta - \delta^*}{\theta} \qquad (7\text{-}62)$$

Head assumed that the dimensionless entrainment velocity E/V_e depends only on H_1 and that H_1, in turn, is a function of $H = \delta^*/\theta$. Cebeci and Bradshaw fit several sets of experimental data with the following formulas [1]:

$$\frac{1}{V_e}\frac{d}{dx}(V_e \theta H_1) = 0.0306(H_1 - 3)^{-0.6169} \qquad (7\text{-}63)$$

$$\begin{aligned} H_1 &= 3.3 + 0.8234(H - 1.1)^{-1.287} && \text{for} \quad H \leq 1.6 \\ &= 3.3 + 1.5501(H - 0.6778)^{-3.064} && \text{for} \quad H > 1.6 \end{aligned} \qquad (7\text{-}64)$$

Equations 7-34, 7-63, and 7-64 represent three equations among the four unknowns θ, H, H_1, and c_f. Head completes the set with the Ludwieg–Tillman skin-friction "law",

$$c_f = 0.246 \times 10^{-0.678H} \mathrm{Re}_\theta^{-0.268} \qquad (7\text{-}65)$$

which was derived simply by fitting data available from various experimental studies with a formula of product form, $f(H)g(\mathrm{Re}_\theta)$. According to White [10], it is accurate to $\pm 10\%$.

7.10. TRANSITION FROM LAMINAR TO TURBULENT FLOW

In the case of flow past an airfoil, the boundary layer starts out as laminar at the stagnation point, with a finite thickness given approximately by equation 7-44. Sooner or later, all boundary layers become unstable, and any small disturbance initiates transition to the erratically unsteady condition known as turbulence. Transition starts at a particular value of the Reynolds number based on the distance x from the start of the boundary layer, the Re_x defined in equation 7-38. For a boundary layer on a smooth flat plate, the critical value of Re_x is about 2.8×10^6, depending on the turbulence in the onset flow. In fact, the value of the transition Reynolds number depends on many factors; see Schlichting [11] for an extensive discussion of them. Among the more important of these factors are the pressure gradient imposed on the boundary layer by the inviscid flow and surface roughness. Transition is hastened—that is, the transition Reynolds number Re_x is lowered—by both surface roughness and a positive value of dp/dx.

For incompressible flows without heat transfer, Michel [12] examined a variety of data and concluded that, for airfoil-type applications, transition should be expected when

$$\mathrm{Re}_\theta > 1.174\left(1 + \frac{22{,}400}{\mathrm{Re}_x}\right)\mathrm{Re}_x^{0.46} \tag{7-66}$$

Although it may not be obvious at first sight, this formula does account somewhat for the effect of pressure gradient, because the momentum thickness grows more rapidly in a positive pressure gradient (along with the displacement thickness, as was discussed earlier in Section 7.6). However, it does not include the effect of surface roughness, which is also very important [11]. Thus it is not useful for every application, but, being based on data taken on airfoils, it should be good for wing analysis.

Unfortunately, the actual growth of Re_θ with Re_x turns out to be very nearly parallel to the growth of the value given by equation 7-66, so that the seemingly minor scatter of experimental data about Michel's curve fit can amount to a large uncertainty as to the x at which transition takes place. Therefore, you may have to bypass the use of equation 7-66 and "impose" transition at some judiciously chosen point along the airfoil. For example, you should not expect the boundary layer to remain laminar much past the minimum pressure point; a positive pressure gradient substantially decreases the stability of a laminar boundary layer, even at relatively low Reynolds numbers.

Actually, we should not speak of a "point" at which transition takes place; transition to turbulence is not an instantaneous process. Rather, over a certain length of the airfoil, the flow is intermittently laminar and turbulent. The distance between the point in front of which the flow is always laminar and the point aft of which the flow is always turbulent can be much larger than the length over which the flow is purely laminar. Since the skin friction usually increases dramatically thru the transition region, see Fig. 7.11, it would be useful to have an integral method that could describe the transition process. Unfortunately, no such method exists. Thus, it is necessary in calculating boundary-layer growth by an integral method to adopt the fiction of a transition point. After using a method like Thwaites's up to the transition point, however it is located, one switches to a method like Head's for the turbulent part of the boundary layer.

As in most integral methods for calculating turbulent boundary layers, Head's set of equations includes two first-order ordinary differential equations. Therefore, two initial conditions are required to start the calculations. This is no problem when, as in the Stanford competition, one is trying the method against experimental data. There should be enough data to provide any number of initial conditions. However, initial data are not so easily obtained when starting the calculations in the laminar part of the flow. A method like that of Thwaites can give data on θ, H, and c_f up to the start of transition, but H and c_f change so radically during transition that only the initial turbulent value of θ can be taken from the laminar calculations.

To get another condition, I suggest you just guess the starting value of H. This is not so crazy as it may sound. The shape factor for a turbulent boundary layer usually lies in a rather narrow range—in a mild pressure gradient, it is seldom out of the range 1.3 to 1.4—and, as you will find in running the program INTGRL that implements this method, the boundary-layer parameters downstream of transition soon forget the starting value chosen.

7.11. BOUNDARY-LAYER SEPARATION

The neat strategy we have outlined—determining the flow about a body as if the fluid were nonviscous, and then patching in a boundary layer near the body surface—is based crucially on the boundary layer experiencing the same pressure gradient in the direction along the body surface as does the outer inviscid flow. This, in turn, depends on the boundary-layer thickness being small compared to the characteristic length for flow variations along the surface. This picture can be destroyed if the pressure gradient imposed by the outer flow is too strongly positive. The boundary layer "separates," in a sense to be made precise below, and the characteristic length for flow variations along the body surface becomes of the same order as the boundary-layer thickness.

To explain, recall from equation 7-14 that, within the boundary layer, the flow is nearly parallel to the body surface. Also, because of the no-slip condition, the flow speed in the boundary layer is generally small compared to V_e, the speed of the outer flow. Therefore, if $dp/dx > 0$, the flow in the boundary layer is decelerated much more (absolutely as well as percentagewise) than is the outer flow.

To illustrate, simplify the equation of motion 7-18 to

$$\rho u \frac{du}{dx} = \frac{1}{2} \frac{d}{dx} \rho u^2 = -\frac{dp}{dx}$$

Then the rate of change in $\frac{1}{2}\rho u^2$ (and hence u^2) is the same all through the boundary layer. For this case, if we look at the point near the surface where $u = 0.1 V_e$, say, that velocity is reduced to zero when the outer flow velocity is reduced by only 1%. Although this local application of Bernoulli's equation (for that is what we are doing) considerably exaggerates the phenomenon, it is, nevertheless, a real possibility that if the pressure increases along the body surface (thus reducing V_e), then there will appear a locus of points within the boundary layer at which $u = 0$. Since the slower parts of the flow (near the wall) will decelerate to zero speed sooner than will the flow in the outer part of the boundary layer, the locus will begin at the wall, as shown in Fig. 7.14.

Even though v is small within the boundary layer, it is of course large compared to u near this locus, and the streamlines there become normal to the surface. Since $u < 0$ on the "downstream" side of the locus, the streamline pattern will be as

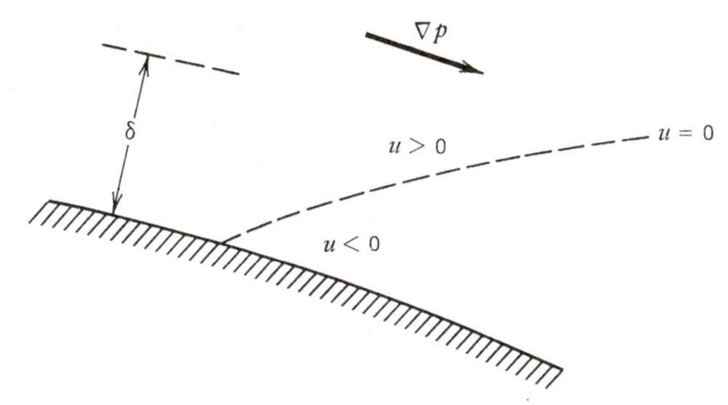

FIG. 7.14. Flow reversal in a positive pressure gradient.

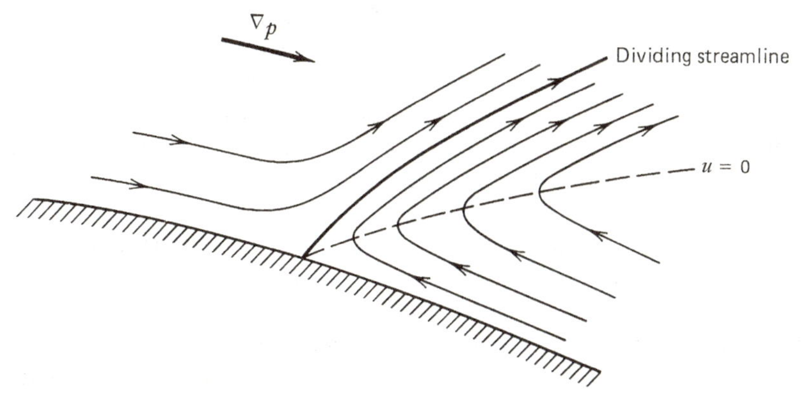

FIG. 7.15. Boundary-layer separation in a positive pressure gradient.

sketched in Fig. 7.15. In particular, there must be one streamline—a *dividing streamline*—which separates streamlines that reverse their direction at the $u = 0$ locus from those that start further upstream. From the geometry, we see that this streamline comes out from the wall. The flow is therefore said to *separate* from the solid boundary; the streamlines no longer follow the wall.

Before examining the influence of boundary-layer separation on airfoil performance, some general remarks as to its characteristics are in order. The "Bernoulli model" used above to argue for the possibility of flow reversal implies that if the pressure increases at all in the direction of the flow, separation is immediate and inevitable. This is not actually the case. A boundary layer in a positive pressure gradient will resist separation in proportion to its ability to transfer momentum from its outer regions to the slower flow closer to the surface. Thus a turbulent boundary layer, with its higher rate of momentum transport, can survive a higher pressure gradient without separating than can a laminar flow. On the other hand, the thicker the boundary layer, the harder it is to get momentum to the slower parts of the layer, and the easier it is to separate.

Returning to Fig. 7.15, we can deduce a characteristic of the velocity field in the vicinity of the separation point that is useful in detecting separation in both experimental and computational investigations. The no-slip condition applies both

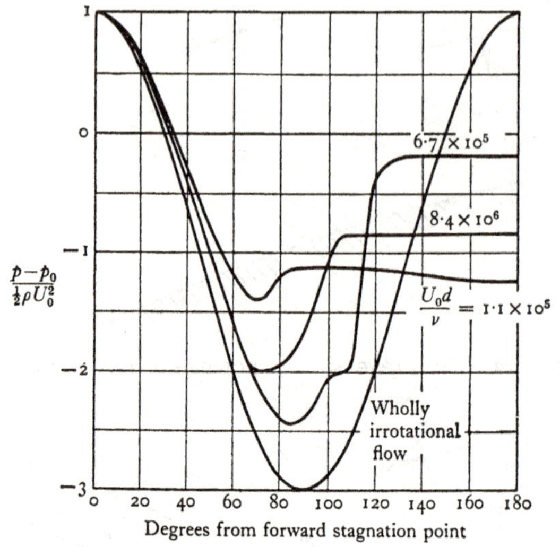

FIG. 7.16. Comparison of measured pressure distributions on a circular cylinder with potential-flow result. Reprinted with permission from Ref. [6].

upstream and downstream of separation, so $u = 0$ at $y = 0$. However, at small values of y, $u > 0$ upstream of separation but $u < 0$ downstream. Thus the value at $y = 0$ of $\partial u/\partial y$ (and so, from equation 7-24, the wall shear stress[7]) is positive upstream of separation, negative downstream, and hence zero at separation.

This fact is, as noted, useful in locating the separation point. For example, in Thwaites's method discussed above, laminar separation can be predicted to occur when $l(\lambda)$ vanishes, since l is defined in equation 7-39 to be proportional to the wall shear stress. According to the correlation formulas 7-43, this will happen when $\lambda \doteq -0.0842$. A similar criterion is not possible in Head's method for turbulent boundary layers, however, since the skin friction given by the Ludwieg–Tillman skin-friction formula (7-65) can never vanish. According to Kline *et al.* [13], one may predict turbulent separation by examining the development of the shape factor H, which is typically about 2.4 at the start of separation.[8]

One immediate consequence of boundary-layer separation is that the boundary layer is no longer "thin" once it separates, in the sense that the characteristic length in the x direction is then as small as is that for changes in the y direction. The pressure in the external flow can no longer be calculated with the viscosity of the fluid ignored completely; you must consider the interaction of the external flow and the boundary layer. The direction of the change in pressure from its inviscid value is that of the displacement effect of an attached boundary layer. The magnitude of the change can be considerably larger, since separation effectively entails a large displacement thickness; see Fig. 7.16, which compares the pressure distributions

[7] Even for turbulent flow. As noted above, the turbulent part of the shear stress, $-\rho\overline{u'v'}$, vanishes at the surface because of the no-slip condition.

[8] Turbulent separation, like transition, is a process that occurs over a finite region, within which the boundary layer is sometimes separated and sometimes attached.

observed on a circular cylinder at various Reynolds number with theoretical results obtained under the assumption of irrotational inviscid flow. Thus, boundary-layer separation substantially increases form drag.

To summarize, if the pressure increases too rapidly in the direction of the external flow, the flow within the boundary layer may back up, which creates all sorts of aerodynamic problems. Even if it does not separate the boundary layer, a positive pressure gradient will substantially thicken it, and alter the pressure distribution on the body in such a way as to produce drag. Also, as noted previously, a positive dp/dx destabilizes a laminar boundary layer; transition to turbulence is often fixed by the location of the minimum pressure point on the surface as much as by the Reynolds number. Positive pressure gradients are therefore called *adverse pressure gradients*, and their control is a major goal of aerodynamic design. A *streamlined body* is one for which the streamlines closely follow the body contour, meaning that separation is minimized so as to maximize performance.

7.12. AIRFOIL PERFORMANCE CHARACTERISTICS

The arguments made in the preceding section, especially in the last paragraph, furnish a basis for understanding the performance of real airfoils in a real (slightly viscous) fluid. Figure 7.17 through 7.24 display experimental data [14, 15] on the performance of several well-known airfoils, including members of the NACA four- and five-digit families discussed in Chapter 1. Each chart shows a cross section of the airfoil, its section lift and moment coefficients (as functions of the angle of attack), and its section drag coefficient (as a function of c_l), for a range of Reynolds numbers.

Let us first consider the variation with angle of attack of lift and drag. As indicated in Fig. 7.25, increasing the angle of attack moves the forward stagnation point rearward on the lower surface of the airfoil, so that the flow must accelerate substantially to negotiate the leading edge. The resultant pressure distribution is sharply spiked near the leading edge (i.e., there are large negative values of C_p), as can be seen in Figs. 7.26 to 7.28. The sharper the spike, the steeper the adverse pressure gradient imposed on the boundary layer aft of the minimum pressure point. This increases the form drag and may also lead to boundary-layer separation. If the separation occurs in the laminar flow near the leading edge, early transition to turbulence may follow, with a consequent increase in drag due to friction. On the other hand, because of the higher rate of momentum transport within a turbulent boundary layer, the separated flow may reattach, as sketched in Fig. 7.29, so that the result is simply an effective thickening of the upper surface near the leading edge.

Increasing the radius of curvature of the leading edge reduces the magnitude of the pressure peak and so avoids or at least delays leading-edge separation. If it can be delayed past the point at which the boundary layer becomes turbulent, so much the better, because of the higher resistance to separation of a turbulent boundary layer. Still, the minimum upper-surface pressure coefficient inevitably becomes more negative as the angle of attack increases.

For a sufficiently large angle of attack, if separation does not occur near the minimum pressure point, it will take place near the trailing edge, where the pressure

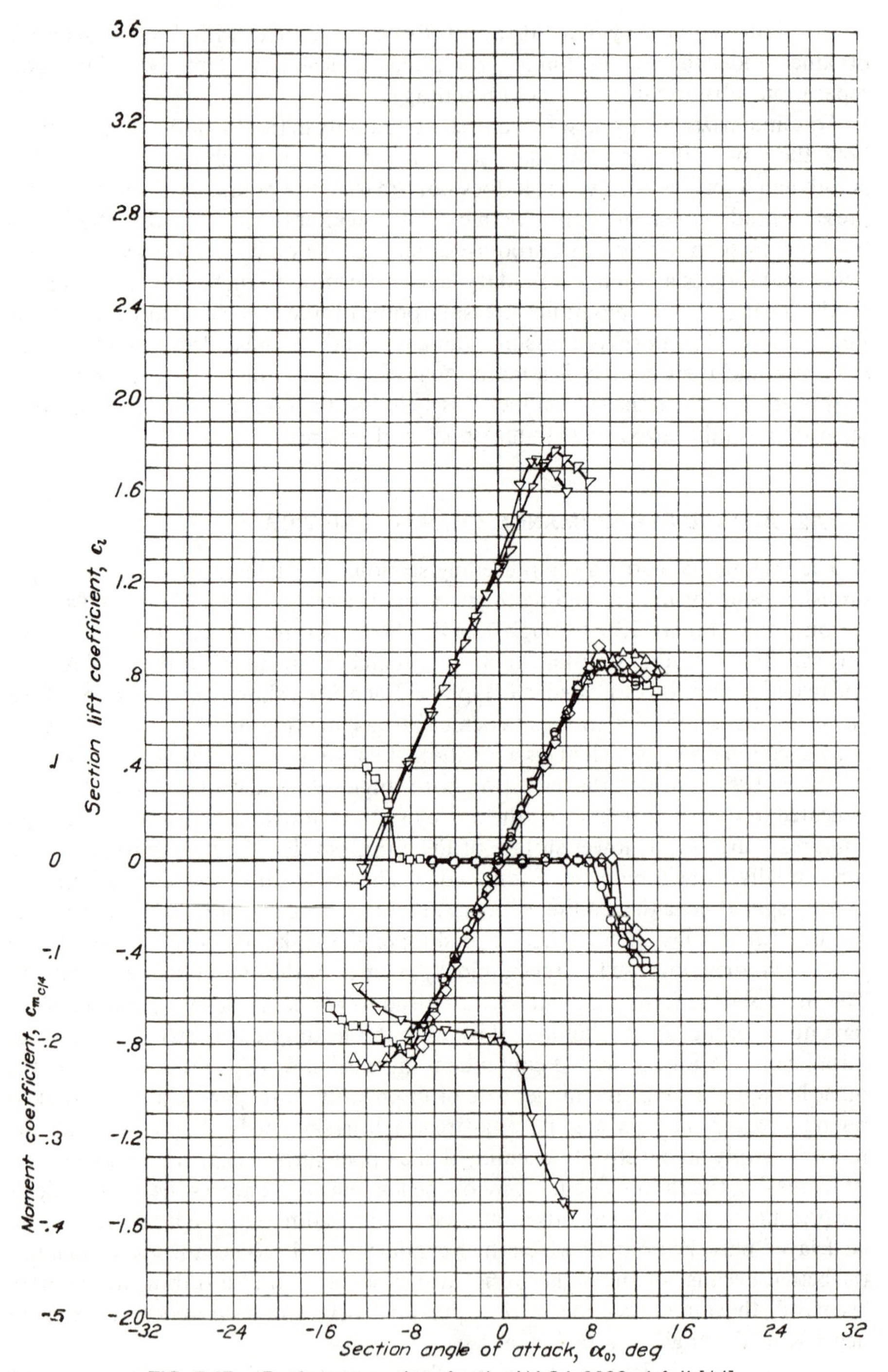

FIG. 7.17. Performance data for the NACA 0006 airfoil [14].

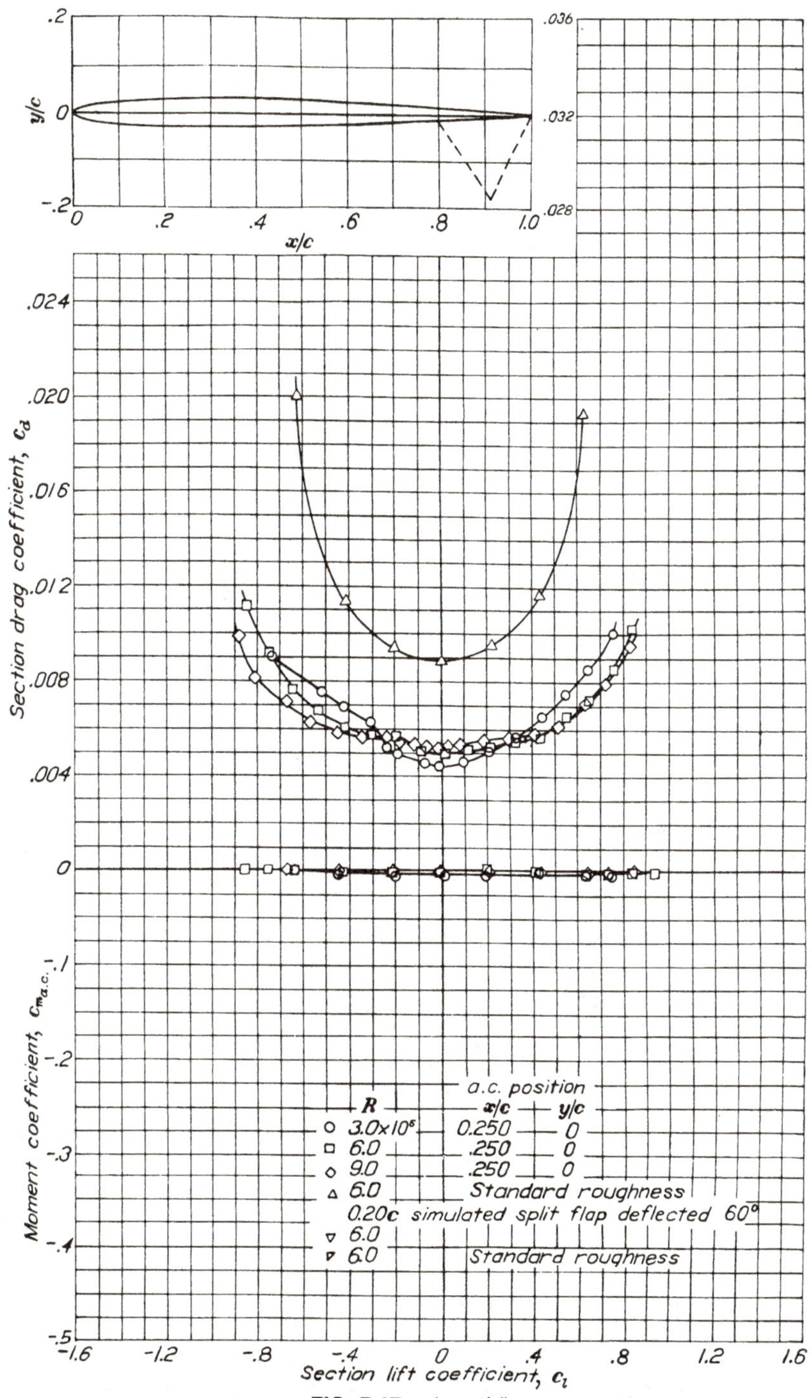

FIG. 7.17. (cont'd)

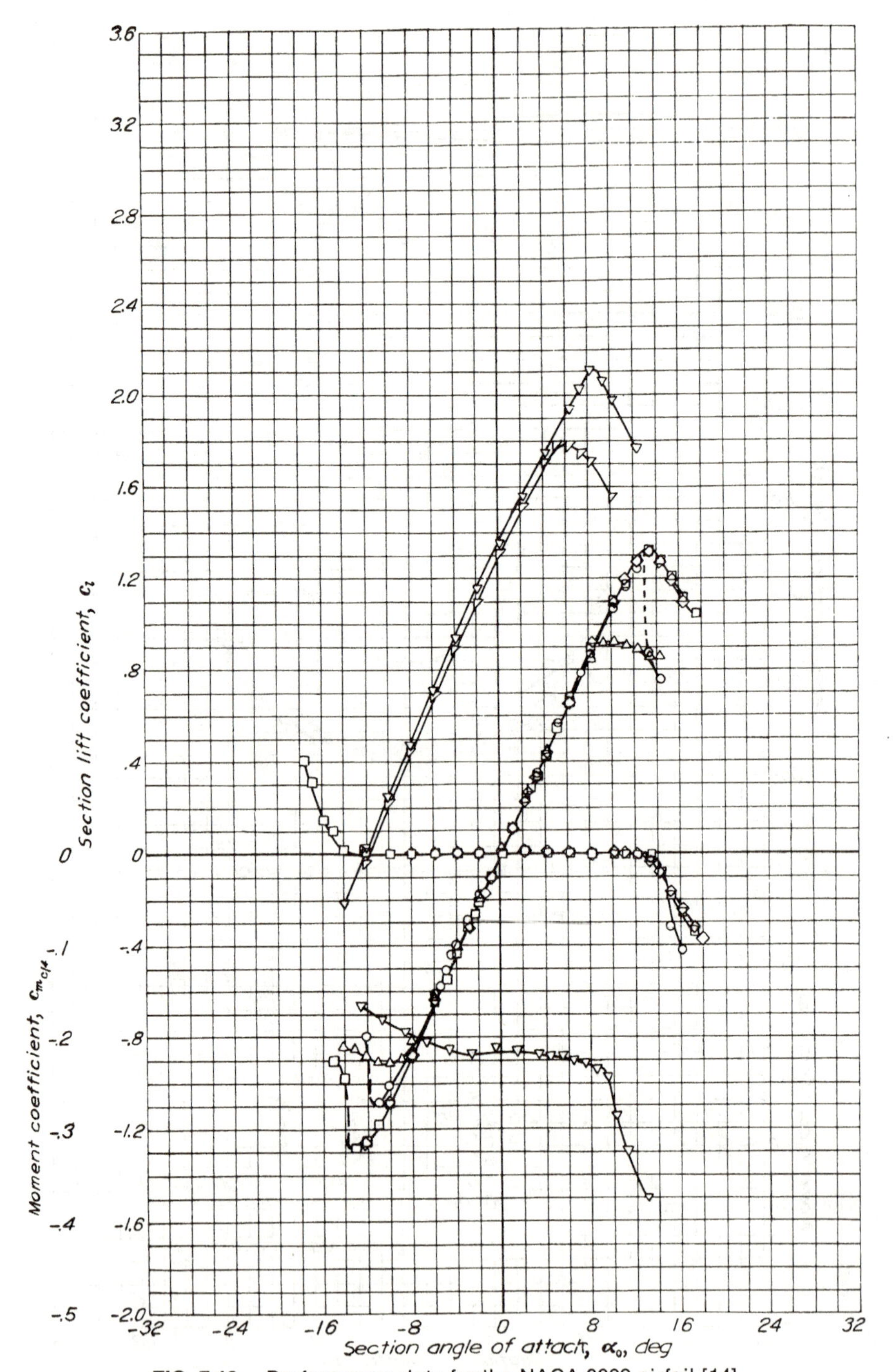

FIG. 7.18. Performance data for the NACA 0009 airfoil [14].

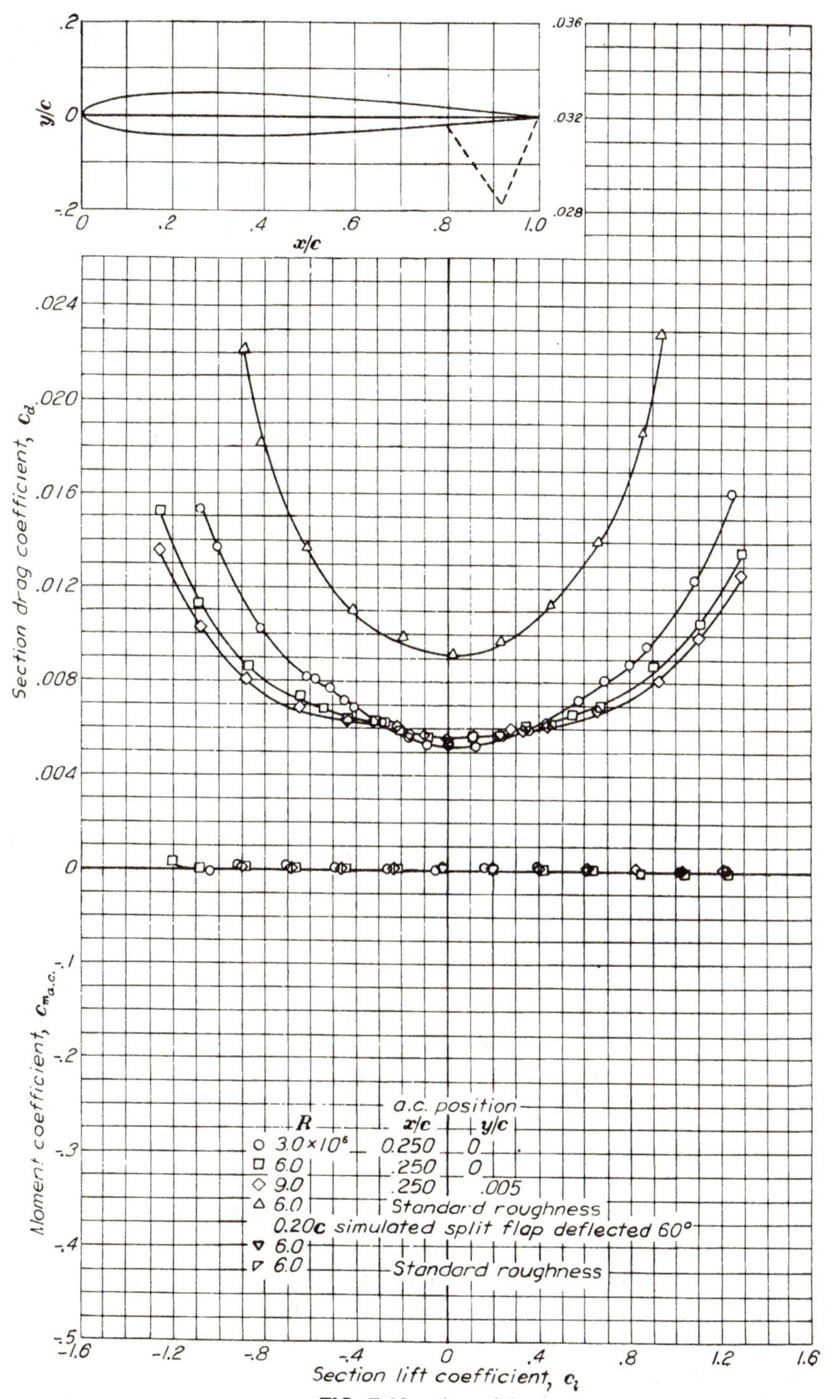

FIG. 7.18. (cont'd)

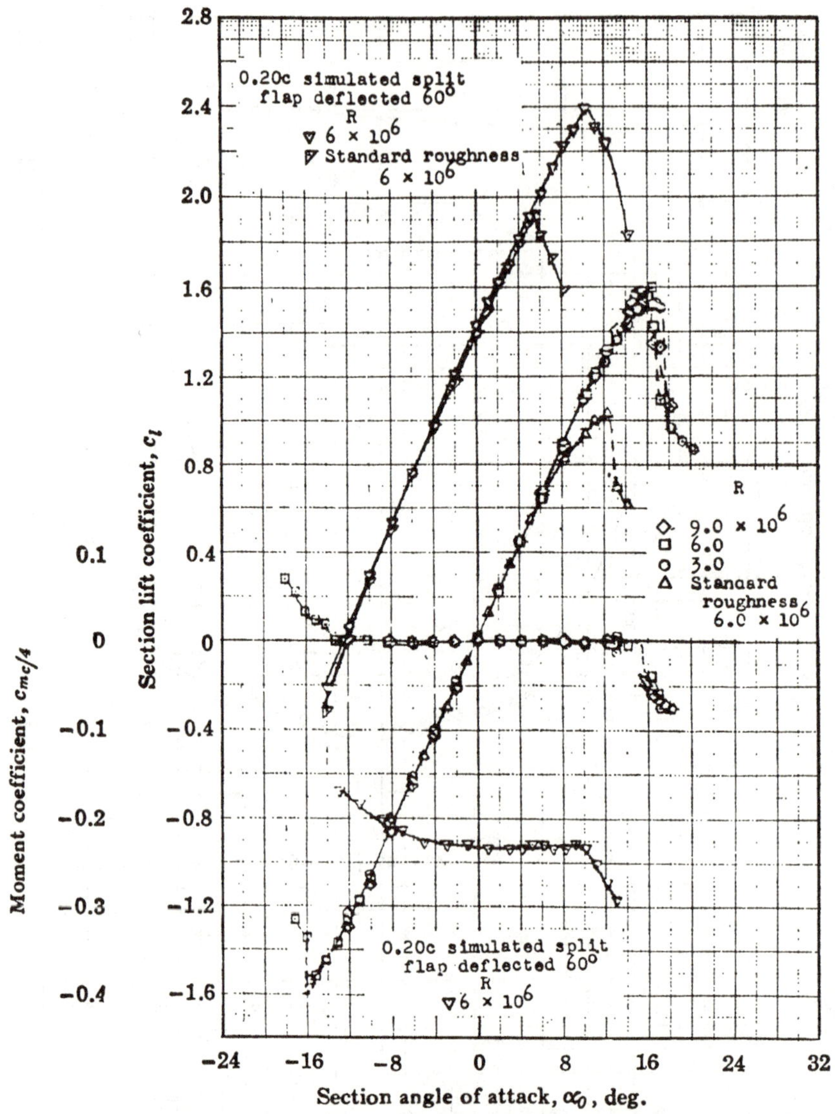

FIG. 7.19. Performance data for the NACA 0012 airfoil [14].

gradient is also positive and the boundary layer, being thicker, is less resistant to separation. Massive separation (as distinguished from a leading-edge separation followed by reattachment) limits the maximum lift coefficient developed by an airfoil. The decrease in c_l with further increase in angle of attack is often precipitous; see, for example, Fig. 7.22. This is called *stall*; note that it is another adverse effect of boundary-layer separation.

Now let us try to relate the differences in performance among airfoils to their geometry. The NACA 0006, 0009, and 0012 airfoils (Figs. 7.17 through 7.19) are all

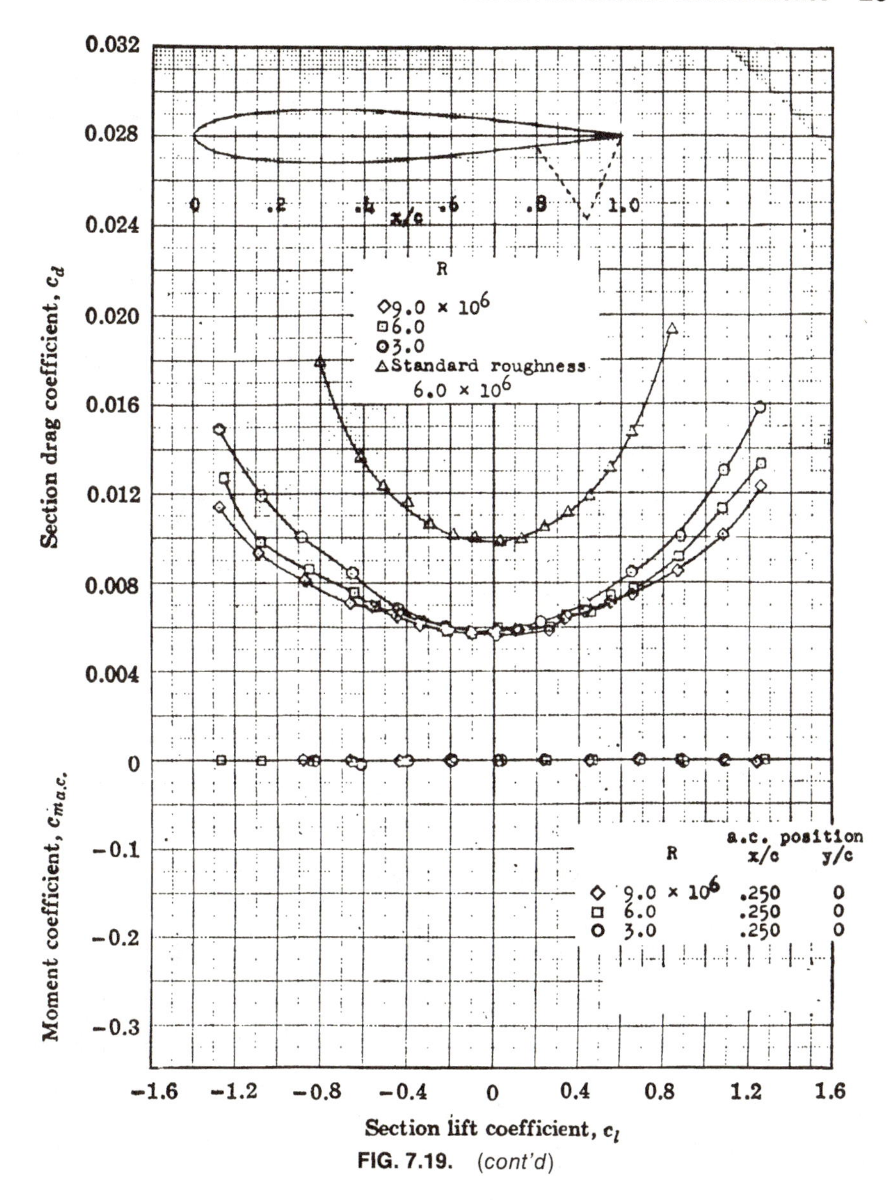

FIG. 7.19. (*cont'd*)

uncambered and differ in their thickness distribution only by scale factors. The NACA 0012, 2412, and 4412 (Figs. 7.19 to 7.21) have similar distributions of thickness and their maximum camber at the same position but differ in the amount of camber. The NACA 23012 (Fig. 7.22) differs from the XX12 airfoils in having its maximum camber relatively close to the leading edge. Like the 0012, the 65_1–012 (Fig. 7.23) is a symmetric 12% thick airfoil but has a much sharper leading edge, and its maximum thickness is closer to midchord. The final example, the relatively modern *GA*(*W*)-1 airfoil (Fig. 7.24), is about 17% thick and much more highly

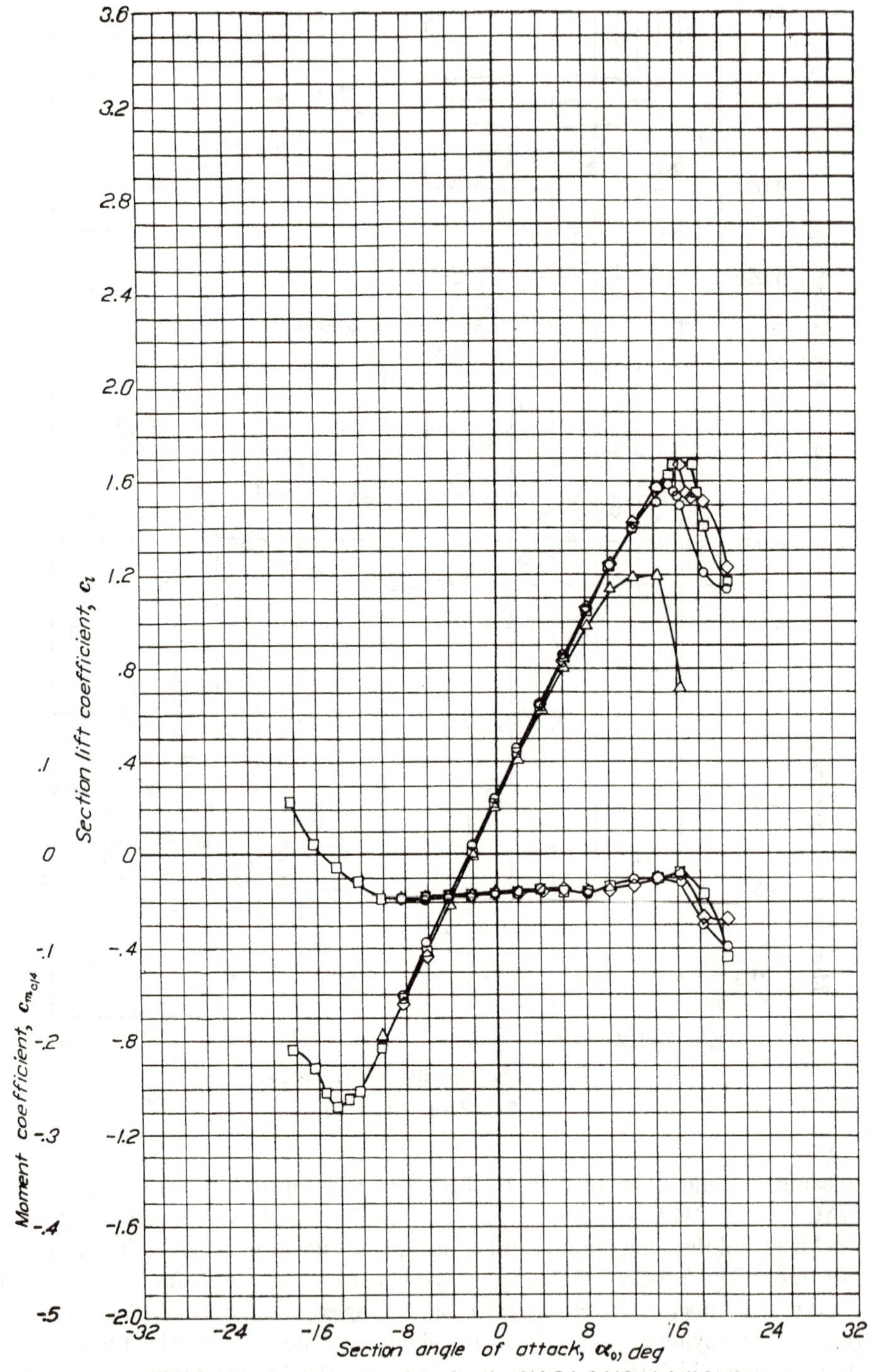

FIG. 7.20. Performance data for the NACA 2412 airfoil [14].

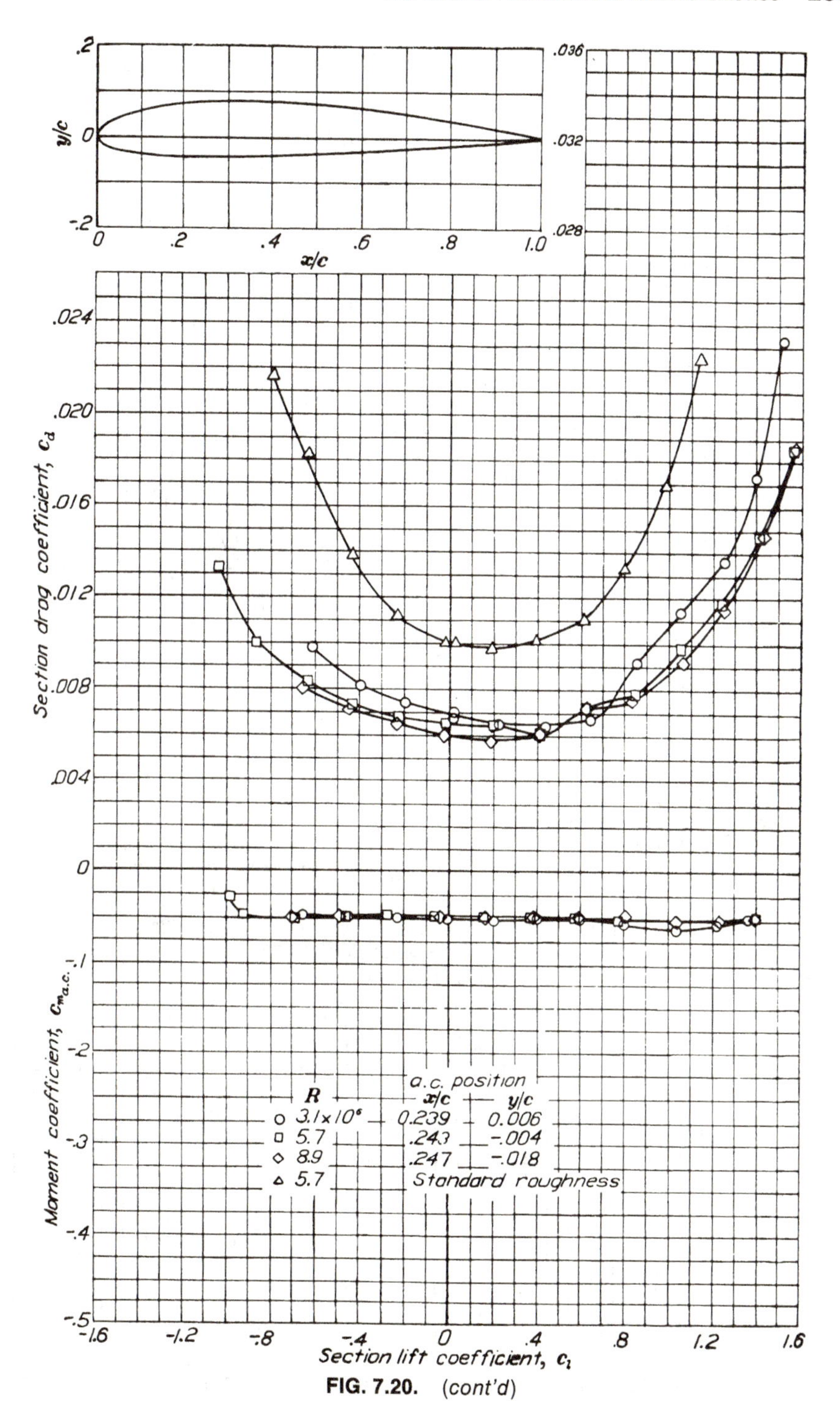

FIG. 7.20. (cont'd)

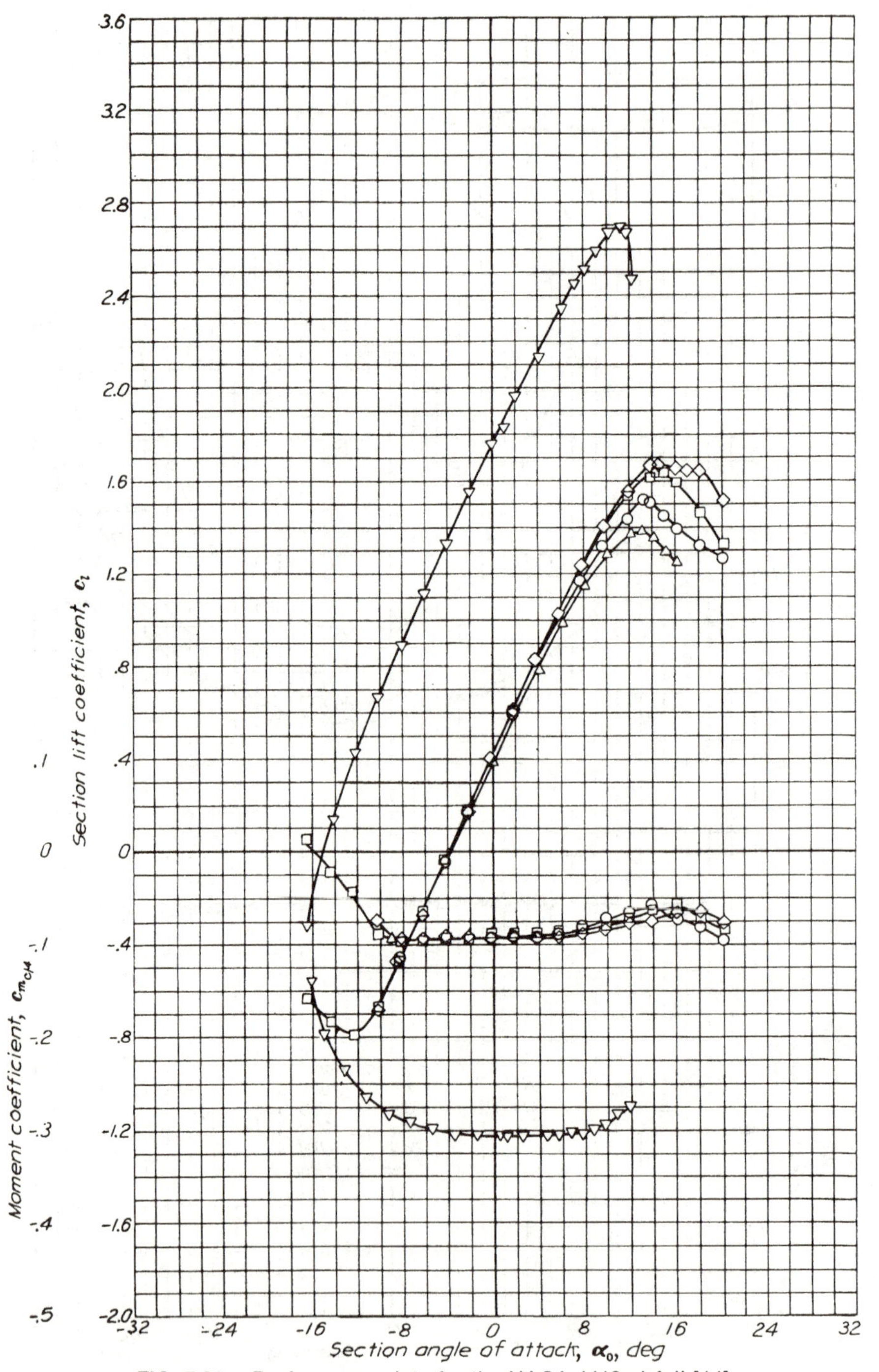

FIG. 7.21. Performance data for the NACA 4412 airfoil [14].

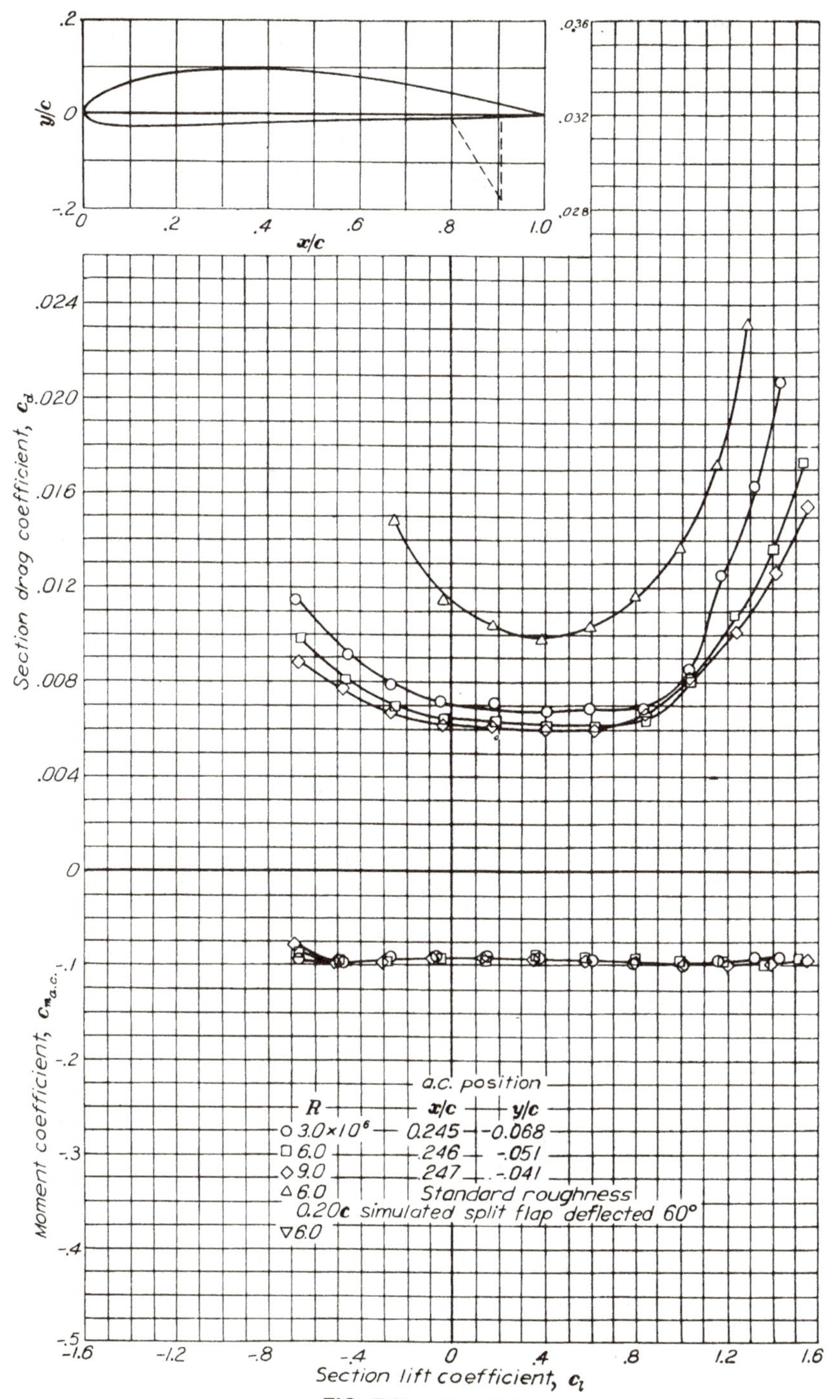

FIG. 7.21. (cont'd)

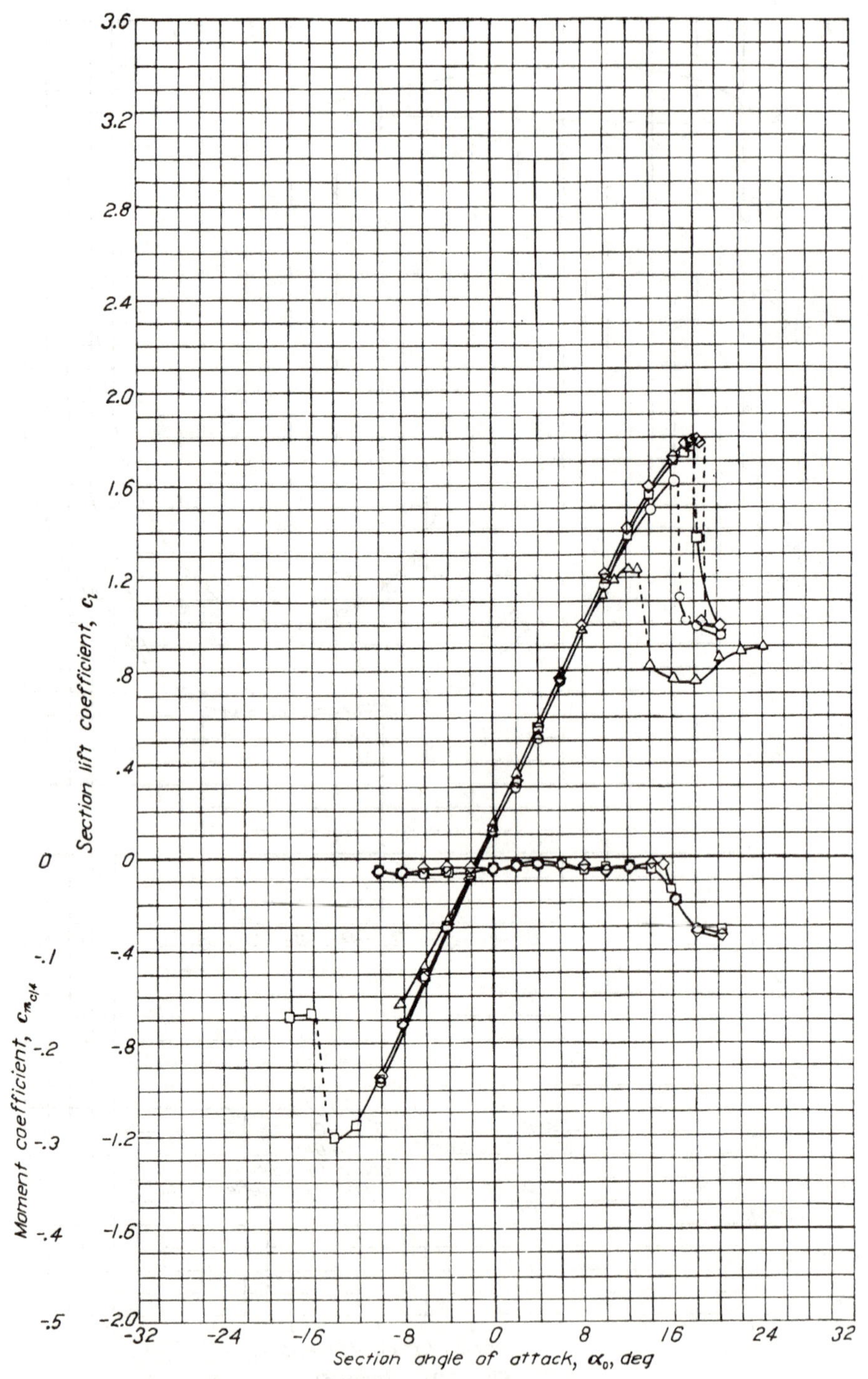

FIG. 7.22. Performance data for the NACA 23012 airfoil [14].

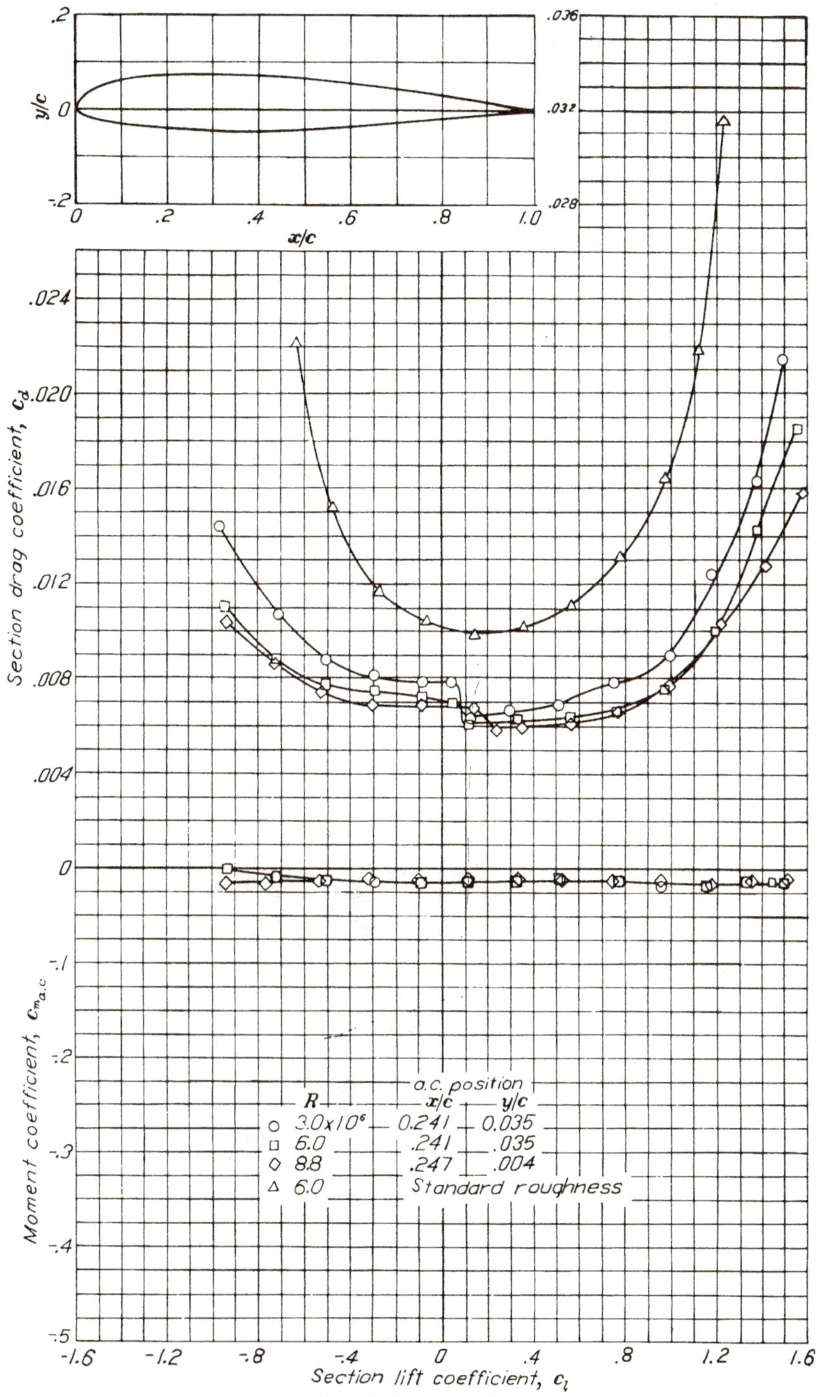

FIG. 7.22. (cont'd)

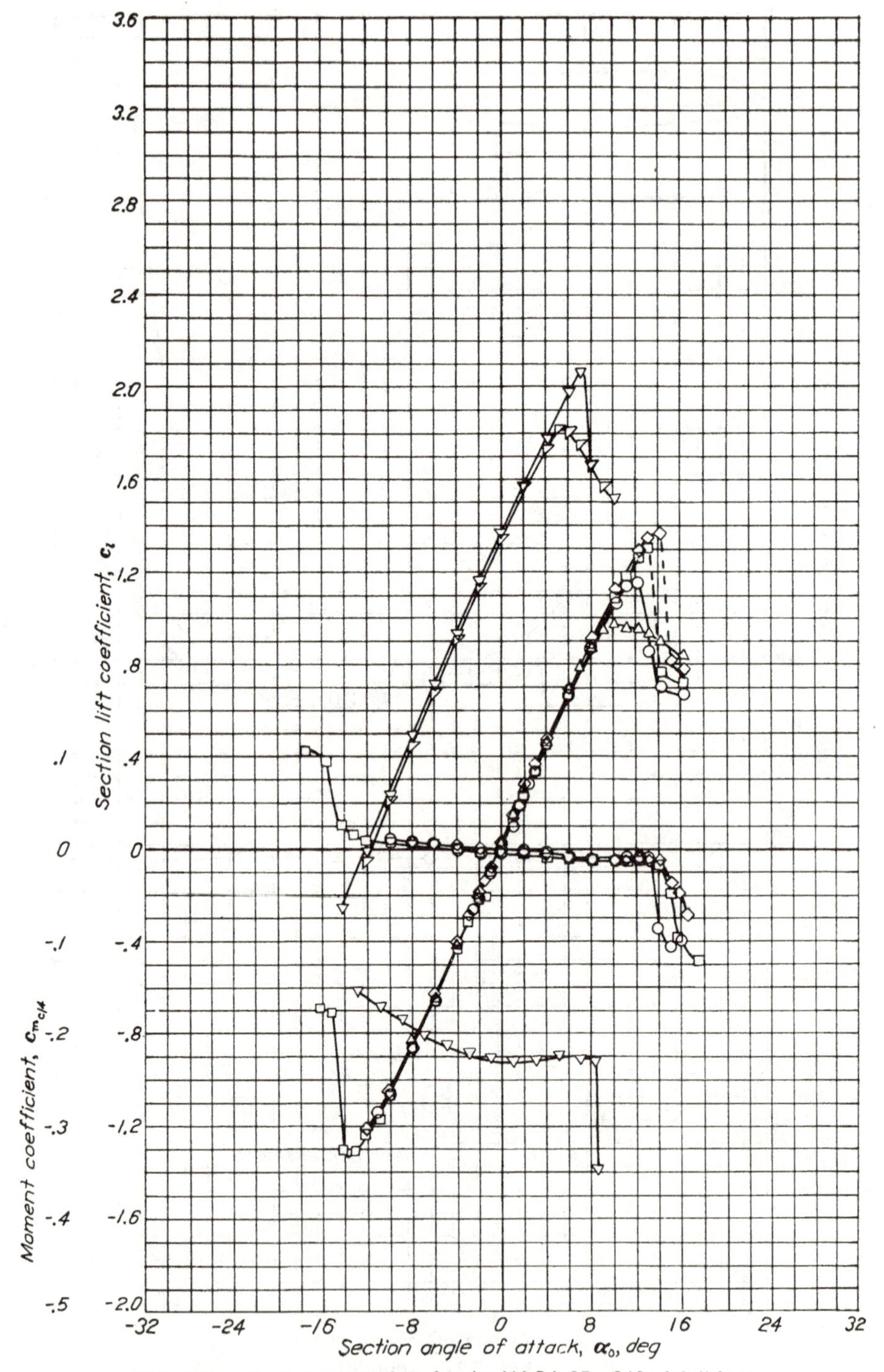

FIG. 7.23. Performance data for the NACA 65_1–012 airfoil [14].

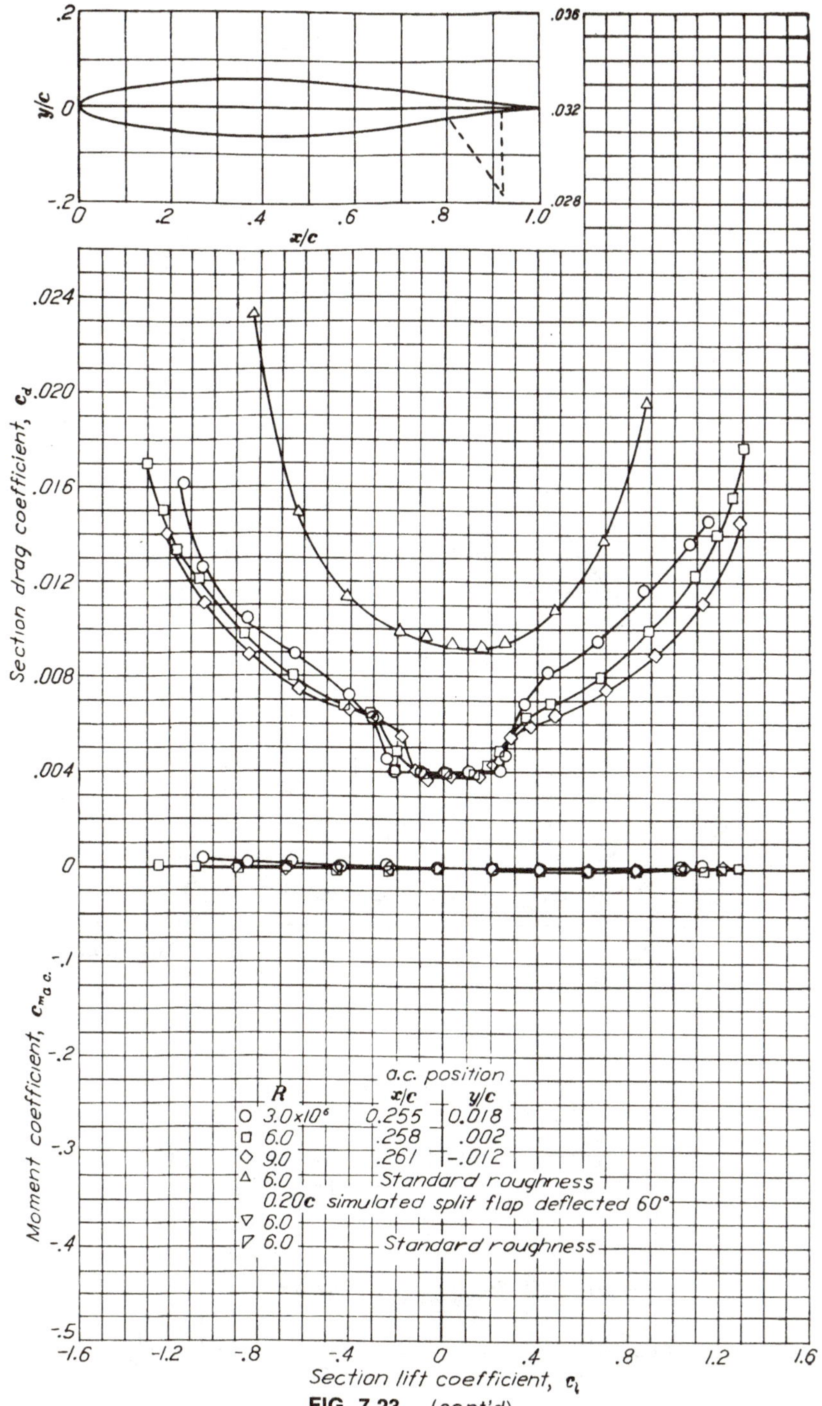

FIG. 7.23. (cont'd)

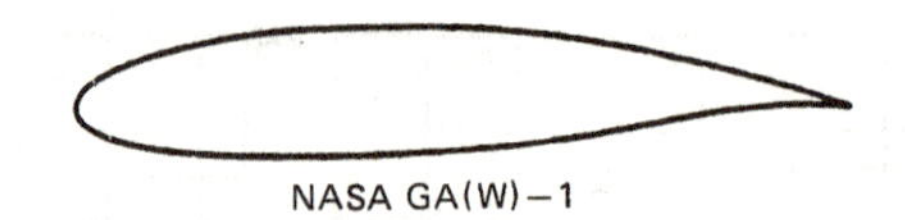

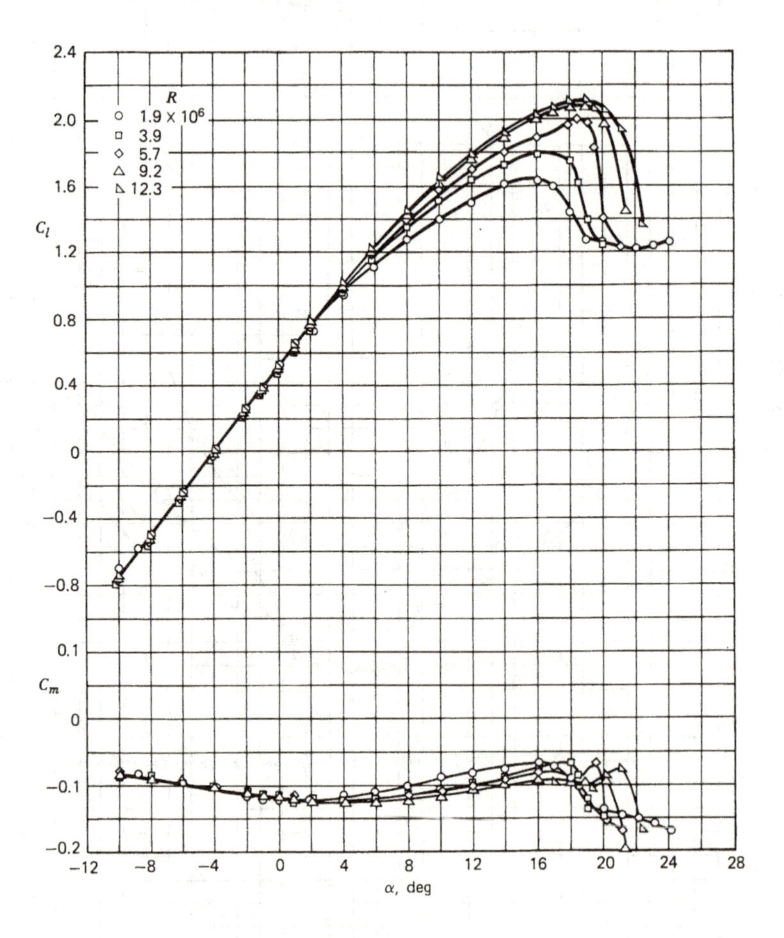

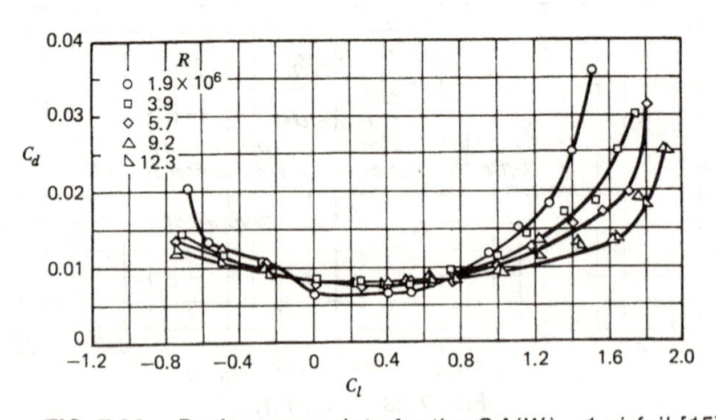

FIG. 7.24. Performance data for the $GA(W) - 1$ airfoil [15].

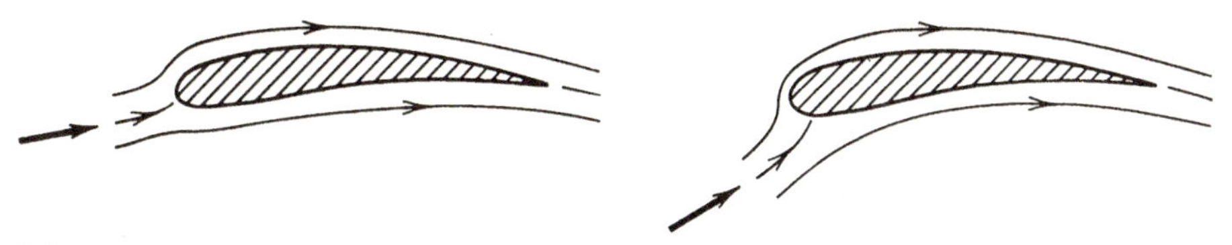

FIG. 7.25. Effect of angle of attack on the adverse pressure gradient on the suction side of a lifting airfoil.

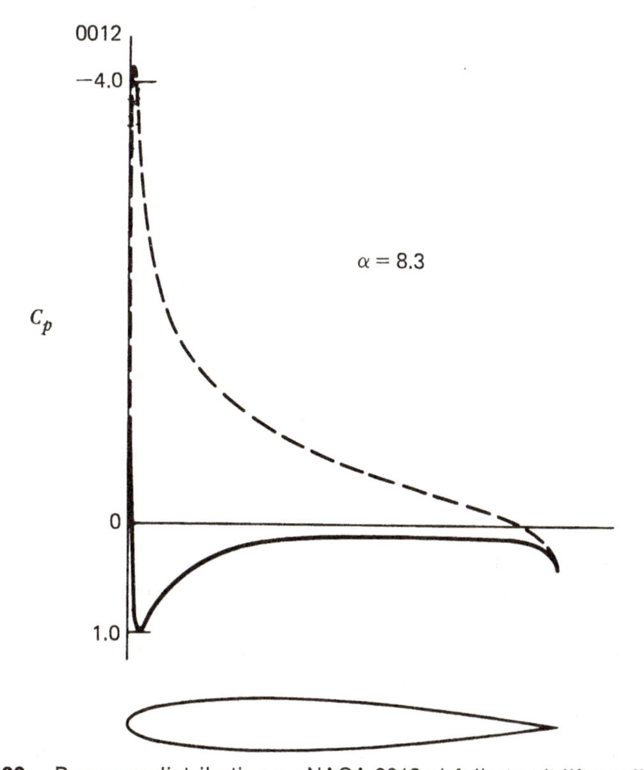

FIG. 7.26. Pressure distribution on NACA 0012 airfoil at unit lift coefficient.

cambered near its trailing edge than are the other airfoils whose performance charts are included here.

The performance characteristics on which you should focus are as follows.

1. Maximum section lift coefficient $c_{l_{max}}$.
2. Minimum section drag coefficient $c_{d_{min}}$.
3. c_l at which $c_d = c_{d_{min}}$.
4. The moment about the aerodynamic center c_{mac} (actually c_m about the quarter chord in most of the cases).
5. The angle of zero lift α_{L0} and the angles of attack at which $c_l = c_{l_{max}}$ and $c_d = c_{d_{min}}$.

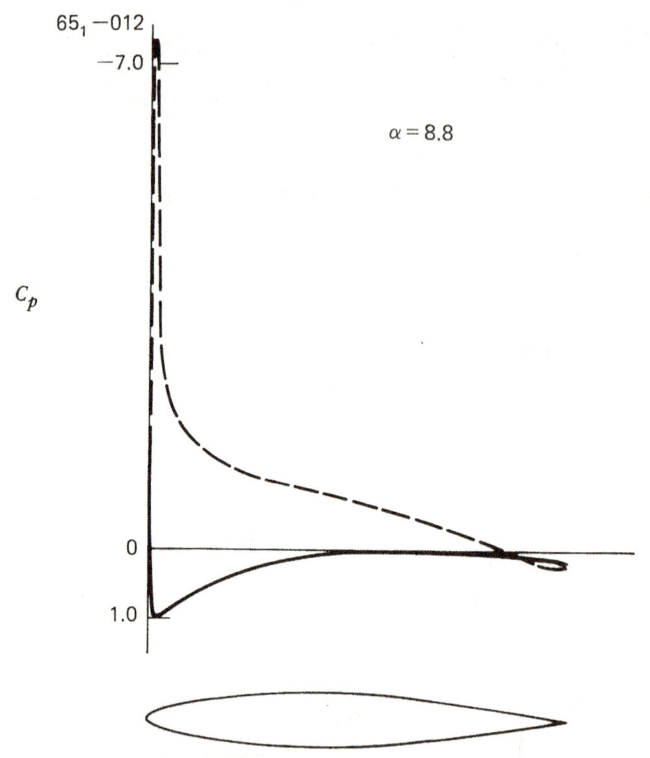

FIG. 7.27. Pressure distribution on NACA 65_1–012 airfoil at unit lift coefficient.

The last two items are easy enough to understand; I just want you to be aware of their magnitudes. You should be alarmed, for example, if asked to increase the angle of attack to 30°.

As can be seen from Figs. 7.19 through 7.21, the amount of camber has very little effect on the maximum lift coefficient. However, both the amount and distribution of thickness have rather profound effects in this respect, as does the Reynolds number. You will examine the effect of the overall amount of thickness in problem 8; I will discuss the other influences here.

What limits the lift coefficient is boundary-layer separation, which becomes more likely as the boundary layer thickens and as the pressure gradient is made more adverse. Increasing the Reynolds number decreases the thickness of the boundary layer (see equation 7-15) and so increases the pressure gradient it can take without separating. The effect of the thickness distribution can be understood by examining the inviscid pressure distribution on the airfoil; the steeper the adverse pressure gradient, the more likely is separation. The pressure distributions shown in Figs. 7.26–7.28 were calculated by program PANEL at the angle of attack for which the airfoil's theoretical lift coefficient was 1.0. Sure enough, the airfoil whose pressure gradient is most gentle (steep) has the best (worst) $c_{l_{\max}}$.

Most of the airfoils for which data are presented in Figs. 7.17 to 7.24 have about the same $c_{d_{\min}}$. There is a small increase of $c_{d_{\min}}$ with thickness ratio, see also Fig. 7.9,

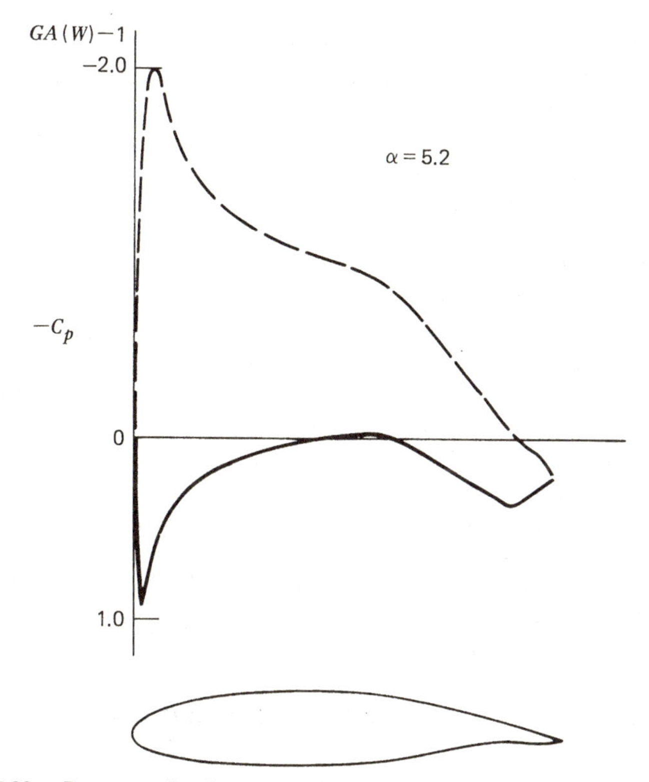

FIG. 7.28. Pressure distribution on $GA(W)-1$ airfoil at unit lift coefficient.

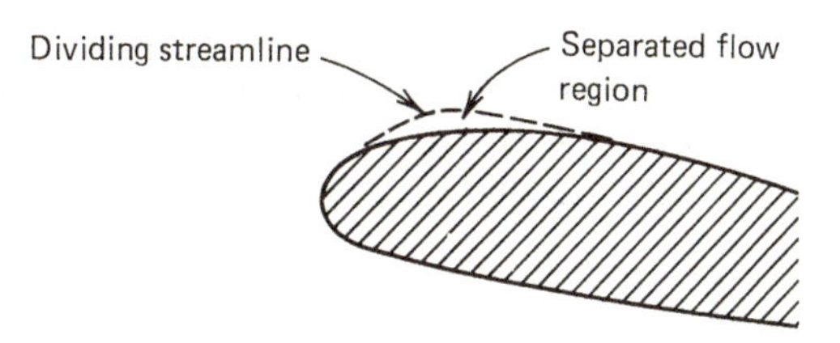

FIG. 7.29. Reattachment of the separated flow near a leading edge.

but the most interesting c_d versus c_l curve is that of the 65_1–012 airfoil. By sharpening the leading edge and moving the maximum thickness point rearward the minimum pressure point at small angles of attack is also moved rearward. An adverse pressure gradient, even if too weak to separate the boundary layer, will almost always trigger transition to turbulence. Thus the so-called *laminar-flow airfoils*, of which the 65_1–012 is an example, achieve relatively low drag coefficients by postponing transition to the higher skin-friction coefficients of the turbulent boundary layer. As the angle of attack strays from the design value (0° for the 65_1–012), the drag advantage disappears, because the minimum pressure point pops forward, just as it does for any airfoil at angle of attack, and brings with it the transition point. This explains the "bucket" in the c_d versus c_l curve, which is characteristic for this family of airfoils. In fact, the subscript "1" in 65_1–012 indicates the width of the drag bucket, the range in c_l in tenths above and below the design value in which favorable pressure gradients exist on both upper and lower surfaces. Unfortunately, these airfoils do not perform as well on aircraft as in the wind tunnel, because it is so hard to maintain a smooth clean surface in the real world, and because transition to turbulence is hastened by surface roughness.

The remaining performance parameter we will discuss is the lift coefficient at which the minimum drag is achieved. Here is where camber is helpful. You can see that the 0012, 2412, and 4412 airfoils all have about the same $c_{d_{min}}$ but achieve it at successively higher lift coefficients. What causes the drag to increase as c_l is varied is the adverse pressure gradient that develops on the "suction" side of the airfoil near the leading edge and which, as described above, increases the rate of growth of the displacement thickness and so the form drag. Which side is the suction side depends on the direction of the onset flow. For sufficiently high angles of attack, the upper surface is the troublesome one, whereas at lower (perhaps negative) angles it is the lower surface. Clearly, the angle of attack midway between the values that cause misbehavior of the boundary layer on the upper and lower surfaces—which should be close to the α at which $c_d = c_{d_{min}}$—increases with the slope of the camber line at the leading edge. So, then, does the lift coefficient at $c_{d_{min}}$. You will have a chance to confirm this line of reasoning in problem 9.

7.13. THE DEVELOPMENT OF CIRCULATION ABOUT A SHARP-TAILED AIRFOIL

Boundary-layer separation does have one good point. While it increases drag and limits the maximum lift, it can also be regarded as the source of the circulation that provides lift in the first place.

To show this, we need a result of nonviscous flow theory known as *Kelvin's theorem.* First recall from equation 2-35 that, in a two-dimensional nonviscous flow, the angular velocity of a fluid particle is conserved:

$$\frac{D\omega}{Dt} = 0$$

Consider then a closed contour C that always consists of the same fluid particles; that is, which drifts with the flow. As indicated in Fig. 7.30, we will call the contour

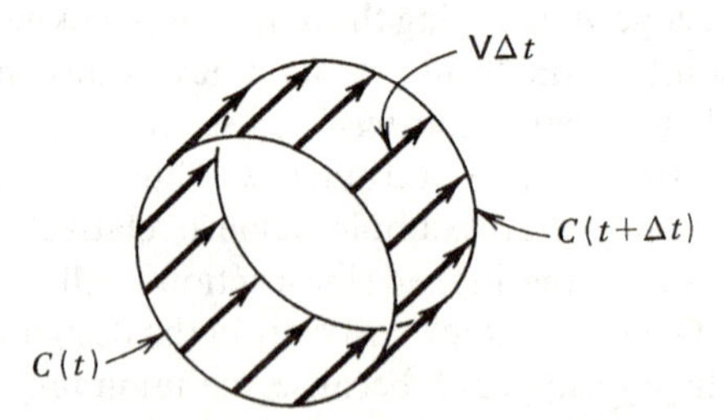

FIG. 7.30. Contour for analysis of the circulation around a closed contour that drifts with the fluid.

$C(t)$, its motion from t to $t + \Delta t$ being described by the vectors $\mathbf{V}\Delta t$. Since, from equation 2-34,

$$\boldsymbol{\omega} = \tfrac{1}{2}\,\text{curl}\,\mathbf{V}$$

we have, from Stokes's theorem (equation 2-25),

$$\Gamma(t) \equiv \oint_{C(t)} \mathbf{V} \cdot \mathbf{dl} = \int_{S(t)} \text{curl}\,\mathbf{V} \cdot \hat{\mathbf{k}}\,ds$$

$$= 2\int_{S(t)} \omega\,ds \qquad (7\text{-}67)$$

where $S(t)$ is the plane surface bounded by $C(t)$ and $\hat{\mathbf{k}}$ is a unit vector out of the plane, whereas $\Gamma(t)$ is the circulation around $C(t)$. Now, in an incompressible flow, the volume of a fluid particle remains constant. In the two-dimensional case under consideration, this implies that the cross-sectional area of a fluid particle remains constant. Since its angular velocity is also constant, so is the contribution of every fluid particle within $C(t)$ to the surface integral in equation 7-67. But because C drifts with the fluid, no particle crosses the contour C. Thus the value of the integral is constant, and we are led to (at least in two-dimensional flow, but it's also true in the three-dimensional case)

Kelvin's Theorem *The circulation around a closed contour which drifts with the fluid is constant in time.*

Now let's look at what happens when an airfoil is put into motion from rest. Suppose, for the sake of the argument, that it is accelerated instantaneously to its ultimate speed. Immediately after the body is set in motion, the circulation around the airfoil is what it was an instant before, namely, zero. Then the rear stagnation point is probably on the upper surface, upstream from the trailing edge, as sketched in Fig. 7.31. To reach the stagnation point, the lower-surface flow must go around the sharp trailing edge. This requires a large acceleration. In potential theory, the velocity becomes infinite at a sharp convex corner (see Appendix C). Thus the flow experiences a very substantial adverse pressure gradient as it approaches the rear stagnation point, which separates the boundary layer between the trailing edge and the rear stagnation point. A strong vortex—a region of high circulation—is then formed in the separated-flow region, as shown in Fig. 7.32.

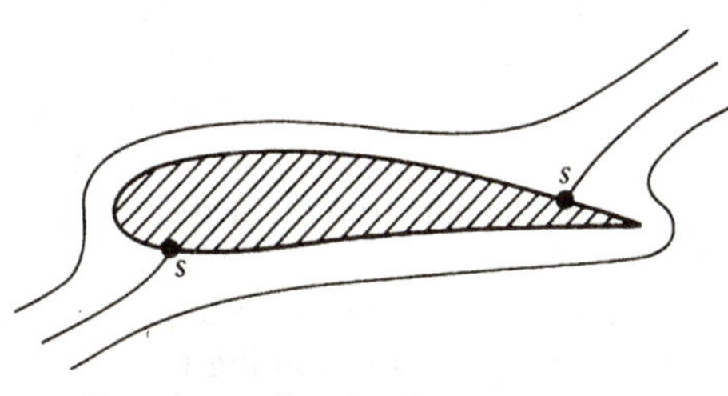

FIG. 7.31. Streamline pattern immediately after an airfoil is put into motion from rest.

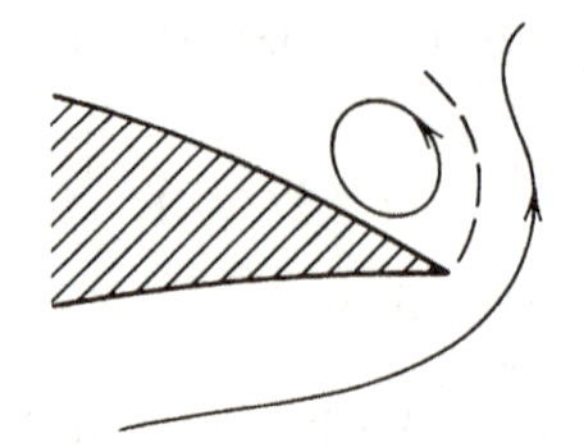

FIG. 7.32. Peeling off a vortex at the trailing edge.

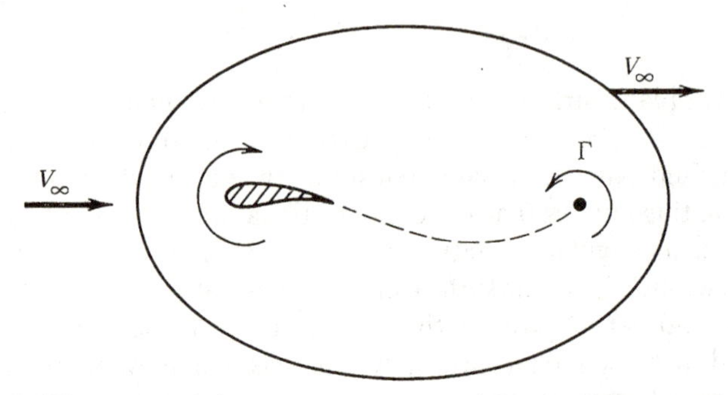

FIG. 7.33. Drifting away of the starting vortex and the development of circulation.

Consider, now, what happens to this vortex. Specifically, look at a contour consisting of the fluid particles that surround this vortex at its formation and that subsequently move with the local fluid velocity. As this contour moves away from the body, the inviscid-flow result of Kelvin's theorem becomes accurate, so that the circulation around the contour becomes constant. Since the vortex is then "staying inside" the contour, we describe this state of affairs by saying that *the vortex itself drifts with the fluid.*

Now construct a contour large enough to enclose both the airfoil and the spun-off vortex, as shown in Fig. 7.33. The circulation around this contour also remains constant, namely, zero. For this to be so, *the airfoil must develop a circulation equal and opposite to that of the drifting vortex.* The origin of the circulation of the airfoil [and, hence, according to the Kutta–Joukowski formula (4-33), its lift], therefore, is the shedding of a vortex at its sharp trailing edge, which is induced by boundary-layer separation.

It may be argued that the flow along the upper surface could also separate as it approaches the stagnation point and so generate a clockwise vortex that cancels out the one described above, as sketched in Fig. 7.34. However, because the velocity is so high at the sharp trailing edge, the counterclockwise vortex generated by the flow that comes around the trailing edge is much stronger. This is a good reason for having a sharp trailing edge; if the net strength of the vortices spun off were zero, so would be the lift of the airfoil.

So long as there is a stagnation point upstream of the trailing edge, the flow continues to separate as it navigates the trailing edge and tries to approach the stagnation point. Thus, the airfoil continues to spin off vortices and so increase its

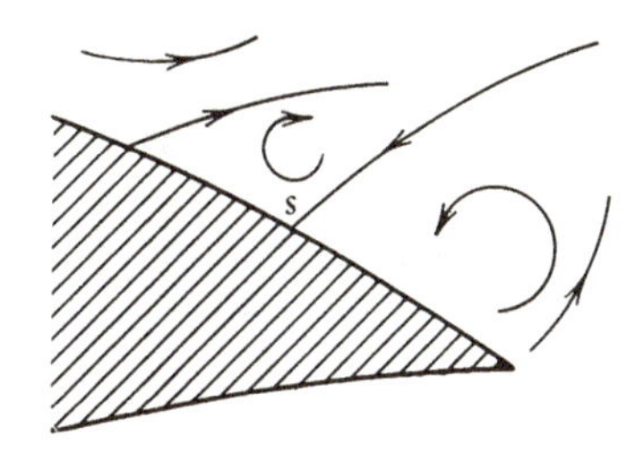

FIG. 7.34. The possibility of a counter vortex.

circulation. This has the effect of adding a clockwise tangential component to the velocity on the airfoil surface, which moves the stagnation point toward the trailing edge.

If the stagnation point reaches the trailing edge, the flow achieves an equilibrium. That is, if the trailing edge is itself a stagnation point, the flow no longer needs to accelerate around a sharp corner and then slow down abruptly at an upstream stagnation point. Of course, the flow may still separate as it approaches the stagnation point, but now it is possible for the separation to be equally severe on both the upper and lower surfaces. Any vortices spun off the two surfaces now cancel each other out; the net circulation of the wake stabilizes, and so does the (equal and opposite) circulation of the airfoil.

This leads to a more precise version of the Kutta condition:

The circulation about an airfoil is such that the vorticity developed on the upper surface and dumped into the wake is equal and opposite to that spun off from the lower surface.

To an excellent approximation, this is equivalent to the requirement that the trailing edge be a stagnation point so far as the inviscid flow is concerned.

7.14. COMPUTATION OF BOUNDARY LAYER GROWTH ALONG AN AIRFOIL

We will now tie together much of the content of this chapter by examining a computer program that calculates the boundary-layer growth on an airfoil. To perform such a calculation, one needs, first of all, the distribution $V_e(x)$ of the velocity of a nonviscous fluid along the airfoil surface. The program listed in Section 7.17, called INTGRL, is set up to calculate the boundary-layer growth on an ellipse of specified thickness ratio τ. For that family of bodies, the potential-flow solution is known in closed form; see equation 3-82. Program INTGRL takes advantage of this fact by using a function subroutine (VE) to set $V_e(x)$. However, the program is easily generalized to cases in which the inviscid flow data are generated by another computer program, such as DUBLET or PANEL. This generalization will be discussed below.

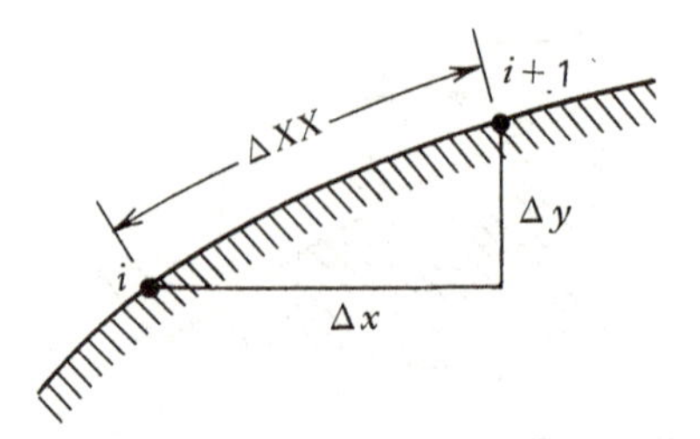

FIG. 7.35. Definition of boundary-layer coordinates by Pythagorean theorem.

The "x" of the boundary-layer equations is not a Cartesian coordinate but is measured along the airfoil surface, with $x = 0$ locating the stagnation point. Program INTGRL contains two function subroutines, X(I) and Y(I), which give, respectively, the Cartesian x and y coordinates of the Ith point at which the boundary layer is to be calculated and at which the potential-flow velocity is VE(I). These points are distributed along the surface of the ellipse by a cosine law, which concentrates computation points near the stagnation points, where V_e varies most rapidly. The first point (I = 1) is the stagnation point. Thus the "x" of the boundary-layer equations, called XX in INTGRL, is set to zero at the first point. Its change from one point to the next is calculated from the Pythagorean theorem; see Fig. 7.35.

The next step is to determine the gradient of the external flow velocity at the calculation points. This is done by fitting a parabola to the values of $V_e(x)$ at three successive points; that is, by finding constants α, β, and γ such that

$$\mathrm{VE} = \alpha + \beta \mathrm{XX} + \gamma \mathrm{XX}^2$$

at three adjacent points. We then differentiate the formula with respect to XX, and evaluate the result at the point under study. See problem 10 for the results. Generally, the three interpolation points are the point at which dV_e/dx is being estimated and the points just upstream and just downstream. At the first and last points, we must instead interpolate data at the first three and last three points, respectively.

As noted in Section 4.8, inviscid flow results can be scaled with the size of the body and the magnitude of the onset flow and so are generally presented in dimensionless form, namely,

$$\tilde{V} \equiv V/V_{\text{ref}}$$
$$= f(\tilde{x}, \tilde{y})$$

where

$$\tilde{x} \equiv \frac{x}{L}; \qquad \tilde{y} \equiv \frac{y}{L}$$

Thus the function VE of program INTGRL is referenced to the onset flow velocity V_∞, and the coordinates yielded by the function subroutines X and Y are made dimensionless by half the chord of the ellipse (since the origin is at the midpoint of the ellipse, the stagnation points are put at $x = \pm 1$, $y = 0$). Accordingly, we must put the boundary-layer equations in compatibly dimensionless form. With

$$\tilde{V}_e \equiv V_e/V_{\text{ref}}$$
$$\tilde{x} \equiv x/L$$
$$\tilde{\theta} \equiv \theta/L \qquad (7\text{-}68)$$

equation 7-42 becomes

$$\text{Re}\,\frac{d}{d\tilde{x}}(\tilde{\theta}^2 \tilde{V}_e^6) = 0.45\tilde{V}_e^5 \qquad (7\text{-}69)$$

in which Re is the Reynolds number based on V_{ref} and L,

$$\text{Re} = \rho V_{\text{ref}} L/\mu \qquad (7\text{-}70)$$

Since l, λ, and H are already dimensionless, equation 7-43 can be used as is. However, it is convenient to rewrite equations 7-39 and 7-40 as

$$l = \tfrac{1}{2}\,\text{Re}\tilde{V}_e\tilde{\theta}\,c_f \qquad (7\text{-}71)$$

$$\lambda = \text{Re}\,\tilde{\theta}^2\,\frac{d\tilde{V}_e}{d\tilde{x}} \qquad (7\text{-}72)$$

The dimensionless equations 7-69 through 7-72 can also accommodate the case when V_e and x are defined in dimensional terms; just let $V_{\text{ref}} = 1$ m/sec and $L = 1$ m, for example.

The dimensionless forms of the equations of Head's method are similar enough to equations 7-34 and 7-63 to 7-65 that they need not be rewritten here; basically, you just replace dimensional variables like V_e and θ with their tilded (dimensionless) counterparts. The various scale factors cancel out of each equation.

When you run program INTGRL, you are asked to input the number of points at which you want the boundary layer calculated (NX) and the Reynolds number based on the reference length and velocity (RE). Then the boundary-layer calculations are ready to start. The value of θ at the stagnation point is obtained from equation 7-44, whose dimensionless version is

$$\tilde{\theta}(0) = \sqrt{\frac{0.075}{\text{Re}\,\tilde{\dot{V}}_0}}$$

where

$$\tilde{\dot{V}}_0 \equiv \frac{d\tilde{V}_e}{d\tilde{x}}(0)$$

Thwaites's method is then used to calculate θ, H, and c_f up to transition. Specifically, the differential equation 7-69 is formally integrated from the (I − 1)th point to the Ith,

$$\text{Re}\,\tilde{V}_e^6\tilde{\theta}\Big|_{\text{I}-1}^{\text{I}} = 0.45\int_{\text{XX(I}-1)}^{\text{XX(I)}} \tilde{V}_e^5\,d\tilde{x} \qquad (7\text{-}73)$$

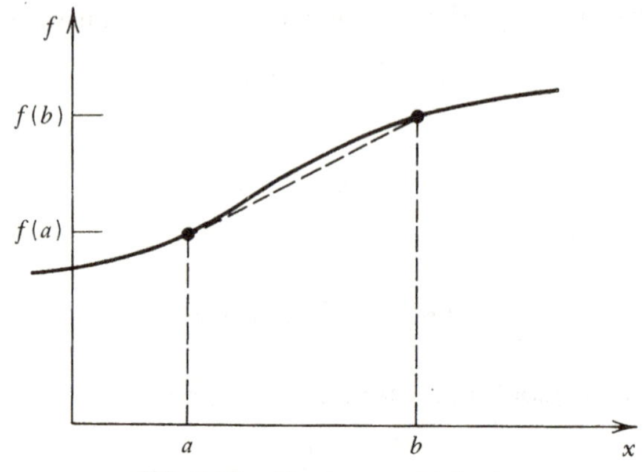

FIG. 7.36. The trapezoidal rule.

and the value of the integral approximated by the *trapezoidal rule*. As the name is intended to imply, this procedure approximates the variation of the integrand between XX(I − 1) and XX(I) as linear, so that the integral is just the area of a trapezoid. For the case illustrated in Fig. 7.36, the trapezoidal rule gives

$$\int_a^b f(x)\,dx \approx \tfrac{1}{2}[f(a) + f(b)](b - a) \tag{7-74}$$

However, a linear approximation of the integrand of equation 7-73 is totally inadequate near $\tilde{x} = 0$ (when I = 2). There

$$\tilde{V}_e^5 \approx \tilde{V}_0^5\, \tilde{x}^5$$

Thus the trapezoidal-rule approximation of equation 7-73 is used only for I = 3, 4, ... The correct consistent result for θ_2 is simply $\theta_2 = \theta_1$.

Thwaites's method is used until one of three things happens:

1. The NXth (last) point at which data are to be calculated is reached. The program then stops.
2. The quantity $\lambda < -0.0842$, indicating laminar separation. The program outputs an appropriate message and then stops.
3. Transition to turbulence is predicted, whether by Michel's method or because the point on the body where transition is known (or guessed) to occur is reached.

In the last case, the program now shifts to Head's method for turbulent boundary layers. It begins by asking for the value of H at transition.[9] As noted previously, the response should be in the range 1.3 to 1.4. The calculation proceeds until turbulent separation is predicted, or the NXth point is reached, at which time a message is

[9] Note that the index I at which the boundary layer first goes turbulent is stored for future reference, as ITRANS.

printed out and the user is invited to restart the turbulent calculations, with another H. If the H input is less than 1.0, then the program stops.

To integrate the two differential equations 7-34 and 7-63, a second-order *Runge–Kutta method* is used. The way this works can be seen by considering the problem

$$\frac{dy}{dx} = f(x, y) \tag{7-75}$$

Suppose we have an approximation Y_i to y at x_i and seek a corresponding approximation Y_{i+1} to $y(x_{i+1})$. Calculate

$$\begin{aligned} f_1 &\equiv f(x_i, Y_i) \\ Y_1^* &\equiv Y_i + (x_{i+1} - x_i)f_1 \\ f_2 &\equiv f(x_{i+1}, Y_1^*) \end{aligned} \tag{7-76}$$

quantities that are illustrated in Fig. 7.37, and then set

$$Y_{i+1} = Y_i + (x_{i+1} - x_i)\tfrac{1}{2}(f_1 + f_2) \tag{7-77}$$

Note that this is almost like the trapezoidal rule discussed earlier, which would differ only in replacing $f_1 + f_2$ by

$$f(x_i, Y_i) + f(x_{i+1}, Y_{i+1})$$

That would be nice, but we don't know Y_{i+1} until equation 7-77 is applied. However, it can be shown that equation 7-77 is just as accurate as a trapezoidal rule, anyway. Both methods give an error that, for sufficiently small $\Delta x \equiv x_{i+1} - x_i$, is proportional to $(\Delta x)^2$ (hence the "second-order" part of the name of the method).

As noted previously, the Ludwieg–Tillman formula (7-65) always gives a positive c_f, so that we cannot predict separation by looking for a zero of c_f. The simplest alternative criterion for turbulent separation is based on the computed value of H. Thus, when $H > 2.4$, INTGRL declares that the boundary layer has separated [13]. This criterion is usually adequate, since H tends to increase rapidly near separation anyway.

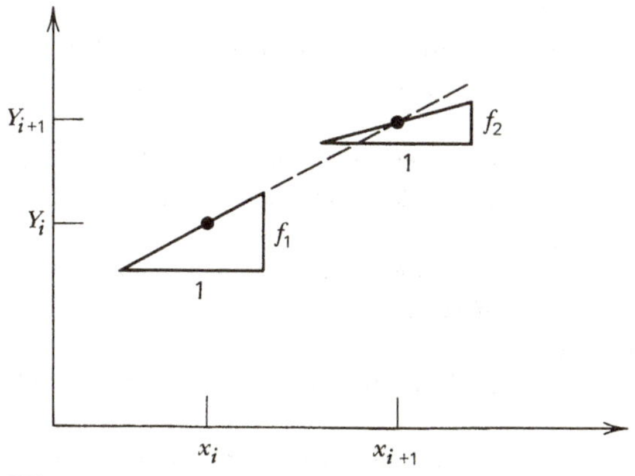

FIG. 7.37. The second-order Runge–Kutta method.

To adapt INTGRL to the case where the data on $V_e(x)$ are computed numerically, let the generator of those data output them on a file or tape, in the form

$$\begin{array}{lll} \text{NX} & & \\ \text{X}_1 & \text{Y}_1 & \text{VE}_1 \\ \text{X}_2 & \text{Y}_2 & \text{VE}_2 \\ \vdots & \vdots & \vdots \\ \text{X}_{\text{NX}} & \text{Y}_{\text{NX}} & \text{VE}_{\text{NX}} \end{array}$$

Delete from INTGRL the three function subroutines, and replace the lines in the main program that input TAU and NX with appropriate "read" statements, such as

```
      READ(1,2000) NX
      DO 50 I = 1, NX
   50 READ(1,2010) X(I), Y(I), VE(I)
```

Add a common block in which to store the vectors **X**, **Y**, and **VE**, and insert that block into subroutine DERIVS as well as the main program. And that's it!

7.15. REFERENCES

1. Cebeci, T., and Bradshaw, P., *Momentum Transfer in Boundary Layers*. McGraw-Hill/Hemisphere, Washington, D.C. (1977).
2. Thwaites, B. (ed.), *Incompressible Aerodynamics*. Clarendon Press, Oxford (1960), p. 179.
3. Von Kármán, T., "Über laminare und turbulente reibung," *ZAMM* **1,** 233–252 (1921).
4. Pohlhausen, K., "Zur näherungsweisen Integration der Differential-gleichung der laminare Reibungsschicht," *ZAMM* **1,** 252–268 (1921).
5. Thwaites, B., "Approximate Calculation of the Laminar Boundary Layer," *Aeronaut. Q.* **1,** 245–280 (1949).
6. Batchelor, G. K., *An Introduction to Fluid Dynamics*. Cambridge University Press, Cambridge (1970), p. 336.
7. Dhawan, S., "Direct Measurements of Skin Friction," NACA Report 1121 (1953).
8. Coles, D., "The Law of the Wake in the Turbulent Boundary Layer," *J. Fluid Mech.* **1,** 191–226 (1956).
9. Coles, D., "The Young Person's Guide to the Data," *Stanford Conference on Computation of Turbulent Boundary Layers*. Stanford University Press, Stanford (1969), Vol. II, pp. 1–19.
10. White, F. M., *Viscous Fluid Flow*. McGraw-Hill, New York (1974).
11. Schlichting, H., *Boundary-Layer Theory*, translated by J. Kestin. McGraw-Hill, New York (1968), 6th ed.
12. Michel, R., "Etude de la Transition sur les Profiles d'Aile," ONERA Report 1/1578A (1951).
13. Kline, S. J., Bardina, J. G., and Stawn, R. C., "Correlation of the Detachment of Turbulent Boundary Layers," *AIAA J.* **21,** 68–73 (1983).

14. Abbott, I. H., and von Doenhoff, A. E., *Theory of Wing Sections*. Dover, New York (1959).
15. McGhee, R. J., and Beasley, W. D., "Low-Speed Aerodynamic Characteristics of a 17-Percent Thick Wing Section Designed for General Aviation Application," NASA TN D-7428 (1973).

7.16. PROBLEMS

1. Consider a steady, nonviscous, isentropic compressible flow. From the isentropic flow relationship

$$\frac{p}{\rho^\gamma} = \frac{p_\infty}{\rho_\infty^\gamma}$$

show that changes in density and pressure in the flow field can be estimated from

$$\frac{\Delta\rho}{\rho_\infty} \sim \frac{1}{\gamma}\frac{\Delta p}{p_\infty} \sim M_\infty^2$$

where M_∞ is the free-stream Mach number

$$M_\infty \equiv \frac{V_\infty}{a_\infty}, \qquad a_\infty \equiv \sqrt{\frac{\gamma p_\infty}{\rho_\infty}}$$

Then show that the continuity equation 2-31 may be reduced to its incompressible-flow version if $M_\infty^2 \ll 1$.

2. Estimate the variation of the wall shear-stress coefficient

$$c_f \equiv \frac{\tau_w}{\frac{1}{2}\rho V_e^2}$$

with Reynolds number $\rho V_e L/\mu$ for a steady laminar incompressible flow. That is, use the concept of a characteristic length to estimate the value of τ_w from equation 7-24, and use equation 7-15 to estimate δ/L.

3. Some inviscid-flow methods (DUBLET, for example) satisfy flow tangency by requiring the stream function to be constant on the body surface. Show that the effect of boundary-layer growth on the pressure distribution can be accounted for in such a method by requiring instead that

$$\psi = -V_e\delta^* + \text{constant}$$

on the body surface.

4. Integrate Thwaites's differential equation 7-69 analytically for the case

$$\tilde{V}_e = \tilde{x}^m$$

Plot against m, for $-0.09 < m < 1.0$, the quantities

$$\frac{\mathrm{Re}_\theta}{\sqrt{\mathrm{Re}_x}}, \qquad H, \qquad \sqrt{\mathrm{Re}_x}\, c_f$$

In Chapter 10, we shall discuss the numerical solution of the differential boundary-layer equations under this velocity distribution, which leads to the well-known *similar solutions* (which constituted part of the data base Thwaites used to construct his method).

5. Use Thwaites's method to estimate the *entrance length* of a two-dimensional duct; that is, the distance required for the flow to become "fully developed," so that the flow is unidirectional and the velocity profile is the same as given by the Poisseuille flow solution of Section 6.15 along the rest of the duct. Take the flow to be uniform across the entrance to the duct, as shown in the sketch. Approximate the variation of the velocity at the edge of the boundary layer to be linear from the entrance to the point where the flow is fully developed. That is, in view of problem 11 of Chapter Six, set $V_e = V_0\left(1 + \frac{1}{2}\frac{x}{L}\right)$. Compare your results with the distance required for the boundary layer on a flat plate to become as thick as the half-width of the duct, and discuss the difference.

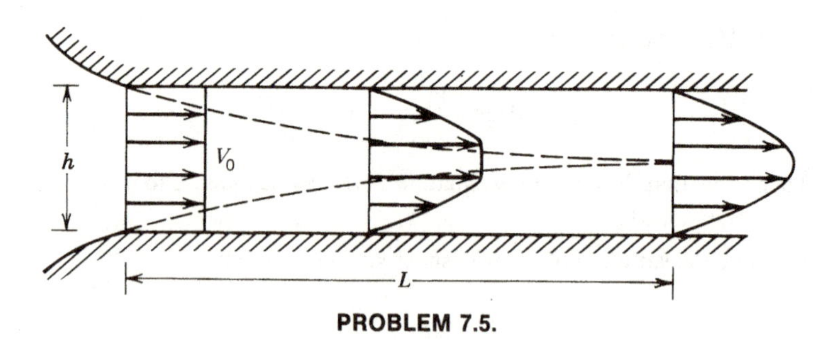

PROBLEM 7.5.

6. If

$$u = u_0, \qquad v = 0 \qquad \text{when} \quad 0 < t < \frac{T}{3}$$

and

$$u = 0, \qquad v = v_0 \qquad \text{when} \quad \frac{T}{3} < t < T$$

and u, v are periodic, with period T, compute $\bar{u}$, $\bar{v}$, $\overline{uv}$, and $\overline{u'v'}$.

7. Explain the effect of thickness on the maximum lift coefficients of the NACA 00XX airfoils whose performance characteristics are shown in Figs. 7.17 through 7.19. *Hint:* Run PANEL and look at their pressure distributions at about 6° angle of attack. However, you ought to be able to guess the effect of thickness on these pressure distributions.

8. Explain the effect of thickness ratio on the minimum drag coefficients of the 00XX airfoils. *Hint:* Look at their pressure distributions at $\alpha = 0°$.

9. Use PANEL to examine the pressure distributions on a cambered NACA $XX12$ airfoil at various angles of attack. Estimate the lift coefficient at which the drag coefficient should achieve its minimum.

10. Show that if $V_i \equiv V_e(x_i)$ for $i = 1, 2, 3$,

$$\frac{dV_e}{dx}(x_1) \approx \frac{1}{x_3 - x_2}\left[a(V_2 - V_1) - \frac{V_3 - V_1}{a}\right]$$

where

$$a \equiv \frac{x_3 - x_1}{x_2 - x_1}$$

Hint: Follow the procedure outlined in Section 7.14, namely,

(a) Find an approximate formula for $V_e(x)$ by fitting the parabola $V_e(x) \approx \alpha + \beta x + \gamma x^2$ to the data $V_e = V_1, V_2, V_3$ at $x = x_1, x_2, x_3$.

(b) Differentiate the parabolic formula with respect to x and evaluate the result at $x = x_1$.

11. Run program INTGRL to explore the effect of thickness ratio and Reynolds number on the behavior of the boundary layer. Take $0.05 \le \tau \le 0.5$ and $4.0 \le \log_{10} \mathrm{Re} \le 7.0$.

(a) Plot the values of x at which you get laminar separation, transition to turbulence (according to Michel's formula), and turbulent separation against $\log_{10} \mathrm{Re}$, one plot for each τ.

(b) For a case in which you do get transition to turbulence, plot H, c_f, and θ versus x. Show the effect of different choices for H at transition.

(c) The "point" at which you get transition is rather sensitive to the values calculated for θ. Try successively larger values of NX (for some particular τ and a fairly large Re) and discuss the convergence of your results.

12. Modify INTGRL to accept output from DUBLET, and repeat problem 11 for an NACA 0012 airfoil.

13. Modify INTGRL to calculate the boundary-layer development on a flat plate. Plot the total drag per unit width for one side of the plate against the Reynolds number based on plate length, for $\mathrm{Re}_L = 10^4$ to 10^9.

7.17. COMPUTER PROGRAM

```
C      PROGRAM INTGRL
C
C         INTEGRAL METHOD FOR CALCULATION OF BOUNDARY-LAYER
C         GROWTH ON AN AIRFOIL, STARTING AT A STAGNATION POINT
C
C         THWAITES'S METHOD USED FOR LAMINAR-FLOW REGION
C         MICHEL'S METHOD USED TO FIX TRANSITION
C         HEAD'S METHOD USED FOR TURBULENT-FLOW REGION
C
```

```
      COMMON /NUM/ PI,NX
      DIMENSION     YY(50)
      COMMON        XX(100),VGRAD(100),THETA(100)
      COMMON /REY/ RE
      COMMON /BOD/ TAU
      REAL          LAMBDA,L
      PI          =  3.1415926535
C
C                   INPUT THICKNESS RATIO OF ELLIPSE
C
      PRINT,      ' INPUT TAU
      READ,       TAU
C
C                   INPUT NUMBER OF STATIONS ALONG AIRFOIL
C
      PRINT,      ' INPUT NX
      READ,       NX
C
C
C                   FIND DISTANCES BETWEEN NODES ALONG SURFACE
C
      XX(1)       =  0.0
      DO 100      I = 2,NX
      DX          =  X(I) - X(I-1)
      DY          =  Y(I) - Y(I-1)
  100 XX(I)       =  XX(I-1) + SQRT(DX*DX + DY*DY)
C
C                   FIND VELOCITY GRADIENT AT NODES
C
      V1          =  VE(3)
      X1          =  XX(3)
      V2          =  VE(1)
      X2          =  XX(1)
C     VE(NX+1)    =  VE(NX-2)
      XX(NX+1)    =  XX(NX-2)
      DO 110      I = 1,NX
      V3          =  V1
      X3          =  X1
      V1          =  V2
      X1          =  X2
      V2          =  VE(NX-2)
      IF (I .LT. NX)        V2 = VE(I+1)
      X2          =  XX(I+1)
      FACT        =  (X3 - X1)/(X2 - X1)
      VGRAD(I)    =  ((V2 - V1)*FACT - (V3 - V1)/FACT)/(X3 - X2)
  110 CONTINUE
C
C                   SELECT TRANSITION CRITERION
C
      PRINT,      ' DO YOU WANT TO FIX TRANSITION LOCATION (F)
      PRINT,      ' OR USE MICHELS CRITERION (M)
      READ 1030, IANS
      IF (IANS .EQ. 'M')      GO TO 120
      PRINT,      ' INPUT TRANSITION LOCATION
      READ,   XTRANS
C
C                   INPUT REYNOLDS NUMBER BASED ON REFERENCE V AND L
C
  120 PRINT,      ' INPUT RE
      READ,       RE
      PRINT 1000
```

```
C
C                    LAMINAR-FLOW REGION
C
      THETA(1)    =  SQRT(.075/RE/VGRAD(1))
      I           =  1
  200 LAMBDA      =  THETA(I)**2*VGRAD(I)*RE
      IF (LAMBDA .LT. -.0842)   GO TO 400
      CALL THWATS(LAMBDA,H,L)
      CF          =  2.*L/RE/THETA(I)
      IF (I .GT. 1)     CF = CF/VE(I)
      PRINT 1010, X(I),Y(I),VE(I),VGRAD(I),THETA(I),H,CF
      I           =  I + 1
      IF (I .GT. NX)         STOP
      DTH2VE6     =  .225*(VE(I)**5 + VE(I-1)**5)*(XX(I) - XX(I-1))/RE
      THETA(I)    =  SQRT(((THETA(I-1)**2)*(VE(I-1)**6) + DTH2VE6)
     +               /(VE(I)**6))
      IF (I .EQ. 2)      THETA(2) = THETA(1)
C
C                    TEST FOR TRANSITION
C
      IF (IANS .EQ. 'M')      GO TO 210
      IF (X(I) .LT. XTRANS)   GO TO 200
      GO TO 300
  210 REX         =  RE*XX(I)*VE(I)
      RET         =  RE*THETA(I)*VE(I)
      RETMAX      =  1.174*(1. + 22400./REX)*REX**.46
      IF (RET .LT. RETMAX)    GO TO 200
C
C                    TURBULENT-FLOW REGION
C
  300 ITRANS      =  I
  310 PRINT,      ' INPUT H AT TRANSITION',
      READ,       H
      IF (H .LT. 1.0)        STOP
      I           =  ITRANS
      YY(2)       =  H1OFH(H)
      YY(1)       =  THETA(I-1)
  320 DX          =  XX(I) - XX(I-1)
      CALL RUNGE2(I-1,I,DX,YY,2)
      THETA(I)    =  YY(1)
      H           =  HOFH1(YY(2))
      RTHETA      =  RE*VE(I)*THETA(I)
      CF          =  CFTURB(RTHETA,H)
      PRINT 1020, X(I),Y(I),VE(I),VGRAD(I),THETA(I),H,CF
      IF (H .GT. 2.4) GO TO 410
      I           =  I + 1
      IF (I .LE. NX)         GO TO 320
      STOP
  400 PRINT,      ' LAMINAR SEPARATION'
      STOP
  410 PRINT,      ' TURBULENT SEPARATION'
      GO TO 310
 1000 FORMAT(///,9X,'X',8X,'Y',7X,'VE',6X,'VDOT',5X,'THETA',8X,'H',
     +           8X,'CF',/)
 1010 FORMAT(' L',F10.5,F9.5,2F9.4,F11.7,F9.4,F10.6)
 1020 FORMAT(' T',F10.5,F9.5,2F9.4,F11.7,F9.4,F10.6)
 1030 FORMAT(A1)
      END
C=================================================================
```

```
      SUBROUTINE THWATS(LAMBDA,H,L)
C
C                   THWAITES'S CORRELATION FORMULAS
C
      REAL            L,LAMBDA
      IF (LAMBDA .LT. 0.0)   GO TO 100
      L               =  .22 + LAMBDA*(1.57 - 1.8*LAMBDA)
      H               =  2.61 - LAMBDA*(3.75  - 5.24*LAMBDA)
      GO TO 200
  100 L               =  .22 + 1.402*LAMBDA + .018*LAMBDA/(.107 + LAMBDA)
      H               =  2.088 + .0731/(.14 + LAMBDA)
  200 RETURN
      END
C===============================================================
      FUNCTION H1OFH(H)
C
C                   HEAD'S CORRELATION FORMULA FOR H1(H)
C
      IF (H .GT. 1.6)         GO TO 100
      H1OFH        =  3.3 + .8234*(H - 1.1)**(-1.287)
      RETURN
  100 H1OFH        =  3.3 + 1.5501*(H - .6778)**(-3.064)
      RETURN
      END
C=======================================================
      FUNCTION HOFH1(H1)
C
C                   INVERSE OF H1(H)
C
      IF (H1 .LT. 3.3)        GO TO 110
      IF (H1 .LT. 5.3)        GO TO 100
      HOFH1        =  1.1 + .86*(H1 - 3.3)**(-.777)
      RETURN
  100 HOFH1        =  .6778 + 1.1536*(H1 - 3.3)**(-.326)
      RETURN
  110 HOFH1        =  3.0
      RETURN
      END
C=========================================================
      FUNCTION CFTURB(RTHETA,H)
C
C                   LUDWIEG-TILLMAN SKIN FRICTION FORMULA
C
      CFTURB       =  .246*(10.**(-.678*H))*(RTHETA)**(-.268)
      RETURN
      END
C========================================================
      SUBROUTINE DERIVS(I)
C
C                   SET DERIVATIVES OY VECTOR Y
C
      COMMON /RNK/ YT(50),YP(50)
      COMMON        XX(100),VGRAD(100),THETA(100)
      COMMON       /REY/ RE
      H1           =  YT(2)
      IF (H1 .LE. 3.)   RETURN
      H            =  HOFH1(H1)
      RTHETA       =  RE*VE(I)*YT(1)
      YP(1)        =  - (H + 2.)*YT(1)*VGRAD(I)/VE(I)
     +                + .5*CFTURB(RTHETA,H)
```

```
      YP(2)          =  - H1*(VGRAD(I)/VE(I) + YP(1)/YT(1))
     +                  +  .0306*(H1 - 3.)**(-.6169)/YT(1)
      RETURN
      END
C=====================================================
      SUBROUTINE RUNGE2(IO,I1,DX,YY,N)
C
C                    2ND-ORDER RUNGE-KUTTA METHOD FOR SYSTEM
C                    OF N FIRST-ORDER EQUATIONS
C
      DIMENSION     YS(50),YY(50)
      COMMON /RNK/ YT(50),YP(50)
      INTVLS        =  I1 - IO
      IF (INTVLS .LT. 1)      GO TO 200
      DO 130        I = 1,INTVLS
      DO 100        J = 1,N
  100 YT(J)         =  YY(J)
      CALL DERIVS(IO + I - 1)
      DO 110        J = 1,N
      YT(J)         =  YY(J) + DX*YP(J)
  110 YS(J)         =  YY(J) + .5*DX*YP(J)
      CALL DERIVS(IO + 1)
      DO 120        J = 1,N
  120 YY(J)         =  YS(J) + .5*DX*YP(J)
  130 CONTINUE
  200 RETURN
      END
C==============================================================
      FUNCTION X(I)
C
C                    X FOR ELLIPSE
C
      COMMON /NUM/ PI,NX
      X             =  - COS(PI*(I - 1)/FLOAT(NX - 1))
      RETURN
      END
C==============================================================
      FUNCTION Y(I)
C
C                    Y FOR ELLIPSE
C
      COMMON /NUM/ PI,NX
      COMMON /BOD/ TAU
      Y             =  SIN(PI*(I - 1)/FLOAT(NX - 1))*TAU
      RETURN
      END
C==============================================================
      FUNCTION VE(I)
C
C                    VE FOR ELLIPSE
C
      COMMON /BOD/ TAU
      VE             =  (1. + TAU)*SQRT((1. - X(I)**2)
     +                    /(1. - (1.-TAU**2)*X(I)**2))
      RETURN
      END
```

CHAPTER 8

PANEL METHODS

In Chapter 4, we described and implemented the classic panel method devised by Smith and Hess [1], the so-called "Douglas Neumann" method based on constant-strength distributions of sources and vortices on straight-line panels. This method made it possible, for the first time, to analyze flows past bodies of realistic geometry. No longer was the aerodynamic analyst forced to ignore (or at least to fudge) the facts that wings have thickness, that flaps are often deflected more than a few degrees, that fuselages are not always axisymmetric, and that wing-body junctions are smoothed over with fillets. It can be said that this method marked the birth of computational aerodynamics, a field that has thrived ever since as computers get larger and faster and analysts get ever more clever. It is not unusual now to read claims that "computers should [soon] begin to supplant wind tunnels in the aerodynamic design and testing process" [2].

Some of the progress that has been made since 1960 is in the design of panel methods themselves. This chapter takes a broader and deeper look at the variety of panel methods now in use.

8.1. MATHEMATICAL FOUNDATIONS: GREEN'S IDENTITY

One of the most important virtues of panel methods is their near-universal applicability. Any incompressible irrotational flow can be represented by a distribution of sources and vortices (or doublets) over its bounding surfaces.

The proof rests on the divergence theorem, equation 2-17

$$\int_{\mathscr{V}} \nabla \cdot \mathbf{U}\, d\mathscr{V} = -\int_{S} \hat{\mathbf{n}} \cdot \mathbf{U}\, dS \tag{8-1}$$

Here $\mathscr{V}$ is a region in space (a flow field in our case), S is the boundary of $\mathscr{V}$, and $\hat{\mathbf{n}}$ is a unit vector normal to S and directed into $\mathscr{V}$.[1] Since equation 8-1 is basically a three-dimensional version of the "fundamental theorem of calculus"

$$\int_a^b \frac{df}{dx}\,dx = f(b) - f(a)$$

it is important that $\mathbf{U}$ be a continuous function of position inside $\mathscr{V}$. This has implications as to the design of the surface S that bounds $\mathscr{V}$, which are hinted at in Fig. 8.1 and will be detailed later.

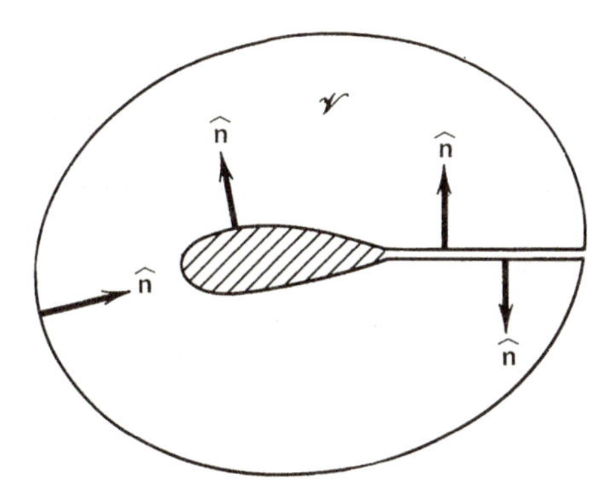

FIG. 8.1. Control volume for application of divergence theorem.

The vector $\mathbf{U}$ to which we shall apply equation 8-1 is defined by

$$\mathbf{U} = \phi\nabla\phi_s - \phi_s\nabla\phi \tag{8-2}$$

where ϕ is the velocity potential of the flow in $\mathscr{V}$, so that

$$\mathbf{V} = \nabla\phi \tag{8-3}$$

is the fluid velocity in $\mathscr{V}$, whereas ϕ_s is the potential of a source of unit strength at some arbitrary point P in $\mathscr{V}$:

$$\phi_s = \frac{1}{2\pi}\ln r \qquad \text{in two dimensions}$$

$$= -\frac{1}{4\pi}\frac{1}{r} \qquad \text{in three dimensions} \tag{8-4}$$

In either case, r is the distance from P to the point at which $\mathbf{U}$ is to be evaluated, as shown in Fig. 8.2. As will be seen shortly, this choice of $\mathbf{U}$ leads to a formula for ϕ *at* P in terms of data on the bounding surface S.

If the function $\mathbf{U}$ defined by equation 8-2 is to be continuous in $\mathscr{V}$, so must ϕ, $\nabla\phi$, ϕ_s, and $\nabla\phi_s$. But ϕ_s and its derivatives blow up at P. Therefore, before integrating

[1] In equation 2-17 and in mathematics texts, the unit normal is directed out of $\mathscr{V}$. Aerodynamicists prefer to have the unit normal outward from solid surfaces, which necessitates the minus sign in equation 8-1.

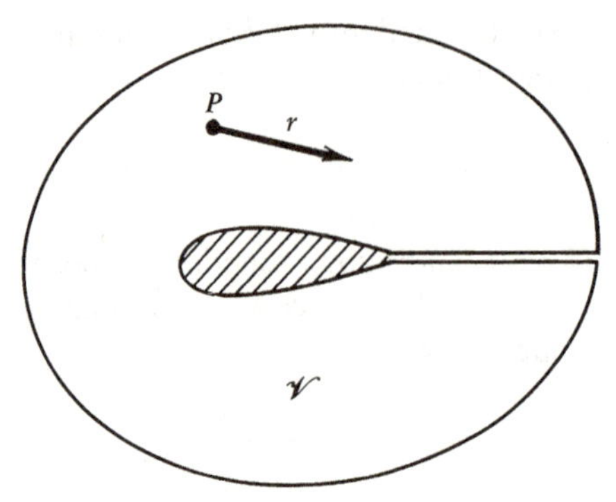

FIG. 8.2. Nomenclature for source potential.

over $\mathscr{V}$, we will carve out a small circle (in two dimensions) or sphere (in three dimensions) centered at P and of radius ε, as shown in Fig. 8.3. Let $\mathscr{V}_\varepsilon$ be the part of $\mathscr{V}$ outside that excluded region, and S_ε be the surface of the circle or sphere.

For the same reason, we must exclude from $\mathscr{V}$ such entities as the vortex sheet shed by a wing of finite span, since the velocity $\nabla\phi$ is discontinuous across such a sheet. This can be done by the device illustrated in Figs. 8.1 to 8.3, namely, by including in the bounding surface S a two-sided component that sandwiches the vortex sheet. Then $\nabla\phi$ can be assigned its proper values on either side of the sheet while remaining continuous in $\mathscr{V}$ (and $\mathscr{V}_\varepsilon$).

A similar device is necessary in two-dimensional problems, for which the velocity $\nabla\phi$ is continuous off the body—an airfoil in steady flow has no wake—but ϕ is not. For example, recall from problem 5 of Chapter 2 the solution for uniform flow past a circular cylinder with circulation Γ:

$$\phi = V_\infty r \cos\theta\left(1 + \frac{a^2}{r^2}\right) - \frac{\Gamma}{2\pi}\theta$$

Since θ is not a single-valued function of position—on the positive x-axis, for example, θ could be 0, 2π, -4π, and so on—neither is ϕ. To make ϕ single valued, we can restrict the range of θ to any interval of 2π. This amounts to bounding $\mathscr{V}$, the

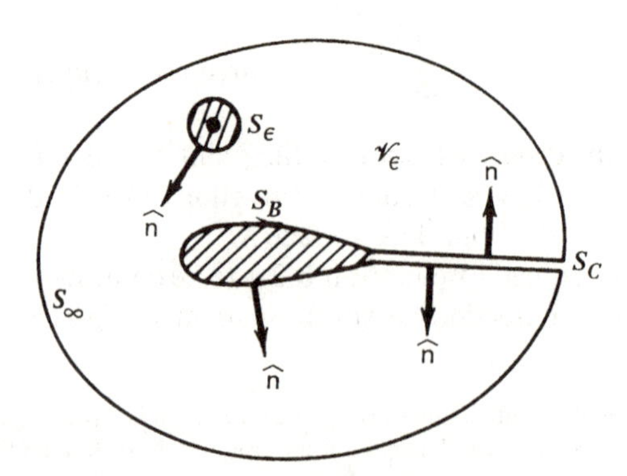

FIG. 8.3. Nomenclature for derivation of Green's identity.

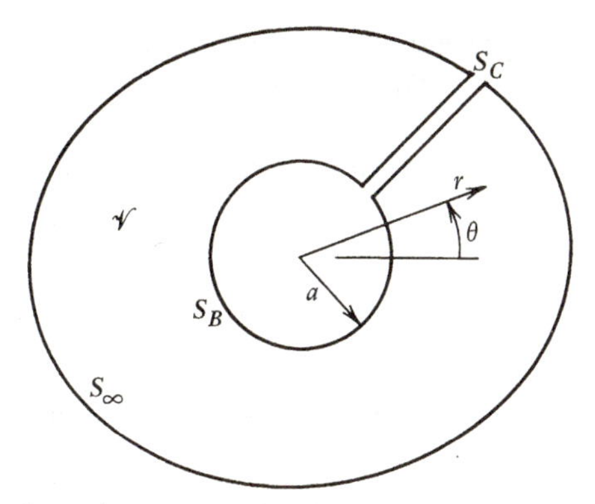

FIG. 8.4. Need for a branch cut to make the potential for flow past a cylinder with circulation single valued.

region in which we want to discuss ϕ, not only by the surface of the body (S_B in Fig. 8.4) and by a surface S_∞ far from the body, but by a surface S_C that connects S_B and S_∞.

More generally, consider flow past an airfoil with circulation Γ. From the definition (2-51),

$$\Gamma = \oint_C \mathbf{V} \cdot \mathbf{dl}$$

where C is a closed curve around the airfoil, as shown in Fig. 8.5. Also by definition, the difference in ϕ between two points P_1 and P_2 is

$$\phi(P_2) - \phi(P_1) = \int_{P_1}^{P_2} \nabla\phi \cdot \mathbf{dl} = \int_{P_1}^{P_2} \mathbf{V} \cdot \mathbf{dl}$$

Now if one goes from P_1 to P_2 on the part of C above the airfoil, and returns to P_1 on the other part of C, one gets

$$\phi(P_1) - \phi(P_1) = \oint_C \mathbf{V} \cdot \mathbf{dl} = \Gamma \tag{8-5}$$

which shows ϕ to be multivalued; the value of ϕ at P_1 is not the same at the beginning and end of the trip. Again, the solution is to insert a "branch cut" in the

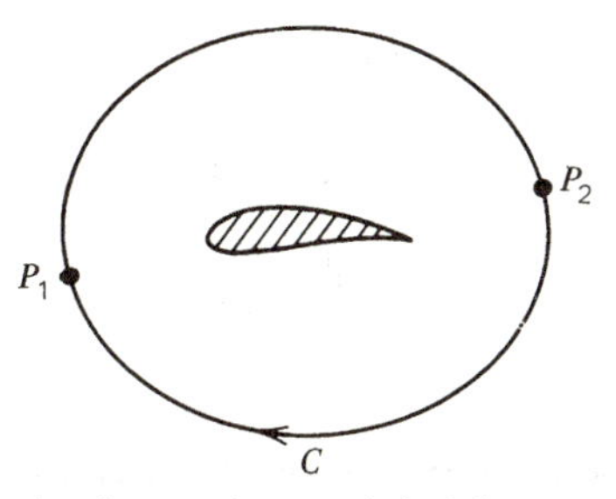

FIG. 8.5. Multivaluedness of potential of flow past a lifting airfoil.

domain of interest that prevents circumnavigation of the airfoil, which is what was done in Figs. 8.1 to 8.3. This makes the region $\mathscr{V}$ "simply connected," in mathematical parlance.

In summary, the surface S generally has three components:

1. S_B, the surface(s) of the body(ies) immersed in the flow.
2. S_∞, a surface far from S_B.
3. S_C, a two-sided surface that runs between S_B and S_∞ and which sandwiches discontinuities in ϕ and/or $\nabla\phi$.

Now the source potential ϕ_s satisfies Laplace's equation 2-38

$$\nabla^2\phi_s = 0$$

everywhere in $\mathscr{V}_\varepsilon$, just as does ϕ. Therefore, the divergence of the vector $\mathbf{U}$ defined in equation 8-2 is

$$\begin{aligned}\nabla\cdot\mathbf{U} &= \nabla\phi\cdot\nabla\phi_s + \phi\nabla\cdot\nabla\phi_s - \nabla\phi_s\cdot\nabla\phi - \phi_s\nabla\cdot\nabla\phi \\ &= \phi\nabla^2\phi_s - \phi_s\nabla^2\phi \\ &= 0 \qquad \text{in } \mathscr{V}_\varepsilon \end{aligned} \tag{8-6}$$

Thus, applying equation 8-1 to the region $\mathscr{V}_\varepsilon$ gives

$$\int_{\mathscr{V}_\varepsilon} \nabla\cdot\mathbf{U}\, d\mathscr{V} = 0 = -\int_{S+S_\varepsilon} \hat{\mathbf{n}}\cdot(\phi\nabla\phi_s - \phi_s\nabla\phi)\, dS \tag{8-7}$$

where we observe from Fig. 8.3 that the surface bounding $\mathscr{V}_\varepsilon$ is comprised of S, the surface bounding $\mathscr{V}$, and S_ε, the surface of the small sphere or cylinder surrounding P. It is convenient to separate the integrals over these two surfaces and to rewrite equation 8-7 as

$$\int_{S_\varepsilon} \hat{\mathbf{n}}\cdot(\phi\nabla\phi_s - \phi_s\nabla\phi)\, dS = -\int_{S} \hat{\mathbf{n}}\cdot(\phi\nabla\phi_s - \phi_s\nabla\phi)\, dS \tag{8-8}$$

Interesting things happen when we let ε, the radius of the sphere or cylinder surrounding P, go to zero. Then ϕ and $\nabla\phi$ approach their values at P,[2] ϕ_P and $\mathbf{V}_P$, say, so the left side of equation 8-8 becomes

$$\int_{S_\varepsilon} \hat{\mathbf{n}}\cdot(\phi\nabla\phi_s - \phi_s\nabla\phi)\, dS \approx \phi_P \int_{S_\varepsilon} \hat{\mathbf{n}}\cdot\nabla\phi_s\, dS - \mathbf{V}_P\cdot\int_{S_\varepsilon} \hat{\mathbf{n}}\phi_s\, dS$$

The first integral on the right is just the volume rate of flow through S_ε, which equals the strength of the source inside, namely, unity. As for the second integral, r is constant (ε) on S_ε, and so, from equation 8-4, is ϕ_s. Taking ϕ_s outside the integral leaves

$$\int_{S_\varepsilon} \hat{\mathbf{n}}\, dS$$

[2] But only because we have been careful to define S so that ϕ and $\nabla\phi$ are continuous functions of position inside $\mathscr{V}$.

which is zero by symmetry, so that equation 8-8 becomes, in the limit, what is known as *Green's identity*:

$$\phi_P = \int_S [(\hat{\mathbf{n}} \cdot \nabla\phi)\, \phi_s - \phi(\hat{\mathbf{n}} \cdot \nabla\phi_s)]\, dS \tag{8-9}$$

This remarkable formula gives the value of ϕ at any point P in $\mathscr{V}$, a region in which ϕ is a continuous solution of Laplace's equation 2-38, in terms of the values of ϕ and $\hat{\mathbf{n}} \cdot \nabla\phi$ on the boundary of $\mathscr{V}$.

Our derivation of equation 8-9 was necessarily rather mathematical. The result is important enough to deserve some physical interpretation. First, look at

$$\int_S (\hat{\mathbf{n}} \cdot \nabla\phi)\phi_s\, dS$$

The quantity ϕ_s, according to equations 8-4, depends only on the distance r between P and the dS whose contribution to the integral is under consideration. Thus, although introduced as the potential of a source of unit strength at point P, evaluated at a point on S that is a distance r away, it could also be taken as the potential of a source of unit strength at dS, evaluated at P. With this interpretation, the integral can be called the potential of a source distribution on S whose strength per unit area is $\hat{\mathbf{n}} \cdot \nabla\phi$, the component normal to S of the local fluid velocity.

The integral

$$\int_S \phi\hat{\mathbf{n}} \cdot \nabla\phi_s\, dS$$

has a similar interpretation. The gradient of a scalar is defined as the vector whose magnitude and direction are those of the maximum rate of change of the scalar. Thus, $\hat{\mathbf{n}} \cdot \nabla\phi_s$ is the rate of change of ϕ_s in the direction of $\hat{\mathbf{n}}$ at the element dS. This can be represented as follows. As shown in Fig. 8.6, let Q_1 and Q_2 be points a distance δ apart, on either side of dS, and arranged so that

$$\overrightarrow{Q_2Q_1} = \delta\hat{\mathbf{n}}$$

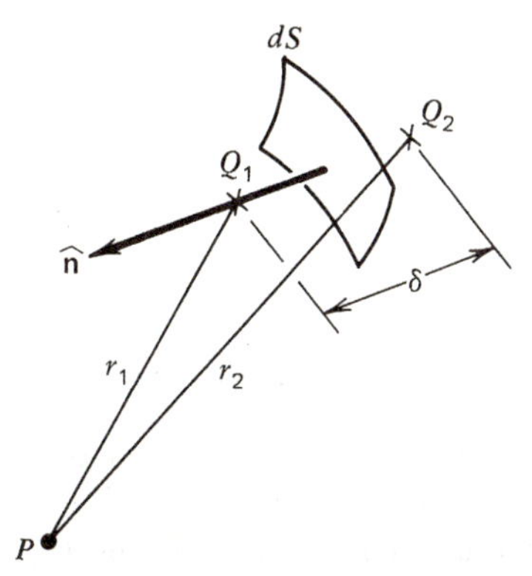

FIG. 8.6. Interpretation of second integral in equation 8-9 as a doublet distribution.

Let ϕ_1 and ϕ_2 be the values at Q_1 and Q_2, respectively, of the potential of a unit-strength source at P. Then

$$\hat{\mathbf{n}} \cdot \nabla\phi_s = \lim_{\delta \to 0} \frac{\phi_1 - \phi_2}{\delta}$$

However, as noted above, ϕ_1 and ϕ_2 can also be regarded as the values at P of the potentials due to unit-strength sources at Q_1 and Q_2. Then $\phi_1/\delta - \phi_2/\delta$ is the difference between the potentials at P of two sources of strength $1/\delta$ at Q_1 and Q_2. As $\delta \to 0$, they coalesce into what is called a *doublet*, whose strength is defined to be the product of the source strength and the distance between the sources (see problem 4 of Chapter 2), or, in this case, unity. From the viewpoint of an observer at P, the second part of the integral in equation 8-9 is, therefore, the potential of a doublet distribution over the surface S. The axes of the doublets are normal to S, and the strength per unit area of the distribution is ϕ, the local velocity potential.

Thus we have shown, as was advertised, that the velocity potential of any irrotational flow can be represented by a distribution of sources and doublets over its bounding surfaces. The strength of the source and doublet distributions per unit area are, respectively, the boundary values of the normal derivative of ϕ and of ϕ itself.

There are many interesting and useful variations on the theme of equation 8-9. First, let us separate out the integrals over the part of S far from whatever bodies are in the flow. On S_∞,

$$\phi \approx V_\infty(x \cos \alpha + y \sin \alpha) \tag{8-10}$$

so the integrals over S_∞ ought to be linear in the constants $V_\infty \cos \alpha$ and $V_\infty \sin \alpha$. Then, by inspection, we see that equation 8-9 can be rewritten

$$\phi_P = V_\infty(x_P \cos \alpha + y_P \sin \alpha) + \int_{S_B + S_C} [(\hat{\mathbf{n}} \cdot \nabla\phi)\phi_s - \phi(\hat{\mathbf{n}} \cdot \nabla\phi_s)]\, dS \tag{8-11}$$

For, when P is far from S_B and S_C, ϕ ought to satisfy equation 8-10, which the form of equation 8-11 ensures.

A related variation is to take the ϕ in the above derivation to be the *perturbation potential*, so

$$\mathbf{V} = \mathbf{V}_\infty + \nabla\phi \tag{8-12}$$

instead of equation 8-3. Then ϕ vanishes on S_∞, so that equation 8-9 can be used with S replaced by $S_B + S_C$.

8.2. POTENTIAL-BASED PANEL METHODS

Equations 8-9 and 8-11 are widely used as bases for very powerful panel methods, capable of treating a wide variety of potential flow problems. This approach to aerodynamics was pioneered by Morino and Kuo [3] and improved by Johnson and Rubbert [4].

To begin with, observe that the normal component of $\mathbf{V}\phi$ is known on the body surface, whether ϕ is the total or perturbation potential. From the flow tangency condition (3-5)

$$\hat{\mathbf{n}} \cdot \mathbf{V} = 0 \tag{8-13}$$

on solid bodies so, from equation 8-3,

$$\hat{\mathbf{n}} \cdot \nabla\phi = 0$$

on S_B if ϕ is the total velocity potential, whereas from equation 8-12,

$$\hat{\mathbf{n}} \cdot \nabla\phi = -\hat{\mathbf{n}} \cdot \mathbf{V}_\infty$$

on S_B if ϕ is the perturbation potential. Since it simplifies the writing, we'll take ϕ to be the total potential, but note that many widely-used panel methods are instead based on the perturbation potential.[3]

The normal component of $\nabla\phi$ can also be eliminated from the integrals over the two-sided surface S_C. The velocity component tangential to S_C is discontinuous across S_C in some cases (e.g., when S_C is the vortex sheet that is shed from a wing of finite span), but the normal component is always continuous. Since the unit normals are equal but opposite on opposite sides of S_C, as shown in Fig. 8.3, the contribution of any element dS of S_C to

$$\int_{S_C} (\hat{\mathbf{n}} \cdot \nabla\phi)\phi_S \, dS$$

is cancelled by that of the corresponding element on the opposite side of S_C.

However, ϕ *can* be discontinuous across S_C. The best we can do with its contribution to the integral over S_C is to integrate over both sides at once. Then only the jump in ϕ shows up, and equation 8-11 can be written

$$\phi_P = V_\infty(x_P \cos\alpha + y_P \sin\alpha) - \int_{S_B} \phi(\hat{\mathbf{n}} \cdot \nabla\phi_s) \, dS - \int_{S_C} \Delta\phi(\hat{\mathbf{n}} \cdot \nabla\phi_s) \, dS \tag{8-14}$$

Here

$$\Delta\phi = \phi^+ - \phi^-$$

where, as shown in Fig. 8.7, ϕ^+ is the value of ϕ on the side of S_C for which $\hat{\mathbf{n}}$ goes into the fluid, ϕ^- is the value of ϕ on the opposite side, and it is to be understood now that we integrate over only one side of S_C.

The jump in ϕ across S_C can often be related to values of ϕ on S_B. The relation is especially simple in two-dimensional problems, for which the two-sided surface S_C was introduced to keep ϕ single valued in the region $\mathscr{V}$ bounded by S_C, S_B, and S_∞. If

[3] Maskew [5] claims that the use of the perturbation potential yields better-conditioned algebraic equations.

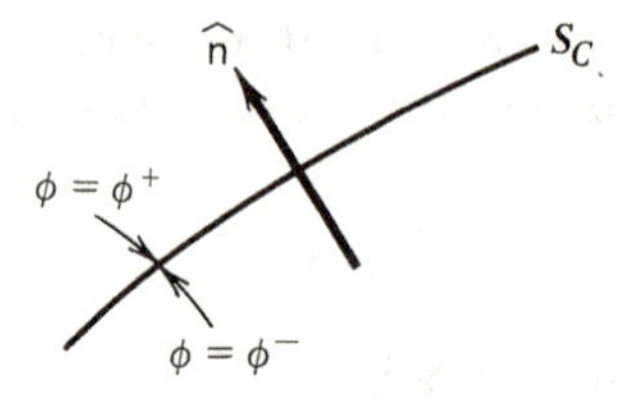

FIG. 8.7. Potential jump across connecting surface S_C.

P_1 and P_2 are the intersections of the two sides of S_C with S_B, as shown in Fig. 8.8, we generally have

$$\phi\Big|_{P_1} \neq \phi\Big|_{P_2}$$

Rather, as shown in equation 8-5,

$$\phi\Big|_{P_2} - \phi\Big|_{P_1} = \oint_{P_1}^{P_2} \nabla\phi \cdot \mathbf{dl} = \Gamma \tag{8-15}$$

the circulation about the airfoil. However, ϕ is continuous so long as we stay *inside* $\mathscr{V}$; in particular, it is continuous as we move along the boundary of $\mathscr{V}$,[4] including the transitions between S_C and S_B at P_1 and P_2. Thus the jump in ϕ across S_C at its junction with S_B is seen to be simply the difference between the values of ϕ at P_2 and P_1, or, from equation 8-15, Γ.

FIG. 8.8. Equating potential jump across connecting surface to circulation about airfoil.

In fact, the jump in ϕ across S_C has the value Γ all along S_C. For, in a two-dimensional flow, the velocity is continuous off the body.[5] Thus the component of velocity tangential to S_C is continuous across S_C. Since that velocity component equals the rate of change of ϕ along S_C, $\partial\phi/\partial t$, say, we have

$$\Delta\frac{\partial\phi}{\partial t} = \frac{\partial}{\partial t}\Delta\phi = 0$$

Therefore, the jump in ϕ is constant along S_C. Since that jump equals Γ at the intersection of S_C with S_B,

$$\Delta\phi = \Gamma \tag{8-16}$$

[4] Barring the appearance of a concentrated vortex, to which a discontinuity in doublet strength corresponds. See below.

[5] In three-dimensional flows, we have seen discontinuities in **V** off the body, at the vortex sheet shed from a lifting wing.

all along S_C, and equation 8-14 now simplifies to

$$\phi_P = V_\infty(x_P \cos\alpha + y_P \sin\alpha) - \int_{S_B} \phi(\hat{\mathbf{n}} \cdot \nabla\phi_s)\, dS - \Gamma \int_{S_C} \hat{\mathbf{n}} \cdot \nabla\phi_s\, dS \quad (8\text{-}17)$$

with Γ being related to values of ϕ on S_B by equation 8-15.

At last we have done enough mathematics, and can turn to the development of a panel method based on our mathematics. Specifically, equation 8-17 will be used as an integral equation for the values of ϕ on S_B, by letting the point P approach different points on S_B. As in the Smith–Hess method discussed in Chapter 4, we discretize the problem of finding ϕ on S_B by assuming a particular functional form for its variation on S_B and find the parameters involved in the chosen functional form by evaluating equation 8-17 at a sufficient number of "control points" on S_B. Once the parameters are determined, equation 8-17 can be used to evaluate ϕ anywhere in the flow field. The velocity field can then be found from equation 8-3 and the pressure field from Bernoulli's equation 2-37.

We shall discuss two such schemes. The first is a two-dimensional version of the method developed by Morino and Kuo [3], whereas the second, hinted at in their paper, may be relatively unexplored.

In both methods, the body is approximated by an N-sided polygon, as shown in Fig. 8.9. That is, as in the Smith–Hess method, the body is defined by N points ("nodes"), which are connected by straight-line "panels," which together comprise the surface S_B. Also as in Chapter 4, the nodes will be numbered in clockwise order, starting at the trailing edge, and the jth panel is the one between nodes j and $j + 1$. There are then N panels and the two which meet at the trailing edge are the panels numbered 1 and N. The two-sided surface S_C will be taken to run along the positive x axis, from the trailing edge (node 1) to $+\infty$, and will be called the $(N + 1)$st panel. See Fig. 8.9.

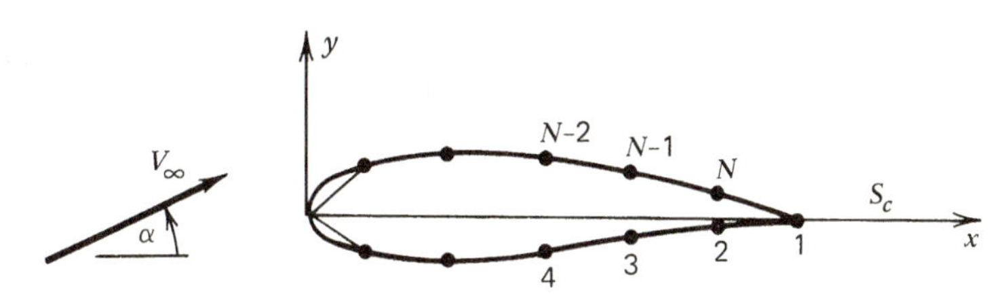

FIG. 8.9. Nomenclature for panels on airfoil.

8.2.1. Constant-Potential Method

Now to define the two-dimensional version of the Morino–Kuo method. Here we take ϕ to be constant on each of the N panels,

$$\phi = \phi_j$$

on panel j, and so, in accordance with equation 8-15, set

$$\Gamma = \phi_N - \phi_1 \quad (8\text{-}18)$$

There are then N unknowns $(\phi_1, \phi_2, \ldots, \phi_N)$, which we determine by evaluating equation 8-17 at the midpoint of each panel. With

$$\bar{x}_i \equiv \tfrac{1}{2}(x_i + x_{i+1}), \qquad \bar{y}_i \equiv \tfrac{1}{2}(y_i + y_{i+1})$$

these equations take the form

$$\phi_i = V_\infty(\bar{x}_i \cos\alpha + \bar{y}_i \sin\alpha) - \sum_{j=1}^{N} \phi_j \int_{\text{panel } j} \hat{\mathbf{n}} \cdot \nabla\phi_s \, dS - (\phi_N - \phi_1) \int_{S_C} \hat{\mathbf{n}} \cdot \nabla\phi_s \, dS$$

$$\text{for } i = 1, \ldots, N \tag{8-19}$$

As in Chapter 4, the integrals in equations 8-19 are most easily evaluated in coordinates oriented to the particular panel. When considering the contribution of the jth panel to equations 8-19, let (x^*, y^*) be the local coordinates of the midpoint of the ith panel, (x_i, y_i), and $(\xi, 0)$ the local coordinates of a typical point on the jth panel, as shown in Fig. 8.10. The value at $(\xi, 0)$ of $\hat{\mathbf{n}} \cdot \nabla\phi_s$ is the y^* component of the velocity due to a unit-strength source at (x^*, y^*). We can thus use equation 3-7 to obtain

$$\hat{\mathbf{n}} \cdot \nabla\phi_s \bigg|_{\xi,0} = -\frac{1}{2\pi} \frac{y^*}{(x^* - \xi)^2 + y^{*2}} \tag{8-20}$$

Then equations 8-19 may be written

$$\phi_i = V_\infty(\bar{x}_i \cos\alpha + \bar{y}_i \sin\alpha) + \frac{1}{2\pi} \sum_{j=1}^{N} \phi_j \int_0^{l_j} \frac{y^*}{(x^* - \xi)^2 + y^{*2}} \, d\xi$$

$$+ \frac{1}{2\pi}(\phi_N - \phi_1) \int_0^{\infty} \frac{y^*}{(x^* - \xi)^2 + y^{*2}} \, d\xi \tag{8-21}$$

These are the same integrals as those shown in Section 4.8 to be the angles subtended at (x_i, y_i) by the panel over which the integration takes place (see equation 4-88). Thus equations 8-19 can be put in the form

$$\sum_{j=1}^{N} A_{ij}\phi_j = b_i \qquad \text{for } i = 1, \ldots, N \tag{8-22}$$

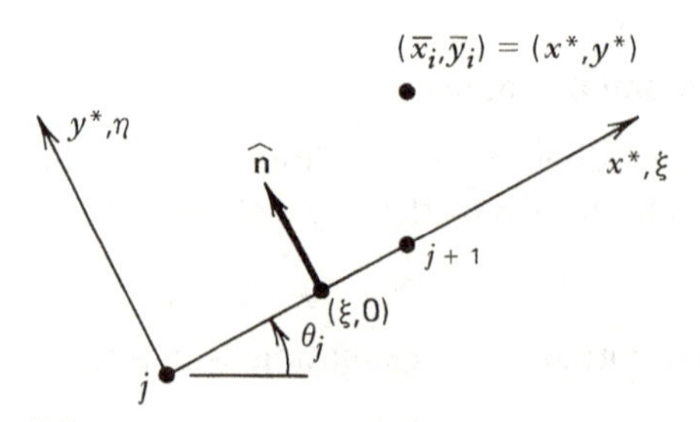

FIG. 8.10. Local panel-fixed coordinates.

with

$$b_i = -V_\infty(\bar{x}_i \cos\alpha + \bar{y}_i \sin\alpha)$$

$$\begin{aligned} A_{ij} &= \frac{1}{2\pi}\beta_{i1} - \delta_{i1} - \frac{1}{2\pi}\beta_{iN+1} && \text{if } j = 1 \\ &= \frac{1}{2\pi}\beta_{iN} - \delta_{iN} + \frac{1}{2\pi}\beta_{iN+1} && \text{if } j = N \\ &= \frac{1}{2\pi}\beta_{ij} - \delta_{ij} && \text{otherwise} \end{aligned} \tag{8-23}$$

Here

$$\begin{aligned} \delta_{ij} &\equiv 1 && \text{if } i = j \\ &\equiv 0 && \text{otherwise} \end{aligned}$$

is the so-called *Kronecker delta*, whereas the angles β_{ij} and β_{iN+1} are defined in Fig. 8.11.

The setting up and solution of equations 8-22 can be accomplished by a program much like the one introduced in Chapter 4. The only subroutines of PANEL that must be changed are COFISH, VELDIS, and FANDM, as follows.

COFISH sets the coefficients of the linear equations 8-22. In PANEL, there are $N + 1$ such equations (flow tangency at the middle of N panels, plus a Kutta condition). Now there are only N, so that all references to A(I, KUTTA) (KUTTA $= N + 1$) may be eliminated. The structure of COFISH is otherwise the same, with the object being to calculate A(I, J), for I, J $= 1, \ldots, N$ (N = NODTOT in PANEL). The quantity β_{ij} in equations 8-23 is nothing more than the FTAN of PANEL's COFISH. FLOG is not needed. From Fig. 8.11, since the coordinates of node 1 are $(c, 0)$, where c is the airfoil chord,

$$\beta_{iN+1} = \tan^{-1}\frac{\bar{y}_i}{c - \bar{x}_i}$$

The Kronecker deltas of equations 8-23 are most easily incorporated by adding, after the DO loop that runs over J but before the end of the one that runs over I, the

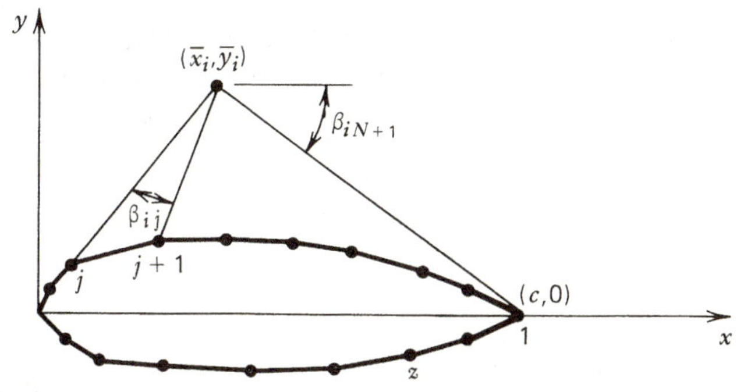

FIG. 8.11. Definitions of angles appearing in results for potential due to doublet panels.

statement

$$\text{A(I, I)} = \text{A(I, I)} - 1.0$$

Finally, in the unlabeled common block that contains the A matrix, KUTTA (which is the number of equations in PANEL) should be replaced by something equal to NODTOT (but not NODTOT itself, which is already used in another common block in PANEL).

Subroutine VELDIS computes the velocity and pressure distributions on the body. In PANEL, the velocity at the midpoint of each panel is calculated by adding the onset flow velocity to the velocities induced by the sources and vortices on every panel. A similar procedure could be used here. In view of equation 8-3, the velocity at any point P could be determined by differentiating equation 8-17 with respect to the coordinates of P. The integrals over S_B in the resultant expression could be approximated just as were those of equation 8-17 in equations 8-19. In fact, replacing ϕ_i in equations 8-21 with ϕ_P, we could get the same result by differentiating equations 8-21 with respect to the coordinates of P. The resultant integrals would not be any harder to evaluate than were those of equations 8-21.

However, for reasons to be discussed below, such a procedure almost always gives terrible results. The following alternative, which has the additional advantage of simplicity, is recommended.

The solution of equations 8-22 yields the values of the potential on the panels, under the assumption that the potential is constant on each panel, though different from one to the next. Since the velocity is just the gradient of the potential, an estimate for the velocity at the nodes can be obtained by taking ϕ_i to be the potential at the midpoint of the ith panel and approximating the velocity at the ith node by

$$V_i = \frac{\phi_i - \phi_{i-1}}{d} \tag{8-24}$$

where d, as shown in Fig. 8.12, is simply the distance between the midpoints of the ith and $(i - 1)$th panels. If you want the velocity at the panel midpoints instead, fit a quadratic polynomial to the values of ϕ at three panel midpoints and differentiate it with respect to the coordinate that is tangential to the panel; see problem 10 of Chapter 7.

Once the velocity is known at a node, the pressure coefficient there can be calculated from Bernoulli's equation 3-2′. Subroutine FANDM, which computes the lift and moment in PANEL by summing the forces and moments on individual panels, can then be adapted to the constant-potential method by replacing the

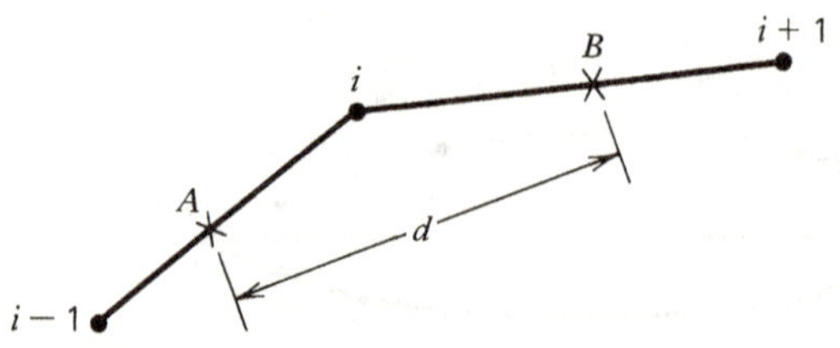

FIG. 8.12. Computation of nodal velocity in constant-potential panel method.

calculated values of CP(I), the pressure coefficient at the middle of the Ith panel, with the average of the pressure coefficients calculated at the ends of the panel. An alternative calculation of the lift coefficient can be based on the Kutta–Joukowski formula (4-33). From the definition (4-60) and equation 8-18,

$$c_l = 2\frac{\Gamma}{V_\infty c} = 2\frac{\phi_N - \phi_1}{V_\infty c} \tag{8-25}$$

Results obtained in this way are in pretty good agreement with those obtained from PANEL and converge at about the same rate for airfoils that are not too thin. However, for the extreme case of a cusped trailing edge (like the Joukowski airfoils discussed in Chapters 3 and 4), the constant-potential method gives very poor accuracy. It often happens that a computational method works well for some problems and not for others. Such a method lacks *robustness.*

8.2.2. Linear-Potential Method

Let us turn now to the second panel method that we promised to base on equation 8-17. The fundamental difference is that the potential will be assumed to vary linearly over each panel; for example,

$$\phi = \phi_j + \frac{\xi}{l_j}(\phi_{j+1} - \phi_j) \tag{8-26}$$

on panel j, where ξ is, as in equations 8-21, the distance from the jth node on the jth panel, and ϕ_j is now the value of ϕ at the jth node. Note the change in nomenclature: in equations 8-21 and 8-22, ϕ_j is the (constant) value of ϕ on the jth panel.

Algebraic equations for the ϕ_j's are derived almost as before and can be put in a form like equations 8-22. However, a little more caution is required to identify and then resolve an apparent singularity. Thus we return to equation 8-17, and first observe that now

$$\Gamma = \phi_{N+1} - \phi_1$$

where ϕ_1 and ϕ_{N+1} are the potentials on the lower and upper sides, respectively, of the trailing edge. Next, we again approximate S_B by a polygon whose nodes are points on the body surface, as shown in Fig. 8.9, and break up the integral over S_B into integrals over the N panels. Using equation 8-20 for $\hat{\mathbf{n}} \cdot \nabla\phi_s$, we thus obtain

$$\phi_P = V_\infty(x_P \cos\alpha + y_P \sin\alpha) + \frac{1}{2\pi}\sum_{j=1}^{N}\int_{\text{panel } j} \phi \frac{y^*}{(x^* - \xi)^2 + y^{*2}}\, d\xi$$
$$+ \frac{1}{2\pi}(\phi_{N+1} - \phi_1)\int_0^\infty \frac{y^*}{(x^* - \xi)^2 + y^{*2}}\, d\xi \tag{8-27}$$

in which (x^*, y^*) are now the coordinates of point P. On introducing equation 8-26 for the value of ϕ on the jth panel, we obtain integrals that are of the same form as

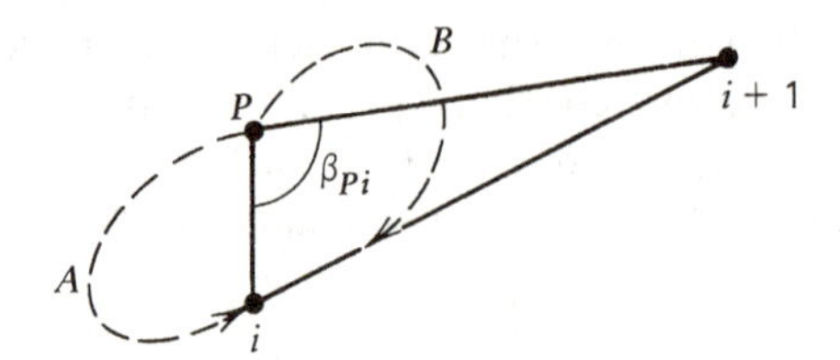

FIG. 8.13. Ambiguity in definition of β_{P_i} at nodes.

those evaluated in Chapter 4 and thus get

$$\begin{aligned}\phi_P = &\ V_\infty(x_P\cos\alpha + y_P\sin\alpha)\\ &+\frac{1}{2\pi}\sum_{j=1}^{N}\left[\phi_j + \frac{x^*}{l_j}(\phi_{j+1}-\phi_j)\right]\beta_{Pj} + \frac{y^*}{l_j}(\phi_{j+1}-\phi_j)\ln\frac{r_{Pj+1}}{r_{Pj}}\\ &+\frac{1}{2\pi}(\phi_{N+1}-\phi_1)\beta_{PN+1}\end{aligned} \tag{8-28}$$

in which, in accordance with our previous nomenclature, β_{Pj} is the angle subtended at point P by the jth panel (or by the connecting surface S_C when $j = N + 1$), while r_{Pj} is the distance from P to the jth node.

Equation 8-28 is turned into a set of algebraic equations for the $N + 1$ nodal values of ϕ by letting P become the ith node, $i = 1, 2, 3, \ldots, N + 1$. However, it is not enough simply to replace the subscript "P" by "i" in equation 8-28. Consider the contribution of the ith and $(i - 1)$th panels to ϕ_P, the terms for which $j = i$ and $(i - 1)$, respectively. Since $r_{ii} = 0$, these terms contain logarithmic singularities. Also, the quantity β_{ij} is indeterminate when $j = i$ or $i - 1$. If P approaches the ith node along path A in Fig. 8.13, $\beta_{Pi} \to 0$, while $\beta_{Pi} \to \pi$ if P approaches the ith node along path B.

Let us consider in detail, therefore, the effect on equation 8-28 of letting the point P approach the ith node. As noted above, the troublesome terms of equation 8-28 are those for which $j = i$ or $i - 1$. Thus we write it

$$\begin{aligned}\phi_P = &\frac{1}{2\pi}\left[\phi_{i-1} + \frac{x^*_{i-1}}{l_{i-1}}(\phi_i - \phi_{i-1})\right]\beta_{Pi-1} + \frac{1}{2\pi}\frac{y^*_{i-1}}{l_{i-1}}(\phi_i - \phi_{i-1})\ln\frac{r_{Pi}}{r_{Pi-1}}\\ &+\frac{1}{2\pi}\left[\phi_i + \frac{x^*_i}{l_i}(\phi_{i+1}-\phi_i)\right]\beta_{Pi} + \frac{1}{2\pi}\frac{y^*_i}{l_i}(\phi_{i+1}-\phi_i)\ln\frac{r_{Pi+1}}{r_{Pi}} + \text{regular terms}\end{aligned} \tag{8-29}$$

Here (x^*_{i-1}, y^*_{i-1}) locates point P in coordinates fixed to the $(i - 1)$th panel, for example, and the rest of the nomenclature is defined in Fig. 8.14.

As $P \to$ node i, the panel-oriented coordinates of point P behave as follows:

$$x^*_i,\ y^*_i,\ y^*_{i-1} \to 0$$

$$x^*_{i-1} \to l_{i-1}$$

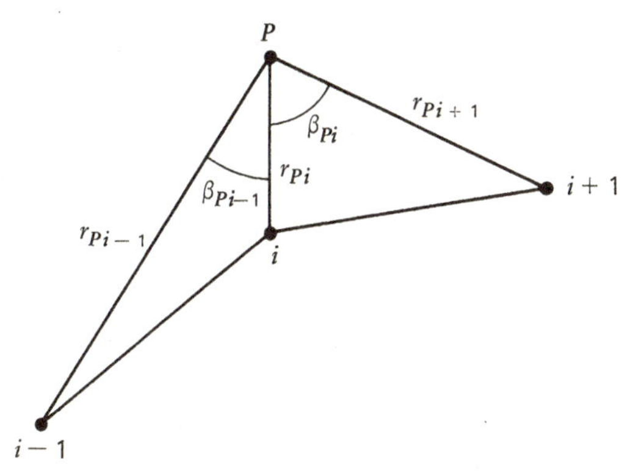

FIG. 8.14. Removal of ambiguity in linear-potential panel method.

Thus equation 8-29 becomes

$$\phi_P \to \frac{1}{2\pi}(\phi_{i-1} + \phi_i - \phi_{i-1})\beta_{Pi-1} + 0 + \frac{1}{2\pi}(\phi_i + 0)\beta_{Pi} + 0 + \text{regular terms}$$

$$= \frac{1}{2\pi}\phi_i(\beta_{Pi-1} + \beta_{Pi}) + \text{regular terms}$$

While β_{Pi} and β_{Pi-1} are individually indeterminate as $P \to i$, their sum can be seen from Fig. 8.14 to approach a unique limit, namely, the angle subtended at the ith node by the $(i-1)$th and the $(i+1)$th nodes. The results for the trailing-edge nodes are somewhat different; see problem 7.

Since the potentially singular log terms have disappeared in the limiting process, we can get around the difficult terms of equation 8-28 as follows. Replace the subscript "P" with "i." However, when $i = j$ or $j + 1$, just zero out β_{ij} and $\ln r_{ij}$. When the summation over j is complete, compensate for these deletions by adding to the right side of equation 8-28 the quantity

$$\frac{1}{2\pi}\phi_i(\beta_{ii-1} + \beta_{ii})$$

where the sum of the two angles can be calculated from (compare equation 4-87)

$$\beta_{ii-1} + \beta_{ii} = \tan^{-1}[(y_i - y_{i+1})(x_i - x_{i-1}) - (x_i - x_{i+1})(y_i - y_{i-1}),$$
$$(x_i - x_{i+1})(x_i - x_{i-1}) + (y_i - y_{i+1})(y_i - y_{i-1})]$$

Having identified and then removed a real problem with the use of equation 8-28, I now must tell you that they are insufficient to determine the $N + 1$ unknowns ϕ_i. Applying equation 8-28 at nodes 1 and $N + 1$ (the two sides of the trailing edge) yields the same result, so that only N equations of the form of (8-28) are independent. It will be shown in the following subsection that a method based on a linear doublet distribution on the panels, unlike Morino's constant-doublet-strength method, needs an explicit Kutta condition, which then completes the set of equations governing the nodal values of the doublet strength. As in the Douglas

Neumann method [1], this may take the form of equating the tangential velocities at the midpoints of the two trailing-edge panels, in which case we require (see equation 8-31)

$$\frac{\phi_{N+1} - \phi_N}{l_N} = \frac{\phi_1 - \phi_2}{l_1} \tag{8-30}$$

Possibly because (as we have shown) the potential is continuous in the method under discussion, there is considerably more freedom in how one uses the solution for the ϕ_i's to derive the velocity and pressure fields than with Morino's method. The most obvious (and simplest) procedure is to differentiate equation 8-26, the formula for the assumed linear variation of the potential along the panel, and to take the result as the tangential velocity at the panel midpoint $(\bar{x}_i, \bar{y}_i)$:

$$V_t(\bar{x}_i, \bar{y}_i) \approx \frac{\phi_{i+1} - \phi_i}{l_i} \tag{8-31}$$

Another approach I have used is to calculate the velocities at the nodes by fitting both linear and quadratic formulas to the potential at the node under consideration and its two nearest neighbors. In the linear case, for which the formula is

$$\phi = \alpha + \beta t$$

where t is the distance tangential to the body surface, the coefficients α and β must be determined in the "least-squares" sense; that is, so as to minimize the sum of the squares of the errors in fitting the formula to the three data:

$$E^2 \equiv (\phi_{i+1} - \alpha - \beta t_{i+1})^2 + (\phi_i - \alpha - \beta t_i)^2 + (\phi_{i-1} - \alpha - \beta t_{i-1})^2 \tag{8-32}$$

Finally, I have computed the velocity at the panel midpoints by substituting the assumed form of the potential distribution over the panels (equation 8-26) into the derivatives of equation 8-17 and then integrating over the panels (i.e., by using equations 8-34, given below). This last approach, which fails so dismally when the potential is assumed constant over the panels, gives quite acceptable results when the potential is taken to vary linearly over the panels. At least for the velocity away from the ends of the panels; although (as we have shown) the potential derived from the piecewise-linear model is continuous at the nodes, the velocity is not.

On the other hand, the results obtained at panel midpoints by integrating the linear potential formula over the panels are no better than those obtained by the much simpler formula (8-31), which is therefore preferred. Since, when the velocities and pressure are computed at the panel midpoints, forces and moments can be calculated from subroutine FANDM of program PANEL as is, the only subroutines of PANEL that have to be replaced to generate a program that implements the linear potential method are COFISH and VELDIS.

8.2.3. Equivalent Vortex Distributions

Panel methods based on doublet distributions are extremely powerful and widely used in the aerospace industry. However, the velocity field of a doublet is sufficiently more complicated than are those of sources and vortices that it is relatively difficult

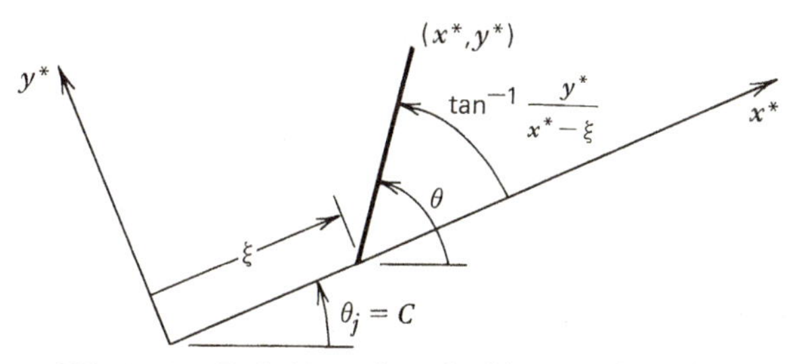

FIG. 8.15. Definition of angle θ in equation 8-33.

to become comfortable with these methods. Thus it is useful to recognize that, to every doublet distribution, there corresponds an equivalent vortex distribution. This equivalence also helps to explain such items as the inability of the constant-doublet-strength model to yield good velocity fields by summing the velocities induced by the individual panels and its lack of an explicit Kutta condition.

Consider the typical integral in equation 8-27, which may be applied to both methods considered above by stipulating the appropriate variation of ϕ over the panels. Since

$$\int \frac{y^*}{(x^* - \xi)^2 + y^{*2}}\, d\xi = \tan^{-1} \frac{y^*}{x^* - \xi} + C$$

we can, by choosing the constant C to be the inclination of the jth panel with respect to the horizontal θ_j, equate this integral to the angle θ defined in Fig. 8.15, which is the angle between the horizontal and the ray that extends from the point $(\xi, 0)$ on the panel to the field point (x^*, y^*). Then the integrals of equation 8-27 may be integrated by parts as follows:

$$\frac{1}{2\pi} \int_{\text{panel } j} \phi \frac{y^*\, d\xi}{(x^* - \xi)^2 + y^{*2}} = \frac{1}{2\pi} \phi\theta \Bigg|_{\xi=0}^{\xi=l_j} - \frac{1}{2\pi} \int_0^{l_j} \frac{\partial \phi}{\partial \xi}\, \theta\, d\xi \qquad \text{(8-33)}$$

The integral on the left is the velocity potential due to a doublet distribution of strength $\phi(\xi)$ on the jth panel. Since $-\theta/2\pi$ is, from equation 2-52, the velocity potential of a unit-strength vortex, we have thus shown that a doublet distribution on a linear panel is equivalent to a distribution of vortices over that panel, plus a pair of concentrated vortices, one at either end of the panel. The strength of the vortex distribution is just the derivative of the strength of the doublet distribution, whereas the point vortices have strengths equal in magnitude to the strength per unit length of the doublet distribution at the ends of the panel.

In the case of constant doublet strength, therefore, the distributed vortices do not appear in equation 8-33; then $\partial\phi/d\xi = 0$. However, because of the discontinuity in potential from one panel to another, neighboring panels contribute concentrated vortices of opposite but not necessarily equal strength at the node they have in common. Specifically, the net strength of the vortex at the ith node is $\phi_i - \phi_{i-1}$. However, the first and Nth panels contribute vortices of strength ϕ_1 and $-\phi_N$, respectively, at the trailing-edge node, which exactly cancel the vortex associated

FIG. 8.16. Velocity field in constant-potential approximation.

with the connecting surface S_C [the "$(N + 1)$th panel"]. Thus the equivalent vortex system consists of concentrated vortices at every node but the first.

This picture explains the poor results obtained for the velocity field by summing the contributions of the individual panels. The problem is that the velocity field due to a piecewise-constant doublet distribution is just too singular to be relied upon. The velocity field due to one of these vortices has little effect on the tangential velocity at points nearby, but its influence at points on the other side of the airfoil is disastrous, as can be seen from Fig. 8.16.

With the understanding that its velocity field is due to a collection of discrete vortices, the constant-potential method is seen to be very much like the vortex–lattice method discussed in Chapter 5, the difference being that, in the vortex–lattice method, the vortex strengths are determined by asking that the velocity field be tangent to the body at specified control points. In the method under discussion, flow tangency is used less directly, in the sense that the integral equation 8-17, though it was derived from equation 8-11 only by setting $V_n = 0$ on S_B, is still basically a statement about the potential field rather than the velocity field and its tangency to the body surface. Like the constant-potential method, vortex–lattice methods can be made to yield good results for the pressure distribution on wings with thickness, but, because of the proximity of the vortices on opposite surfaces, some "massaging" of the data is needed to make them work. Rubbert [6] got fairly good pressure distributions (except near sharp trailing edges) by smearing out the concentrated vortices into constant-strength vortex distributions before calculating the vortex-induced velocity field. That is, the vortex strength on each panel was defined to be the average of the strengths of the vortices at the ends of the panel, divided by the panel length. This is not unlike the procedure recommended in Section 8.2.1 for use in the constant-potential method, namely, computing the nodal velocities by taking differences among the results obtained for the panel potentials.

The vortex-based interpretation of the velocity field also sheds light on another interesting feature of Morino's method, namely, its lack of an explicit Kutta condition. As noted above, the equivalent vortex system is comprised of concentrated vortices at all nodes except the trailing edge. Therefore, the velocity is continuous at the trailing edge (though not at any other node), and so the Kutta condition is satisfied automatically.

The vortex system associated with the linear-potential method is quite different: Because the potential is continuous at the nodes, the nodal vortices cancel one another (including those at the trailing-edge node), and the only terms that remain when equation 8-33 is substituted into (8-27) are the continuous vortex distributions on the panels. The strengths of these distributions, being derivatives of the assumed linear potential distributions, are constant on each panel. Thus the velocity field can be taken from the analysis of Section 4.8, in which constant-strength vortex panels were considered, and the velocity associated with the potential given by equation 8-28 has components

$$u_P = V_\infty \cos\alpha + \sum_{j=1}^{N} \frac{\gamma_j}{2\pi}\left(\beta_{Pj}\cos\theta_j - \ln\frac{r_{Pj+1}}{r_{Pj}}\sin\theta_j\right)$$

$$v_P = V_\infty \sin\alpha + \sum_{j=1}^{N} \frac{\gamma_j}{2\pi}\left(\beta_{Pj}\sin\theta_j + \ln\frac{r_{Pj+1}}{r_{Pj}}\cos\theta_j\right) \tag{8-34}$$

where θ_j is, as shown in Fig. 8.10, the inclination of the jth panel, and

$$\gamma_j \equiv \frac{\phi_{j+1} - \phi_j}{l_j} \tag{8-35}$$

is the derivative of the doublet strength on the jth panel and so its vortex strength. Thus equations 8-34 can be compared with equation 4-95; just set $\theta_i = 0$ to get u_P, and $\theta_i = \pi/2$ to get v_P.

As the point P approaches the trailing edge, the terms $j = 1$ and $j = N$ of equations 8-34 exhibit logarithmic singularities, which would seem to show the need for an explicit Kutta condition in the linear-potential method. The singularities cancel if

$$\gamma_1 \sin\theta_1 = \gamma_N \sin\theta_N$$

$$\gamma_1 \cos\theta_1 = \gamma_N \cos\theta_N \tag{8-36}$$

where θ_1 and θ_N are the inclinations to the x axis of the two panels that meet at the trailing edge, as shown in Fig. 8.17. Unfortunately, this is one more equation than we need. Fortunately, it is one more equation than seems to be necessary. To demand that the velocity field be completely free from singularities at the trailing edge is overly strict; after all, the velocity components given by equations 8-34 are also singular at all the other nodes. Specifically, the velocity components tangential to the panels are finite, but discontinuous from one panel to the next, whereas the normal velocity component is logarithmically singular at both ends of every panel. It seems to be enough, therefore, to equate the tangential velocities on the last two panels, as was done in equation 8-30 (see equation 8-31), or, in terms of the equivalent vortex strength introduced in equation 8-35, to require

$$\gamma_1 + \gamma_N = 0 \tag{8-37}$$

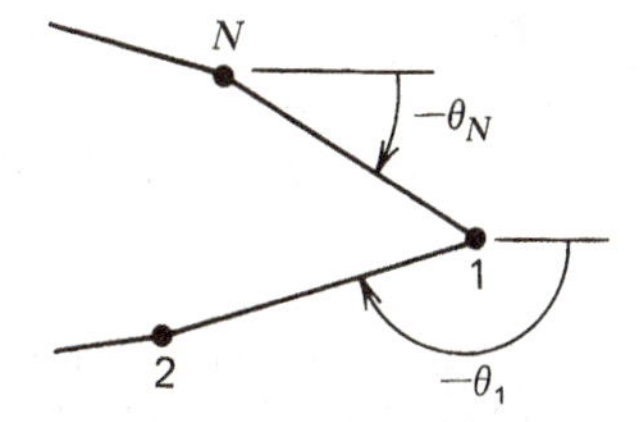

FIG. 8.17. Nomenclature for study of Kutta condition.

instead of equations 8-36. That is, if the normal velocity component is so poorly behaved in any case, the Kutta condition ought to be confined to the tangential component. This is consistent with Section 4.3's "second corollary" of the Kutta condition, that the tangential velocities on either side of the trailing edge approach one another at the trailing edge.

This reasoning has been confirmed by numerical experiments, in which I solved (in a least-squares sense, see the next section for details) the overdetermined system consisting of equations 8-28, evaluated at nodes 1 through N, and equations 8-36. The results were little different from those I got by supplementing the N equations 8-28 with the single Kutta condition (8-31). In fact, the simpler approach was slightly more accurate when compared with exact solutions for flows past ellipses and Joukowsky airfoils.

8.3. VORTEX-BASED PANEL METHODS

In the preceding section, we looked at panel methods in which the potential on the panels—the strength of the doublet distribution thereon—was the major unknown. However, the results of those methods became easier to interpret when the doublet distribution was replaced by an equivalent vortex distribution.

In particular, a continuous distribution of doublets over straight-line panels was shown to be equivalent to a distribution of vortices over the same panels, with the vortex strength being the derivative of the doublet strength. Since the doublet strength is the value of the potential on the surface, the vortex strength is the tangential velocity on the surface.

This is also true when the surface on which the doublets are distributed is curved, so that equation 8-17 can be written (for the two-dimensional case)

$$\phi_P = V_\infty(x_P \cos\alpha + y_P \sin\alpha) + \int_{S_B} \gamma \phi_v \, dS \tag{8-38}$$

Here ϕ_v is the potential of a unit strength vortex

$$\phi_v = \frac{-1}{2\pi}\theta$$

(r, θ) being the polar coordinates of the field point P relative to dS, as shown in Fig. 8.18, whereas γ, the vortex strength, is the tangential velocity component and the derivative of the potential in the direction of $\hat{\mathbf{t}}$, the unit vector tangential to S_B.[6]

Panel methods may be based on equation 8-38 as well as on the equivalent equation 8-17. If the body surface S_B is approximated by the usual straight-line panels, and the vortex strength is taken to be constant on each panel, the velocity field corresponding to equation 8-38 can be taken directly from equation 8-34,

[6] Note that the integrations in equation 8-38 extend only over S_B, and not, as in equation 8-17, over the surface S_C that extends from S_B to infinity. Since the doublet strength on S_C was shown to be constant, its derivative along S_C is zero. Alternatively, note that the velocity tangential to S_C is continuous across S_C (unlike the potential), so that the integrals of $(\hat{\mathbf{t}} \cdot \nabla\phi)\phi_v$ along the two sides of S_C cancel one another.

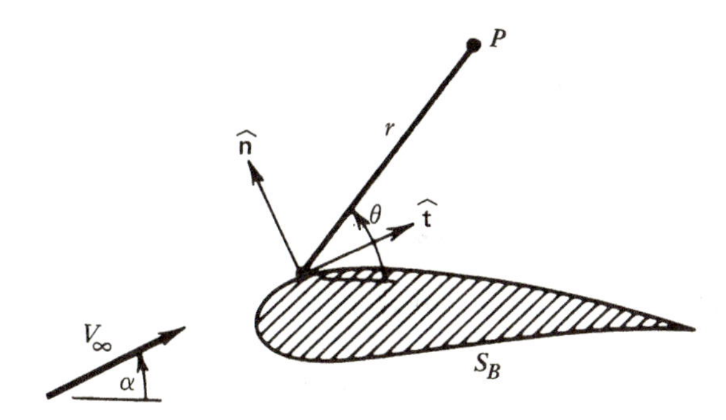

FIG. 8.18. Nomenclature for vortex-based panel method.

which is derived for a piecewise-linear doublet distribution. According to equation 8-33, the strength of the equivalent vortex distribution is the derivative of the doublet strength and so is given by equation 8-35, which was used in equation 8-34.

To determine the vortex strength, we may use the fact (see equations 8-31 and 8-35) that the vortex strength on the ith panel equals the velocity tangential to that panel:

$$\gamma_i = u_i \cos\theta_i + v_i \sin\theta_i \tag{8-39}$$

Here θ_i is the inclination of the ith panel to the horizontal, and u_i and v_i are the (x, y) velocity components at the midpoint of the ith panel, which are obtained by letting the point P of equation 8-34 approach that midpoint. Substituting equations 8-34 into 8-39 yields

$$\gamma_i = V_\infty \cos(\theta_i - \alpha) + \frac{1}{2\pi}\sum_{j=1}^{N} \gamma_j \left[\sin(\theta_i - \theta_j)\ln\frac{r_{ij+1}}{r_{ij}} + \cos(\theta_i - \theta_j)\beta_{ij}\right] \tag{8-40}$$

If equation 8-40 is evaluated at the middle of every panel, we obtain N equations in the N unknowns $\gamma_1, \ldots, \gamma_N$. Flow tangency was implicitly satisfied when we set $\hat{\mathbf{n}} \cdot \nabla\phi = 0$ on the panels, but nothing has yet been done about the Kutta condition. Since a constant vortex distribution is equivalent to a linear doublet distribution, the discussion of the Section 8.2.3 carries over. Thus, to eliminate a discontinuity in the tangential velocity at the trailing edge, we must satisfy equation 8-37. This gives us an overdetermined system of equations: N equations like (8-40), one like (8-37), and only N unknowns.

One way to deal with overdetermined systems is the *least-squares method.* Write equations 8-37 and 8-40 in the form

$$\sum_{j=1}^{N} A_{ij}\gamma_j = b_i \qquad \text{for } i = 1, \ldots, N+1 \tag{8-41}$$

The idea is to find the γ_j's so as to minimize

$$E^2 \equiv \sum_{i=1}^{N+1}\left(\sum_{j=1}^{N} A_{ij}\gamma_j - b_i\right)^2$$

Thus we require

$$\frac{\partial E^2}{\partial \gamma_k} = 0$$

$$= 2 \sum_{i=1}^{N+1} \left(\sum_{j=1}^{N} A_{ij}\gamma_j - b_i \right) A_{ik}$$

and so determine the γ_j's so as to satisfy

$$\sum_{j=1}^{N} \left(\sum_{i=1}^{N+1} A_{ij}A_{ik} \right) \gamma_j = \sum_{i=1}^{N+1} b_i A_{ik} \qquad \text{for } k = 1, \ldots, N \tag{8-42}$$

which is a determinate system.

The extra summations involved in equation 8-42 are rather time consuming. An alternate way to deal with equation 8-41, slightly less accurate but much more efficient, was introduced by Bristow [7]. He simply added a constant error term e to each of the governing equations 8-41, obtaining

$$\sum_{j=1}^{N} A_{ij}\gamma_j + e = b_i \qquad \text{for } i = 1, \ldots, N + 1 \tag{8-43}$$

Since e is unknown, equations 8-43 contain $N + 1$ unknowns and are thus determinate.

It should be noted that, although the linear-potential and constant-vortex methods both yield velocity distributions calculable from equations 8-34, their results are quite different. In the first case, the γ_j's are calculated in terms of the nodal values of the potential, which are governed by equations 8-28 and 8-31. In the vortex-based method, the γ_j's are found so as to satisfy equations 8-40 and 8-37, which, except for the Kutta conditions (8-31) and (8-37), are not at all equivalent.

8.4. SOURCE-BASED PANEL METHODS

The relatively mathematical approach of this chapter has now yielded a number of alternatives to the Smith–Hess panel method introduced in Chapter 4 on a much less formal basis. The question then arises, how does the earlier method fit into the present framework? Although equation 8-11 does give the potential as a distribution of sources and doublets over the body, and we know that the doublet distribution is equivalent to the vortex distribution of equation 8-38, the source and doublet strengths are related in those equations to the boundary values of the normal and tangential components of the velocity, respectively. In the Smith–Hess method, the vortex strength is taken to be constant, and the source strength, which was zero in the methods discussed in the preceding two sections, is the major unknown variable.

To see how the present equations may be made to encompass the earlier method, we return to equation 8-11. Let ϕ^* be a solution of Laplace's equation in the region $\mathscr{V}_B$ that is *inside* the body surface S_B, as shown in Fig. 8.19. If ϕ_s is still the potential of a point source at P, a point *outside* the body surface, and (compare with equation 8-2)

$$\mathbf{U}^* \equiv \phi^* \nabla \phi_s - \phi_s \nabla \phi^* \tag{8-44}$$

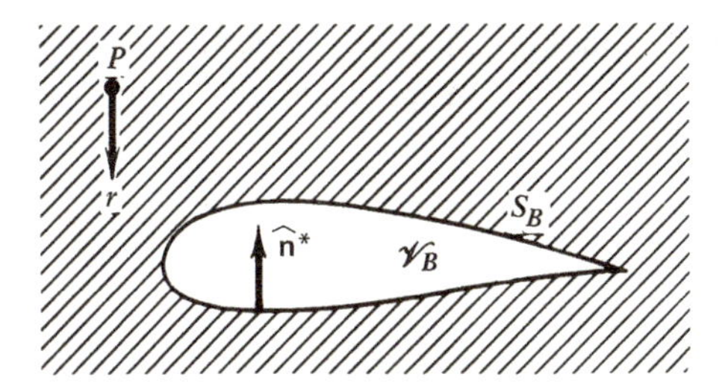

FIG. 8.19. Definition of potential inside airfoil.

then, as we found in equation 8-6,

$$\nabla \cdot \mathbf{U}^* = 0 \qquad \text{in } \mathscr{V}_B$$

Note that this equation holds everywhere in $\mathscr{V}_B$. Since P, the source point, is not in $\mathscr{V}_B$, we do not have to carve anything out of $\mathscr{V}_B$ as we did in deriving equation 8-6. Then, from equations 8-1 and 8-44,

$$0 = \int_{S_B} \hat{\mathbf{n}}^* \cdot (\phi^* \nabla \phi_s - \phi_s \nabla \phi^*)\, dS \tag{8-45}$$

The unit normal $\hat{\mathbf{n}}^*$ in equation 8-45 is directed into $\mathscr{V}_B$ and so is equal and opposite to the $\hat{\mathbf{n}}$ of equation 8-11 at the same point on S_B. Thus, we can let $\hat{\mathbf{n}}^* = -\hat{\mathbf{n}}$ and add equations 8-11 and 8-45 to get

$$\phi_P = V_\infty(x \cos\alpha + y \sin\alpha) + \int_{S_B + S_C} (\sigma\phi_s - \mu\hat{\mathbf{n}} \cdot \nabla\phi_s)\, dS \tag{8-46}$$

in which

$$\sigma \equiv \hat{\mathbf{n}} \cdot \nabla(\phi - \phi^*)$$
$$\mu \equiv \phi - \phi^* \tag{8-47}$$

The inclusion of S_C in the integral of equation 8-46 is justified by letting ϕ^* and $\hat{\mathbf{n}} \cdot \nabla\phi^*$ take on any values at all on S_C; so long as they are the same on both sides of S_C, their contributions cancel. Since the integrals of $(\hat{\mathbf{n}} \cdot \nabla\phi)\phi_s$ over the two sides S_C are already known to cancel, the σ term of equation 8-46 may be deleted when integrating over S_C.

Equation 8-46 thus represents the potential at any point P in the flow field in terms of distributions of sources and doublets over the body surface S_B and of doublets over the connecting surface S_C. This was also true of equation 8-11. Now, however, the source strength σ and doublet strength μ are not simply related to the normal derivative and value of ϕ on the boundary. Rather, according to equation 8-47, they are related to the difference between ϕ and ϕ^*, a function that satisfies Laplace's equation inside S_B but whose value and normal derivative on S_B are otherwise quite arbitrary.

A solution of Laplace's equation is not unique until we specify its value *or* normal derivative at every point on the boundary, and so we have quite a bit of freedom in using equations 8-46 and 8-47. The methods described earlier in this chapter correspond to the choice $\phi^* = 0$ on S_B. The unique solution of Laplace's equation is then

$$\phi^* = 0 \qquad \text{in } \mathscr{V}_B$$

which makes

$$\hat{\mathbf{n}} \cdot \nabla \phi^* = 0 \qquad \text{on } S_B$$

and so equation 8-47 reduces to

$$\sigma = \hat{\mathbf{n}} \cdot \nabla \phi$$

$$\mu = \phi$$

and equation 8-46 reassumes the form of equation 8-11.

To get the Smith–Hess method from equations 8-46 and 8-47 requires a bit more work. Let γ be the circulation about the airfoil, divided by the perimeter of the panels. Then

$$\Gamma = \gamma \oint_{S_B} dS$$

Set

$$\frac{\partial \phi^*}{\partial t} = \frac{\partial \phi}{\partial t} - \gamma \tag{8-48}$$

where $\partial/\partial t$ is the derivative in the direction tangential to S_B. Then

$$\oint_{S_B} \frac{\partial \phi^*}{\partial t} dS = \oint_{S_B} \frac{\partial \phi}{\partial t} dS - \Gamma = 0$$

It can be shown that this condition ensures that ϕ^* is single valued on S_B and so in $\mathscr{V}_B$. Thus, at least in principle, ϕ^* can be determined in $\mathscr{V}_B$, in terms of γ and the tangential velocity $\partial\phi/\partial t$ on S_B. Replacing the doublet distribution in equation 8-46 by its equivalent vortex distribution leads to (see equation 8-38)

$$\phi_P = V_\infty(x_P \cos\alpha + y_P \sin\alpha) + \int_{S_B} \left(\sigma\phi_s + \frac{d\mu}{dt}\phi_v\right) dS \tag{8-49}$$

where, from equations 8-47 and 8-48,

$$\frac{d\mu}{dt} = \frac{\partial}{\partial t}(\phi - \phi^*)$$

$$= \gamma$$

with γ a constant.

Equation 8-49 is thus the potential used in the Smith–Hess method. Of course, ϕ^* is not really determined in executing the method. Thus, although we could apply the flow tangency condition

$$\hat{\mathbf{n}} \cdot \nabla \phi = 0 \qquad \text{on } S_B$$

implicitly and reduce equation 8-47 to

$$\sigma = -\hat{\mathbf{n}} \cdot \nabla \phi^*$$

such a reduction is not useful. Instead σ and γ are determined so as to satisfy flow tangency at the panel midpoints and a Kutta condition, as was described in detail in Chapter 4.

This somewhat contorted approach to the Smith–Hess method shows that the method is as generally applicable as the doublet- and vortex-based methods, since all can be derived from Green's identity (8-9) without restriction on the body shape. It should be emphasized, however, that the *ad hoc* approach of Chapter 4 tells you all you need to know about the use of the Smith–Hess method, besides being less obscured by mathematics. On the other hand, without Green's identity, we would never have gotten the doublet- or vortex-based alternatives.

8.5. COMPARISONS OF SOURCE-, DOUBLET-, AND VORTEX-BASED METHODS

Naturally, with four different panel methods now at our disposal, some comparisons of their performance are called for. For an NACA four-digit airfoil of reasonable thickness (10–12%, say), the source-based Smith–Hess method and the doublet-based linear potential method are usually substantially more accurate than are the other two methods for a given number of panels. On the other hand, the source-based method takes more computer time for a given number of panels than do the alternative procedures. All the methods take about the same time to set up and solve the linear system that approximates the integral equation, but the Smith–Hess method requires an extra integration over the panels to find the tangential velocity on the body, which is about as time consuming as is the setting up of the algebraic system. In the vortex-based method, the tangential velocity is the unknown in the integral equation. In the potential-based methods, it can be determined from simple difference equations like 8-31.

An important characteristic of any numerical method is its rate of convergence. As you increase the number of panels, the error of any of these methods is found to decrease roughly in proportion to $1/N$. Thus they are all "first-order" methods, although, as noted above, their accuracy differs considerably for any given N. Knowing that a method is first-order accurate allows you to extrapolate results obtained with different numbers of panels to a higher degree of accuracy. Just plot the results against $1/N$, as is illustrated in Fig. 8.20, and fit them with a straight line. If the N's used are large enough, the intercept of the straight line at $1/N = 0$ should be an excellent approximation to the exact result.[7]

[7] Note how much better an idea this is than plotting results against N. Then your estimate of the asymptote as $N \to \infty$ depends on what French curve you use.

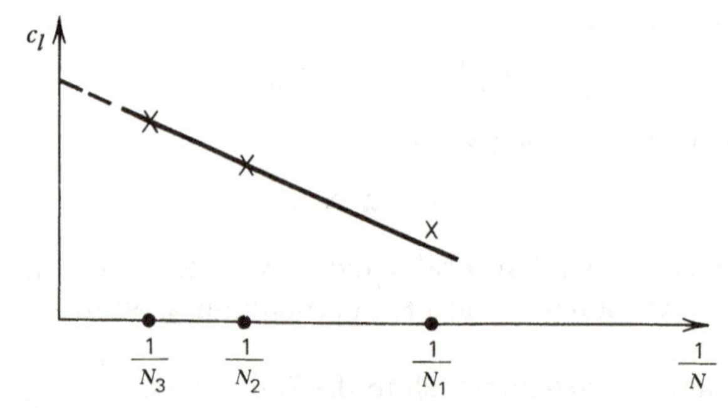

FIG. 8.20. Convergence of panel method results for lift coefficient.

That the constant-potential method has the same order of accuracy as the linear-potential procedure is curious (but true); one would expect a higher order of accuracy from the latter approach. Similarly, methods based on quadratic distributions of doublet strength, although very useful in problems that require fine resolution of the flow field, still converge no more rapidly than do the simpler methods in airfoil- and wing-type problems [5].

Since panel methods were introduced in Chapter 4 as a more accurate alternative to thin-airfoil theory, it is ironic that very thin airfoils are tough for some panel methods to handle. The Smith–Hess method is easily seen to suffer this affliction. If the airfoil thickness goes to zero, the source strength should vanish, and all you have left is a constant-strength vortex distribution, which also disappears when you apply the Kutta condition. The problem carries over when the trailing edge is very sharp, and you often get a spurious negative loading of such an airfoil near its trailing edge; see, for example, Fig. 7.27, which was generated by program PANEL. For the Joukowski airfoil, whose trailing edge is cusped, only the linear-potential method gives really good results. The constant-potential method is particularly bad for such airfoils.

Source- and vortex-based methods may have greater appeal to most aerodynamicists, who learn to develop a better feel for the velocity fields of sources and vortices than for doublets. However, not only are doublet-based methods usually more efficient (less computing time for a given level of accuracy), but they have such substantial advantages in three-dimensional problems that we have had to learn to use them whether we like them or not. The problem with using vortices to generate lift in three dimensions is that a vortex distribution must satisfy a "Helmholtz condition" [8] in order to generate an irrotational flow. For a concentrated vortex, this condition is that the vortex not end in the fluid. The strength of distributed vortices, which is related to the tangential velocity component on the surface, is a vector in three-dimensional problems, and must satisfy a continuity equation to meet the Helmholtz condition. These constraints make it considerably more difficult to parameterize the strength of a vortex distribution than that of the equivalent doublet distribution. Also, since the vortex strength is a vector, the number of unknowns associated with a given number of panels may be significantly larger with vortices than with doublets.

8.6. REFERENCES

1. Hess, J. L., and Smith, A. M. O., "Calculations of Potential Flow About Arbitrary Bodies," *Prog. Aeronaut. Sci.* **8**, 1–139 (1966).
2. Chapman, D. A., Mark, H., and Pirtle, M. W., "Computers vs. Wind Tunnels for Aerodynamic Flow Simulations," *Aeronaut. Astronaut.* **35**, 22–30 (1975).
3. Morino, L. and Kuo, C. C., "Subsonic Potential Aerodynamics for Complex Configurations: A General Theory," *AIAA J.* **12**, 191–197 (1974).
4. Johnson, F. T., and Rubbert, P., "Advanced Panel-Type Influence Coefficient Methods Applied to Subsonic Flows." *AIAA Paper*, 77–641 (1977).
5. Maskew, B., "Prediction of Subsonic Aerodynamic Characteristics: A Case for Low-Order Panel Methods," *J. Aircr.* **19**, 157–163 (1982).
6. Rubbert, P., "Theoretical Characteristics of Arbitrary Wings by a Non-Planar Vortex Lattice Method," Boeing Document *D*6-9246 (1962).
7. Bristow, D. R., "Recent Improvements in Surface Singularity Methods for the Flowfield Analysis about Two-Dimensional Airfoils," *AIAA Paper* 77–641 (1977).
8. Milne-Thomson, L. M., *Theoretical Aerodynamics*, 4th ed. Dover, New York (1973) Sec. 9.3.
9. Abbott, I., and von Doenhoff, A. E., *Theory of Airfoil Sections*. Dover, New York (1959).

8.7. PROBLEMS

1. Verify the result of equation 8-6 by writing out the divergence of **U** and the gradients of ϕ and ϕ_s in Cartesian coordinates; for example,

$$\nabla\phi = \hat{\mathbf{i}}\frac{\partial\phi}{\partial x} + \hat{\mathbf{j}}\frac{\partial\phi}{\partial y} + \hat{\mathbf{h}}\frac{\partial\phi}{\partial z}$$

$$\nabla\cdot\mathbf{U} = \frac{\partial U_x}{\partial x} + \frac{\partial U_y}{\partial y} + \frac{\partial U_z}{\partial z}$$

2. Find the coefficients α and β required to minimize the quantity E^2 defined in equation 8-32.

3. **A major project:** Program one of the three panel methods discussed in this chapter (constant doublet, linear doublet, and constant vortex). Compare its performance with the source-based method on which program PANEL is based for an NACA 4412 airfoil. That is, plot the lift and moment coefficients at some angle of attack against $1/N$, where N is the number of panels, and against $1/T$, where T is the computer time used.

4. If you programmed a doublet-based method in problem 4, revise it to calculate the tangential velocities at the panel midpoints by summing the velocities contributed by the panels; that is, by representing the velocity field as due to discrete vortices at the nodes or by constant-strength vortex panels, as is appropriate. Compare your results with those obtained by differentiating the potential distribution along the panels.

5. When the point P is far from the airfoil, ϕ_v in equation 8-38 can be approximated as $\Theta/2\pi$, where (R, Θ) are the polar coordinates of P relative to some *fixed* point within

the airfoil, rather than to dS. Then equation 8-38 becomes

$$\phi_p \approx V_\infty(x_p \cos\alpha + y_p \sin\alpha). - \frac{\Gamma}{2\pi}\Theta$$

where Γ is the circulation about the airfoil. Use the integral momentum theorem for a control volume bounded by the surfaces S_B and S_∞, equation 2-12, to verify the Kutta–Joukowski theorem, equation 4-33.

6. One of the virtues of a numerical method is that it can deal with inputs that are numerical rather than analytical. Most airfoil designs are unlike the NACA four- and five-digit series in that their shapes are defined numerically, that is, by giving the coordinates of selected points on the body. Revise PANEL and/or the method you programmed in problem 3 to deal with numerically defined geometries. That is, replace the bulk of subroutine SETUP by reading in NODTOT and [X(I), Y(I), I = 1, NODTOT] from a "tape" (file). Test the program by constructing such a file for one of the airfoils whose coordinates are given in Ref. [9].

7. Show that the limit of equation 8-28 as the point P approaches either node at the trailing edge (node 1 or $N + 1$) is

$$0 = \bar{V}_\infty(x_1 \cos\alpha + y_1 \sin\alpha) - \frac{1}{2\pi}\phi_1\delta - \frac{1}{2\pi}\phi_{N+1}\beta + \text{regular terms}$$

in which δ and β are angles subtended at the trailing edge node by the trailing edge panels, as shown in the sketch.

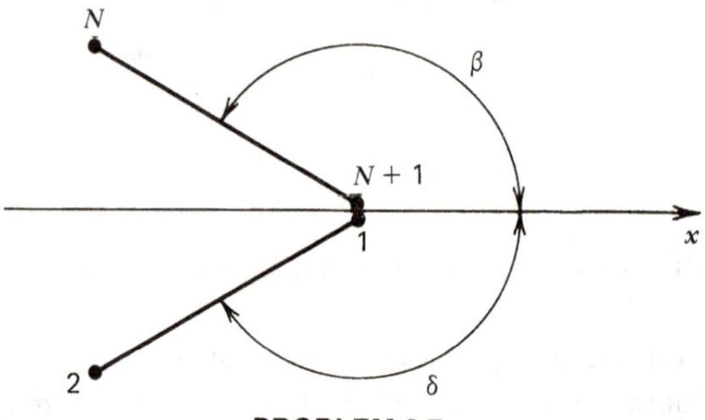

PROBLEM 8.7.

8. As shown in Section 7.2.2, the displacement effect of the boundary layer on the pressure distribution can be determined by solving an irrotational flow problem with surface normal velocity V_n not zero, but prescribed in terms of the boundary-layer solution. Revise one of the panel methods described in this chapter to accommodate a known, nonzero value of $\partial\phi/\partial n$ on S_B.

CHAPTER 9

FINITE DIFFERENCE METHODS

The panel methods described in Chapters 4 and 8 are definitely "numerical" methods. It's hard to imagine trying to solve any problem with less than 20 panels, which is far more than you could hope to handle in a computerless calculation. However, the panel method becomes numerical only when, after much analysis, you obtain a set of simultaneous algebraic equations, whose setting up and solution requires the use of a computer. The problems to be discussed in the last two chapters will also be solved numerically, but the method to be employed—the finite-difference method—is brought in much earlier in the process, and plays a much more prominent part in the solution. This chapter introduces you to the finite-difference method.

9.1. BOUNDARY-VALUE PROBLEMS IN ONE DIMENSION

To gain an easy introduction to the finite-difference method, we will first look at the relatively simple problem

$$\frac{d^2y}{dx^2} + p(x)\frac{dy}{dx} + q(x)y = r(x) \tag{9-1}$$

$$y(0) = V_0 \tag{9-2}$$

$$\frac{dy}{dx}(1) = S_1 \tag{9-3}$$

When $r = 0$ and p and q are constant, you should know that the problem is easily solved by assuming the solution to have the exponential form

$$y = e^{\lambda x} \tag{9-4}$$

On substituting this form into equation 9-1, we obtain a quadratic equation for the parameter λ. Since equation 9-1 is linear and homogeneous, its general solution can be expressed as the superposition of two solutions of the form (9-4) that correspond to the two roots (λ_1 and λ_2, say) of the quadratic:

$$y = A_1 e^{\lambda_1 x} + A_2 e^{\lambda_2 x} \tag{9-5}$$

The constants A_1 and A_2 are then found so that equation 9-5 satisfies the boundary conditions (9-2) and (9-3).

However, when p, q, and r vary with x, it is usually difficult to solve equation 9-1 in terms of functions you can find on your calculator. Even though equation 9-1 is linear and doesn't *look* formidable, you may have to solve it numerically. That is not to say a numerical solution is always inferior to an analytical one. If you only have to solve the problem for a few specific values of the data V_0 and S_1, a numerical solution can give all the results you need just as efficiently as one in "closed form." However, an analytical solution is more flexible and may help you to obtain much more easily results such as the maximum value of the output solution as a function of your input data.

In any case, our purpose here is to solve equations 9-1 to 9-3 by a finite-difference method. *The first step in any numerical method is to discretize the problem*; that is, to reduce it to a finite number of unknowns. Equation 9-1 governs the behavior of a continuous function; its description over the interval (0, 1) involves an infinite amount of information, in the sense that there are an infinite number of points in the interval. One way around this problem is to approximate the solution by a function with a finite number of parameters, for example, a polynomial, and to devise a method for determining those parameters so that the function approximately satisfies the equations and boundary conditions.

In the finite-difference method, we discretize the problem by seeking approximations to $y(x)$ only at a finite number of ordered points $x_1, x_2, \ldots, x_N$, with $0 = x_1 < x_2 < \cdots < x_N = 1$. The unknowns

$$Y_i \approx y(x_i) \qquad i = 1, \ldots, N \tag{9-6}$$

are determined by solving simultaneously a set of algebraic equations that approximate the differential equation and the boundary conditions. Specifically, as we shall now show, these equations are derived by approximating the derivatives that appear in equations 9-1 and 9-3 at the "nodes" x_i in terms of the nodal values of y, which are in turn approximated by the Y_i's.

Approximating the derivatives is simplest (and usually most accurate) when the nodes are equally spaced:

$$x_i = (i - 1)\Delta x \tag{9-7}$$

Then, by Taylor-series expansion,

$$y(x_{i+1}) = y + \Delta x\, y' + \frac{(\Delta x)^2}{2} y'' + \frac{(\Delta x)^3}{6} y''' + O(\Delta x)^4 \tag{9-8}$$

and

$$y(x_{i-1}) = y - \Delta x\, y' + \frac{(\Delta x)^2}{2} y'' - \frac{(\Delta x)^3}{6} y''' + O(\Delta x)^4 \tag{9-9}$$

where the primes indicate differentiation and the quantities on the right sides of equations 9-8 and 9-9 are understood to be evaluated at x_i. By the "O" symbol is meant that, if

$$F(z) = O(z^n) \qquad \text{as } z \to 0$$

(read, "F is of the order of z^n as $z \to 0$"), then

$$\lim_{z \to 0} \left| \frac{F(z)}{z^n} \right| < \infty \tag{9-10}$$

Dropping terms of order $(\Delta x)^2$ from equations 9-8 and 9-9 yields forward-difference and backward-difference approximations to $y'(x_i)$, respectively, whose accuracy you will examine in problem 1. For present purposes, we prefer to subtract equations 9-8 and 9-9 and solve for $y'(x_i)$ to get a *central-difference approximation to* dy/dx:

$$\frac{dy}{dx}(x_i) = \frac{y(x_{i+1}) - y(x_{i-1})}{2\Delta x} + O(\Delta x)^2 \tag{9-11}$$

A similar approximation to the second derivative is obtained by adding equations 9-8 and 9-9:

$$\frac{d^2y}{dx^2}(x_i) = \frac{y(x_{i+1}) - 2y(x_i) + y(x_{i-1})}{(\Delta x)^2} + O(\Delta x)^2 \tag{9-12}$$

Since the derivation of equations 9-11 and 9-12 is rather formal, Figs. 9.1 and 9.2 have been prepared to help you understand the results. In particular, note from Fig. 9.2 that

$$\frac{y(x_{i+1}) - y(x_i)}{\Delta x} \qquad \text{and} \qquad \frac{y(x_i) - y(x_{i-1})}{\Delta x}$$

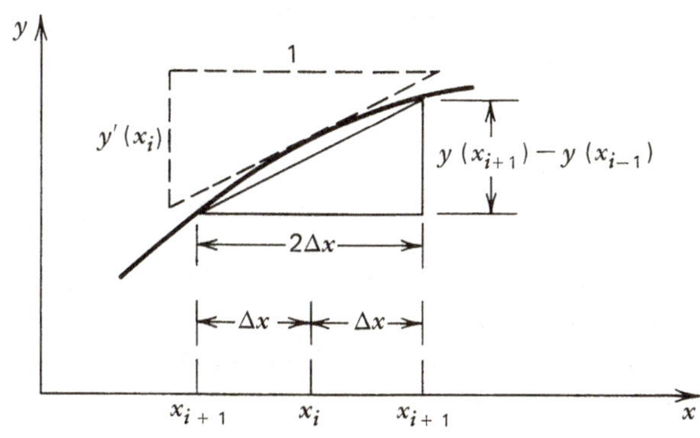

FIG. 9.1. Interpretation of central-difference approximation to first derivative.

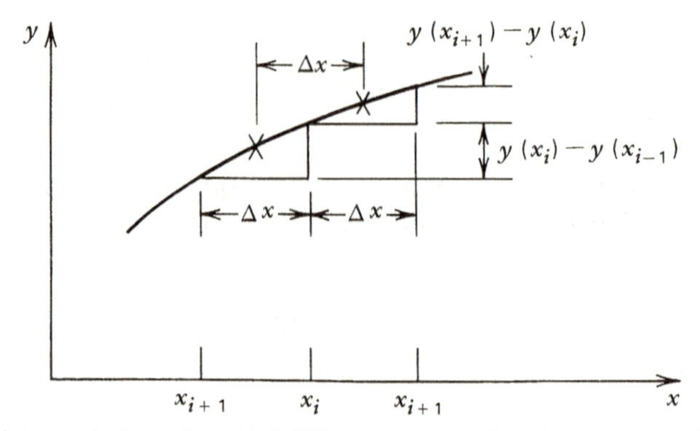

FIG. 9.2. Interpretation of central-difference approximation to second derivative.

are central-difference approximations to $y'(x_i + \Delta x/2)$ and $y'(x_i - \Delta x/2)$, respectively. Thus, by an extension of equation 9-11,

$$y''(x_i) \approx \frac{y'(x_i + \Delta x/2) - y'(x_i - \Delta x/2)}{\Delta x}$$

$$\approx \frac{y(x_{i+1}) - y(x_i)}{(\Delta x)^2} - \frac{y(x_i) - y(x_{i-1})}{(\Delta x)^2}$$

in agreement with equation 9-12.

The differential equation 9-1 can now be approximated at x_i in terms of values of y at that node and its neighbors by substituting therein for the derivatives of y from equations 9-11 and 9-12:

$$\frac{y(x_{i+1}) - 2y(x_i) + y(x_{i-1})}{(\Delta x)^2} + p(x_i)\frac{y(x_{i+1}) - y(x_{i-1})}{2\Delta x} + q(x_i)y(x_i) = r(x_i) + O(\Delta x)^2 \tag{9-13}$$

An equation for the finite-difference approximation to $y(x)$ is then obtained by letting $y(x_j) \rightarrow Y_j$ in equation 9-13 and dropping the order-of-magnitude symbol that represents the errors incurred in approximating the derivatives:

$$\frac{Y_{i+1} - 2Y_i + Y_{i-1}}{(\Delta x)^2} + p_i\frac{Y_{i+1} - Y_{i-1}}{2\Delta x} + q_i Y_i = r_i \tag{9-14}$$

Here, of course, $p_i \equiv p(x_i)$ and so forth.

Equation 9-14 is a *finite-difference equation*, specifically, a finite-difference approximation to equation 9-1. It may be applied at every node x_i that lies between the endpoints $x = 0$ and $x = 1$; that is, for $i = 2, \ldots, N - 1$. Thus it gives us $N - 2$ equations in the N unknowns $Y_1, Y_2, \ldots, Y_N$.

The two equations we lack can be obtained from the boundary conditions (9-2) and (9-3). The former clearly implies

$$Y_1 = V_0 \tag{9-15}$$

but the latter requires more work. If we base a finite-difference approximation to dy/dx at $x = 1$ on equation 9-11, we bring in $y(x_{N+1})$, which is beyond the interval of interest. However, we can also invoke the finite-difference approximation to the differential equation at $x = 1$ and eliminate Y_{N+1} between them. That is, from equations 9-3 and 9-11,

$$Y_{N+1} = Y_{N-1} + 2\Delta x\, S_1$$

which, when plugged into equation 9-14 (with $i = N$), yields

$$2\frac{Y_{N-1} - Y_N}{(\Delta x)^2} + q_N Y_N = r_N - p_N S_1 - 2\frac{S_1}{\Delta x} \tag{9-16}$$

Alternatively, we could use a *backward-difference approximation* to $y'(1)$. From equation 9-9,

$$y'(x_N) = \frac{y(x_N) - y(x_{N-1})}{\Delta x} + O(\Delta x) \tag{9-17}$$

which suggests that equation 9-3 be approximated by

$$\frac{Y_N - Y_{N-1}}{\Delta x} = S_1 \tag{9-18}$$

However, the error term in equation 9-17 is of first order in Δx, whereas those of equations 9-11 and 9-12, on which equation 9-16 is based, are of second order. Thus, as $\Delta x \to 0$, the difference between equation 9-16 and the differential equations on which it is based vanishes much more rapidly than does the difference between equations 9-18 and 9-13. In this sense, equation 9-16 is more accurate than is equation 9-18. As might then be expected (but see the next section for a more detailed discussion), a finite-difference solution based on equation 9-16 does turn out to be closer to the exact solution of equations 9-1 to 9-3 than does one which uses equation 9-18. You will verify this in problem 4. In the meantime, we will continue along the more accurate course.

Significantly, equations 9-14 to 9-16 contain no more than three unknowns each, and those have consecutive indexes. Therefore, when written in matrix form, the coefficient matrix is *tridiagonal*:

$$\begin{bmatrix} B_1 & C_1 & & & & \\ A_2 & B_2 & C_2 & & & \\ & A_3 & B_3 & C_3 & & \\ & & \ddots & \ddots & \ddots & \\ & & & A_{n-1} & B_{n-1} & C_{n-1} \\ & & & & A_n & B_n \end{bmatrix} \begin{bmatrix} Y_1 \\ Y_2 \\ Y_3 \\ \vdots \\ Y_{n-1} \\ Y_n \end{bmatrix} = \begin{bmatrix} D_1 \\ D_2 \\ D_3 \\ \vdots \\ D_{n-1} \\ D_n \end{bmatrix} \tag{9-19}$$

That is, aside from the main diagonal (on which the entries are $B_1, B_2, \ldots, B_N$) and the two adjacent diagonals, all the entries are zero. This leads to a very neat algorithm for determining the Y_i's, as we'll see shortly. First let's identify the A_i's, B_i's, and C_i's. From equation 9-15,

$$B_1 = 1$$
$$C_1 = 0$$
$$D_1 = V_0$$

From equation 9-14,

$$A_i = \frac{1}{(\Delta x)^2} - \frac{p_i}{2\Delta x}$$

$$B_i = -\frac{2}{(\Delta x)^2} + q_i \qquad \text{for } i = 2, \ldots, N-1$$

$$C_i = \frac{1}{(\Delta x)^2} + \frac{p_i}{2\Delta x}$$

$$D_i = r_i$$

Finally, from equation 9-16,

$$A_N = \frac{2}{(\Delta x)^2}$$

$$B_N = -\frac{2}{(\Delta x)^2} + q_N$$

$$D_N = r_N - p_N S_1 - 2\frac{S_1}{\Delta x}$$

To solve equation 9-19, the famous and very useful *Thomas algorithm* can be used. All it is, is Gauss elimination, specialized for a tridiagonal system of equations. Thus, we multiply the first equation

$$B_1 Y_1 + C_1 Y_2 = D_1$$

by (A_2/B_1) and subtract it from the second equation

$$A_2 Y_1 + B_2 Y_2 + C_2 Y_3 = D_2$$

so as to eliminate the Y_1 term. This yields an equation of the form

$$B'_2 Y_2 + C_2 Y_3 = D'_2$$

where

$$B'_2 = B_2 - C_1 A_2/B_1$$
$$D'_2 = D_2 - D_1 A_2/B_1$$

Multiplying this equation by (A_3/B'_2) and subtracting it from the third equation eliminates Y_2 from the latter, so it becomes

$$B'_3 Y_3 + C_3 Y_4 = D'_3$$

This *Gauss reduction* phase of the calculations is nicely summarized in FORTRAN:

```
    DO 100    I = 2, N
    B_I = B_I - C_{I-1} A_I / B_{I-1}
100 D_I = D_I - D_{I-1} A_I / B_{I-1}
```

Note that we have dropped the primes on the B_i's and D_i's; once those coefficients are changed, we need store only the revised values.

At the conclusion of this phase of the algorithm, the linear system (9-20) has been reduced to bidiagonal form:

$$\begin{bmatrix} B_1 & C_1 & & & & \\ & B_2 & C_2 & & & \\ & & B_3 & C_3 & & \\ & & & \ddots & \ddots & \\ & & & & B_{N-1} & C_{N-1} \\ & & & & & B_N \end{bmatrix} \begin{bmatrix} Y_1 \\ Y_2 \\ Y_3 \\ \vdots \\ Y_{N-1} \\ Y_N \end{bmatrix} = \begin{bmatrix} D_1 \\ D_2 \\ D_3 \\ \vdots \\ D_{N-1} \\ D_N \end{bmatrix}$$

Now we can solve the last equation

$$B_N Y_N = D_N$$

for Y_N, use that result to solve the preceding equation

$$B_{N-1} Y_{N-1} + C_{N-1} Y_N = D_{N-1}$$

for Y_{N-1}, and so forth. Again, the process is easily written in FORTRAN:

```
    Y_N = D_N / B_N
    DO 200    J = 2, N
    I = N - J + 1
200 Y_I = (D_I - C_I Y_{I+1}) / B_I
```

This is called the *back-substitution* phase of the algorithm.

The Thomas algorithm goes so fast that numerical methods are often specially structured to reduce any algebraic systems involved to tridiagonal form, even if that entails some otherwise unnecessary iterations to solve the whole problem. The method which will be described in Section 10.2.1 for solving the Falkner–Skan equation is a case in point.

9.2. CONVERGENCE AND ORDER OF ACCURACY

This is a good point at which to discuss in general terms a problem that the finite-difference method just described shares with all numerical methods, including those discussed previously in this book, namely, *convergence*. Finite-difference methods for solving differential equations approximate the derivatives in the differential equation by differences among the function values at the nodes. To quantify these approximations, define the *local truncation error* of a *difference* equation as the quantity that is required to balance the equation when the exact solution of the corresponding *differential* equation is substituted therein for its finite-difference approximation. In the case of equation 9-14, the local truncation error T_i is thus defined by

$$\frac{y_{i+1} - 2y_i + y_{i-1}}{(\Delta x)^2} + p_i \frac{y_{i+1} - y_{i-1}}{2\Delta x} + q_i y_i = r_i + T_i$$

in which $y_i \equiv y(x_i)$. Substituting from the Taylor-series-based equations 9-12 and 9-13 yields

$$\frac{d^2 y}{dx^2}(x_i) + p(x_i)\frac{dy}{dx}(x_i) + q(x_i)y(x_i) + O(\Delta x)^2 = r(x_i) + T_i$$

Then, since y is defined to be the exact solution of equation 9-1 at x_i [as well as elsewhere in the interval (0, 1)], the local truncation error satisfies

$$T_i = O(\Delta x)^2$$

which is also implied by the definition of T_i and equation 9-13.

However, all this means is that the difference between the differential and difference *equations* is of order $(\Delta x)^2$. The more important question is, what is the difference between their *solutions*? As $\Delta x \to 0$, does $Y_i \to y(x_i)$?

This is the question of *convergence*, which must be asked of any numerical method. It is most convincingly answered simply by redoing the calculations with different numbers of nodes and comparing the results. If you do this with the program LINBVP that implements the finite-difference method described above (and is listed in Section 9.7), you will find that it does indeed converge, and quite rapidly in the example problem.

Once we are sure a method converges, we want to know its rate of convergence, or *order of accuracy*. In the case of the finite-difference methods under discussion, a method is said to be *pth-order accurate* if

$$y(x_i) - Y_i = O(\Delta x)^p$$

Knowing that the error of one method is of fourth order in the interval whereas that of another is only of second order suggests that the first one may require fewer intervals and (perhaps) less work to achieve a given accuracy than the second one, as indicated by Fig. 9.3. Thus information on the order of accuracy can be useful in selecting a method. Also, as was indicated in Chapter 8, you can use a knowledge of the order of accuracy of a method to extrapolate results obtained numerically to the limit of zero interval size (or infinite number of iterations, or whatever). Suppose it is known that the error is of first order in Δx. Let $Y(x; \Delta x)$ be the numerical solution obtained at x using the step size Δx, and $y(x)$ the exact solution at x. Then, as you can see from Fig. 9.4, data obtained with different step sizes Δx_1 and Δx_2 can be extrapolated to

$$y(x) \approx \frac{Y(x; \Delta x_2)\Delta x_1 - Y(x; \Delta x_1)\Delta x_2}{\Delta x_1 - \Delta x_2}$$

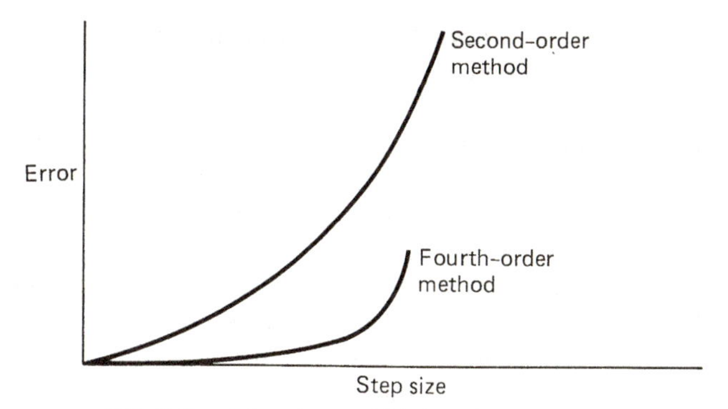

FIG. 9.3. Advantage of high-order accuracy.

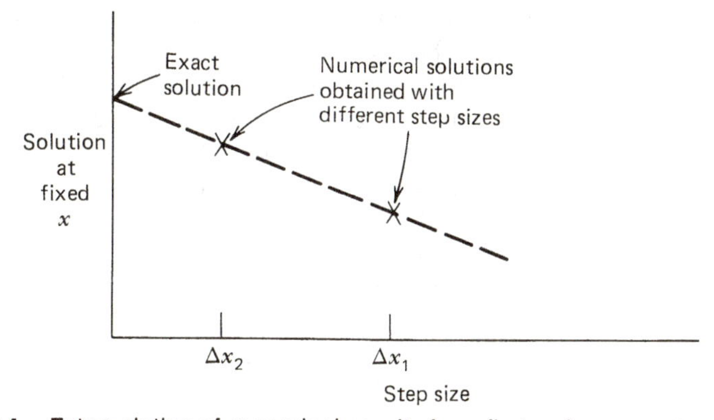

FIG. 9.4. Extrapolation of numerical results from first-order-accurate method.

If, instead, the error varies quadratically with Δx, we can proceed as follows. If

$$Y(x; \Delta x) \simeq y(x) + C(\Delta x)^2$$

then

$$Y\left(x; \frac{\Delta x}{2}\right) \simeq y(x) + \frac{C(\Delta x)^2}{4}$$

would be the result obtained at x with half the original mesh size. Although the constant C is unknown, the term involving C can be eliminated from these two expressions, and

$$y(x) \approx \frac{4}{3} Y\left(x; \frac{\Delta x}{2}\right) - \frac{1}{3} Y(x; \Delta x) \tag{9-20}$$

would be expected to be a better approximation to $y(x)$ than either of its constituent numerical solutions.

Combining numerical solutions obtained with different step sizes in an attempt to extract results of greater accuracy than any of them is called *Richardson extrapolation.* It is built into the program LINBVP discussed below.

Again, the order of accuracy of a numerical method is best established by running the method with different numbers of nodes. Of course, this is expensive, and any means for predicting the order of accuracy is appreciated. In the case of finite-difference methods, the error of the solution is usually of the same order of magnitude in the step size as the local truncation error, the local difference between the differential and finite-difference equations, which is usually easy to predict by making Taylor-series expansions.

However, there are cases in which the local truncation error underestimates the order of accuracy. In particular, if, as seems reasonable, the nodes x_i are concentrated in parts of the domain where the solution is expected to vary relatively rapidly, the uneven spacing reduces the accuracy of three-point approximations to d^2y/dx^2 like equation 9-12 to first order. However, as will be discussed in detail in Chapter 10 and will be tested by you in problem 3, the error of finite-difference solutions based on such approximations often continues to be of second order even when the uneven spacing makes the local truncation error of first order.

Thus it is always advisable to test the accuracy of a finite-difference method numerically. Also, knowing a method is second-order accurate tells you only its rate of convergence, not the actual magnitude of the error. The difference between Y_i and $y(x_i)$ could be $100(\Delta x)^2$ just as well as $10^{-6}(\Delta x)^2$. Two methods with the same order of accuracy can therefore differ widely in their rate of convergence.

The foregoing ideas are built into a FORTRAN program called LINBVP, which solves the linear boundary-value problem

$$-\frac{d}{dx}\left(p\frac{dy}{dx}\right) + qy = r$$

$$y(0) = 1$$

$$\frac{dy}{dx}(1) = 0 \tag{9-21}$$

This can be reduced to an equation like 9-1 by expanding the first derivative and dividing by $-p$. The advantage of equation 9-21 is that its solution can be guaranteed to be unique if

$$p(x) > 0 \qquad \text{and} \qquad q(x) \geq 0$$

in the interval (0, 1). These criteria are met by the functions built into LINBVP, namely,

$$p = 1 + x^2, \qquad q = x^2(1 + x^2)$$

The exact solution was specified to be

$$y = 1 - x^2 + x^4/4$$

which meets the boundary conditions of equation 9-21, and then the function $r(x)$ was defined so that the differential equation would be satisfied. Thus the output of LINBVP includes a comparison of the numerical solution of the exact solution.

Program LINBVP also demonstrates the worth of Richardson extrapolation.

The program is set up to repeat the entire solution, with the interval halved each time. When you start it up, you are asked to "INPUT MOUT, NHALF," where NHALF is the number of times the interval will be halved and MOUT is the number of points at which data will be output, regardless of the current calculation interval; it is one more than the number of intervals used in the first solution. After each solution (except the first), equation 9-20 is used to extrapolate data obtained at the output points with the interval just used and with the interval used in the previous solution. The results are listed under "YXTRAP." You should find that they converge to the exact solution ("YEXACT") much faster than do the raw results (labeled just plain "Y").

9.3. INCOMPRESSIBLE POTENTIAL FLOW PAST A THIN SYMMETRIC AIRFOIL

Let's now apply the finite-difference method to the problem solved in Chapter 3 with source distributions; the flow at zero angle of attack past a symmetric airfoil. Let ϕ be the *perturbation potential* of the velocity field, so that

$$\mathbf{V} = \mathbf{V}_\infty + \nabla\phi$$

where $\mathbf{V}_\infty = V_\infty\hat{\mathbf{i}}$ is the onset flow velocity. Then ϕ satisfies the Laplace equation

$$\frac{\partial^2\phi}{\partial x^2} + \frac{\partial^2\phi}{\partial y^2} = 0 \tag{9-22}$$

in the flow field, and boundary conditions

$$\nabla\phi \to 0 \tag{9-23}$$

far from the airfoil and

$$\hat{\mathbf{n}}\cdot\mathbf{V} = \hat{\mathbf{n}}\cdot\nabla\phi + \hat{\mathbf{n}}\cdot\mathbf{V}_\infty = 0 \tag{9-24}$$

on the airfoil.

The first step of a finite-difference method is to discretize the problem. Here we will look for approximations to ϕ only at the intersections of *mesh lines* parallel to the x and y axes, as sketched in Fig. 9.5. Let our unknowns be

$$\Phi_{ij} \approx \phi(x_i, y_j)$$

We shall not require the mesh lines $x = x_i$ and $y = y_j$ to be equally spaced. This complicates somewhat the approximation of the derivatives in equation 9-22. As you will show in problem 5, the following five-point approximation is valid at the typical point (x_i, y_j):

$$\frac{\dfrac{\Phi_E - \Phi_O}{h_E} - \dfrac{\Phi_O - \Phi_W}{h_W}}{\frac{1}{2}(h_E + h_W)} + \frac{\dfrac{\Phi_N - \Phi_O}{h_N} - \dfrac{\Phi_O - \Phi_S}{h_S}}{\frac{1}{2}(h_N + h_S)} = 0 \tag{9-25}$$

Here, for brevity, we replace Φ_{ij} with Φ_O, and let Φ_E, Φ_N, Φ_W, and Φ_S be the approximations to the potential to the east, north, west, and south of the central

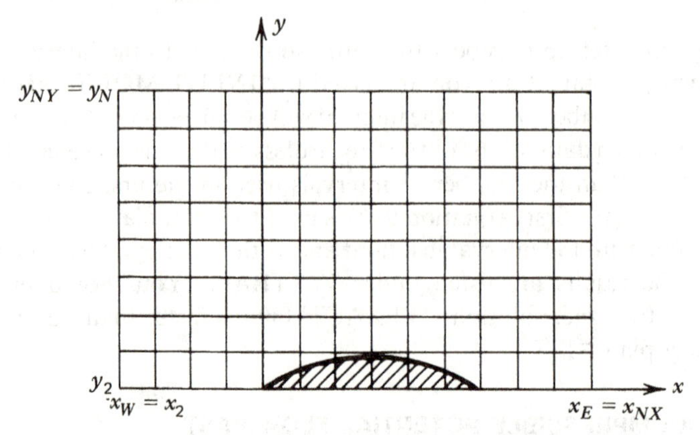

FIG. 9.5. Grid for finite-difference analysis of symmetric flow past a symmetric airfoil.

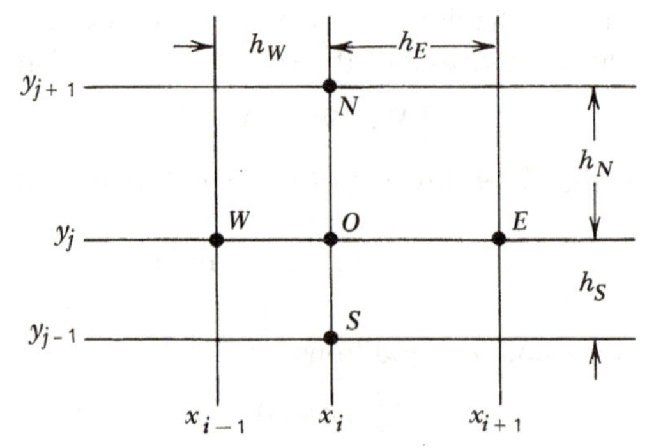

FIG. 9.6. Nomenclature for finite-difference approximation to Laplace equation at central point *O*.

point, respectively. Similarly, the *h*'s are distances from the central point to its nearest neighbors, as shown in Fig. 9.6.

The discretization of the boundary conditions (9-23) and (9-24) requires some discussion. Equation 9-23 is meant to be applied infinitely far from the airfoil. This, of course, is impossible. The easiest solution is to apply it at some large but finite distance from the airfoil. The question that then arises—how large is "large"—is best answered experimentally. Let the values of x and y at which equation 9-23 is applied be parameters input to the program (x_W, x_E, and y_N in Fig. 9.5), and run the program for various values of these parameters. If, on increasing the boundaries of the solution domain by 50%, say, you find that the pressure distribution on the airfoil changes very little, then you can be reasonably confident that you've set the boundary conditions far enough away from the airfoil.[1]

[1] With some deeper analysis, you can do much better. Murman and Cole [1], for example, analyzed the way in which their potential decayed to its far-field value, their results having a form similar to that derived in problem 5 of Chapter 8. Using such a result to stipulate the far-field behavior generally allows you to bring in the boundaries of the computational domain much closer to the airfoil, which saves computational storage space and, because the number of points at which results are found is reduced, computing time.

Since a velocity potential consistent with the far-field behavior demanded by equation 9-23 is

$$\phi \to 0 \qquad \text{far from airfoil}$$

we could simply set the values of ϕ accordingly on the boundaries of the solution domain. However, since the potential is only a compact way of storing information about the more physically interesting velocity field, it is preferable to set boundary conditions on $\mathbf{V}$, or on $\nabla\phi = \mathbf{V} - V_\infty\hat{\mathbf{i}}$. Specifically, to ensure overall conservation of mass, we set the values of the component of $\mathbf{V}$ normal to the boundaries of the computational domain.[2] Referring again to Fig. 9.5, we require

$$\frac{\partial\phi}{\partial x} = u - V_\infty = 0 \qquad \text{at } x = x_W \qquad \text{and} \qquad x = x_E$$

$$\frac{\partial\phi}{\partial y} = v = 0 \qquad \text{at } y = y_N \tag{9-26}$$

which are not hard to discretize accurately. If

$$X_2 = x_W, \qquad X_{NX} = x_E, \qquad Y_{NY} = y_N \tag{9-27}$$

we can use the central-difference approximations (see equation 9-12)

$$\frac{\Phi_{3j} - \Phi_{1j}}{X_3 - X_1} = 0; \qquad \frac{\Phi_{NX+1,j} - \Phi_{NX-1,j}}{X_{NX+1} - X_{NX-1}} = 0$$

$$\frac{\Phi_{iNY+1} - \Phi_{iNY-1}}{Y_{NY+1} - Y_{NY-1}} = 0 \tag{9-28}$$

in which the lines $x = X_1, x = X_{NX+1}$, and $y = Y_{NY+1}$ are outside the computational domain and are defined by

$$X_2 - X_1 = X_3 - X_2$$

for example.

Note that we do not impose any far-field boundary condition beneath the airfoil. That is because we intend to take advantage of the symmetry of the problem about the x axis by solving for ϕ only in the region $y \geq 0$. Thus we instead set boundary conditions on the x axis off the airfoil. Specifically, we note

$$v = \frac{\partial\phi}{\partial y} = 0 \qquad \text{at } y = 0 \text{ off the airfoil} \tag{9-29}$$

Which brings us to the airfoil, the flow tangency condition, and a major difficulty of the finite-difference method. Up to this point, the method has been quite straightforward. However, the discretization of equation 9-24 is complicated if, as is implied by Fig. 9.5, the body surface crosses mesh lines at points other than mesh line

[2] Strictly speaking, these boundary conditions determine ϕ only up to an additive constant. However, we will use an iterative method that will be determinate for the velocity field, although the level of the potential will depend on what guess is made for the solution initially.

intersections. To be sure, one can approximate $\hat{\mathbf{n}} \cdot \nabla\phi$ at point O in Fig. 9.7 in terms of values of ϕ at O and nearby points like A, B, and C, but the geometry required to locate the intersections of the surface normal with the appropriate mesh lines is very complicated. Also, as with the one-sided difference approximation to the boundary condition (9-3) in the one-dimensional problem discussed in the previous section, using a first-order approximation to the derivatives of ϕ at point O would spoil the accuracy of the whole solution. With a major increase in complexity, it would be possible to get formal second-order accuracy, but then the solution sometimes blows up [2].

Two remedies are commonly used. The more accurate is a finite-difference mesh that conforms to the body, as shown in Fig. 9.8. Although it helps somewhat if the mesh lines are orthogonal to the body, that is not essential. The effort spent on generating nonorthogonal body-conforming meshes for use in finite-difference (and related) methods in computational fluid mechanics may surprise you; see the papers in Ref. [3], for example. Some even solve (numerically) nonlinear partial differential equations, just to get meshes they can use to solve other equations!

Here, as befits an introductory text, we shall follow a simpler procedure, approximating the boundary condition (9-24) as we did in the thin-airfoil theory of Chapter 3. Using the same arguments that led to equation 3-29, we thus require

$$v = \frac{\partial \phi}{\partial y} = V_\infty \frac{dY}{dx} \qquad \text{at } y = 0 \text{ on the airfoil} \tag{9-30}$$

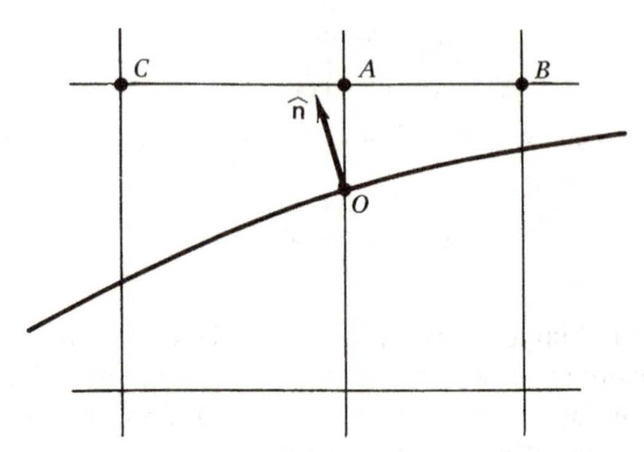

FIG. 9.7. Approximating normal derivative in terms of function values at mesh points.

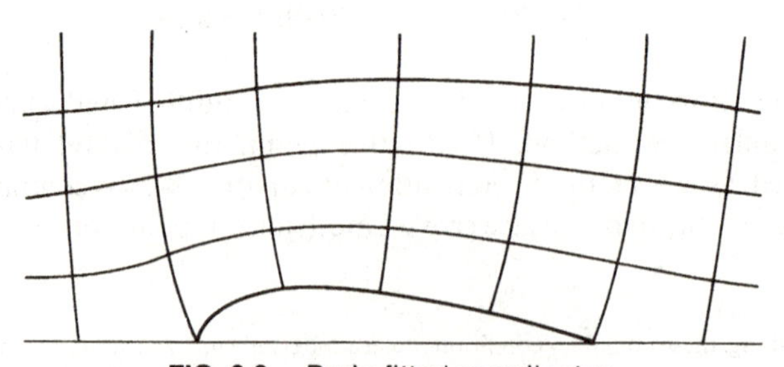

FIG. 9.8. Body-fitted coordinates.

where $Y(x)$ is the half-thickness of the airfoil. This is as easy to discretize as were the far-field boundary conditions (9-26); if we let

$$Y_2 = 0$$
$$Y_2 - Y_1 = Y_3 - Y_2 \tag{9-31}$$

then equations 9-29 and 9-30 are approximated by

$$\frac{\Phi_{i3} - \Phi_{i1}}{Y_3 - Y_1} = V_\infty \frac{dY}{dx} \qquad \text{for } 0 < x_i < c$$
$$= 0 \qquad \text{otherwise} \tag{9-32}$$

Of course, as observed in Chapters 3 and 4, the thin-airfoil approximation (9-30) costs you quite a bit of accuracy, especially near stagnation points. However, when the geometry is as complicated as it is for real aircraft, body-conforming meshes are so hard to construct that approximations are sometimes necessary for the finite-difference method to be usable. This is the main advantage of the panel method, its ability to describe accurately flows past realistic geometries with relative ease. Unfortunately, as we'll see in Chapter 11, it is not applicable to some very important classes of flows, in particular to transonic flows.

9.3.1. Direct Methods

The set of equations to be solved now consists of finite-difference equations like (9-25) [one for each of the $(NX - 1) \times (NY - 1)$ mesh points], and $2(NX - 1) + 2(NY - 1)$ boundary conditions like (9-28) and (9-32). The number of equations and unknowns is considerable. Since we want to put a good number of mesh points on the airfoil and must have enough fore and aft of the airfoil to approximate the flow field far away, to have 30 constant-x mesh lines is by no means excessive, and even a 30×10 mesh is rather coarse. To put equations 9-25, 9-28, and 9-32 into the generic form

$$\sum_j A_{ij} x_j = b_i \tag{9-33}$$

is therefore possible, but not very attractive, if you then intend to plug the A matrix into some standard linear-equation solver. Storing the coefficients of a 300×300 matrix is beyond the capacity of all but the biggest computers. And what if the problem were three dimensional!

Fortunately, there are alternatives. After all, none of the equations we have to solve has more than five unknowns, regardless of the number of mesh points. Thus there should be (and are) ways to take advantage of the *sparseness* of our coefficient matrix (i.e., the fact that nonzero elements of the matrix are few and far between).

One possibility is based on the observation that equations 9-25, 9-28, and 9-32 each involve unknowns only on (at most) three neighboring columns (or rows) of the mesh. Therefore, if we collect the unknowns on the line $x = x_i$ for example, into a vector

$$\{\mathbf{\Phi}_i\} \equiv \begin{Bmatrix} \Phi_{i1} \\ \Phi_{i2} \\ \vdots \\ \Phi_{iNY} \\ \Phi_{iNY+1} \end{Bmatrix} \tag{9-34}$$

matrices $[A_i]$, $[B_i]$, $[C_i]$, and $\{d_i\}$ can be defined so that the algebraic equations associated with points on the line $x = x_i$ can be collected into

$$[A_i]\{\mathbf{\Phi}_{i-1}\} + [B_i]\{\mathbf{\Phi}_i\} + [C_i]\{\mathbf{\Phi}_{i+1}\} = \{d_i\} \tag{9-35}$$

with $[A_1] = [C_{NY+1}] = 0$. These equations can themselves be collected into the *block-tridiagonal* form

$$\begin{bmatrix} B_1 & C_1 & & & & \\ A_2 & B_2 & C_2 & & & \\ & A_3 & B_3 & C_3 & & \\ & & \ddots & \ddots & \ddots & \\ & & & A_{NY} & B_{NY} & C_{NY} \\ & & & & A_{NY+1} & B_{NY+1} \end{bmatrix} \begin{bmatrix} \mathbf{\Phi}_1 \\ \mathbf{\Phi}_2 \\ \mathbf{\Phi}_3 \\ \vdots \\ \mathbf{\Phi}_{NY} \\ \mathbf{\Phi}_{NY+1} \end{bmatrix} = \begin{bmatrix} d_1 \\ d_2 \\ d_3 \\ \vdots \\ d_{NY} \\ d_{NY+1} \end{bmatrix} \tag{9-36}$$

that is, into a matrix that *looks* tridiagonal but whose elements are themselves matrices. The vectors $\{\mathbf{\Phi}_i\}$ can be found by a procedure remarkably like the Thomas algorithm that was applied to equation 9-19, the key difference being that divisions by diagonal elements B_i become multiplications by inverses of matrices $[B_i]$. Such a procedure will be implemented in Chapter 10 in connection with the "box method" for solving the Falkner–Skan equation, for which it works very well. However, in the present case, it is hard to fit even a moderately fine two-dimensional grid into the computer's core storage, whereas a three-dimensional problem almost certainly entails using tapes and discs for storage. Going out of core is time-consuming (i.e., expensive), so that alternative methods that require less storage are worth looking at.

9.3.2. Iterative Methods

In particular, *iterative methods*—methods in which one guesses the solution and then improves it in successive trials—are very effective in reducing the storage requirements of finite-difference methods. One such method is *relaxation by points.* It starts with the observation (again) that each of equations 9-25, 9-28, and 9-32 has at most five unknowns and also that the coefficient of Φ_O in equation 9-25, the finite-difference approximation to the Laplace equation at (x_i, y_j), is larger than the coefficients of all the other unknowns.[3] Thus, suppose we have approximations $\tilde{\Phi}_{ij}$

[3] This may not be obvious from the clumsy form of equation 9-25. If you check it out, you will see that the absolute value of the coefficient of Φ_O in fact exactly equals the sum of the absolute values of the other coefficients.

for the unknowns Φ_{ij}. A new (and hopefully improved) approximation for Φ_{ij} might be obtained by "solving" the equation associated with the point x_i, y_j for that unknown and replacing the unknowns other than Φ_{ij} by whatever approximations we may have available.

For example, suppose we have an evenly spaced, square mesh,

$$h_E = h_W = h_N = h_S = h \tag{9-37}$$

in which case equation 9-25 reduces to

$$\frac{1}{h^2}(\Phi_E + \Phi_W + \Phi_N - \Phi_S - 4\Phi_O) = 0 \tag{9-38}$$

Given approximations $\tilde{\Phi}_N$, $\tilde{\Phi}_E$, $\tilde{\Phi}_W$, and $\tilde{\Phi}_S$ for the Φ's at the points surrounding the central point, the new approximation to Φ_O, Φ_O^*, say, is then

$$\Phi_O^* = \tfrac{1}{4}(\tilde{\Phi}_E + \tilde{\Phi}_W + \tilde{\Phi}_N + \tilde{\Phi}_S) \tag{9-39}$$

for all points (x_i, y_j) away from the boundaries. A similar approach may be used for the equations associated with the boundaries, equations 9-28 and 9-32.

After using equations like (9-39) to update the tilded approximations for the Φ_{ij}'s for all points (x_i, y_j) in the domain of interest, we must decide whether to continue the iterations or to accept the answers we have. If the changes $\{\Phi_{ij}^* - \tilde{\Phi}_{ij}\}$ are sufficiently small, the new approximations Φ_{ij}^* may be presumed to be close enough to the solutions Φ_{ij} of the finite-difference equations. If not, the Φ_{ij}^*'s just calculated become the $\tilde{\Phi}_{ij}$'s of the next approximation.

An iterative process, it is seen, thus has three stages:

1. The initial guess. Usually, the better the guess, the better the performance of the method. Thus it is sometimes worthwhile to solve the problem on a relatively coarse mesh, just to get (by interpolation) a good first guess for use on a fine mesh.
2. An algorithm to change the current guess into an improved one.
3. A criterion for stopping the iterations.

To turn the $\tilde{\Phi}$'s into Φ^*'s by use of equation 9-39 requires that we store only two pieces of data for each x_i, y_j, namely, $\tilde{\Phi}_{ij}$ and Φ_{ij}^*. However, we could do even better, so far as storage is concerned, by discarding the old approximation to Φ_{ij} as soon as we calculate a new one; that is, by overwriting $\tilde{\Phi}_{ij}$ with Φ_{ij}^*. Then we need only store the current approximation to Φ_{ij}, and so cut the storage requirements to the bare minimum. This is called the *method of successive displacements or Gauss–Seidel method*. The method requiring storage of both the old and new approximations is known as the *method of simultaneous displacements*, or *Jacobi's method*.

If we continue to distinguish between old and new approximations to the Φ_{ij}'s by tildes and asterisks, the form taken by the equation that defines Φ_{ij}^* will differ from equation 9-39 in that some of the data on the right side will have come from the previous (tilded) approximation and the rest from the new (starred) approximation. The results obtained depend on the order in which you sweep through the mesh and update your approximations to the Φ_{ij}'s. If you go from left to right and down to up,[4] the data at points to the left and below (x_i, y_j) will have been updated by the time you compute Φ^*_{ij}, and the formula to use would be[5]

$$\Phi^*_O = \tfrac{1}{4}(\tilde{\Phi}_E + \Phi^*_W + \tilde{\Phi}_N + \Phi^*_S) \tag{9-40}$$

Before getting down to programming details, I want to consider one more variation on the iterative method. Although the use of equation 9-40 rather than equation 9-39 relieves the storage requirement, that does not necessarily make it the method of choice. Another criterion in selecting a computational method is speed (i.e., cost). In an iterative method, speed is controlled by the amount of work required per trial, and the number of trials required for the approximate solutions $\tilde{\Phi}_{ij}$ and Φ^*_{ij} to converge to one another (and so, presumably, to the solution of equations 9-38). There is not much to choose between equations 9-39 and 9-40 so far as work per sweep through the mesh is concerned, but there are substantial differences in the rates of convergence of the two methods. Unfortunately, these differences depend somewhat on the problem being attacked; see problem 6. However, Stein and Rosenberg [4] showed that, at least for a certain class of matrices $[A]$ (which includes the coefficient matrix of equations 9-38), the Jacobi (equation 9-39) and Gauss–Seidel (equation 9-40) methods either both converge or both diverge. When they do converge, the Gauss–Seidel method does so faster. Since it also entails less storage, equation 9-40 is generally preferred to equation 9-39.

Still, the convergence rate of the Gauss–Seidel method is painfully slow, and it gets worse when you try to refine the mesh. What it needs is an adjustable parameter that can be varied to improve the rate of convergence; a dial you can twist to tune in the method. This is what it gets in the *successive overrelaxation method*, in which the change in the approximation to Φ_{ij} is taken to be a factor ω times the change you would make in the Gauss–Seidel method:

$$\begin{aligned}\Phi^*_O &= \tilde{\Phi}_O + \omega(\Phi^*_O - \tilde{\Phi}_O)_{\text{Gauss–Seidel}} \\ &= \frac{\omega}{4}(\tilde{\Phi}_E + \Phi^*_W + \tilde{\Phi}_N + \Phi^*_S) + (1-\omega)\tilde{\Phi}_O \end{aligned} \tag{9-41}$$

in which we have used equation 9-40. The method is called "successive" because, like the Gauss–Seidel method, it is used to update the approximations to Φ_{ij} at successive points while the iteration is in process, and "overrelaxation" because iterative methods are called *relaxation methods* and the best value of the parameter ω in equation 9-41 is generally greater than one. The solution at each point should be coupled to the values of Φ at each of its neighbors, but, in the Gauss–Seidel

[4] Since these are the directions of increasing i and j, this order makes some sense and is given the name *lexicographical order*.

[5] As you may realize already, but will certainly see when we get to the program that implements the iterative method, it is not really necessary to distinguish between the tilded and starred approximations once you get down to computing. The only Φ_{ij}'s stored are the current approximations thereto, which are partly tilded and partly starred.

process, the Φ's at two of those neighbors are frozen at previously computed values. Thus the change called for in the Gauss–Seidel method is underdone.

With the right choice of the overrelaxation parameter ω, dramatic increases in the convergence rate are possible. For example, suppose we have to solve the Laplace equation on an $N \times N$ mesh (N mesh lines in both the x and y directions). The Jacobi and Gauss–Seidel methods both converge. Their rate of convergence can be measured by examining the root-mean-squared errors

$$\tilde{E} \equiv \left[\sum_{ij} (\Phi_{ij} - \tilde{\Phi}_{ij})^2\right]^{1/2}, \qquad E^* \equiv \left[\sum_{ij} (\Phi_{ij} - \Phi^*_{ij})^2\right]^{1/2} \tag{9-42}$$

It can be shown that [5]

$$1 - \frac{E^*}{\tilde{E}} = O\left(\frac{1}{N^2}\right) \tag{9-43}$$

Thus the error is reduced in each iteration by a factor of order N^{-2}, so that the number of iterations required to reduce the error below a given tolerance is of the order of N^2. For the successive overrelaxation method. however, ω can be chosen so that

$$1 - \frac{E^*}{\tilde{E}} = O\left(\frac{1}{N}\right) \tag{9-44}$$

and the number of trials required for convergence is of order N^1. As the mesh is refined, therefore, the SOR (successive overrelaxation) method proves much more effective than the Jacobi and Gauss–Seidel methods.

It is also much faster than the direct (noniterative) method. With N^2 mesh points, the SOR method requires the use of N^2 equations like (9-41) in each trial. Since the number of trials required for convergence is of order N, the total number of operations required is of the order of N^3. If you count the number of operations required to solve M linear equations by Gauss elimination, it turns out to be of order M^3. Thus, if the N^2 equations like (9-33) were simply solved by a standard linear-equation solver, the work required would go like N^6. Organizing the coefficients into a block-tridiagonal matrix, as in equation 9-36, does save a lot of work. There are N rows of matrices within the matrix, so that the reduction process entails $N - 1$ inversions of the matrices $[B_i]$. This part of the process dominates the operation count. Since the matrices $[B_i]$ are $N \times N$, these inversions each require of the order of N^3 operations, and the total operation count is of order N^4. That is about the same as the work required in the Jacobi and Gauss–Seidel methods but about N times as much as that required in the optimized SOR method.

Since the Gauss–Seidel method is, from equation 9-41, simply an SOR method with $\omega = 1$, it is worth spending some effort to determine the optimum value of ω. This you will do by trial and error in problem 7. However, for the Laplace equation, the optimum relaxation factor can also be determined analytically, which is done in Appendix H. The results of that analysis are incorporated in program THINAIR, which implements the SOR method for the thin-airfoil problem under discussion.

Other noteworthy features of program THINAIR are

1. You have to specify the size of the domain, that is, the values of x and y on which you will apply the far-field boundary conditions (9-26). This should be done experimentally; that is, you should repeat this calculation with different specifications for x_W, x_E, and y_N until satisfied that the solution on the airfoil would be insensitive to further increases in the size of the domain.
2. The mesh lines are not evenly spaced, but are concentrated near the leading and trailing edges of the airfoil. This slows the convergence for a given number of mesh lines, since the rate of convergence is controlled by the size of the smallest grid. However, it improves the resolution of the details of the pressure distribution.[6]
3. The principal output of the program is the pressure distribution on the airfoil. From equation 3-35, the pressure coefficient on a thin airfoil is

$$C_p = -2\left(1 - \frac{u}{V_\infty}\right)\bigg|_{y=0}$$

$$= -\frac{2}{V_\infty}\frac{\partial \phi}{\partial x}\bigg|_{y=0}$$

   In program THINAIR, this is calculated at points midway between the mesh lines that intersect the airfoil surface by a central-difference formula.
4. Velocities are normalized with V_∞, and lengths with the airfoil chord, so that the airfoil lies on the x axis between $x = 0$ and $x = 1$.

9.4. INITIAL-VALUE PROBLEMS: THE HEAT EQUATION

The problems considered so far in this chapter are called *boundary-value problems.* A solution is sought that satisfies a differential equation in a particular domain (an interval on the x axis or a region in the x–y plane) and that meets certain conditions ("boundary conditions") on the entire boundary of the domain (at both ends of the interval or on a curve that surrounds the domain).

New phenomena appear when we apply finite-difference methods to *initial-value problems.* These are characterized (mathematically) as having supporting conditions on only part of the boundary of the domain. In problems with one independent variable, data on the solution of the differential equation are given at one point, or one end of the interval of interest. In two variables, the "initial data" are given on a line or curve that does not surround the domain. Physically, as the name implies, initial-value problems usually arise because of a time dependence of the solution, and the initial data describe the state of the system at the initial time of observation. However, as we shall see in Chapters 10 and 11, there are also cases in which one of the spatial directions is "time like," and initial data are prescribed at a particular value of the corresponding coordinate. This is not so strange; it only means there are

[6] A better way of achieving this resolution is to transform the coordinates mathematically so that equally spaced lines in the transformed coordinate space are concentrated near appropriate points in the physical space; see Ref. [3].

cases in which the flow properties at a point are influenced only by what goes on upstream of that point. What is perhaps stranger is the case in which flow properties are also affected by what goes on downstream of the point in question. Such is the behavior of an incompressible inviscid flow; recall the symmetry of the velocity fields of sources and vortices, for example (sections 2.6.2 and 2.6.3).

As our prototypical initial-value problem, we consider the one-dimensional *heat equation*

$$\frac{\partial^2 u}{\partial x^2} = \frac{\partial u}{\partial t} \tag{9-45}$$

so called because it governs the temperature distribution in a one-dimensional medium. For boundary conditions, take

$$\begin{aligned} u &= 0 \qquad \text{at } x = 0 \\ u &= 0 \qquad \text{at } x = 1 \end{aligned} \tag{9-46}$$

and let the initial condition be

$$u = g(x) \qquad \text{at } t = 0 \tag{9-47}$$

The problem is then to determine $u(x, t)$ for $t > 0$ and $0 < x < 1$.

Before we attempt a numerical solution of the heat equation, let us look at its analytical solution, with a method called *separation of variables*. The same technique will be found useful later in analyzing its various numerical solutions and will help to explain their strengths and weaknesses.

To begin with, we seek a solution in product form,

$$u(x, t) = X(x)T(t) \tag{9-48}$$

This is not the most general form, but it will be seen below that solutions of this form can be superposed to meet any given data. After substituting the product solution into the heat equation 9-45 and dividing through by XT, we get

$$\frac{1}{X}\frac{d^2X}{dx^2} = \frac{1}{T}\frac{dT}{dt}$$

The left side is independent of t, and the right side of x. Hence, neither side is variable, and both sides equal the same constant, $-a^2$, say, and the problems of determining $X(x)$ and $T(t)$ "separate":

$$\frac{d^2X}{dx^2} = -a^2X \tag{9-49}$$

$$\frac{dT}{dt} = -a^2T \tag{9-50}$$

Both equations are linear, with constant coefficients, and so have solutions of exponential form. For equation 9-49, the exponentials are complex and so are conveniently rearranged into sine and cosine form:

$$X(x) = A \cos ax + B \sin ax \tag{9-51}$$

As for equation 9-50, it yields the solution

$$T(t) = Ce^{-a^2t} \tag{9-52}$$

Thus far the constants A, B, C, and a are arbitrary. Since we are only interested in the product of X and T, we may as well set $C = 1$; no generality is lost. If we require the product solution to satisfy the boundary conditions (9-46),

$$T(t)\,X(0) = 0 = e^{-a^2t}(A + 0)$$

$$T(t)\,X(1) = 0 = e^{-a^2t}(A \cos a + B \sin a)$$

we see that $A = 0$ and

$$B \sin a = 0$$

If $B = 0$ along with A, equation 9-51 shows $X(x)$ to be zero for all x. Rejecting this trivial solution, we require that $\sin a = 0$ and so that a be an integral multiple of π:

$$a = m\pi, \qquad m = 0, \pm 1, \pm 2, \pm 3, \ldots \tag{9-53}$$

Then we write

$$XT = B_m e^{-m^2\pi^2 t} \sin m\pi x \tag{9-54}$$

the subscript m on B indicating that it may depend on m, which could be any integer.

There remains the initial condition (9-47). Although no single term of the form (9-54) is liable to suffice, we can sum such terms over the integer m:

$$u(x, t) = \sum_{m=1}^{\infty} B_m e^{-m^2\pi^2 t} \sin m\pi x \tag{9-55}$$

Since $\sin m\pi x = -\sin(-m\pi x)$, it is enough to sum over positive values of m. Then the constants B_m should be determined so as to satisfy

$$g(x) = u(x, 0) = \sum_{m=1}^{\infty} B_m \sin m\pi x \tag{9-56}$$

which is a problem of finding the coefficients of a Fourier series; see Appendix E.

However, the important thing is the form of the solution (9-55). It consists of an infinite number of terms, each sinusoidal in x and decaying with time. The higher the value of m, the higher the frequency and the more rapidly the term is damped. Physically, this means that any rapid variations in temperature are quickly smoothed out; *diffused* is the technical term.

9.4.1 An Explicit Finite-Difference Method

Although we have an analytic solution of the heat equation at hand, it is not all that useful, unless only the first few terms of the potentially infinite series are needed. In any case, our interest in the heat equation is to try out different finite-difference methods.

Lay a rectangular mesh on the (x, t) plane, as shown in Fig. 9.9, spacing the mesh lines equally in both x and t directions for convenience. We then seek approxi-

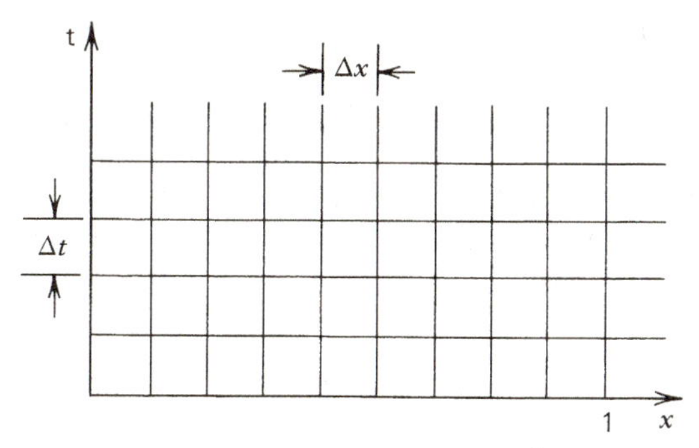

FIG. 9.9. Grid for finite-difference analysis of heat equation.

mations U_{ij}, say, to the value of $u(x_i, t_j)$, where

$$x_i = (i - 1)\Delta x$$

$$t_j = (j - 1)\Delta t \tag{9-57}$$

A finite-difference approximation to the heat equation 9-45 at the mesh point (x_i, t_j) can then be derived by approximating the derivatives therein by differences among the values of u at nearby mesh points. An obvious approximation for $\partial^2 u/\partial x^2$ is the usual central-difference formula (see equation 9-12)

$$\left.\frac{\partial^2 u}{\partial x^2}\right|_{ij} = \frac{u_{i+1,j} - 2u_{ij} + u_{i-1,j}}{\Delta x^2} + O(\Delta x)^2 \tag{9-58}$$

in which $u_{kl} = u(x_k, t_l)$. A central difference approximation is also a possibility for $\partial u/\partial t$,

$$\left.\frac{\partial u}{\partial t}\right|_{ij} = \frac{u_{ij+1} - u_{ij-1}}{2\Delta t} + O(\Delta t)^2 \tag{9-59}$$

but then it is hard to get the calculation started, since the finite-difference equation obtained by substituting equations 9-58 and 9-59 into the heat equation 9-45 involves three time levels, and we have initial data (9-47) only at one time level.[7]

A vastly simpler algorithm results if we use the forward-difference approximation:

$$\left.\frac{\partial u}{\partial t}\right|_{ij} = \frac{u_{ij+1} - u_{ij}}{\Delta t} + O(\Delta t) \tag{9-60}$$

On substituting equations 9-58 and 9-60 into 9-45, and defining the finite-difference approximations U_{ij} by asking that they satisfy the resulting equation without the error terms, we get

$$\frac{U_{i+1j} - 2U_{ij} + U_{i-1j}}{(\Delta x)^2} = \frac{U_{ij+1} - U_{ij}}{\Delta t}$$

[7] Also, using equation 9-59 for the time derivative turns out to lead to unconditional *instability*, a term that will be defined below.

which we solve for U_{ij+1}:

$$U_{ij+1} = U_{ij} + \frac{\Delta t}{(\Delta x)^2}(U_{i+1j} - 2U_{ij} + U_{i-1j}) \tag{9-61}$$

This equation may be used for $j = 1, 2, 3, \ldots$ and for $2 \le i \le N - 1$, where $x_N = (N - 1)\Delta x = 1$, the right-hand boundary of the domain of interest. The finite-difference approximations to the boundary conditions (9-46) and the initial condition (9-47) are obviously

$$U_{1j} = 0 \qquad \text{for } j > 1 \tag{9-62}$$

$$U_{Nj} = 0 \qquad \text{for } j > 1 \tag{9-63}$$

$$U_{i1} = g(x_i) \qquad \text{for } i = 1, \ldots, N \tag{9-64}$$

The algorithm for determining a finite-difference approximation to the solution of equations 9-45 through 9-47 is then as follows:

1. Use equation 9-64 for initial values.
2. To obtain the solution at t_j, $j = 2, 3, 4, \ldots$.
   (a) Use equation 9-62 for U_{1j}.
   (b) Use equation 9-61, with j reduced by 1, for $U_{2j}, \ldots, U_{N-1,j}$.
   (c) Use equation 9-63 for U_{Nj}.

Surely no problem of equal importance has a simpler or more transparent solution.

The program used to implement this method, HEATEX, is set up to handle boundary conditions somewhat more general than (9-46), namely,

$$a_0 u - b_0 \frac{\partial u}{\partial x} = c_0 \qquad \text{at } x = 0$$

$$a_1 u + b_1 \frac{\partial u}{\partial x} = c_1 \qquad \text{at } x = 1$$

with the coefficients being constants input by the user, such that

$$a_0 b_0,\ a_1 b_1 \ge 0$$

Violations of these conditions lead to an instability that, since the heat equation governs the temperature distribution is a one-dimensional medium, can be interpreted as a consequence of violating the second law of thermodynamics. Heat could flow from cold to hot without any work being done.

9.4.2. Stability

The behavior of the solution generated by program HEATEX will be found to depend critically on the parameter

$$r \equiv \frac{\Delta t}{(\Delta x)^2} \tag{9-65}$$

When $r < 1/2$, the solution proceeds in time as you might expect, if you think of

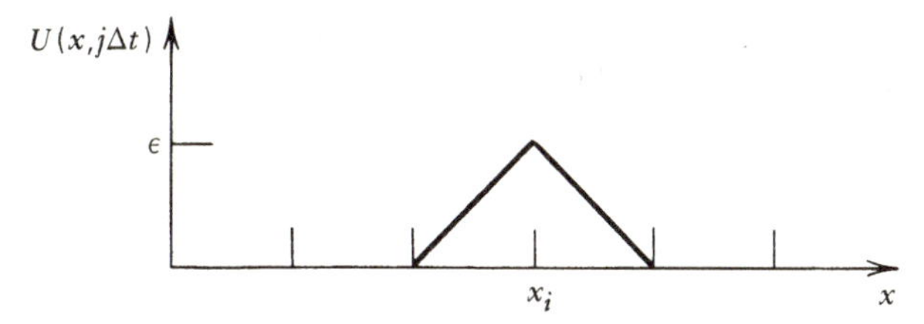

FIG. 9.10. Local excess in temperature.

nature as trying to diffuse all temperature gradients. However, when $r > 1/2$, all hell breaks loose. The solution begins to exhibit oscillatory spatial variations, which grow disastrously in amplitude as time marches on. This is the phenomenon of *instability*. It is something to be contended with in any initial-value problem, including the boundary-layer equations to be discussed in Chapter 10 and the transonic flow equations of Chapter 11.

A simple explanation of the phenomenon can be obtained by looking at a special case. Suppose the solution at the jth time level exhibited a little blip near x_i, as shown in Fig. 9.10:

$$U_{i+1j} = U_{i-1j} = 0, \qquad U_{ij} = \varepsilon$$

Then, according to the finite-difference approximation (9-61), the solution at the same x at the next time level is

$$U_{ij+1} = (1 - 2r)\varepsilon$$

so that, if $r > 1/2$, $U_{ij+1} < 0$. In terms of the physical interpretation of the heat equation, this would mean that a local excess in temperature would lead subsequently to a local valley in the temperature distribution, which makes no sense whatsoever. What has happened is that we have extrapolated too far in time our local estimate of $\partial u/\partial t$.

A more rigorous explanation of instability can be obtained by solving the finite-difference equations with the separation-of-variables method that led to the solution (9-55) of the heat equation. So doing will give us an expression of the finite-difference solution very similar in form to equation 9-55, thus enabling a discussion of the relation of the numerical solution to the continuous solution. Of course, the results of the finite-difference method are actually obtained by following the algorithm described after equation 9-64.

Thus, assume a solution of equations 9-61 to 9-64 of the form

$$U_{ij} = X_i T_j \tag{9-66}$$

in which X_i is a function of the x index i only and T_j of j only. Substituting this into equation 9-61 and dividing by U_{ij}, we get

$$\frac{T_{j+1} - T_j}{T_j} = r\left(\frac{X_{i+1} - 2X_i + X_{i+1}}{X_i}\right)$$

r being given by equation 9-65. Since the left side is independent of i, and the right of

j, neither side is variable. The constant to which they are equal is conveniently (though not obviously) set equal to

$$2r(\cos \alpha - 1)$$

Then

$$X_{i+1} + X_{i-1} = 2 \cos \alpha \, X_i$$

$$T_{j+1} = [1 - 2r(1 - \cos \alpha)] T_j$$

These are *linear difference equations* with constant (with respect to the independent variables i and j) coefficients. As is the case with linear *differential* equations with constant coefficients, their general solutions are of exponential form:

$$\begin{aligned} X_i &= Ae^{\lambda x_i} \\ &= Ae^{\lambda(i-1)\Delta x} \\ &= A\rho^{i-1} \end{aligned}$$

where $\rho = e^{\lambda \Delta x}$. On substituting this assumed solution form into the equation for X_i, a quadratic equation for ρ is derived, whose two solutions are $e^{\pm i\alpha}$, where $i^2 = -1$. The general solution, which must be a linear combination of these two solutions, is then conveniently written as

$$X_i = A \cos(i - 1)\alpha + B \sin(i - 1)\alpha \tag{9-67}$$

Similarly, we find

$$T_j = C[1 - 2r(1 - \cos \alpha)]^{j-1} \tag{9-68}$$

as you can also show by recursion (if $T_{j+1} = KT_j$, then $T_j = KT_{j-1}$, and $T_{j+1} = K^2 T_{j-1} = K^3 T_{j-2} = \cdots$).

Because we are interested only in the product $X_i T_j$, we can set $C = 1$. Insisting that $X_i T_j$ meet the boundary condition (9-62) yields $A = 0$, whereas equation 9-63 then implies

$$0 = X_N = B \sin(N - 1)\alpha$$

Since setting B equal to 0 would make X_i identically zero, we require

$$\sin(N - 1)\alpha = 0$$

and so

$$\alpha = \frac{m\pi}{N - 1} = m\pi\Delta x \qquad m = 0, \pm 1, \pm 2, \pm 3, \ldots$$

since $\Delta x = 1/(N - 1)$. To satisfy the initial condition (9-64) requires a series of product solutions, and so we set

$$U_{ij} = \sum_{m=1}^{N-1} B_m [1 - 2r(1 - \cos m\pi\Delta x)]^{j-1} \sin(i - 1)m\pi\Delta x \tag{9-69}$$

where the B_m's should be determined so as to satisfy equation 9-64.

As was advertised, equation 9-69 is as similar in its form as in its derivation to the continuous solution (9-55) of the heat equation. To heighten the similarity, we use equations 9-57 to rewrite equation 9-55 as

$$u(x_i, t_j) = \sum_{m=1}^{\infty} B_m(e^{-m^2\pi^2\Delta t})^{j-1} \sin(i-1)m\pi\Delta x \tag{9-70}$$

The dependence of the two results on the spatial coordinate is then seen to be identical.

Equations 9-69 and 9-70 show that both the continuous and finite-difference solutions can be expressed as a sum of terms, each sinusoidal in x and proportional to the $(j-1)$th power of what we shall call a *damping factor*. The differences between equations 9-69 and 9-70 are two:

1. The numerical solution contains only $(N-1)$ terms, since, at the mesh points x_i, all other integral values of m yield linearly dependent terms.
2. The damping factor for the terms of the continuous solution (9-70), that is, the factor by which the mth term of the sum changes from the jth time level to the $(j+1)$th,

$$e^{-m^2\pi^2\Delta t}$$

decreases monotonically with the index m. The behavior of the corresponding factor for the numerical solution (9-69)

$$1 - 2r(1 - \cos m\pi\Delta x)$$

depends on r. When $r > \frac{1}{2}$ and m is close to N, the factor is less than -1, and "damping factor" is a misnomer; the higher-frequency terms increase in magnitude with j, and we observe what we have called *instability*. When $r < \frac{1}{2}$, the damping factor is less than 1 in magnitude, and the solution is stable. Note, however, that if $1/4 < r < 1/2$, and m is close to N, the damping factor is negative, in which case the terms decay in magnitude but oscillate in sign. Thus the decay of the higher-frequency terms is poorly represented by the numerical solution if we choose Δt close to the limit permitted by stability considerations.

9.4.3. Convergence

The importance of stability is only partially due to the observation that, without it, your results are simply garbage. More positively, there is *Lax's equivalence theorem* (see, for example, Richtmyer and Morton [6]):

Given a properly posed initial-value problem and a finite-difference approximation to it that satisfies the consistency condition, stability is the necessary and sufficient condition for convergence as the mesh is refined.

By "properly posed" is meant that the exact problem has a unique solution that is stable to small perturbations of the initial data. The "consistency condition" is simply that the local truncation error, the local difference between the differential and finite-difference equations, tend to zero as the mesh and time step are refined. In fact, one can add the corollary that the error of a stable solution is of the same order of magnitude in the mesh spacing as is the local truncation error.

It is easy to prove the sufficiency part of the Lax theorem for the case under discussion. First, recall the precise definition of the local truncation error; it is the quantity that must be added to one side of the finite-difference equation for U_{ij} when U_{ij} is replaced by the corresponding value $u(x_i, t_j)$ of the exact solution of the associated differential equation. Letting

$$u_{ij} = u(x_i, t_j)$$

we thus have, from equations 9-60 and 9-65,

$$u_{ij+1} = u_{ij} + r(u_{i+1,j} - 2u_{ij} + u_{i-1j}) + \Delta t\, T_{ij} \tag{9-71}$$

where T_{ij} is the local truncation error (recall that the difference equation derived by approximating the derivatives in the heat equation with equations 9-58 and 9-60 was multiplied through by Δt in obtaining equation 9-61). Then, calling the error of the solution

$$e_{ij} \equiv u(x_i, t_j) - U_{ij}$$

we subtract equation 9-61 from 9-71 and see that e_{ij} also satisfies (9-71). At the boundaries, $U_{ij} = u_{ij}$, so

$$e_{ij} = e_{Mj} = 0$$

Now let

$$\varepsilon_j \equiv \max_{i=1,\dots,N} |e_{ij}|$$

$$\tau_j \equiv \max_{i=1,\dots,N} |T_{ij}| \tag{9-72}$$

Then, from equation 9-71 (with u_{ij} replaced by e_{ij}),

$$\varepsilon_{j+1} = \max_{i=1,\dots,N} |re_{i+1j} + (1-2r)e_{ij} + re_{i-1j} + \Delta t\, T_{ij}|$$

Since the absolute value of a sum is less than or equal to the sum of the absolute values,

$$\varepsilon_{j+1} \le \max_{i=1,\dots,N} \{|re_{i+1j}| + |(1-2r)e_{ij}| + |re_{i-1j}| + \Delta t|T_{ij}|\}$$

For stability, $1 - 2r \ge 0$. In that case, the preceding inequality can be written

$$\varepsilon_{ij} \le \max_{i=1,\dots,N} \{r|e_{i+1j}| + (1-2r)|e_{ij}| + r|e_{i-1j}| + \Delta t|T_{ij}|\}$$

$$\le (r + 1 - 2r + r)\varepsilon_j + \Delta t\, \tau_j$$

in which we have used the definitions (9-72). Thus

$$\varepsilon_{j+1} \le \varepsilon_j + \Delta t \tau_j$$

But then

$$\varepsilon_j \le \varepsilon_{j-1} + \Delta t \tau_{j-1}$$

so

$$\varepsilon_{j+1} \le \varepsilon_{j-1} + \Delta t(\tau_j + \tau_{j-1})$$

Continuing, we find

$$\varepsilon_{j+1} \le \varepsilon_1 + \Delta t(\tau_j + \tau_{j-1} + \cdots + \tau_1)$$

But $U_{ij} = u_{ij}$ at $j = 1$ ($t = 0$), so $\varepsilon_1 = 0$ and

$$\varepsilon_{j+1} \le j\Delta t \max_{k=1,\ldots,j} \tau_k$$

Since $j\Delta t = t_{j+1}$, we have shown that (see equations 9-72)

$$\max_{i=1,\ldots,N} |U_{ij+1} - u_{ij+1}| \le t_{j+1} \max_{\substack{i=1,\ldots,N \\ k=1,\ldots,j}} |T_{ik}|$$

Thus, if the local truncation error vanishes as Δx and $\Delta t \to 0$, so does the error of the finite-difference solution. Further, from the Taylor-series expansions (9-58) and (9-60) that led to the finite-difference equation 9-61, we see that

$$T_{ij} = O(\Delta t, \Delta x^2)$$

Hence the error of the solution is of the same order of magnitude:

$$U_{ij} - u(x_i, t_j) = O(\Delta t) + O(\Delta x)^2 \tag{9-73}$$

9.4.4. The Crank–Nicolson Method

Because the method used in program HEATEX is, according to equation 9-73, only first-order accurate in Δt, it seems improvable. However, as noted earlier, simply using the central-difference approximation (9-59) leads to a method that is unconditionally unstable, as you will show in problem 11. The problem is that the finite-difference equation is then second order in time and so has solutions that have nothing to do with the heat equation.

However, there is a way to get second-order accuracy for the time derivative, without introducing more than two time levels. We can set up a finite-difference approximation to the heat equation at points midway between the nodes at which data were stored, as is illustrated in Fig. 9.11. The central-difference approximation to $\partial u/\partial t$ at $(x_i, t_{j+1/2})$

$$\frac{u_{ij+1} - u_{ij}}{\Delta t}$$

involves data only at two adjacent time levels. To get a second-order accurate approximation for $\partial^2 u/\partial x^2$ at the same point, simply average the usual approximations at (x_i, t_{j+1}) and (x_i, t_j). By Taylor-series expansion about $(x_i, t_{j+1/2})$, it is

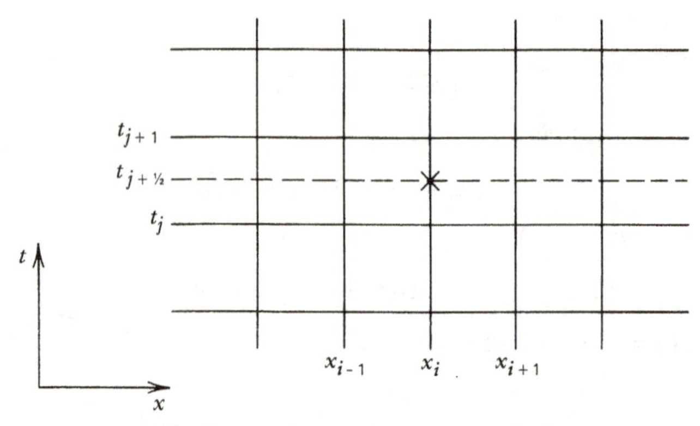

FIG. 9.11. Crank–Nicolson method.

easy to show, for example, that

$$\frac{u_{ij} + u_{ij+1}}{2} = u_{ij+1/2} + O(\Delta t)^2$$

Thus we get the finite-difference method of Crank and Nicolson [7]:

$$\frac{1}{2}\frac{U_{i+1,j+1} - 2U_{ij+1} + U_{i-1,j+1}}{(\Delta x)^2} + \frac{1}{2}\frac{U_{i+1,j} - 2U_{ij} + U_{i-1,j}}{(\Delta x)^2} = \frac{U_{ij+1} - U_{ij}}{\Delta t} \qquad (9\text{-}74)$$

Equation 9-74 contains the values of U at three adjacent nodes at the $(j + 1)$th time level, and so, in contrast with equation 9-61, cannot be solved explicitly for an approximation to the solution at the advanced time level. Rather, we must solve simultaneously the $N - 2$ equations 9-74 (with $i = 2, \ldots, N - 1$) and the boundary conditions (9-62) and (9-63) for the N unknowns $U_{1j+1}, \ldots, U_{Nj+1}$. For this reason, the Crank–Nicolson scheme is classified as *implicit*, while the method based on equation 9-61 is termed *explicit.*

However, the form of equations 9-74 makes their solution relatively easy. Since each equation contains only three unknowns and since their indexes are consecutive, their coefficient matrix is tridiagonal. Specifically, after multiplying equation 9-74 through by Δt, rearranging, and using the definition (9-65), we have to solve

$$\begin{bmatrix} 1 & & & & & & \\ -\frac{r}{2} & 1+r & -\frac{r}{2} & & & & \\ & -\frac{r}{2} & 1+r & -\frac{r}{2} & & & \\ & & \cdot & \cdot & \cdot & & \\ & & & \cdot & \cdot & \cdot & \\ & & & & \cdot & \cdot & \cdot \\ & & & & -\frac{r}{2} & 1+r & -\frac{r}{2} \\ & & & & & & 1 \end{bmatrix} \begin{bmatrix} U_{1j+1} \\ U_{2j+1} \\ U_{3j+1} \\ \cdot \\ \cdot \\ \cdot \\ U_{N-1,j+1} \\ U_{Nj+1} \end{bmatrix} = \begin{bmatrix} d_1 \\ d_2 \\ d_3 \\ \cdot \\ \cdot \\ \cdot \\ d_{N-1} \\ d_N \end{bmatrix}$$

Here

$$d_1 = 0$$

$$d_i = \frac{r}{2}(U_{i+1,j} + U_{i-1,j}) + (1 - r)U_{ij} \qquad \text{for } i = 2, \ldots, N-1$$

$$d_N = 0$$

Because the Thomas algorithm described earlier requires two sweeps through the system, roughly twice as much work per time step is required in the implicit method as in the explicit. Of course, if we get higher accuracy with the Crank–Nicolson method, the extra work may be worthwhile.

Because the Crank–Nicolson scheme is implicit, its accuracy is not so easily predicted as is the case with the explicit scheme. However, it can be shown that, as is usual, the error of the Crank–Nicolson solution is of the same order of magnitude in the mesh spacings as is the local truncation error, namely, $O(\Delta t)^2$ and $O(\Delta x)^2$. As for stability, we can still apply the separation-of-variables technique. The analysis is similar in form and results to the one we made earlier of the explicit scheme. Anticipating that the i-dependent part of the product solution will, as was the case then, be of the same sinusoidal form as was the x-dependent part of the continuous solution, we substitute $T_j \sin m\pi x_i$ into the difference equation 9-74 for U_{ij}, and get

$$\frac{1}{2}\frac{1}{(\Delta x)^2}(T_j + T_{j+1})(\sin m\pi x_{i+1} - 2\sin m\pi x_i + \sin m\pi x_{i-1})$$

$$= \frac{1}{\Delta t}(T_{j+1} - T_j)\sin m\pi x_i$$

which simplifies to

$$[1 - r(\cos m\pi\Delta x - 1)]T_{j+1} = [1 + r(\cos m\pi\Delta x - 1)]T_j$$

This difference equation is easily solved by recursion or by assuming a solution proportional to ρ^{j-1}:

$$T_j = \left[\frac{1 - r(1 - \cos m\pi\Delta x)}{1 + r(1 - \cos m\pi\Delta x)}\right]^{j-1}$$

in which the arbitrary multiplicative constant has already been set to 1. Since the assumed product solution already satisfies the boundary conditions at $x = 0$ and 1, the solution of the finite-difference equations 9-62, 9-63, and 9-74 differs from equation 9-69 only in the damping factor:

$$U_{ij} = \sum_{m=1}^{N-1} B_m \left(\frac{1 - r(1 - \cos m\pi\Delta x)}{1 + r(1 - \cos m\pi\Delta x)}\right)^{j-1} \sin(i-1)m\pi\Delta x \tag{9-75}$$

As before, the constants B_m are determined by the initial condition (9-64).

To avoid instability, the damping factors

$$\frac{1 - r(1 - \cos m\pi\Delta x)}{1 + r(1 - \cos m\pi\Delta x)}$$

must be less than 1 in magnitude for every m. Since $1 - \cos m\pi\Delta x \geq 0$, each factor is less than $+1$ for every positive $r = \Delta t/(\Delta x)^2$. Thus we must only check that

$$\frac{1 - r(1 - \cos m\pi\Delta x)}{1 + r(1 - \cos m\pi\Delta x)} \geq -1$$

or

$$1 - r(1 - \cos m\pi\Delta x) \geq -1 - r(1 - \cos m\pi\Delta x)$$

or

$$2 \geq 0$$

which, of course, is always true. Thus the Crank–Nicolson method, in contrast with the explicit method, is stable for any r!

Unconditional stability means you can take as big a time step as you like. This is an important advantage of the Crank–Nicolson method over the explicit method, for which we must take

$$\Delta t < \frac{1}{2}(\Delta x)^2$$

so that, as the spatial mesh is defined, the number of time steps required goes up markedly, even if the solution's variation with time is slow. As noted above, the Crank–Nicolson method also gives superior accuracy; it can be shown that

$$u(x_i, t_j) - U_{ij} = O(\Delta x)^2 + O(\Delta t)^2$$

However, if $r > 1/2$, the damping factor is negative for the highest frequency terms, which, while damped, oscillate in sign from one time step to the next. If m and r are large enough, the magnitude of the damping factor can be close to 1; the magnitude does not decrease monotonically with increasing frequency. This is quite a disparity in behavior from the exact solution (9-70), in which the higher the frequency, the higher the rate of damping. If the initial data $u(x, 0)$ are smooth enough, the weights B_m of the higher-frequency terms are small, and the oscillatory behavior of those terms in the Crank–Nicolson solution is not important. However, if the initial data vary rapidly with x (e.g., if the boundary conditions are discontinuous in time, as in case b of problem 9), the higher-frequency terms are not negligible, and their oscillations spoil the solution. Then the time step must be limited just as in the explicit method, at least until those terms become unimportant.

9.4.5. Backward-Difference Schemes

There are alternative schemes that, like the Crank–Nicolson method, are unconditionally stable but that model the behavior of the higher-frequency terms more accurately. In the *fully implicit method*, the time derivative is approximated by a backward difference:

$$\frac{\partial u}{\partial t}(x_i, t_j) \simeq \frac{u_{ij} - u_{ij-1}}{\Delta t}$$

which leads to the finite-difference equation

$$\frac{U_{ij} - U_{ij-1}}{\Delta t} = \frac{U_{i+1j} - 2U_{ij} + U_{i-1j}}{(\Delta x)^2} \tag{9-76}$$

Since this involves three approximations to u at the advanced time level, the advancement of the solution requires solution of a tridiagonal system, namely,

$$\begin{bmatrix} 1 & & & & & & \\ -r & 1+2r & -r & & & & \\ & -r & 1+2r & -r & & & \\ & & \cdot & \cdot & \cdot & & \\ & & & \cdot & \cdot & \cdot & \\ & & & & \cdot & \cdot & \cdot \\ & & & & -r & 1+2r & -r \\ & & & & & & 1 \end{bmatrix} \begin{bmatrix} U_{1j} \\ U_{2j} \\ U_{3j} \\ \cdot \\ \cdot \\ \cdot \\ U_{N-1j} \\ U_{Nj} \end{bmatrix} = \begin{bmatrix} 0 \\ U_{2j-1} \\ U_{3j-1} \\ \cdot \\ \cdot \\ \cdot \\ U_{N-1,j-1} \\ 0 \end{bmatrix}$$

A product solution $T_j \sin m\pi x_i$ of the finite-difference equation 9-76 satisfies

$$T_j[1 + 2r(1 - \cos m\pi\Delta x)] = T_{j-1}$$

whose solution is

$$T_j = \frac{1}{[1 + 2r(1 - \cos \alpha)]^{j-1}}$$

By following the procedure that led to equations 9-69 and 9-75, we finally determine the solution of the finite-difference equations in the form

$$U_{ij} = \sum_{m=1}^{N-1} B_m[1 + 2r(1 - \cos m\pi\Delta x)]^{-j+1} \sin(i-1)m\pi\Delta x \tag{9-77}$$

It is easy to see that the damping factor in equation 9-77

$$[1 + 2r(1 - \cos m\pi\Delta x)]^{-1}$$

is always positive and less than 1. Therefore, as advertised, the fully implicit method is unconditionally stable, and the damping factor decreases monotonically with the frequency index m. Thus this method is much better suited than the Crank–Nicolson method for problems in which the high-frequency components of the solution are important. Its one disadvantage is that it is accurate only to first order in Δt. Thus the time step, although unlimited by stability considerations, does have to be kept relatively small for accuracy.

This disadvantage can be overcome in two ways:

1. Use a more elaborate backward-difference approximation to the time derivative. For equal time intervals,

$$\frac{\partial u}{\partial t}(x_i, t_j) \simeq \frac{\frac{3}{2}u_{ij} - 2u_{ij-1} + \frac{1}{2}u_{ij-2}}{\Delta t} \tag{9-78}$$

is designed to provide second-order accuracy in Δt and can be shown to be unconditionally stable and to have a damping factor that decreases in magnitude monotonically as the frequency increases. However, it requires storage of data at two time steps and so needs to be supplemented by a special procedure for either starting the calculations or changing the time step.

2. Repeat the calculation with half the time step and use the Richardson extrapolation to manipulate the results to obtain second-order accuracy. This could be done either step by step or by repeating the whole calculation from the initial time. Either way, this alternative requires three times as much computation as either the preceding alternative or the Crank–Nicolson method. It is probably most useful as the supplementary starting procedure required in the second-order backward-difference scheme.

9.5. REFERENCES

1. Murman, E., and Cole, J. D., "Calculation of Plane Steady Transonic Flows," *AIAA J.* **9,** 114–121 (1971).
2. Greenspan, D., *Introductory Numerical Analysis of Elliptic Boundary Value Problems.* Harper & Row, New York (1965).
3. Smith, R. E. (ed.), *Numerical Grid Generation Techniques,* Proceedings of NASA/ICASE Workshop, NASA Conference Pub. 2166 (1980).
4. Stein, P., and Rosenberg, R. L., "On the Solution of Linear Simultaneous Equations by Iteration," *J. London Math. Soc.* **23,** 111–118 (1948).
5. Forsythe, G., and Wasow, W. R., *Finite-Difference Methods for Partial Differential Equations.* Wiley, New York (1960).
6. Richtmyer, R. D., and Morton, K. W., *Difference Methods for Initial-Value Problems,* 2nd ed. Interscience, New York (1967).
7. Crank, J., and Nicolson, P., "A Practical Method for Numerical Evaluation of Solutions of Partial Differential Equations of the Heat-Conduction Type," *Proc. Cambridge Phil. Soc.* **43,** 50–67 (1947).

9.6. PROBLEMS

1. Show that the *forward-difference approximation* to dy/dx

$$\frac{dy}{dx}(x) \approx \frac{y(x+\Delta) - y(x)}{\Delta}$$

and the *backward-difference approximation*

$$\frac{dy}{dx}(x) \approx \frac{y(x) - y(x-\Delta)}{\Delta}$$

are accurate to first order in Δ.

2. Consider the finite-difference approximation

$$\frac{Y_{i+1} - Y_i}{\Delta x} = A\left(x_i + \frac{\Delta x}{2}\right)\frac{Y_i + Y_{i+1}}{2}$$

to the differential equation

$$\frac{dy}{dx} = A(x)y$$

What is its local truncation error?

3. Examine the rate of convergence of the finite-difference method implemented in LINBVP with both evenly and unevenly spaced meshes. For example, try $x_i = [(i-1)/(N-1)]^p$ for various values of p. In particular, what effect does the mesh spacing have on the order of accuracy of the method? Consider also the convergence of the results obtained by Richardson extrapolation.

4. Modify LINBVP to use a first-order approximation to the boundary condition at $x = 1$ (see equation 9-18). Test its rate of convergence and determine its order of accuracy.

5. Verify that equation 9-25 is an appropriate finite-difference approximation to the Laplace equation 9-22 when the mesh lines are arbitrarily spaced. What is the order of magnitude of its local truncation error?

6. Solve these two systems of equations iteratively with both the Jacobi and Gauss–Seidel methods. They are not so large that you must use a computer, but you can if you want to.

   (a)
$$\begin{aligned} u_1 + 2u_2 - 2u_3 &= 1 \\ u_1 + u_2 + u_3 &= 3 \\ 2u_1 + 2u_2 + u_3 &= 5 \end{aligned}$$

   (b)
$$\begin{aligned} 5u_1 + 3u_2 + 4u_3 &= 12 \\ 3u_1 + 6u_2 + 4u_3 &= 13 \\ 4u_1 + 4u_2 + 5u_3 &= 13 \end{aligned}$$

7. Modify program THINAIR by removing its call to OPTRLX, the subroutine that optimizes the overrelaxation parameter automatically. Run the program, a few trials at a time, and see if you can find an optimum factor by trial and error.

8. Compare the results of program THINAIR with those you obtained from thin-airfoil theory in Chapter 3 for an ellipse and/or for the airfoil-shaped body defined by equation 3-42. *Hint*: Do not attempt to satisfy the flow tangency condition at the blunt stagnation point(s), where dY/dx is infinite. Treat such points like ordinary mesh points, at which the only requirement is to satisfy the Laplace equation.

9. The local truncation error of the explicit method for the heat equation 9-61 is $O(\Delta t) + O(\Delta x)^2$. Check whether this is reflected in the numerical solution implemented in program HEATEX by repeating the calculation with $\Delta t \to \Delta t/4$, $\Delta x \to \Delta x/2$ (two cycles) and examining the solution at some fixed x, t. Try it for two cases:

   (a) $u(x,0) = \sin \pi x, \qquad u(0,t) = u(1,t) = 0$

   (b) $u(x,0) = 1, \qquad u(0,t) = \dfrac{\partial u}{\partial x}(1,t) = 0$

10. Repeat problem 9 for the Crank–Nicolson method, which is implemented in program CRANKN. Since its local truncation error is $O(\Delta x)^2 + O(\Delta t)^2$, halve both Δt and Δx when you repeat the calculations.

11. Show that the finite difference scheme for the heat equation obtained by using the central-difference approximation (9-59) for $\partial u/\partial t$

$$\frac{U_{ij+1} - U_{ij-1}}{2\Delta t} = \frac{U_{i+ij} - 2U_{ij} + U_{i-ij}}{(\Delta t)^2}$$

is unconditionally unstable. *Hint*: Use the separation-of-variables method. Assume a solution of the difference equation for the time-dependent part T_j of your product solution in the form

$$T_j = A\rho^j$$

There will be two roots ρ, since the equation for T_j is of second order. For stability, both must be less than 1 in magnitude.

12. Modify CRANKN to use the backward-difference scheme, equation 9-76. Compare its accuracy and stability with the explicit and Crank–Nicolson methods for the two cases described in problem 9.

9.7. COMPUTER PROGRAMS

```
C       PROGRAM LINBVP
C
C          FINITE-DIFFERENCE SOLUTION OF STURM-LIOUVILLE PROBLEM
C
C                       D/DX (- P DY/DX) + Q Y = F
C
C                       Y = 1    AT X = 0
C                       DY/DX = 0    AT X = 1
C
        COMMON /TRID/ A(251),B(251),C(251),D(251),Y(251)
        DIMENSION     YOLD(251),X(251)
C
C                         INPUT NUMBER OF OUTPUT POINTS (MOUT) AND NUMBER OF
C                         TIMES COMPUTATIONAL INTERVAL WILL BE HALVED (NHALF)
C
        PRINT,          ' INPUT MOUT,NHALF',
        READ,           MOUT,NHALF
        M               =  MOUT
        NTOTAL          =  NHALF + 1
        DO 300          ITRY = 1,NTOTAL
C
C                         SET COORDINATES OF GRID POINTS
C
        DO 100          I = 1,M
        X(I)            =  (I - 1)/FLOAT(M - 1)
    100 CONTINUE
        X(M+1)          =  2.*X(M) - X(M-1)
C
C                         SET COEFFICIENTS OF TRIDIAGONAL MATRIX
C
```

```
C                   --  BOUNDARY CONDITIONS
C
      B(1)       =  1.0
      C(1)       =  0.0
      D(1)       =  1.0
      A(M)       =  - 1.0
      B(M)       =  1.0
      D(M)       =  0.0
C
C                   --  POINTS IN BETWEEN
C
      DO 200     I = 2,M
      FACTOR     =  - 2./(X(I+1) - X(I-1))
      XA         =  .5*(X(I-1) + X(I))
      A(I)       =  FACTOR*P(XA)/(X(I) - X(I-1))
      XC         =  .5*(X(I) + X(I+1))
      C(I)       =  FACTOR*P(XC)/(X(I+1) - X(I))
      B(I)       =  Q(X(I)) - A(I) - C(I)
      D(I)       =  F(X(I))
      IF (I .EQ. M)         A(I) = A(I) + C(I)
  200 CONTINUE
      CALL TRIDIG(M)
C
C                   EXTRAPOLATE RESULTS TO 4TH-ORDER ACCURACY
C
      PRINT 2000,M
      JUMP       =  (M - 1)/(MOUT - 1)
      DO 240     I = 1,M,JUMP
      J          =  (I - 1)/JUMP + 1
      IF (ITRY .GT. 1)      YXTRAP = (4.*Y(I) - YOLD(J))/3.
      PRINT 2010,X(I),Y(I),YXTRAP,YEXACT(X(I))
  240 YOLD(J)    =  Y(I)
  300 M          =  2*(M - 1) + 1
 2000 FORMAT(/////,' NUMBER OF GRID POINTS =',I4,//,5X,'X',10X,'Y',
     +           8X,'YXTRAP',6X,'YEXACT',/)
 2010 FORMAT(F8.4,3F12.6)
      STOP
      END
C============================================================
      FUNCTION P(X)
      P          =  1. + X**2
      RETURN
      END
C============================================================
      FUNCTION Q(X)
      Q          =  X**2*(1. - X**2)
      RETURN
      END
C============================================================
      FUNCTION F(X)
      F          =  Q(X)*YEXACT(X) + 2.*(1. - 5.*X**4)
      RETURN
      END
C============================================================
      FUNCTION YEXACT(X)
      YEXACT     =  1. - X**2*(1. - .5*X**2)
      RETURN
      END
C============================================================
```

```
      SUBROUTINE TRIDIG(NEQNS)
C
C                   TRIDIAGONAL SYSTEM SOLVER
C
      COMMON /TRID/ A(251),B(251),C(251),D(251),V(251)
      DO 100      I = 2,NEQNS
      FACTOR      =  A(I)/B(I-1)
      B(I)        =  B(I) - FACTOR*C(I-1)
  100 D(I)        =  D(I) - FACTOR*D(I-1)
      V(NEQNS)    =  D(NEQNS)/B(NEQNS)
      DO 200      L = 2,NEQNS
      I           =  NEQNS + 1 - L
  200 V(I)        =  (D(I) - C(I)*V(I+1))/B(I)
      RETURN
      END
C     PROGRAM THINAIR
C
C        FINITE-DIFFERENCE SOLUTION OF INCOMPRESSIBLE POTENTIAL FLOW
C        PAST A THIN SYMMETRIC AIRFOIL AT ZERO ANGLE OF ATTACK
C
C        SUCCESSIVE OVERRELAXATION WITH OPTIMIZATION
C
      COMMON /UXYN/ PHI(100,50),X(100),Y(50),NX,NY
      COMMON /AIRF/ ILE,ITE
      PI          =  3.141592685
C
C                   INPUT FIELD DIMENSIONS (XW < X < XE, 0 < Y < YN)
C                   AND NUMBERS OF INTERVALS UPSTREAM (NUP),
C                   DOWNSTREAM (NDOWN), ON (NON), AND ABOVE (NABOVE)
C                   AIRFOIL
C
      PRINT, ' INPUT XW,XE,YN',
      READ, XW,XE,YN
      PRINT, ' INPUT NUP,NDOWN,NON,NABOVE',
      READ,  NUP,NDOWN,NON,NABOVE
      ILE         =  NUP + 2
      ITE         =  ILE + NON
      NX          =  ITE + NDOWN
      NY          =  NABOVE + 2
C
C                   SET UP MESH AND MAKE INITIAL GUESS
C
      DO 50       I = 2,ILE
   50 X(I)        =  XW*(1. - COS(PI/2.*(ILE-I)/FLOAT(ILE-2)))
      X(1)        =  2.*X(2) - X(3)
      DO 60       I = ILE,ITE
   60 X(I)        =  .5*(1. - COS(PI*(I-ILE)/FLOAT(ITE-ILE)))
      DO 70       I = ITE,NX
   70 X(I)        =  XE - (XE-1.)*COS(PI/2.*(I-ITE)/FLOAT(NX-ITE))
      X(NX+1)     =  2.*X(NX) - X(NX-1)
      DO 80       J = 2,NY
   80 Y(J)        =  YN*(1. - COS(PI/2.*(J-2)/FLOAT(NY-2)))
      Y(1)        =  2.*Y(2) - Y(3)
      Y(NY+1)     =  2.*Y(NY) - Y(NY-1)
      DO 90       I = 1,NX
      DO 90       J = 1,NY
   90 PHI(I,J)       =  0.0
C
C                   INPUT MAXIMUM NUMBER OF TRIES (MAXTRY)
```

```
C                    AND OVERRELAXATION FACTOR (RELAX)
C
      KOUNT       =  0
      CHGOLD      =  0.0
  100 PRINT,' INPUT MAXTRY,RELAX',
      READ,       MAXTRY,RELAX
      IF (MAXTRY .EQ. 0)     STOP
      PRINT 2000
      DO 240      ITRY = 1,MAXTRY
      KOUNT       =  KOUNT + 1
      CHGRMS      =  0.0
      CHGMAX      =  0.0
      DO 170      I = 2,NX
      DO 160      J = 2,NY
      CALL COFISH(I,J,DPHI)
      PHI(I,J)      =  PHI(I,J) + DPHI*RELAX
      IF (ABS(DPHI) .LE. CHGMAX)  GO TO 160
      CHGMAX      = ABS(DPHI)
      XMAX        =  X(I)
      YMAX        =  Y(J)
  160 CHGRMS            =  CHGRMS + DPHI**2
  170 CONTINUE
      CHGRMS            =  SQRT(CHGRMS/FLOAT(NX*NY))
      IF (CHGRMS .LT. 1.E-6) GO TO 300
      CALL OPTRLX(KOUNT,CHGRMS,RELAX)
      RATIO       =  CHGRMS/CHGOLD
      CHGOLD      =  CHGRMS
  240 PRINT 2010,       KOUNT,CHGMAX,XMAX,YMAX,CHGRMS,RATIO
  300 PRINT, ' WANT CONTOUR PLOT (Y/N)? ',
      READ 1000,  IANS
      IF (IANS .EQ. 1HY)  CALL CONTOUR(40)
      CALL CPOUT(ILE,ITE)
      GO TO 100
 1000 FORMAT(A1)
 2000 FORMAT(///,
     + ' TRIAL  MAX CHANGE     XMAX     YMAX    RMS CHANGE  CHANGE RATIO')
 2010 FORMAT(I5,E11.3,F12.4,F8.4,E11.3,F12.4)
      END
C=================================================================
      SUBROUTINE COFISH(I,J,DPHI)
C
C                    SET COEFFICIENTS OF TRIDIAGONAL SYSTEM
C                    FOR CHANGE OF POTENTIAL IN ITH COLUMN
C
      COMMON /UXYN/ PHI(100,50),X(100),Y(50),NX,NY
C
C                    SET LOCAL COEFFICIENTS OF FINITE-DIFFERENCE EQUATIONS
C
      HN          =  Y(J+1) - Y(J)
      HS          =  Y(J) - Y(J-1)
      HE          =  X(I+1) - X(I)
      HW          =  X(I) - X(I-1)
      CN          =  2./HN/(HN + HS)
      CS          =  2./HS/(HN + HS)
      CE          =  2./HE/(HE + HW)
      CW          =  2./HW/(HE + HW)
      CO          =  - CN - CW - CS - CE
C
```

```
C                    SET POTENTIAL OUTSIDE COMPUTATIONAL SPACE
C
      IF (I .EQ. 2)  PHI(1,J)    =  PHI(3,J)
      IF (I .EQ. NX) PHI(NX+1,J)=  PHI(NX-1,J)
      IF (J .EQ. 2)  PHI(I,1)    =  PHI(I,3) - (Y(3)-Y(1))*FP(I)
      IF (J .EQ. NY) PHI(I,NY+1)=  PHI(I,NY-1)
C
C                    SET GLOBAL COEFFICIENTS OF FINITE-DIFFERENCE EQUATIONS
C
      DPHI          =  - CN*PHI(I,J+1) - CW*PHI(I-1,J) - CS*PHI(I,J-1)
     +                 - CE*PHI(I+1,J) - CO*PHI(I,J)
      DPHI          =  DPHI/CO
      RETURN
      END
C=============================================================
      SUBROUTINE OPTRLX(KOUNT,CHGRMS,RELAX)
      COMMON /RELX/ CHANGE(2),EIGVAL(2)
      CHANGE(2)  =  CHGRMS
      IF (KOUNT .EQ. 1)       GO TO 110
      EIGVAL(2)  =  CHANGE(2)/CHANGE(1)
      IF (KOUNT .EQ. 2)       GO TO 100
      IF (ABS(EIGVAL(2)/EIGVAL(1) - 1.) .GT. .001)  GO TO 100
      IF (EIGVAL(2) .LE. 0.0)GO TO 100
      EIGLOG        =  ALOG(EIGVAL(2))
      RADCAL        =  - EIGLOG*((1./RELAX - .25)*EIGLOG + 2./RELAX - 1.)
      IF (RADCAL .LE. 0.0)    GO TO 100
      RELAX         =  2.*(1. - SQRT(RADCAL))
      PRINT,         ' RELAX =',RELAX
  100 EIGVAL(1)     =  EIGVAL(2)
  110 CHANGE(1)     =  CHANGE(2)
      RETURN
      END
C=================================================================
      FUNCTION FP(I)
C
C                    SLOPE OF AIRFOIL SURFACE
C
      COMMON /UXYN/ PHI(100,50),X(100),Y(50),NX,NY
      COMMON /AIRF/ ILE,ITE
      FP            =  0.0
      IF ((I .LT. ILE) .OR. (I .GT. ITE)) RETURN
      FP            =  .1*(1. - 2.*X(I))
      RETURN
      END
C=============================================================
      SUBROUTINE CONTOUR(NWIDTH)
C
C                    CONTOUR PLOT OF U(X,Y) ON RECTANGLE
C
      COMMON /UXYN/ PHI(100,50),X(100),Y(50),NX,NY
      COMMON /CNTR/ UPLOT(50),ISYM(100)
      DIMENSION        U(100,50)
      JGAP            =  NWIDTH - 8
C
C                    SCALE U BETWEEN MAXIMUM AND MINIMUM VALUES
C
      UMIN          =  0.0
      UMAX          =  0.0
      DO 100        I = 2,NX
      DO 100        J = 2,NY
```

```
      IF (PHI(I,J) .GT. UMAX)   UMAX = PHI(I,J)
      IF (PHI(I,J) .LT.  UMIN)  UMIN =  PHI(I,J)
  100 CONTINUE
      FACTOR       =  10./(UMAX -UMIN)
      DO 110       I = 2,NX
      DO 110       J = 2,NY
  110 U(I,J)       =  (PHI(I,J) - UMIN)*FACTOR
      YMAX         =  Y(NY) - Y(2)
      XMAX         =  X(NX) - X(2)
      DY           =  YMAX/FLOAT(NWIDTH)
      DX           =  DY*1.67
      LINES        =  (XMAX + .5*DX)/DX +1
      PRINT 1000,              UMAX,UMIN,Y(2),JGAP,Y(NY),X(2)
      XLINE        =  X(2)
      DO 200       J = 2,NY
  200 UPLOT(J)     =  U(2,J)
      CALL TERPLT(NWIDTH,DY)
      DO 260       L = 2,LINES
      XLINE        =  XLINE + DX
      DO 210       I = 2,NX
      IF (XLINE .LT. X(I))   GO TO 220
  210 CONTINUE
  220 DO 230       J = 2,NY
  230 UPLOT(J)     =  U(I,J) + (XLINE - X(I))*(U(I-1,J) - U(I,J))
     +                        /(X(I-1) - X(I))
      CALL TERPLT(NWIDTH,DY)
  260 CONTINUE
      PRINT 1010,             X(NX)
 1000 FORMAT(//////////,' SCALED CONTOUR PLOT OF PHI(X,Y)',////,
     +            10X,'MAXIMUM PHI =',E10.3,/,10X,'MINIMUM PHI =',E10.3,
     +            //,5X,'Y =',F6.3,=X,'Y =',F6.3,/,' X =',F6.3)
 1010 FORMAT(' X =',F6.3)
      RETURN
      END
C=======================================================
      SUBROUTINE TERPLT(NWIDTH,DY)
      COMMON /UXYN/ U(100,50),X(100),Y(50),NX,NY
      COMMON /CNTR/ UPLOT(50),ISYM(100)
      YPLOT        =  Y(2)
      ISYM(2)      =  UPLOT(2) + 0.5
      DO 120       J = 2,NWIDTH
      YPLOT        =  YPLOT + DY
      DO 100       K = 2,NY
      IF (YPLOT .LT. Y(K))   GO TO 110
  100 CONTINUE
  110 ISYM(J+1)    =  UPLOT(K) + 0.5 + (YPLOT - Y(K))
     +                     *(UPLOT(K-1) - UPLOT(K))/(Y(K-1) - Y(K))
  120 CONTINUE
      JUP          =  NWIDTH + 1
      DO 190       J = 2,JUP
      INDEX        =  ISYM(J) + 1
      GO TO (130,180,140,180,150,180,160,180,170,180,130), INDEX
  130 ISYM(J)      =  1H0
      GO TO 190
  140 ISYM(J)      =  1H2
      GO TO 190
  150 ISYM(J)      =  1H4
      GO TO 190
  160 ISYM(J)      =  1H6
      GO TO 190
```

```
  170 ISYM(J)    =  1H8
      GO TO 190
  180 ISYM(J)    =  1H
  190 CONTINUE
      PRINT 1100,(ISYM(J), J = 2,JUP)
 1100 FORMAT(10X,110A1)
      RETURN
      END
C=======================================================
      SUBROUTINE CPOUT(ILE,ITE)
C
C                    PRINT AND PLOT PRESSURE DISTRIBUTION ON AIRFOIL
C
      COMMON /UXYN/ PHI(100,50),X(100),Y(50),NX,NY
      COMMON /SKAL/ NZERO,YMULT,YSLOP
      NZERO       =  31
      YMULT       =  40.
      YSLOP       =  .5/YMULT
      PRINT 1000
      ILEP        =  ILE + 1
      DO 100      I = ILEP,ITE
      XP          =  .5*(X(I) + X(I-1))
      U           =  (PHI(I,2) - PHI(I-1,2))/(X(I) - X(I-1))
      CP          =  - 2.*U
      CALL PLOTXY(XP,CP)
  100 CONTINUE
 1000 FORMAT(////,' PRESSURE DISTRIBUTION ON AIRFOIL',//,
     +            4X,'X',8X,'CP',/)
      RETURN
      END
C==============================================================
      SUBROUTINE PLOTXY(X,Y)

      (SEE PROGRAM DUBLET)
C     PROGRAM HEATEX
C
C       EXPLICIT FINITE-DIFFERENCE SOLUTION OF 1-D HEAT EQUATION
C
C                    DU/DT = D2U/DX2
C
C       BOUNDARY CONDITIONS
C
C                    A0 U - B0 DU/DX = C0     AT X = 0
C                    A1 U + B1 DU/DX = C1     AT X = 1
C
      COMMON /SKAL/ NZERO,YMULT
      DIMENSION     U(100),V(100)
      NZERO       =  31
      YMULT       =  20.0
C
C                    READ IN CONSTANTS IN BOUNDARY CONDITIONS
C                    ALL MUST BE NONNEGATIVE
C                    IF B0 IS NEGATIVE, PROGRAM STOPS
C
      PRINT,' INPUT A0,B0,C0,A1,B1,C1',
      READ,       A0,B0,C0,A1,B1,C1
      IF (B0 .LT. 0.0)        STOP
```

```
C
C                    READ IN INTERVAL SIZES FOR COMPUTATIONS (DX,DT)
C                    AND OUTPUT (DXOUT)
C
C                    IF DX IS NEGATIVE, GO BACK TO 100
C
  110 PRINT,' INPUT DX,DT,DXOUT',
      READ,       DX,DT,DXOUT
      IF (DX .LT. 0.0)        STOP
      NX          =  (1. + DX*.1)/DX + 1
      NXOUT       =  (DXOUT + DX*.1)/DX
      RATIO       =  DT/DX**2
C
C                    PRESCRIBE INITIAL VALUES
C
      PI          =  3.1415926535879
      DO 120      I = 1,NX
  120 U(I)        =  SIN((I-1)*PI/FLOAT(NX-1))
      TIME        =  0.0
      PRINT 1020,              TIME
      DO 130      I = 1,NX,NXOUT
      X           =  (I - 1)*DX
      CALL PLOTXY(X,U(I))
  130 CONTINUE
C
C                    READ IN NEXT TIME AT WHICH SOLUTION IS WANTED
C                    AND ADVANCE SOLUTION STEP BY STEP
C                    UNTIL THAT TIME IS REACHED
C
C                    IF TNEXT IS LESS THAN PRESENT TIME, GO BACK TO 110
C
  200 PRINT,' INPUT NEXT TIME' ,
      READ,       TNEXT
      IF (TNEXT .LT. TIME)   GO TO 110
      NT          =  (TNEXT - TIME + DT*.1)/DT
      DO 270      J = 1,NT
      TIME        =  TIME + DT
C
C                    --  SOLUTION AT X = 0
C
      IF (B0 .LT. 1.E-4)     GO TO 210
      V(1)        =  2.*RATIO*(U(2) + C0*DX/B0)       +
     +               (1. - 2.*RATIO*(1. + DX*A0/B0))*U(1)
      GO TO 220
  210 V(1)        =  C0/A0
C
C                    --  SOLUTION AT X = 1
C
  220 IF (B1 .LT. 1.E-4)     GO TO 230
      V(NX)       =  2.*RATIO*(U(NX-1) + C1*DX/B1)       +
     +               (1. - 2.*RATIO*(1. + DX*A1/B1))*U(NX)
      GO TO 240
  230 V(NX)       =  C1/A1
C
C                    --  SOLUTION AT POINTS IN BETWEEN X = 0 AND X = 1
C
  240 IUP         =  NX - 1
      DO 250      I = 2,IUP
  250 V(I)        =  RATIO*(U(I+1) + U(I-1)) + (1. - 2.*RATIO)*U(I)
C
```

```
C                       --  SET UP FOR NEXT CYCLE
C
      DO 260      I  =  1,NX
  260 U(I)        =   V(I)
  270 CONTINUE
C
C                       --  OUTPUT RESULTS
C
      PRINT 1020,                TIME
      DO 280      I  =  1,NX,NXOUT
      X           =   (I - 1)*DX
      CALL PLOTXY(X,U(I))
  280 CONTINUE
      GO TO 200
 1020 FORMAT(/////////,' TIME =',E10.3,//,'     X            U',/)
      END
C=====================================================================
      SUBROUTINE PLOTXY(X,Y)

      (SEE PROGRAM DUBLET)
C     PROGRAM CRANKN
C
C       CRANK-NICOLSON SOLUTION OF 1-D HEAT EQUATION
C
C                       DU/DT = D2U/DX2
C
C       BOUNDARY CONDITIONS
C
C                       A0 U - B0 DU/DX = C0      AT X = 0
C                       A1 U + B1 DU/DX = C1      AT X = 1
C
      COMMON /SKAL/ NZERO,YMULT
      COMMON /TRID/ A(251),B(251),C(251),D(251),V(251)
      DIMENSION       U(100)
      NZERO       =   31
      YMULT       =   20.0
C
C                       READ IN CONSTANTS IN BOUNDARY CONDITIONS
C                       ALL MUST BE NONNEGATIVE
C
      PRINT,' INPUT A0,B0,C0,A1,B1,C1',
      READ,        A0,B0,C0,A1,B1,C1
C
C                       READ IN INTERVAL SIZES FOR COMPUTATIONS (DX,DT)
C                       AND OUTPUT (DXOUT)
C
C                       TO STOP, CHOOSE NEGATIVE DT
C
  110 PRINT,' INPUT DX,DT,DXOUT',
      READ,        DX,DT,DXOUT
      IF (DT .LT. 0.0)       STOP
      NX          =   (1. + DX*.1)/DX + 1
      NXOUT       =   (DXOUT + DX*.1)/DX
      RATIO       =   DT/DX**2
C
C                       PRESCRIBE INITIAL VALUES
C
      PI          =   3.1415926535879
      DO 120      I  =  1,NX
```

```
  120 U(I)         =  SIN((I-1)*PI/FLOAT(NX-1))
      TIME         =  0.0
      PRINT 1020,           TIME
      DO 130       I = 1,NX,NXOUT
      X            =  (I-1)*DX
      CALL PLOTXY(X,U(I))
  130 CONTINUE
C
C                   READ IN NEXT TIME AT WHICH SOLUTION IS WANTED
C                   AND ADVANCE SOLUTION STEP BY STEP
C                   UNTIL THAT TIME IS REACHED
C
C                   IF TNEXT IS LESS THAN PRESENT TIME
C                   SELECT NEW INTERVAL SIZES OR STOP
C
  200 PRINT,' INPUT NEXT TIME' ,
      READ,        TNEXT
      IF (TNEXT .LT. TIME)   GO TO 110
      NT           =  (TNEXT - TIME + DT*.1)/DT
      DO 270       J = 1,NT
      TIME         =  TIME + DT
C
C                  --  SET UP LINEAR SYSTEM
C
C                  -----   EQUATION AT X = 0
C
      IF (BO .LT. 1.E-4)     GO TO 210
      A(1)         =  - RATIO
      B(1)         =  1. + RATIO*(1. + DX*AO/BO)
      D(1)         =  RATIO*(U(2) + 2.*CO*DX/BO)      +
     +                (1. - RATIO*(1. + DX*AO/BO))*U(1)
      GO TO 220
  210 A(1)         =  0.0
      B(1)         =  AO
      D(1)         =  CO
C
C                  -----   EQUATION AT X = 1
C
  220 IF (B1 .LT. 1.E-4)     GO TO 230
      A(NX)        =  - RATIO
      B(NX)        =  1. + RATIO*(1. + DX*A1/B1)
      D(NX)        =  RATIO*(U(NX-1) + 2.*C1*DX/B1)      +
     +                (1. - RATIO*(1. + DX*A1/B1))*U(NX)
      GO TO 240
  230 A(NX)        =  0.0
      B(NX)        =  A1
      D(NX)        =  C1
C
C                  -----   EQUATION AT POINTS IN BETWEEN X = 0 AND X = 1
C
  240 IUP          =  NX - 1
      DO 250       I = 2,IUP
      A(I)         =  - .5*RATIO
      B(I)         =  1. + RATIO
  250 D(I)         =  .5*RATIO*(U(I+1) + U(I-1)) + (1. - RATIO)*U(I)
C
C                  -----   SET UP TRIDIAGONAL EQUATIONS SOLVER
C
      DO 255       I = 1,NX
  255 C(I)         =  A(I)
```

```
      CALL TRIDIG(NX)
C
C                    --  SET UP FOR NEXT CYCLE
C
      DO 260      I = 1,NX
  260 U(I)        =  V(I)
  270 CONTINUE
C
C                    --  OUTPUT RESULTS
C
      PRINT 1020,             TIME
      DO 280      I = 1,NX,NXOUT
      X           =  (I - 1)*DX
      CALL PLOTXY(X,U(I))
  280 CONTINUE
      GO TO 200
 1020 FORMAT(///,' TIME =',E10.3,//,'    X          U',/)
      END
C=========================================================
      SUBROUTINE TRIDIG(NEQNS)

      (SEE PROGRAM LINBVP)

C============================================================
      SUBROUTINE PLOTXY(X,Y)

      (SEE PROGRAM DUBLET)
```

CHAPTER 10

FINITE-DIFFERENCE SOLUTION OF THE BOUNDARY LAYER EQUATIONS

We previously considered the boundary-layer equations in Chapter 7. Semi-empirical integral methods were developed to calculate the main descriptors of boundary-layer growth—the skin friction and the momentum and displacement thicknesses—and to predict separation. The methods apply to either laminar or turbulent boundary layers but not through transition, which was assumed to be instantaneous. Since the length of the transition region is usually greater than the run of laminar flow, this approximation is tolerable only when the Reynolds number is high enough that the flow is almost completely turbulent. A similar comment applies to the necessity for guessing starting conditions for the turbulent part of the calculation.

Nevertheless, for two-dimensional attached boundary layers, integral methods are fast and accurate. Experience is accumulating with integral methods for three-dimensional flows, and it seems likely that, at least for certain narrow classes of problems (including flow over uncluttered wings), they will be very useful.

However, for problems that are strongly three dimensional, we must use methods that give more detailed attention to the velocity distribution within the boundary layer. Specifically, when integral methods fail, we turn to finite-difference methods like those described in this chapter. Such methods must also be used for the still more difficult problems in which the boundary-layer approximation breaks down, such as flows with separation. However, for simplicity, most of this chapter is confined to two-dimensional attached boundary layers.

10.1. STATEMENT OF THE PROBLEM

For incompressible two-dimensional flow, the velocity components u, v in the x, y directions satisfy the continuity equation

$$\frac{\partial u}{\partial x} + \frac{\partial v}{\partial y} = 0 \tag{10-1}$$

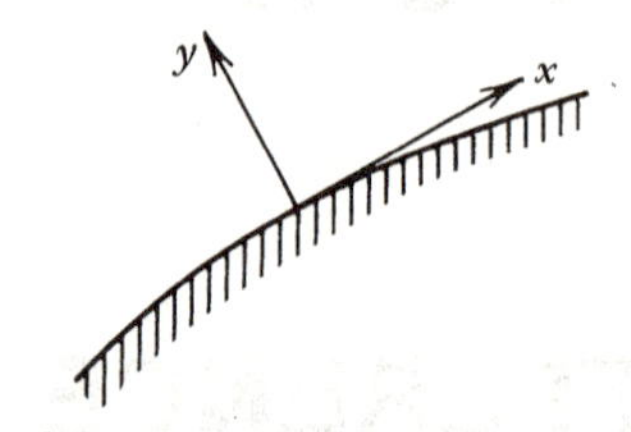

FIG. 10.1. Boundary layer coordinates.

everywhere in the flow field. Within the boundary layer adjacent to a solid surface, the momentum equation in the direction along the surface may be written

$$\rho u \frac{\partial u}{\partial x} + \rho v \frac{\partial u}{\partial y} = -\frac{dp}{dx} + \frac{\partial \tau}{\partial y} \tag{10-2}$$

where, as shown in Fig. 10.1, y is the distance from the surface and x is measured along it. We are interested only in flows that are steady, in the sense that the boundary conditions do not change with time. By selecting an appropriate formula for the shear stress τ, we may apply equation 10-2 to turbulent flows, in which case u, v, p, and τ are understood to be time-averaged values.

The momentum equation in the direction normal to the surface shows that the pressure is nearly constant through a boundary layer and so can be related to the velocity outside the boundary layer $V_e(x)$ by a Bernoulli equation:

$$p + \tfrac{1}{2}\rho V_e^2 = \text{constant} \tag{10-3}$$

Since the boundary layer is so thin on the scale of the inviscid flow outside, and since the flow within the boundary layer is very nearly parallel to the surface, $V_e(x)$ can be taken as the inviscid-flow result for the velocity *on* the body surface and so is known at the start of the boundary-layer analysis. The condition that the boundary-layer velocity field merge into the inviscid field can be written

$$\lim_{y \to \infty} u(x, y) = V_e(x) \tag{10-4}$$

As is usual in boundary-layer analyses, the limit $y \to \infty$ in equation 10-4 means that y takes on values large compared to δ, the scale of flow-property variations in the direction normal to the surface. However, the distance from the surface at which $u(x, y)$ is essentially $V_e(x)$ is supposed to be small compared with L, the scale of flow variations along the surface. That $\delta \ll L$ is the essence of the boundary-layer approximation.

The shear stress τ within the boundary layer is

$$\tau = \mu \frac{\partial u}{\partial y} - \rho \overline{u'v'} \tag{10-5}$$

Here u' and v' are fluctuations in time of the x- and y-velocity components from their mean values u and v, and the overbar indicates an averaging process. Sufficiently close to the start of the boundary layer, $u' = v' = 0$, and the shear stress takes on the Newtonian form of *laminar* flow. Eventually, however, the flow becomes unstable, and $\overline{u'v'} \neq 0$, even if the boundary conditions are independent of time. The flow has become *turbulent*. For turbulent flows, $\overline{u'v'}$ must be related to the mean velocity components u and v either by partial differential equations or, as we shall do later on, by empirically based algebraic formulas.

Besides equation 10-4, the solution of equations 10-1 and 10-2 must meet the no-slip boundary condition

$$u = v = 0 \qquad \text{at } y = 0 \tag{10-6}$$

Also, one must supply an initial condition of the form

$$u(x, y) = u_0(y) \qquad \text{at } x = x_0 \tag{10-7}$$

10.2. SIMILAR SOLUTIONS OF THE LAMINAR INCOMPRESSIBLE BOUNDARY LAYER

While the boundary-layer equations are vastly simpler than the Navier–Stokes equations—pressure is not an unknown and several derivatives are absent—they are still formidable. Because they are nonlinear, superposition of solutions, the basis of most methods for incompressible potential flow is out of the question. Even for laminar flows, if we want details of the velocity distribution through the boundary layer, we must use numerical methods.

As an introduction to such methods, we consider first a class of problems for which the partial differential equations 10-1, 10-2, and 10-5 can be reduced to an ordinary differential equation. These are the *similar solutions*, so called because their velocity profiles are geometrically similar all along the boundary layer, as illustrated in Fig. 10.2:

$$\frac{u}{V_e(x)} = g\left(\frac{y}{\delta(x)}\right)$$

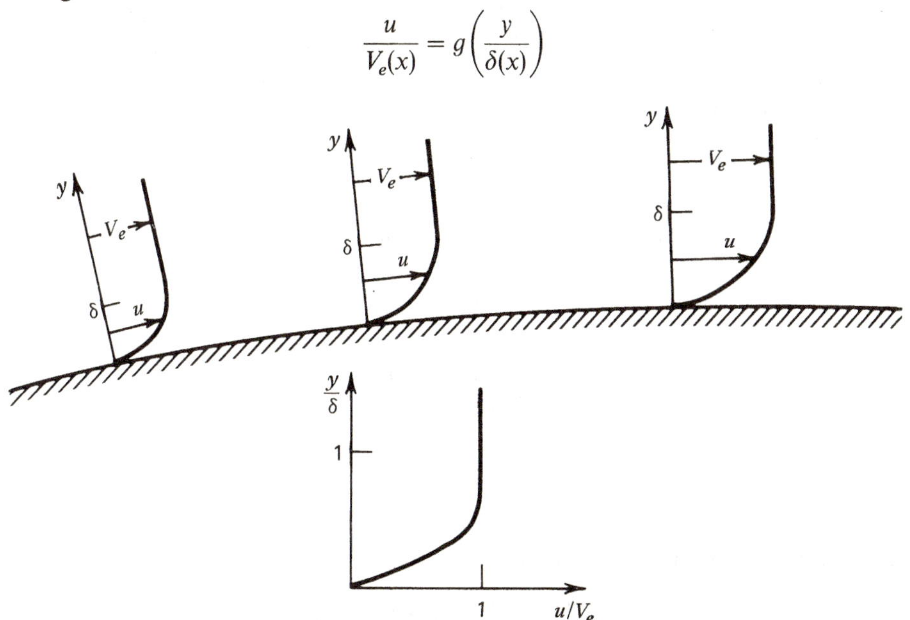

FIG. 10.2. Similar velocity profiles.

or, as will prove more useful,

$$\frac{u}{V_e} = \frac{df}{d\eta}(\eta) \tag{10-8}$$

where

$$\eta \equiv y/\delta(x) \tag{10-9}$$

and $\delta(x)$ is some measure of the boundary-layer thickness, which will be made specific later. Similar solutions exist only for the laminar boundary layer, so that we will drop the turbulent stress contribution $-\rho\overline{u'v'}$ from equation 10-5. Then equation 10-2 can be written

$$\rho u \frac{\partial u}{\partial x} + \rho v \frac{\partial u}{\partial y} = \rho V_e \frac{dV_e}{dx} + \mu \frac{\partial^2 u}{\partial y^2} \tag{10-10}$$

in which we have used the Bernoulli equation 10-3 to eliminate the pressure.

Our first problem is to find the velocity distribution $V_e(x)$, if any, that permits solutions of the form of equation 10-8. Since the velocity distribution is ordinarily prescribed, this is like the inverse approach to solutions of potential-flow problems; we are taking what is ordinarily an input (the V_e distribution) to be an output so as to get a relatively simple and exact solution. As with those inverse solutions, the solutions obtained can be used to evaluate more general methods. The integral method of Thwaites described in Chapter 7 was based, in part, on correlation formulas derived from the similar solutions. Some members of the similar-solution family, moreover, are of great importance in their own right, as we shall see. Also, the equations for boundary-layer growth under an arbitrary pressure gradient can be manipulated into a form remarkably like that which governs the similar solutions, so that the special case under consideration is an excellent test bed for methods capable of treating the general case.

To proceed, consider a change of independent variables from x, y to ξ, η, where

$$\xi = x \tag{10-11}$$

and η is given by equation 10-9. By the chain rule,

$$\begin{aligned}\frac{\partial u}{\partial x} &= \frac{\partial u}{\partial \xi}\frac{\partial \xi}{\partial x} + \frac{\partial u}{\partial \eta}\frac{\partial \eta}{\partial x}\\ &= \frac{\partial u}{\partial \xi} - \frac{y}{\delta^2}\frac{d\delta}{dx}\frac{\partial u}{\partial \eta}\\ \frac{\partial u}{\partial y} &= \frac{1}{\delta}\frac{\partial u}{\partial \eta}\end{aligned}$$

so equations 10-1 and 10-10 become

$$\frac{\partial u}{\partial \xi} - \eta \frac{\dot{\delta}}{\delta}\frac{\partial u}{\partial \eta} + \frac{1}{\delta}\frac{\partial v}{\partial \eta} = 0 \tag{10-12}$$

$$\rho u\left(\frac{\partial u}{\partial \xi} - \eta\frac{\dot{\delta}}{\delta}\frac{\partial u}{\partial \eta}\right) + \frac{\rho v}{\delta}\frac{\partial u}{\partial \eta} = \rho V_e \dot{V}_e + \frac{\mu}{\delta^2}\frac{\partial^2 u}{\partial \eta^2} \tag{10-13}$$

where the dots denote differentiation with respect to ξ.

If the similarity hypothesis (10-8) is valid,

$$\frac{\partial u}{\partial \xi} = \dot{V}_e \frac{df}{d\eta}$$

and equation 10-12 may be written

$$\delta \dot{V}_e \frac{df}{d\eta} - \eta \dot{\delta} V_e \frac{d^2 f}{d\eta^2} + \frac{\partial v}{\partial \eta} = 0 \tag{10-12a}$$

Since

$$\eta \frac{d^2 f}{d\eta^2} = \frac{d}{d\eta}\left(\eta \frac{df}{d\eta}\right) - \frac{df}{d\eta}$$

as you can check by expanding the right side, equation 10-12a may be integrated with respect to η to get

$$\delta \dot{V}_e f - \dot{\delta} V_e(\eta f' - f) + v = c \tag{10-14}$$

in which the prime indicates differentiation with respect to η. Since only the derivative of f is involved in its definition (10-8), its value is so far determinate only to within an additive constant. For convenience, set

$$f = 0 \qquad \text{at } y, \eta = 0 \tag{10-15}$$

Then, evaluating equation 10-14 at $\eta = 0$, we use the no-slip condition (10-6) to get $c = 0$, and equation 10-14 can be solved for v:

$$v = \dot{\delta} V_e(\eta f' - f) - \delta \dot{V}_e f \tag{10-16}$$

Now substitute equations 10-8 and 10-16 into equation 10-13:

$$\rho V_e f'(\dot{V}_e f' - \eta \frac{\dot{\delta}}{\delta} V_e f'') + \frac{\rho}{\delta}[\dot{\delta} V_e(\eta f' - f) - \delta \dot{V}_e f] V_e f''$$

$$= \rho V_e \dot{V}_e + \frac{\mu}{\delta^2} V_e f'''$$

This simplifies to

$$f''' = \rho \frac{V_e \delta}{\mu}\left[\frac{\delta \dot{V}_e}{V_e}(f'^2 - 1 - ff'') - \dot{\delta} ff''\right] \tag{10-17}$$

Thus, for the similarity hypothesis to be valid, $V_e(x)$ and $\delta(x)$ must be such that

$$\rho \frac{V_e \delta}{\mu} \frac{\delta \dot{V}_e}{V_e} = \frac{\rho}{\mu} \delta^2 \dot{V}_e = C_1 \tag{10-18}$$

$$\rho \frac{V_e \delta}{\mu} \dot{\delta} = \frac{1}{2}\frac{\rho}{\mu} V_e \frac{d\delta^2}{d\xi} = C_2 \tag{10-19}$$

in which C_1 and C_2 are constants. Otherwise the equation governing f is not independent of $x = \xi$ after all.

Adding twice equation 10-19 to equation 10-18 gives

$$\frac{\rho}{\mu}\left(\delta^2 \frac{dV_e}{d\xi} + V_e \frac{d\delta^2}{d\xi}\right) = C_1 + 2C_2$$

or

$$\frac{d}{d\xi}(V_e\delta^2) = \frac{\mu}{\rho}(C_1 + 2C_2) \equiv C_3$$

while taking the ratio of equation 10-19 to equation 10-18 gives

$$\frac{\dot{\delta}}{\delta} = \frac{C_2}{C_1}\frac{\dot{V}_e}{V_e}$$

After a little calculus, you find that V_e must have a power-law dependence on ξ,[1]

$$V_e \propto \xi^m \tag{10-20}$$

Then, from equation 10-18,

$$\delta \propto \xi^{(1-m)/2} \propto \left(\frac{\xi}{V_e}\right)^{1/2}$$

Since δ is defined as a distance, we insert a factor of μ/ρ for dimensional consistency and write

$$\delta = \xi\left(\frac{\mu}{\rho V_e \xi}\right)^{1/2} \tag{10-21}$$

Then, from equations 10-18 to 10-21,

$$C_1 = \rho\frac{V_e\delta^2}{\mu}\frac{\dot{V}_e}{V_e} = m$$

$$C_2 = \rho\frac{V_e\delta}{\mu}\dot{\delta} = \frac{1-m}{2}$$

and so equation 10-17 becomes

$$f''' = m(f'^2 - 1) - \left(\frac{m+1}{2}\right)ff'' \tag{10-22}$$

This is the *Falkner–Skan equation* [1]. The boundary conditions on $f(\eta)$ can be collected from equations 10-4, 10-6, 10-8, and 10-15:

$$f = f' = 0 \qquad \text{at } \eta = 0$$

$$f' \to 1 \qquad \text{as } \eta \to \infty \tag{10-23}$$

[1] Actually, on $\xi + C$, where C is some constant, but we can always shift the origin of ξ to make C zero, as is convenient.

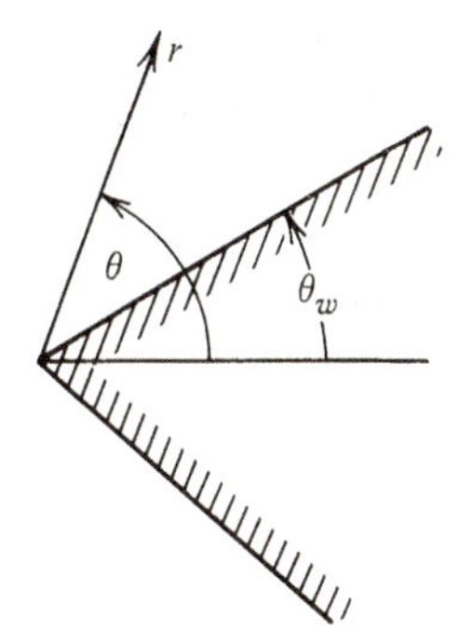

FIG. 10.3. Nomenclature for flow past a wedge.

The exterior-flow velocity distribution called for by equation 10-20 may be generated by potential flow past a wedge. As is shown in Appendix C, the velocity near a sharp wedge of half-included angle θ_w, see Fig. 10.3, behaves like

$$v_r \sim r^{\theta_w/(\pi - \theta_w)}$$

Thus, identifying r with x and ξ, and v_r with V_e, we set

$$m = \frac{\theta_w}{\pi - \theta_w} \tag{10-24}$$

to obtain equation 10-20.

The case $m = 0$, for which

$$V_e = \text{constant}$$

is found in the flow past a thin flat plate aligned with the flow, an easily generated experimental situation. Also of great practical interest is the case $m = 1$, or $\theta_w = \pi/2$, which is the stagnation point flow:

$$V_e = \dot{V}_e x \qquad \text{with } \dot{V}_e = \text{constant} \tag{10-25}$$

The solution for this case is often required as an initial condition for studying boundary-layer growth on airfoils.

Equation 10-22 was first solved for the cases $m = 0$ and $m = 1$ by Blasius [2] and Hiemenz [3], respectively. When $m = -0.0904$, the wall shear stress vanishes; this is the *separation profile*. No solutions exist for m less than this value, and the solutions are not unique for $0 > m > -0.0904$, the adverse-pressure gradient cases.

It may be noted that the function f defined by equation 10-8 is nothing more than a scaled stream function. I avoided this term in the derivation, because similar functions can be used in studying three-dimensional boundary layers [4], for which there is no stream function.

10.2.1. Finite-Difference Methods for the Falkner–Skan Equation

We shall discuss two methods for solving equation 10-22, both of which yield the velocity profile $u/V_e = f'(\eta)$ directly. However, as noted in Chapter 7, the more interesting outputs of a boundary-layer analysis are the displacement and momentum thicknesses and the wall shear stress (skin friction). You will show in problems 1 and 2 that these may be calculated in terms of the solution of equation 10-22 from

$$\frac{\delta^*}{\delta} = \lim_{\eta \to \infty} [\eta - f(\eta)] \tag{10-26}$$

$$\frac{\theta}{\delta} = \frac{2}{3m+1}\left[f''(0) - m\frac{\delta^*}{\delta}\right] \tag{10-27}$$

$$c_f = \frac{2}{\sqrt{\mathrm{Re}_x}} f''(0) \tag{10-28}$$

where

$$\mathrm{Re}_x \equiv \rho \frac{V_e x}{\mu} \tag{10-29}$$

Both methods are finite-difference methods, such as were discussed in Chapter 9. However, the Falkner–Skan equation 10-22 differs in several important respects from the equations solved in Chapter 9:

1. Equation 10-22 is nonlinear. Thus, if we approximate its derivatives by finite differences, we will obtain a system of nonlinear algebraic equations for the values of f at the nodes.
2. Equation 10-22 is third order. To approximate third derivatives requires a difference equation involving at least four unknowns. Even if the equations were linear, the *Thomas algorithm* introduced in Chapter 9 would not be applicable.
3. One of the boundary conditions (10-23) on the solution of equation 10-22 is imposed at infinity. Clearly, we cannot use equally spaced nodes going all the way to infinity; we must confine ourselves to a finite number of unknowns.

The problem last mentioned is most easily dealt with. We simply impose the last of the boundary conditions (10-23) at a finite value of η, say, η_∞:

$$f'(\eta_\infty) = 1 \tag{10-30}$$

However, we must then try different values of η_∞ until we are sure it is large enough to include the asymptotic part of the variation of f' with η; that is, there should be at least a couple of mesh points besides η_∞ at which the computed value of f' is very close to 1.0.

10.2.2. Iterative Solution of Nonlinear Equations

Almost all nonlinear problems require solution by *iteration.* We start with some initial guess for the solution and set up some procedure to improve it. Usually this involves an approximation to the governing equation that *linearizes* the equation on the assumption that the present guess is close to the exact solution.

For example, consider the algebraic problem of finding the root x^* of some nonlinear function $g(x)$:

$$g(x^*) = 0 \tag{10-31}$$

Let $\tilde{x}$ be an approximation to x^*. If $g(x)$ is differentiable, it can be expanded in a Taylor series about $\tilde{x}$:

$$g(x) = g(\tilde{x}) + (x - \tilde{x})g'(\tilde{x}) + O(x - \tilde{x})^2$$

Evaluating this series expression at $x = x^*$, where $g = 0$, and supposing $\tilde{x}$ is sufficiently close to x^* that the quadratic error terms can be ignored, we get

$$x^* \approx \tilde{x} - \frac{g(\tilde{x})}{g'(\tilde{x})} \tag{10-32}$$

If the value of x^* calculated from equation 10-32 is sufficiently close to $\tilde{x}$, we can take it as the solution to equation 10-31. If not, we can let that value of x^* be the $\tilde{x}$ of the next approximation. That is, plug it into the right side of equation 10-32, and repeat the process. This procedure, known as *Newton's method* (due to Sir Isaac), is summarized graphically in Fig. 10.4. There $\tilde{x}_j$ indicates the jth approximation to x^*.

A fundamental question concerning any iterative method is its *convergence*. As we take more and more trials, do we get closer to the true solution? If so, how fast?

Whether Newton's method converges depends on the initial guess. In some cases, a bad initial guess can keep you from converging. However, if the initial guess is "sufficiently accurate"—how accurate depends on the problem—Newton's method can be guaranteed to converge, and "quadratically" at that, by which we mean

$$|x^* - x_j| = O(x^* - x_{j-1})^2 \tag{10-33}$$

That is, the error at any stage of the iteration is of the order of the square of the error at the previous stage.

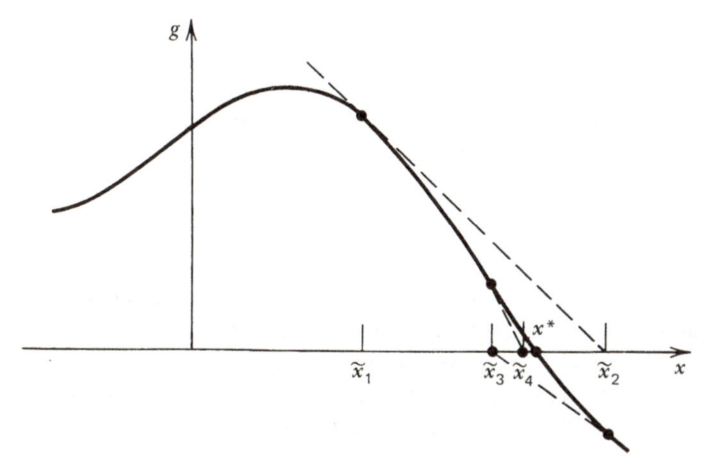

FIG. 10.4. Newton's method for finding a root of $g(x)$.

Example 1 Find a root of

$$g(x) \equiv x^2 - 2$$

Solution The answer, of course, is $x = \pm\sqrt{2}$, so that we will be getting a numerical approximation to one of those numbers. According to equation 10-32, a guess $\tilde{x}$ is improved with the formula

$$x^* \approx \tilde{x} - \frac{\tilde{x}^2 - 2}{2\tilde{x}}$$

$$= \frac{1}{2}\tilde{x} + \frac{1}{\tilde{x}}$$

If our first guess is $\tilde{x} = 1$, the iterative process goes as follows:

$\tilde{x}$	x^*
1.0	1.5
1.5	1.416667
1.416667	1.414216
1.414216	1.414214

The result is now accurate to the places indicated. What results would you get with an initial guess of -1.0?

10.2.3. A Finite-Difference Method Based on a Second-Order Differential Equation

Neither of the two finite-difference methods we shall discuss for the solution of equation 10-22 deals with its third-derivative term directly. In the first method, we decompose equation 10-22 into a pair of differential equations, one first order and the other second order, and iterate for their solution sequentially. Let

$$u^* \equiv \frac{df}{d\eta} \tag{10-34}$$

From equation 10-8, $u^* = u/V_e$. Using equation 10-34 to eliminate the derivatives of f from equation 10-22, we get

$$\frac{d^2u^*}{d\eta^2} = m(u^{*2} - 1) - \frac{m+1}{2} f \frac{du^*}{d\eta} \tag{10-35}$$

Then we can set up a sequential iterative scheme in which the value of $f(\eta)$ corresponding to a guess at $u^*(\eta)$ is calculated from equation 10-34 and then used in equation 10-35, which is of second order in the derivatives of u^*.

It is not hard to set up a finite-difference method to carry out this program. Let the mesh points at which the solution is sought be

$$0 = \eta_1 < \eta_2 < \eta_3 < \cdots < \eta_M = \eta_\infty$$

Let $\tilde{u}(\eta)$ be a given approximation to $u^*(\eta)$. The change in f between adjacent mesh points corresponding to $\tilde{u}$ can be obtained by integrating equation 10-34 (with u^* replaced by $\tilde{u}$) between those mesh points:

$$\tilde{f}(\eta_i) = \tilde{f}(\eta_{i-1}) + \int_{\eta_{i-1}}^{\eta_i} \tilde{u}(\eta)\, d\eta \tag{10-36}$$

If $\tilde{F}_i$ and $\tilde{U}_i$ are the finite-difference approximations to $\tilde{f}(\eta_i)$ and $\tilde{u}(\eta_i)$, respectively, and we approximate the integral in equation 10-36 by the trapezoidal rule of equation 7-74, we get

$$\tilde{F}_i = \tilde{F}_{i-1} + \tfrac{1}{2}(\tilde{U}_i + \tilde{U}_{i-1})(\eta_i - \eta_{i-1}) \tag{10-37}$$

This can be used for $\tilde{F}_2, \tilde{F}_3, \ldots, \tilde{F}_M$, since the value of $\tilde{F}_1$ can be taken from the boundary condition (10-23):

$$\tilde{F}_1 = f(\eta_1) = f(0) = 0 \tag{10-38}$$

Before making a finite-difference approximation to equation 10-35, we will linearize it via Newton's method. Since f is considered known, the only nonlinear term in equation 10-35 is

$$\begin{aligned} u^{*2} &= [\tilde{u} + (u^* - \tilde{u})]^2 \\ &= \tilde{u}^2 + 2(u^* - \tilde{u})\tilde{u} + (u^* - \tilde{u})^2 \end{aligned}$$

Supposing $\tilde{u}$ to be close enough to u^* that the term quadratic in the error $u^* - \tilde{u}$ may be ignored,

$$u^{*2} \approx 2\tilde{u}u^* - \tilde{u}^2$$

we approximate equation 10-35 by

$$\frac{d^2u^*}{d\eta^2} + \frac{m+1}{2}\tilde{f}\frac{du^*}{d\eta} - 2m\tilde{u}u^* = -m(\tilde{u}^2 + 1) \tag{10-39}$$

Equation 10-39 is of the same form as equation 9-1, whose finite-difference solution was discussed in Chapter 9. However, that solution was based on evenly spaced nodes. Any such solution of the Falkner–Skan equation wastes a lot of time and storage. Typically, to obtain accurate results for the wall shear stress and displacement thickness, we have to compute out to $\eta = 5$ to 10, but the velocity profile changes very little for $\eta > 3$. Therefore, it is desirable to concentrate the nodes near the wall and to use relatively wider spacing further out. A common prescription is to increase the spacing between adjacent pairs of nodes by a constant factor, say, ρ:

$$\eta_{i+1} - \eta_i = \rho(\eta_i - \eta_{i-1}) \tag{10-40}$$

This is a linear difference equation with constant coefficients, which can be solved exactly:

$$\eta_i = \eta_2 \frac{\rho^{i-1} - 1}{\rho - 1} \tag{10-41}$$

A finite-difference approximation to second derivatives for unequally spaced nodes was developed in problem 5 of Chapter 9. In the present nomenclature, it is

$$\frac{d^2u^*}{d\eta^2} = \frac{2}{\eta_{i+1} - \eta_{i-1}}\left[\frac{u^*(\eta_{i+1}) - u^*(\eta_i)}{\eta_{i+1} - \eta_i} - \frac{u^*(\eta_i) - u^*(\eta_{i-1})}{\eta_i - \eta_{i-1}}\right] + O(\Delta\eta) \quad (10\text{-}42)$$

To approximate $du^*/d\eta$, we will use something like a central difference formula

$$\frac{du^*}{d\eta} = \frac{u^*(\eta_{i+1}) - u^*(\eta_{i-1})}{\eta_{i+1} - \eta_{i-1}} + O(\Delta\eta) \quad (10\text{-}43)$$

Note that, because of the unequally spaced nodes, both approximations are only first-order accurate, in contrast with the second-order accuracy they would have were the nodes equally spaced; compare equations 9-12 and 9-13.

The finite-difference equation corresponding to equation 10-39 is now obtained by using equations 10-42 and 10-43, less their error terms, to approximate the derivatives of u^*, and replacing $u^*(\eta_i)$, $\tilde{u}(\eta_i)$, and $\tilde{f}(\eta_i)$ by their numerical approximations U_i^*, $\tilde{U}_i$, and $\tilde{F}_i$, respectively. The result is of the form

$$A_i U_{i-1}^* + B_i U_i^* + C_i U_{i+1}^* = D_i \quad (10\text{-}44)$$

where

$$A_i = \left(\frac{2}{h_-} - \frac{m+1}{2}\tilde{F}_i\right)\frac{1}{(h_+ + h_-)}$$

$$C_i = \left(\frac{2}{h_+} + \frac{m+1}{2}\tilde{F}_i\right)\frac{1}{(h_+ + h_-)}$$

$$B_i = -\frac{2}{h_- h_+} - 2m\tilde{U}_i$$

$$D_i = -m(1 + \tilde{U}_i^2) \quad (10\text{-}45)$$

and

$$h_+ \equiv \eta_{i+1} - \eta_i$$

$$h_- \equiv \eta_i - \eta_{i-1} \quad (10\text{-}46)$$

The boundary conditions on u^* are found from equations 10-23, 10-30, and 10-34 to be

$$u^*(0) = 0$$

$$u^*(\eta_\infty) = 1 \quad (10\text{-}47)$$

the finite-difference approximations to which are brought into the form of equation 10-44 by setting

$$B_1 = 1$$

$$C_1 = D_1 = 0$$

$$A_M = 0$$

$$B_M = 1$$

$$C_M = 1 \quad (10\text{-}48)$$

The solution procedure can now be summarized as follows:

1. Select a value of η at which the boundary condition at infinity will be imposed (η_∞), and define nodes between $\eta_1 = 0$ and $\eta_M = \eta_\infty$ by a formula like equation 10-41.
2. Make some initial approximation to the velocity profile. Naturally, it ought to satisfy the boundary conditions (10-47). The simple formula

$$\begin{aligned} \tilde{u} &= \eta \qquad \text{for } 0 < \eta < 1 \\ &= 1 \qquad \text{for } \eta > 1 \end{aligned} \tag{10-49}$$

is adequate.
3. Find the finite-difference approximation to $\tilde{f}$ that corresponds to $\tilde{u}$ from equations 10-37 and 10-38.
4. Solve the tridiagonal system (10-44), whose coefficients are given by equation 10-45 for $i = 2, 3, \ldots, M-1$ and by equation 10-48 for $i = 1$ and M. The Thomas algorithm developed in Chapter 9 can be used.
5. If the solutions U_i^* of equations 10-44 are sufficiently close to the guesses $\tilde{U}_i$, return to step 1 and try another value of η_∞, until you are satisfied that the value chosen is large enough that the results would not change were η_∞ to be still larger. If the sets of numbers U_i^* and $\tilde{U}_i$ are not sufficiently close, take U_i^* as the next guess $\tilde{U}_i$ for $i = 2, \ldots, M-1$, and return to step 3.

As noted above, because the nodes η_i are unequally spaced, the *local truncation error* incurred by using equations 10-42 and 10-43 to approximate the derivatives in equation 10-39 is only of first order in $\Delta\eta$. In contrast, equations 10-44 would have second-order accuracy were the nodes spaced equally. Therefore, as a check on the results obtained with these formulas, consider the following alternative procedure, which would appear *a priori* to be second-order accurate. Introduce a new independent variable z by letting

$$\eta = \eta_2 \frac{\rho^{z/\eta_2} - 1}{\rho - 1} \tag{10-50}$$

Then, if

$$z_i = (i-1)\eta_2$$

so that the nodes are evenly spaced on the z axis, we recover the node spacing on the η axis called for by equation 10-41. Using the chain rule, you can transform equation 10-39 into one in which the derivatives of u^* are with respect to z. For example,

$$\begin{aligned} \frac{du^*}{d\eta} &= \frac{du^*}{dz}\frac{d\eta}{dz} \\ &= \frac{du^*}{dz}\frac{\rho - 1}{\rho^{z/\eta_2}\ln\rho} \end{aligned}$$

The resulting equation is considerably messier than (10-39), but, since the nodes are evenly spaced on the z axis, the difference approximations to the derivatives are simpler and second-order accurate.

I have found experimentally that the solution of those finite-difference equations is essentially the same as that of equations 10-44. Thus, it may be concluded that uneven mesh spacing does not destroy the second-order accuracy of our finite-difference method. The finite-difference equation 10-44 is only a first-order accurate approximation to the differential equation 10-39, but the local truncation error is not reflected in any difference between the solutions of those equations.

10.2.4. A Finite-Difference Method Based on a System of First-Order Equations

Another finite-difference method, somewhat more generally applicable than the one based on equations 10-34 and 10-35, and also more easily adapted to variable mesh spacing, can be devised by working with the equations in the form of a system of first-order equations that can be solved simultaneously. In the case of the third-order equation 10-22, this is done by defining

$$f_1 \equiv f(\eta)$$

$$f_2 \equiv \frac{df}{d\eta}$$

$$f_3 \equiv \frac{d^2f}{d\eta^2} \tag{10-51}$$

Then, from these definitions,

$$\frac{df_1}{d\eta} = f_2 \tag{10-52a}$$

$$\frac{df_2}{d\eta} = f_3 \tag{10-52b}$$

whereas equation 10-22 can be written

$$\frac{df_3}{d\eta} = m(f_2^2 - 1) - \frac{m+1}{2} f_1 f_3 \tag{10-52c}$$

and the boundary conditions (10-23) become

$$\begin{aligned} f_1 = f_2 = 0 \qquad & \text{at } \eta = 0 \\ f_2 \to 1 \qquad & \text{as } \eta \to \infty \end{aligned} \tag{10-53}$$

A quite general method for solving systems of nonlinear equations like (10-52) was devised by Sylvester and Meyer [5] and improved by Keller [6]. It is somewhat more complex than the method described above but, being the basis of the considerable work of Tuncer Cebeci (see Ref. [4], for example), is worth some attention here.

The basic idea is quite simple. As is usual, equations 10-52a–c are first linearized about some guesses $f_1(\eta)$, $f_2(\eta)$, and $f_3(\eta)$ for their solution. The linearized equations are then approximated by finite differences centered at points midway between each pair of adjacent nodes. This can easily be shown to be second-order accurate, regardless of the node spacing (see problem 2 of Chapter 9). The equations are then solved simultaneously.

Let

$$\varepsilon_i(\eta) \equiv f_i(\eta) - \tilde{f}_i(\eta) \qquad \text{for } i = 1, 2, 3 \tag{10-54}$$

be the errors of the current approximations to the f_i's. If the f_i's are eliminated from equations 10-52a–c in favor of the ε_i's, and terms quadratic in the ε_i's then dropped, we get

$$\frac{d\varepsilon_3}{d\eta} - 2m\tilde{f}_2\varepsilon_2 + \frac{m+1}{2}(\tilde{f}_3\varepsilon_1 + \tilde{f}_1\varepsilon_3) = -\left(\frac{d\tilde{f}_3}{d\eta} - m\tilde{f}_2^2 + m + \frac{m+1}{2}\tilde{f}_1\tilde{f}_3\right)$$
$$\frac{d\varepsilon_1}{d\eta} - \varepsilon_2 = -\left(\frac{d\tilde{f}_1}{d\eta} - \tilde{f}_2\right)$$
$$\frac{d\varepsilon_2}{d\eta} - \varepsilon_3 = -\left(\frac{d\tilde{f}_2}{d\eta} - \tilde{f}_3\right) \tag{10-55}$$

in which, because we have dropped quadratic terms, the ε_i's now represent the changes in our estimate of the solution from one iteration to the next. That is, the idea is to solve equations 10-55 for the ε_i's and then to determine the next approximation to the solution by (in accordance with equation 10-54) adding ε_i to $\tilde{f}_i$. If the ε_i's are sufficiently small, we accept the new approximations as the solution of equations 10-52a–c. If not, the new approximations are substituted back into equations 10-55, which are then resolved.

First we need boundary conditions on the solutions of equations 10-55. From equations 10-53 and 10-54, these are

$$\varepsilon_1 = \varepsilon_2 = 0 \qquad \text{at } \eta = 0$$
$$\varepsilon_2 \to 0 \qquad \text{as } \eta \to \infty \tag{10-56}$$

assuming that the current approximation satisfies the boundary condition (10-53).[2] The order of equation 10-55 should be noted; for later convenience, we have listed the ε version of equation 10-52a–c first.

Note that the right sides of equations 10-55 are simply the residuals obtained when the current approximations $\tilde{f}_i$ are substituted into the exact differential equations 10-52a–c. This is a consequence of our writing the linearized versions of equation 10-52a–c in terms of the changes in our estimate of the solution from one iteration to the next, rather than in terms of the new "improved" approximation to

[2] It is always useful to have trial functions that satisfy the boundary conditions. When the boundary conditions are linear in the unknowns, all that is required is to make initial approximations that meet equations 10-53. Equations 10-54 and 10-56 then ensure that all subsequent approximations will also meet them. Similarly, if the initial approximation satisfies the linear differential equations 10-52a and b (a provision that is easy to satisfy), so will all subsequent approximations, and the last two of equations 10-55 become homogeneous.

the f_i's. I recommend that you do likewise in any iterative solution. The magnitudes of these residuals may be examined in each iteration to obtain an alternative criterion for ending the iterations (as noted above, one may also examine the changes in the solution, the ε_i's).[3] Also, it may reduce round-off errors and simplify the programming.

It is convenient to collect equation 10-55 in vector form. Let

$$\{\boldsymbol{\varepsilon}(\eta)\} \equiv \begin{bmatrix} \varepsilon_1 \\ \varepsilon_2 \\ \varepsilon_3 \end{bmatrix} \tag{10-57}$$

and write them as

$$[I^*]\frac{d}{d\eta}\{\boldsymbol{\varepsilon}\} + [g]\{\boldsymbol{\varepsilon}\} = \{\mathbf{r}\} \tag{10-58}$$

Here $\{\mathbf{r}\}$ is a vector whose components are the residuals of the three equations, whereas

$$[I^*] \equiv \begin{bmatrix} 0 & 0 & 1 \\ 1 & 0 & 0 \\ 0 & 1 & 0 \end{bmatrix}$$

$$[g(\eta)] \equiv \begin{bmatrix} \dfrac{m+1}{2} f_3 & -2mf_2 & \dfrac{m+1}{2} f_1 \\ 0 & -1 & 0 \\ 0 & 0 & -1 \end{bmatrix} \tag{10-59}$$

If $\{\mathbf{E}_i\}$ is the finite-difference approximation to $\{\boldsymbol{\varepsilon}(\eta_i)\}$, $[G_i]$ to $[g(\eta_i)]$, and so on, we can approximate the differential equation 10-58 at the midpoint of the interval (η_i, η_{i+1}) by using central-difference approximations to the derivatives

$$\left.\frac{d}{d\eta}\{\boldsymbol{\varepsilon}\}\right|_{i+\frac{1}{2}} = \frac{1}{\eta_{i+1} - \eta_i}\{\mathbf{E}_{i+1} - \mathbf{E}_i\}$$

and averaging the undifferentiated terms[4]

$$\left.[g]\{\boldsymbol{\varepsilon}\}\right|_{i+\frac{1}{2}} \approx \tfrac{1}{2}[G_i]\{\mathbf{E}_i\} + \tfrac{1}{2}[G_{i+1}]\{\mathbf{E}_{i+1}\}$$

[3] As pointed out in the last footnote, equations 10-52a and b are satisfied by every approximation to their solution if they are satisfied by the first approximation. Assuming the initial approximation meets this condition, the only residual that must be examined is that of equation 10-52c (the only nonlinear equation of the three), which simplifies (and so recommends) examination of the residuals as a criterion for deciding when to terminate the iterations.

[4] This may remind you of the Crank–Nicolson method described in Chapter 9. See also problem 2 of Chapter 9.

The residual contains a mixture of differentiated and undifferentiated terms, which are approximated by central differences and averaging, respectively. Let us call the result $\{\mathbf{R}_i\}$, for brevity, and write the finite-difference approximations to equation 10-58 as

$$\left[I^* + \frac{h_i}{2}G_{i+1}\right]\{\mathbf{E}_{i+1}\} - \left[I^* - \frac{h_i}{2}G_i\right]\{\mathbf{E}_i\} = h_i\{\mathbf{R}_i\} \tag{10-60}$$

where

$$h_i \equiv \eta_{i+1} - \eta_i$$

In these equations, i ranges from 1 (if $\eta_1 = 0$) to $N - 1$ (if $\eta_N = \eta_\infty$). Thus equations 10-60 and the boundary conditions (10-56) comprise a set of $3N$ equations in the $3N$ unknowns, the vectors $\{\mathbf{E}_1\}, \{\mathbf{E}_2\}, \ldots, \{\mathbf{E}_N\}$.

It is useful (though perhaps confusing, hopefully only temporarily) to order these equations as follows. First, take the first two of the boundary conditions (10-56), the ones that apply at $\eta = 0$. Then take equation 10-60 for $i = 1, \ldots, N - 1$, and finally the last of the boundary condition (10-56), the one that applies at $\eta = \eta_\infty$. If the unknowns are ordered naturally, from $\{\mathbf{E}_1\}$ to $\{\mathbf{E}_N\}$, the system can be arranged in matrix form as shown below:

$$\begin{bmatrix} 1 & & & & \\ & 1 & & & \\ \left[-I^* + \frac{h_0}{2}G_1\right] & \left[I^* + \frac{h_0}{2}G_2\right] & & & \\ & \left[-I^* + \frac{h_1}{2}G_2\right] & \left[I^* + \frac{h_1}{2}G_3\right] & & \\ & \ddots & \ddots & & \\ \ddots & \ddots & & & \\ & \left[-I^* + \frac{h_{N-1}}{2}G_{N-1}\right] & \left[I^* + \frac{h_{N-1}}{2}G_N\right] & & \\ & & 1 & & \end{bmatrix} \begin{bmatrix} \{\mathbf{E}_1\} \\ \{\mathbf{E}_2\} \\ \vdots \\ \{\mathbf{E}_{N-1}\} \\ \{\mathbf{E}_N\} \end{bmatrix} = h \begin{bmatrix} 0 \\ 0 \\ \{R_1\} \\ \{R_2\} \\ \vdots \\ \{R_{N-2}\} \\ \{R_{N-1}\} \\ 0 \end{bmatrix} \tag{10-61}$$

Note that the blocks corresponding to the finite-difference equation 10-60 start one row above the row on which the vector $\{\mathbf{E}_{i+1}\}$ starts. Also note that the elements are zero except in the 3×3 blocks that either include the main diagonal or are adjacent to the blocks that do. Thus the system can be arranged in the *block-tridiagonal form*:

$$\begin{bmatrix} [B_1] & [C_1] & & & & \\ [A_2] & [B_2] & [C_2] & & & \\ & [A_3] & [B_3] & [C_3] & & \\ & & \ddots & \ddots & \ddots & \\ & & & [A_{N-1}] & [B_{N-1}] & [C_{N-1}] \\ & & & & [A_N] & [B_N] \end{bmatrix} \begin{bmatrix} \{\mathbf{E}_1\} \\ \{\mathbf{E}_2\} \\ \{\mathbf{E}_3\} \\ \vdots \\ \{\mathbf{E}_{N-1}\} \\ \{\mathbf{E}_N\} \end{bmatrix} = \begin{bmatrix} \{\mathbf{D}_1\} \\ \{\mathbf{D}_2\} \\ \{\mathbf{D}_3\} \\ \vdots \\ \{\mathbf{D}_{N-1}\} \\ \{\mathbf{D}_N\} \end{bmatrix} \tag{10-62}$$

in which,

$$[A_i] = \begin{bmatrix} -1 & -\frac{1}{2}h_{i-1} & 0 \\ 0 & -1 & -\frac{1}{2}h_{i-1} \\ 0 & 0 & 0 \end{bmatrix}$$

$$[B_i] = \begin{bmatrix} 1 & -\frac{1}{2}h_{i-1} & 0 \\ 0 & 1 & -\frac{1}{2}h_{i-1} \\ \frac{(m+1)}{4} h_i \tilde{F}_{3i} & -mh_i\tilde{F}_{2i} & -1 + \frac{m+1}{4} h_i \tilde{F}_{1i} \end{bmatrix}$$

$$[C_i] = \begin{bmatrix} 0 & 0 & 0 \\ 0 & 0 & 0 \\ \frac{m+1}{4} h_i \tilde{F}_{3i+1} & -mh_i\tilde{F}_{2i+1} & 1 + \frac{m+1}{4} h_i \tilde{F}_{1i+1} \end{bmatrix}$$

$$\{\mathbf{D}_i\} = \begin{Bmatrix} \frac{h_{i-1}}{2}(\tilde{F}_{2i} + \tilde{F}_{2i-1}) - (\tilde{F}_{1i} - \tilde{F}_{1i-1}) \\ \frac{h_{i-1}}{2}(\tilde{F}_{3i} + \tilde{F}_{3i-1}) - (\tilde{F}_{2i} - \tilde{F}_{2i-1}) \\ \frac{m}{2} h_i(\tilde{F}_{2i}^2 + \tilde{F}_{2i+1}^2) - \frac{m+1}{4} h_i(\tilde{F}_{1i}\tilde{F}_{3i} + \tilde{F}_{1i+1}\tilde{F}_{3i+1}) - \tilde{F}_{3i+1} + \tilde{F}_{3i} - mh_i \end{Bmatrix}$$

$$\text{for } i = 2, \ldots, N-1 \tag{10-63}$$

For $i = 1$ and N, these equations must be revised to account for the boundary condition (10-56). Again, note that the first two elements of the $\{D_i\}$ vectors are zero if the initial approximations satisfy equations 10-52a and b.

The block-tridiagonal system (10-62) may be solved by a procedure that is the analog of the Thomas algorithm for a simple tridiagonal system. Premultiply the first row of matrices by $[B_1]^{-1}$, and write the result

$$\{\mathbf{E}_1\} + [C_1']\{\mathbf{E}_2\} = \{\mathbf{D}_1'\} \tag{10-64}$$

where

$$[C_1'] = [B_1]^{-1}[C_1], \qquad \{\mathbf{D}_1'\} = [B_1]^{-1}\{\mathbf{D}_1\}$$

Now multiply equation 10-64 by $[A_2]$, and subtract the result from the second row of the system (10-62). This knocks out the $\{\mathbf{E}_1\}$ term and leaves

$$[B'_2]\{\mathbf{E}_2\} + [C_2]\{\mathbf{E}_3\} = \{\mathbf{D}'_2\} \tag{10-65}$$

where

$$[B'_2] = [B_2] - [A_2][C_1]$$

$$\{\mathbf{D}'_2\} = \{\mathbf{D}_2\} - [A_2]\{\mathbf{D}'_1\}$$

We then premultiply equation 10-65 by $[B'_2]^{-1}$, and then by $[A_3]$, subtract the result from the third row of equation 10-62, and so on. As with the Thomas algorithm, the process is easily summarized in FORTRAN:

$$\begin{aligned}
&\text{DO 100} \qquad I = 2, N \\
&[C_{I-1}] = [B_{I-1}]^{-1}[C_{I-1}] \\
&\{\mathbf{D}_{I-1}\} = [B_{I-1}]^{-1}\{\mathbf{D}_{I-1}\} \\
&[B_I] = [B_I] - [A_I][C_{I-1}] \\
100&\{\mathbf{D}_I\} = \{\mathbf{D}_I\} - [A_I]\{\mathbf{D}_{I-1}\}
\end{aligned}$$

This leaves the matrix (10-62) in the following form,

$$\begin{bmatrix}
[I] & [C_1] & & & & \\
 & [I] & [C_2] & & & \\
 & & [I] & [C_3] & & \\
 & & & \ddots & \ddots & \\
 & & & & [I] & [C_{N-1}] \\
 & & & & & [B_N]
\end{bmatrix}
\begin{bmatrix}
\{\mathbf{E}_1\} \\ \{\mathbf{E}_2\} \\ \{\mathbf{E}_3\} \\ \vdots \\ \{\mathbf{E}_{N-1}\} \\ \{\mathbf{E}_N\}
\end{bmatrix}
=
\begin{bmatrix}
\{\mathbf{D}_1\} \\ \{\mathbf{D}_2\} \\ \{\mathbf{D}_3\} \\ \vdots \\ \{\mathbf{D}_{N-1}\} \\ \{\mathbf{D}_N\}
\end{bmatrix}$$

which we solve for $\{\mathbf{E}_N\}$

$$\{\mathbf{E}_N\} = [B_N]^{-1}\{\mathbf{D}_N\}$$

and then execute a back-substitution process:

$$\begin{aligned}
&\text{DO 200} \qquad J = 2, N \\
&I = N - J + 1 \\
200&\{\mathbf{E}_I\} = \{\mathbf{D}_I\} - [C_I]\{\mathbf{E}_{I+1}\}
\end{aligned}$$

Note that this process requires that the matrices $[C_i]$ and $\{\mathbf{D}_i\}$ be stored. However, the $[A_i]$'s and $[B_i]$'s can be destroyed immediately after creation.

The programming required for this method is much more intricate than for any of the others we have discussed thus far. However, as is generally the case, once you have done it for one problem, it is not hard to revise it for another. Thus it would not be hard to generalize the program FSKBOX that implements this procedure, listed under Section 10.8, to the Keller–Cebeci "box method" for nonsimilar boundary layers [4].

10.3. TRANSFORMATION OF THE LAMINAR BOUNDARY-LAYER EQUATIONS FOR ARBITRARY PRESSURE GRADIENTS

As was noted in the preceding section, the methods that solve the ordinary differential equation that governs the similar solutions of the boundary-layer equations can readily be adapted to the partial differential equations that apply when the external velocity distribution does not follow the power law (10-20). To do so requires that equations 10-1 and 10-10 first be transformed by introducing variables much like those defined in equations 10-8, 10-9, and 10-11:

$$\xi \equiv x$$

$$\eta \equiv y/\delta(x)$$

$$u = V_e \frac{\partial f}{\partial \eta}(\xi, \eta) \tag{10-66}$$

The only difference is that f is now allowed to depend on ξ as well as on η. Thus, introducing the new independent variables ξ and η into equations 10-1 and 10-10 leads us back to equations 10-12 and 10-13. However, when u is eliminated from equation 10-12 with the third of equations 10-66, the result is

$$\delta \dot{V}_e \frac{\partial f}{\partial \eta} + \delta V_e \frac{\partial^2 f}{\partial \eta \partial \xi} - V_e \dot{\delta} \frac{\partial}{\partial \eta}\left(\eta \frac{\partial f}{\partial \eta} - f\right) + \frac{\partial v}{\partial \eta} = 0$$

which may be integrated to give

$$v = V_e \dot{\delta}\left(\eta \frac{\partial f}{\partial \eta} - f\right) - \delta \dot{V}_e f - \delta V_e \frac{\partial f}{\partial \xi} \tag{10-67}$$

a result that differs from equation 10-16 in its last term (the constant of integration is zero, as before, because we set $f = 0$ at $\eta = 0$, just as in equation 10-15).

Now substituting for u and v in equation 10-13 from equations 10-66 and 10-67, we get, at first,

$$\rho V_e \frac{\partial f}{\partial \eta}\left(\dot{V}_e \frac{\partial f}{\partial \eta} + V_e \frac{\partial^2 f}{\partial \xi \partial \eta} - \eta \frac{\dot{\delta}}{\delta} V_e \frac{\partial^2 f}{\partial \eta^2}\right)$$
$$+ \frac{\rho}{\delta}\left[V_e \dot{\delta}\left(\eta \frac{\partial f}{\partial \eta} - f\right) - \delta \dot{V}_e f - \delta V_e \frac{\partial f}{\partial \xi}\right] V_e \frac{\partial^2 f}{\partial \eta^2} = \rho V_e \dot{V}_e + \frac{\mu}{\delta^2} V_e \frac{\partial^3 f}{\partial \eta^3}$$

which simplifies to

$$f''' = \frac{\rho V_e \delta}{\mu}\left[\frac{\delta \dot{V}_e}{V_e}(f'^2 - 1 - ff'') - \dot{\delta} ff'' + \delta\left(f' \frac{\partial f'}{\partial \xi} - \frac{\partial f}{\partial \xi} f''\right)\right] \tag{10-68}$$

As before, the prime indicates differentiation with respect to η; now, however, it is partial differentiation.

Equation 10-68 differs from equation 10-17 only in its last two terms. The boundary conditions are also the same as for the similar solutions, equations 10-23. But, because of the ξ derivatives in equation 10-68, we now need an initial condition on the whole velocity profile at some starting value of ξ:

$$f(\xi_0, \eta) = f_0(\eta) \tag{10-69}$$

The initial data are usually taken from some similar solution of the Falkner–Skan equation 10-22, the cases $m = 0$ (flat plate) and $m = 1$ (stagnation point) being especially common. In any case, the idea is then to solve equation 10-68 at successively larger values of ξ, starting of ξ_0. The problem of finding $f(\xi, \eta)$ is thus an *initial-value problem*, like (in this respect) the problem of solving the heat equation 9-45.

Much of what we learned in Chapter 9 by applying various finite-difference methods to the heat equation can be carried over when those methods are applied to other initial-value problems, including the boundary-layer equations. In particular, the following generalizations can be made:

1. *Explicit methods*, in which present data are extrapolated to advance the solution from one time level to the next point by point, are simple but subject to instabilities unless the time step is severely limited.
2. The *Crank–Nicolson method* is unconditionally stable and accurate to higher order in the time step than any other "one-step method" (a method in which the solution can be advanced to the next time level just by using the solution at the current time). However, if the spatial variation of the solution is not sufficiently smooth, this accuracy can be realized only by limiting the time step to nearly the same extent as in the explicit method. For larger time steps, the higher-frequency components of the solution oscillate from one time step to the next, and their amplitude decays much more slowly than it should.
3. *Fully implicit methods*, in which the time derivative is approximated by backward differences, are also unconditionally stable, but much better than the Crank–Nicolson method in dealing with high-frequency components of the solution. To obtain accuracy comparable to that which the Crank–Nicolson yields for smooth solutions, a two-step method must be used.

These conclusions can be applied to the integration of the boundary-layer equation 10-68 by noting that ξ is the time-like direction and η is like the spatial variable of the heat equation. Because data on the external flow velocity V_e are generally available only at discrete values of ξ—for example, they may have been generated by a panel method—it is important to choose a method whose "time" step is not limited by stability consideration; $\Delta\xi$ will be governed by the ξ's at which V_e is known. To be able to deal with velocity profiles that vary rapidly and to maximize accuracy, we therefore choose a two-step backward-difference method. For the spatial (η) differencing, we will use the simpler of the methods with which we studied the similar solutions. Thus, as we did in equations 10-34 and 10-35, we split equation 10-68 into the simultaneous equations

$$u^{*\prime\prime} = \frac{\rho V_e \delta}{\mu}\left[\frac{\delta \dot{V}_e}{V_e}(u^{*2} - 1 - fu^{*\prime}) - \dot{\delta} f u^{*\prime} + \delta\left(u^* \frac{\partial u^*}{\partial \xi} - \frac{\partial f}{\partial \xi} u^{*\prime}\right)\right]$$

$$u^* = \frac{\partial f}{\partial \eta} \tag{10-70}$$

We must now define δ. Ideally, we want to do so such that, if we confine the calculations to the strip $0 < y/\delta < \eta_\infty$, we will include all of the active part of the boundary layer without wasting too many nodes in the region where u^* is not noticeably different from 1.0. Mellor and Herring [7] use the displacement thickness for δ, but this complicates the iterative procedure, since $\delta^*(\xi)$ is one of the unknowns. Here we follow Keller and Cebeci [8] in using the same scaling as we did for the similar solutions, equation 10-21, in which $\xi = 0$ is the point at which the calculations are supposed to start. Then our general solution will merge smoothly into a similar solution when that is appropriate, a point which is important because the similar solutions usually furnish the initial data.

From equation 10-21, we see that

$$\dot{\delta} = \frac{1}{2}\left(\frac{\mu}{\rho V_e \xi}\right)^{1/2}\left(1 - \frac{\xi \dot{V}_e}{V_e}\right)$$

and so reduce equation 10-70 to

$$u^{*\prime\prime} = m(u^{*2} - 1) - \frac{m+1}{2} f u^{*\prime} + \xi u^* \frac{\partial u^*}{\partial \xi} - \xi \frac{\partial f}{\partial \xi} u^{*\prime} \tag{10-71}$$

in which the similarity between equations 10-71 and 10-35 has been heightened by defining

$$m(\xi) \equiv \frac{\xi \dot{V}_e}{V_e} \tag{10-72}$$

Our initial and boundary conditions are found from equations 10-23 and 10-69 to be

$$\begin{aligned} f = u^* &= 0 && \text{at } \eta = 0 \\ u^* &\to 1 && \text{as } \eta \to \infty \\ u^* &= u_0^*(\eta) && \text{at } \xi = 0 \end{aligned} \tag{10-73}$$

As with the similar solutions, we will iterate for the solution of these equations at a given ξ by integrating the second of equations 10-70 to find the f corresponding to a guess for $u^*(\xi, \eta)$, and then solving a quasilinearized version of equation 10-71. Before quasilinearizing, it is convenient to eliminate the derivative of u^* with respect to ξ by approximating it with a second-order backward difference formula. Let ξ indicate the station at which u^* is currently sought, and ξ_1, ξ_2 the stations immediately upstream, at which u^* is already known, say as $u_1^*(\eta)$ and $u_2^*(\eta)$, respectively. Then, expanding u_1^* and u_2^* in Taylor series in ξ at constant η about $u^*(\xi, \eta)$, we get

$$u_1^* = u^* + (\xi_1 - \xi)\frac{\partial u^*}{\partial \xi} + \frac{(\xi_1 - \xi)^2}{2}\frac{\partial^2 u^*}{\partial \xi^2} + O(\xi_1 - \xi)^3$$

$$u_2^* = u^* + (\xi_2 - \xi)\frac{\partial u^*}{\partial \xi} + \frac{(\xi_2 - \xi)^2}{2}\frac{\partial^2 u^*}{\partial \xi^2} + O(\xi_2 - \xi)^3$$

Eliminating $\partial^2 u^*/\partial \xi^2$ between the two expressions and dropping the error terms yields a second-order accurate approximation to $\partial u^*/\partial\xi$ at ξ, which we write[5]

$$\Delta\xi \frac{\partial u^*}{\partial \xi} \approx \beta u^*(\xi, \eta) - u^{**} \tag{10-74}$$

where

$$\Delta\xi = \xi_1 - \xi_2$$

$$\beta = \frac{\xi_2 - \xi}{\xi_1 - \xi} - \frac{\xi_1 - \xi}{\xi_2 - \xi}$$

$$u^{**} = \frac{\xi_2 - \xi}{\xi_1 - \xi} u_1^* - \frac{\xi_1 - \xi}{\xi_2 - \xi} u_2^* \tag{10-75}$$

Substituting in equation 10-71 for $\partial u^*/\partial\xi$ from equation 10-74, we obtain

$$u^{*''} \simeq m(u^{*^2} - 1) - \frac{m+1}{2} f u^{*'} + \frac{\xi u^*}{\Delta\xi}(\beta u^* - u^{**}) - \xi \frac{\partial f}{\partial \xi} u^{*'}$$

in which, except for u^{**}, all quantities are understood to be evaluated at ξ. The quasilinearized version of this equation is

$$u^{*''} \simeq m(2\tilde{u}u^* - \tilde{u}^2 - 1) - \frac{m+1}{2} \tilde{f} u^{*'}$$

$$+ \frac{\xi}{\Delta\xi}(2\beta\tilde{u}u^* - \beta\tilde{u}^2 - u^{**}u^*) - \xi \frac{\partial \tilde{f}}{\partial \xi} u^{*'} \tag{10-76}$$

where $\tilde{u}(\eta)$ is the current approximation to $u^*(\xi, \eta)$; compare equation 10-39.

It remains to form finite-difference approximations to the η derivatives (those symbolized by primes) in equation 10-76. Because of its similarity to equation 10-39, the resulting algebraic equations are of the same form as equation 10-44,

$$A_j U_{j-1}^* + B_j U_j^* + C_j U_{j+1}^* = D_j \tag{10-77}$$

but the coefficients are now given by

$$A_j = \frac{2}{(h_+ + h_-)h_-} - \left(\frac{m+1}{2} \tilde{F}_j + \xi \frac{\partial \tilde{F}_j}{\partial \xi}\right) \frac{1}{h_+ + h_-}$$

$$B_j = -\frac{2}{h_- h_+} - 2m\tilde{U}_j - \frac{\xi}{\Delta\xi}(2\beta\tilde{U}_j - U_j^{**})$$

$$C_j = \frac{2}{(h_+ + h_-)h_+} + \left(\frac{m+1}{2} \tilde{F}_j + \xi \frac{\partial \tilde{F}_j}{\partial \xi}\right) \frac{1}{h_+ + h_-}$$

$$D_j = -m(\tilde{U}_j^2 + 1) - \frac{\xi}{\Delta\xi} \beta \tilde{U}_j^2 \tag{10-78}$$

instead of equations 10-45. Here h_+ and h_- are defined in equations 10-46.

[5] See problem 10 of Chapter 7.

To evaluate $\tilde{F}_j$, we use the same trapezoidal rule we used in equation 10-37,

$$\tilde{F}_j = \tilde{F}_{j-1} + \tfrac{1}{2}h_-(\tilde{U}_j + \tilde{U}_{j-1}) \tag{10-79}$$

along with the boundary condition (10-38). The second-order backward-difference approximation to $\partial\tilde{F}_j/\partial\xi$ may be written in a form identical with that of equation 10-74:

$$\Delta\xi \frac{\partial \tilde{F}_j}{\partial \xi} \approx \beta\tilde{F}_j - F_j^{**} \tag{10-80}$$

with, as in equation 10-75,

$$F_j^{**} \approx \frac{\xi_2 - \xi}{\xi_1 - \xi} F_{1j} - \frac{\xi_1 - \xi}{\xi_2 - \xi} F_{2j} \tag{10-81}$$

in which F_{1j}, F_{2j} are the values of F_j at the previous two stations. In view of the boundary conditions (10-73),

$$F_1^{**} = 0 \tag{10-82}$$

whereas, from equations 10-34, 10-75, and 10-81,

$$U^{**} = \frac{\partial F^{**}}{\partial \eta}$$

Using the trapezoidal rule here, too, gives

$$F_j^{**} = F_{j-1}^{**} + \tfrac{1}{2}h_-(U_j^{**} + U_{j-1}^{**}) \tag{10-83}$$

From equations 10-78 to 10-83, we see that the coefficients of equation 10-77 can be calculated if we store $\tilde{U}_j$ and U_j^{**}. The boundary conditions (10-73) on u^* can also be brought into the form of equation 10-77, with coefficients given by equation 10-48.

Thus, as was the case with equation 10-44, equation 10-77 can be used as the basis for an iterative determination of the velocity profile at ξ provided we already know the velocity profiles at two upstream stations (and so can calculate U^{**} from equation 10-75). As was noted in Chapter 9, it is characteristic of methods based on second-order backward differences that they need a special procedure to start the calculations. The program listed in Section 10.8, BDYLAY, cheats on this point; the first step is taken with a first-order, one-step backward-difference method, as follows. For this step, the ξ_1 of equation 10-75 is just zero. Define a fictitious point ξ_2 upstream of ξ_1 by

$$\xi_2 = -\xi \tag{10-84a}$$

so that, as shown in Fig. 10.5, ξ_1 (the origin) is equidistant from ξ and ξ_2, and set

$$u_2^* = u_1^* = u^*(\xi_0, \eta) \tag{10-84b}$$

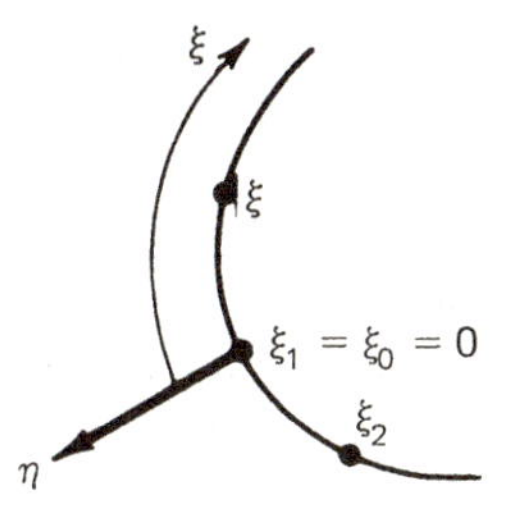

FIG. 10.5. Nomenclature for adaptation of second-order backward difference formula to start of boundary-layer calculation.

Then, from equation 10-75,

$$\Delta\xi = \xi - \xi_1$$

$$\beta = \tfrac{3}{2}$$

$$u^{**} = \tfrac{3}{2}u_1^*$$

and so equation 10-74 becomes

$$\Delta\xi \frac{\partial u^*}{\partial \xi} = \tfrac{3}{2}(u^* - u_1^*)$$

This will reduce to the correct first-order backward-difference formula

$$(\xi - \xi_1)\frac{\partial u^*}{\partial \xi} = u^*(\xi, \eta) - u^*(\xi_1, \eta)$$

if, at this initial step, we replace the first of equations 10-75 by

$$\Delta\xi = \tfrac{3}{2}(\xi - \xi_1) \tag{10-84c}$$

10.3.1. Program BDYLAY

You should be able to see where equations 10-84a–c are used in program BDYLAY (note that $\xi_0 = 0$). Although their usage implies that the velocity profile at the first step is accurate only to first order in the change in ξ, the extra error committed there should tend to disappear as the calculations proceed downstream, just as do local hot spots in a medium whose temperature distribution is governed by the heat equation. As for the calculation of the initial data $u^*(\xi_0, \eta)$, it is assumed in BDYLAY that they are to be taken from one of the similar solutions. The easiest way to reduce equations 10-78 to their similar-solution counterparts (10-45) is to set $\Delta\xi$ and U^{**} arbitrarily but to take ξ as zero. That is what is done in BDYLAY.

Much of BDYLAY should remind you of the programs INTGRL and FALKSK, which illustrated the integral methods for calculating boundary-layer growth developed in Chapter 7 and the method described earlier in this chapter for determining the similar solutions, respectively. The program is set up to calculate the boundary layer on an ellipse at zero angle of attack, for which case the physical coordinates of the body and the inviscid velocity distribution are supplied in function subroutines. These can be removed when those data are instead to be read from a file, just as we discussed in Chapter 7 in connection with INTGRL. The external velocity gradient $\dot{V}_e$ is calculated from a procedure borrowed from INTGRL (or from the method used to derive equation 10-74 above) and then used to calculate the velocity-gradient parameter m from equation 10-72. At $\xi = 0$, equation 10-72 is indeterminate, since the flow stagnates there. However, the proper value of m at the initial (stagnation) point is easily shown to be 1.0, so that BDYLAY simply initializes $m = 1.0$ and avoids using equation 10-72 until the second point along the boundary layer. The displacement thickness and skin-friction coefficient are calculated as in FALKSK, but the momentum thickness must now be calculated by integrating equation 7-30 numerically (a trapezoidal rule is used in BDYLAY). However, the values output for δ^* and θ are scaled by the reference length L in BDYLAY, rather than by δ as was the case with FALKSK.

Separation is detected in BDYLAY (and the calculations terminated) when $U_j^* < 0$ for some j.

10.4. TURBULENT BOUNDARY LAYERS

Now it is time to face reality and to consider the calculation of boundary layers that are turbulent. For the Reynolds numbers encountered in aeronautics, transition to turbulence is inevitable, and in fact most of the boundary layer is usually turbulent. Equations 10-1 and 10-2 then apply to the time-averaged velocity field, but the turbulent shear stress $-\rho u'v'$ cannot be dropped from equation 10-5 and must be counted as an additional unknown. You can generate differential equations for the turbulence stresses by multiplying the Navier–Stokes equations by the velocity components before averaging but that multiplies the number of unknowns more than it adds to the number of equations [9]. Various means for "closing" the system of equations—for reducing the number of unknowns to the number of equations—have been proposed. Sooner or later, you either approximate some of the unknowns in terms of others by empirical formulas, or, if there aren't enough data to suggest an appropriate correlation formula, you simply drop some of the extra unknowns.

We shall choose the least sophisticated alternative, relating the turbulent stress to mean-flow properties through empirically based algebraic formulas. The more complicated methods, in which one must solve simultaneously sets of partial differential equations not much simpler in form than equation 10-2, take more calculation time without, at least for boundary layers, being much more accurate or powerful. It is to be hoped that this is not the last word on the subject and that the multiequation models will improve in accuracy, ease of use, and especially ability to handle situations beyond the capability of the model described here.

Specifically, as in most practical analyses, we shall work in terms of a quantity called the *eddy viscosity* ε, defined so that

$$-\overline{u'v'} \equiv \varepsilon \frac{\partial u}{\partial y} \tag{10-85}$$

Then equation 10-5 can be written

$$\tau = \rho(\nu + \varepsilon)\frac{\partial u}{\partial y} \tag{10-86}$$

where ν is the *kinematic viscosity* μ/ρ. The difference between ν and ε is that ν is a fluid property, independent of the flow, whereas ε is not. Prandtl [10] hypothesized that

$$\varepsilon = l^2\frac{\partial u}{\partial y} \tag{10-87}$$

where l is the "mixing length" introduced in Chapter 7, whose value, according to equation 7-56, is

$$l = \kappa y \tag{10-88a}$$

κ being the same constant that shows up in the *law of the wall*, equation 7-57, which fits experimentally observed velocity profiles remarkably well when κ is about 0.4. Prandtl's reasoning was based on a model of turbulent mixing in which eddies act like large fluid particles, exchanging momentum through collisions just like the hard spheres of kinetic theory. The model has no basis in fact, but equation 10-87 is at least dimensionally correct.

Dimensional reasoning hardly ever leads to unique results. An important modification of equation 10-88 was made by van Driest [11], who sought to account for the laminar sublayer. Taking a cue from the form of the exact solution for laminar flow past an oscillating flat plate,[6] he wrote

$$l = \kappa y(1 - e^{-y^+/A^+}) \tag{10-88b}$$

where A^+, a damping constant, is generally taken to be 24 to 26, whereas

$$y^+ \equiv \frac{yu_\tau}{\nu} \tag{10-89}$$

$$u_\tau \equiv \sqrt{\frac{\tau_w}{\rho}} \tag{10-90}$$

On the basis of available experimental data, Cebeci and Smith [13] let A^+ depend on the local pressure gradient, as follows:

$$A^+ = 26\left\{1 - 11.8\frac{\nu V_e}{u_\tau^3}\dot{V}_e\right\}^{-1/2} \tag{10-91}$$

Also in the interests of improving the agreement with experimental data, they modify equation 10-87 by introducing an "intermittency factor" γ, which accounts for the gradual development of a fully turbulent boundary layer after the start of transition:

$$\varepsilon = l^2\left|\frac{\partial u}{\partial y}\right|\gamma \qquad \text{in inner layer} \tag{10-87a}$$

[6] See Pai [12], p. 62, for example.

The formula they use for γ is

$$\gamma = 1 - \exp\left(-\frac{\mathrm{Re}_t^{0.66}(x - x_t)V_e}{1200\, x_t^2} \int_{x_t}^{x} \frac{dx}{V_e} \right) \tag{10-92}$$

where x_t is the value of x at the start of transition, and Re_t is the Reynolds number based on V_e and x_t.

Equation 10-87a is used only in the inner layer and below. In the outer layer, Cebeci and Smith take

$$\varepsilon = 0.0168\, V_e \delta^* \gamma \qquad \text{in outer layer} \tag{10-87b}$$

The position within the boundary layer at which one switches from equation 10-87a to 10-87b is simply the point where equation 10-87a yields a value for ε equal to the constant given by 10-87b, so that the variation of ε through the boundary layer is as sketched in Fig. 10.6.

Naturally, other possibilities exist. Mellor and Herring [7] use

$$\varepsilon = \nu\phi(\kappa y^+) + V_e \delta^* \Phi\left(\kappa \frac{y u_\tau}{\delta^* V_e} \right) \tag{10-93}$$

where

$$\phi(x) = \frac{x^4}{x^3 + (6.9)^3} \tag{10-94}$$

and

$$\begin{aligned} \Phi(x) &= 0 && \text{for } x \le 0.016 \\ &= 0.016 - x && \text{for } x > 0.016 \end{aligned} \tag{10-95}$$

Thus their $\varepsilon(y)$ is similar in form to the Cebeci–Smith formulas.

Now let us see how the eddy viscosity may be taken into account in the computation of the boundary layer. Equation 10-68 was derived by substituting equations 10-66 and 10-67 into 10-2, with p given by equation 10-3 and τ by the laminar version of 10-5, and then multiplying by $\delta^2/\mu V_e$. If we do this with the $\partial\tau/\partial y$

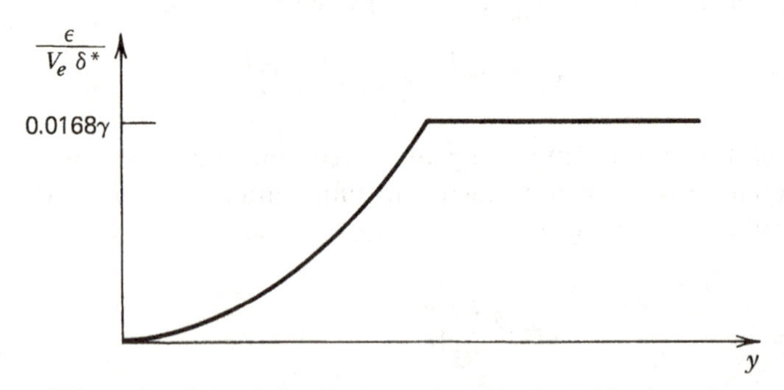

FIG. 10.6. Cebeci–Smith approximation for eddy viscosity.

term of equation 10-2 when τ is given by equation 10-86, we get

$$\frac{\delta^2}{\mu V_e}\left[\frac{\mu V_e}{\delta^2} f''' + \frac{\rho}{\delta}\frac{\partial}{\partial \eta}\varepsilon\left(\frac{V_e}{\delta} f''\right)\right] = f''' + \frac{1}{\nu}\frac{\partial}{\partial \eta}(\varepsilon f''')$$

Thus f''' in equation 10-68 is replaced by this expression, and $u^{*''}$ in equation 10-71 by

$$u^{*''} + \frac{1}{\nu}\frac{\partial}{\partial \eta}(\varepsilon u^{*'})$$

Let the additional term in equation 10-71 be called E:

$$E \equiv \frac{1}{\nu}\frac{\partial}{\partial \eta}(\varepsilon u^{*'}) \tag{10-96}$$

In the outer layer, where ε is the constant given by equation 10-87b, we have simply

$$E = \frac{\varepsilon}{\nu} u^{*''} \qquad \text{in outer layer} \tag{10-97a}$$

In the inner layer, we must use equation 10-87a in 10-96, so

$$E = \frac{\gamma V_e}{\delta \nu}\left[u^{*'}|u^{*'}|2l\frac{\partial l}{\partial \eta} + l^2\frac{\partial}{\partial \eta}(u^{*'}|u^{*'}|)\right]$$

Since $\eta = y/\delta$,

$$\frac{1}{\delta}\frac{\partial l}{\partial \eta} = \frac{\partial l}{\partial y}$$

Also,

$$\begin{aligned}\frac{\partial}{\partial \eta} u^{*'}|u^{*'}| &= \frac{\partial}{\partial \eta} u^{*'}\sqrt{(u^{*'})^2} \\ &= u^{*''}\sqrt{(u^{*'})^2} + \frac{(u^{*'})^2 u^{*''}}{\sqrt{(u^{*'})^2}} \\ &= 2|u^{*'}|u^{*''}\end{aligned}$$

Thus

$$E = 2\frac{\gamma V_e l}{\nu}|u^{*'}|\left(u^{*'}\frac{\partial l}{\partial y} + \frac{l}{\delta}u^{*''}\right) \qquad \text{in inner layer} \tag{10-97b}$$

After adding equation 10-97a and b to the left side of equation 10-71, the next step is to linearize the result about some guess u for u^*. In the inner layer, the additional term E is nonlinear in u^* and so requires some special attention. If

$$u^* = \tilde{u} + e$$

then

$$\begin{aligned}|u^{*\prime}| &= \sqrt{(\tilde{u}' + e')^2} \\ &\simeq \sqrt{\tilde{u}'^2 + 2e'\tilde{u}'} \\ &= |\tilde{u}'|\sqrt{1 + \frac{2e'}{\tilde{u}'}} \\ &\approx |\tilde{u}'|\left(1 + \frac{e'}{\tilde{u}'}\right)\end{aligned}$$

and so,

$$\begin{aligned}|u^{*\prime}|u^{*\prime} &\cong |\tilde{u}'|\left(1 + \frac{e'}{\tilde{u}'}\right)(\tilde{u}' + e') \\ &\cong |\tilde{u}'|(\tilde{u}' + 2e') \\ &\cong |\tilde{u}'|(2u^{*\prime} - \tilde{u}') \\ |u^{*\prime}|u^{*\prime\prime} &\cong |\tilde{u}'|\left(1 + \frac{e'}{\tilde{u}'}\right)(\tilde{u}'' + e'') \\ &\cong |\tilde{u}'|\left(\tilde{u}'' + e'\frac{\tilde{u}''}{\tilde{u}'} + e''\right) \\ &= |\tilde{u}'|\left(\frac{\tilde{u}''}{\tilde{u}'}u^{*\prime} + u^{*\prime\prime} - \tilde{u}''\right)\end{aligned}$$

Thus, in the inner layer, the extra term defined by equation 10-96 has the quasilinearized form

$$E \approx \frac{2\gamma V_e l}{\nu}|\tilde{u}'|\left\{(2u^{*\prime} - \tilde{u}')\frac{\partial l}{\partial y} + \frac{l}{\delta}\left(\frac{\tilde{u}''}{\tilde{u}'}u^{*\prime} + u^{*\prime\prime} - \tilde{u}''\right)\right\} \qquad \text{in inner layer} \tag{10-97c}$$

Note, from equation 10-87a, that the factor outside the curly brackets in equation 10-97c is just $2\delta\varepsilon/\nu l$.

Equations 10-97a and c thus describe the term that must be added to the left side of equation 10-76 when the boundary layer is turbulent. On approximating the η derivatives with differences, an equation of the form of (10-77) is still obtained, but the coefficients given by (10-78) must be changed, as follows:

$$A_j = \left(\frac{2}{h_-}S - P\right)\Big/(h_+ + h_-)$$

$$B_j = -\frac{2}{h_-h_+}S - 2m\tilde{U}_j - \frac{\xi}{\Delta\xi}(2\beta\tilde{U}_j - U_j^{**})$$

$$C_j = \left(\frac{2}{h_+} S + P\right) \Big/ (h_+ + h_-)$$

$$D_j = -m(\tilde{U}_j^2 + 1) - \frac{\xi}{\Delta\xi} \beta \tilde{U}_j^2 + T$$

Here S is the coefficient of $u^{*''}$ in the quasilinearized equation, or

$$S = 1 + 2\frac{\varepsilon}{\nu} \qquad \text{in inner layer}$$

$$= 1 + \frac{\varepsilon}{\nu} \qquad \text{in outer layer}$$

P is the coefficient of $u^{*'}$,

$$P = \frac{m+1}{2}\tilde{F}_j + \xi\frac{\partial \tilde{F}_j}{\partial \xi} + 2\frac{\varepsilon}{\nu}\left(\frac{\tilde{u}''}{\tilde{u}'} + \frac{2\delta}{l}\frac{\partial l}{\partial y}\right) \qquad \text{in inner layer}$$

$$= \frac{m+1}{2}\tilde{F}_j + \xi\frac{\partial \tilde{F}_j}{\partial \xi} \qquad \text{in outer layer}$$

and T is the negative of the part of E that is known in terms of u:

$$T = 2\frac{\varepsilon}{\nu}\left(\tilde{u}'' + \frac{\delta}{l}\frac{\partial l}{\partial y}\tilde{u}'\right) \qquad \text{in inner layer}$$

$$= 0 \qquad \text{in outer layer}$$

Equations 10-87a and b are to be used for ε, and (10-88b) for l. Note, from equations 10-88b and 10-89, that

$$\frac{\delta}{l}\frac{\partial l}{\partial y} = \delta\frac{\partial}{\partial y}[\ln \kappa + \ln y + \ln(1 - e^{-y^+/A^+})]$$

$$= \frac{\delta}{y} + \frac{\delta u_\tau}{A^+ \nu}\frac{e^{-y^+/A^+}}{1 - e^{-y^+/A^+}}$$

$$= \frac{1}{\eta}\left(1 + \frac{y^+}{A^+}\frac{1}{e^{y^+/A^+} - 1}\right)$$

Another problem in the calculation of turbulent boundary-layer growth that requires a modification of program BDYLAY is the addition of points to the domain. You should have found in your study of the similar solutions, problem 3, that you need to apply the asymptotic boundary condition $u^* \to 1$ at a larger value of η when the pressure gradient is adverse than when it is favorable. Thus, in calculating the boundary-layer growth on an airfoil, the value of η_∞ should increase with distance from the stagnation point. However, the increase is small enough to be accounted for simply by using a value of η_∞ that is constant and about twice as large as it needs to be near the stagnation point.

Once you get past transition, however, the fact that the δ introduced in equation 10-21 was based on an analysis of laminar boundary layers catches up with you. It is not a very good scale factor for a turbulent boundary layer. Thus you will have to test the solution at each station to see if the value used for η_∞ is large enough, as follows. With $\eta_\infty = \eta_N$, if

$$|U_N^* - U_{N-1}^*| > 10^{-4}$$

for example, N should be increased by 1 and the calculations redone.

The recipe given above for modifying BDYLAY so that it can handle turbulent boundary layers is complete, but hardly simple. Good luck.

10.5. SEPARATED FLOWS

Many flows of aerodynamic interest feature pressure gradients strong enough to separate the boundary layer. As noted in Chapter 7, the boundary-layer approximation fails in the vicinity of a separation point; the characteristic length for flow property variations along the surface is no longer controlled by the geometry of the body, and may become as small as the boundary-layer thickness.

Separated flows ought, therefore, to be analyzed with the Navier–Stokes equations. This has been done, for example, by Briley and McDonald [14]. As in this chapter, the differential equations are approximated by finite-difference equations. Setting up such equations is not so hard, but solving them is. The boundary-layer equations allow information to propagate downstream, but not upstream. Therefore, if we know their solution at one station along the body, we can determine the solution at all downstream stations by a single forward sweep through the finite-difference mesh. However, the double derivatives with respect to x in the Navier–Stokes equations allow communication of information both upstream and downstream. Thus their solution requires specification of boundary conditions downstream as well as upstream, and the finite-difference equations for all points in the mesh must be solved simultaneously. In mathematical parlance, the Navier–Stokes equations are *elliptic* and so are formulated as a *boundary-value problem*, whereas the boundary-layer equations are *parabolic* and are associated with an *initial-value problem.*

Since the Navier–Stokes equations are so much more difficult to solve than are the boundary-layer equations, there has been much interest in exploring the degree to which solutions of the boundary-layer equations approximate Navier–Stokes solutions for separated flows. It has been found that, at least in some cases of practical interest, boundary-layer-based solutions are quite satisfactory. One such case is the small separation bubble that appears near the leading edge of an airfoil at the angle of attack [15]. Since their foundation is somewhat shaky once the flow separates, the boundary-layer equations cannot be recommended for use in every case, but it must be expected that the limits on their applicability will become more firmly established and that these limits will be found to encompass a substantial set of interesting aerodynamic problems. If it turns out that they cannot be used for flows with massive separation, such as the flow past a stalled airfoil, that will not be so serious, so long as they can predict the angle of attack at which stall occurs.

Therefore, we shall discuss in outline how the boundary-layer equations can be used for separated flows. It is important to note at the outset that the procedures outlined so far in this chapter and in Chapter 7 are inadequate. If the flow separates, the interaction between the boundary-layer growth and the body-surface distribution is much stronger than those procedures can handle. Indeed, if you take the pressure distribution as given (from an inviscid calculation or otherwise), the calculations blow up when you reach the separation point.

However, as was first shown by Catherall and Mangler [16], the boundary-layer equations can be solved past separation if the pressure distribution along the boundary layer is regarded as one of the outputs of the solution, and the displacement-thickness distribution is specified instead. That is, the boundary condition as $y \to \infty$ is taken to be[7]

$$\psi \to u(y - \delta^*)$$

instead of equation 10-4, which is used only to deduce the value of $V_e(x)$. Because δ^* and V_e have their usual input/output roles reversed in such a calculation, it is called the *inverse boundary-layer method.*

Of course, the displacement-thickness distribution is not known before the boundary layer is calculated. It must be determined so that the inviscid surface velocity distribution (corrected for the displacement effect of the boundary layer as discussed in Section 7.2.2) matches the value of V_e calculated from equation 10-4. If these distributions are called $V_{e,\text{inv}}$ and $V_{e,\text{bl}}$, respectively, Pletcher [15] suggests that the trial distribution of δ^* be multiplied by $V_{e,\text{bl}}/V_{e,\text{inv}}$ in the next iteration. He further recommends making two or three iterations through the inviscid calculation—set

$$\delta^* = \frac{V_{e,\text{bl}}}{V_{e,\text{inv}}} (\delta^*)_{\text{trial}}$$

and then correct the value of $V_{e,\text{inv}}$ for that distribution of δ^*—for each solution of the boundary-layer equations.

As noted above, the boundary-layer equations allow information to propagate only downstream. Specifically, information is convected with the local x component of velocity. Because separation is accompanied by flow reversal within the boundary layer, there are parts of the flow in which information propagates "upstream"; that is, against the direction of the flow outside the boundary layer. For stability, the finite-difference equations that are used to approximate the boundary-layer equations should reflect the direction of propagation that is locally appropriate. Thus forward differences should be used to approximate x derivatives when $u < 0$, instead of the backward differences used previously in this chapter. If the finite-difference equations are solved by marching through the mesh in the $+x$ direction, we must have available guesses for the velocities at points "downstream" (with

[7] Compare equation 10-26; note from equation 10-66 that f is a dimensionless stream function, specifically,

$$f = \psi / V_e \delta$$

respect to the external flow) of points at which $u < 0$. These guesses will, one hopes, be improved in subsequent iterations.[8]

Yet another difference between the solution of the direct and inverse boundary-layer problems is in the treatment required of the finite-difference equations. Pletcher [15] uses equations similar in form to our equations 10-70; specifically, equations for u and the stream function ψ. In the direct problem solved by program BDYLAY, these equations may be solved sequentially, with the value for ψ (whose dimensionless form is our f) determined from a guess at u (u^*, in our formulation) and then used in what becomes a second-order equation for u. He finds that these equations must be solved simultaneously in the inverse problem or else the displacement thickness exhibits spurious oscillations. This considerably complicates the programming of the solution, to the point where the approach outlined in Section 10.2.4 begins to look attractive.

Calculations have been carried out for the practically important case in which the boundary layer is laminar at separation, becomes turbulent, and then reattaches [15]. The primary obstacle to accuracy in such cases is not computational; all the tools needed have been laid out in this chapter. What are needed are better data on the process of transition and on the growth of a turbulent boundary layer that is separated and reattaching.

10.6. REFERENCES

1. Falkner, V. M., and Skan, S. W., "Some Approximate Solutions of the Boundary Layer Equations," *Philos. Mag.* **12,** (1931); British ARC Report 1314 (1930).
2. Blasius, H., "Grenzschichten in Flussigkeiten mit Kleiner Reibung," *Z. Math. Phys.* **56,** 1 (1908); NACA TM 1256.
3. Hiemenz, K., "Die Grenzschicht in einem in den Gleichformigen Flussigkeitsstrom eingetauchten gerade Kreiszylinder," *Dingl. Polytech. J.* **326,** 321 (1913).
4. Cebeci, T., and Bradshaw, P., *Momentum Transfer in Boundary Layers.* McGraw-Hill/Hemisphere, Washington, D. C. (1977).
5. Sylvester R. J., and Meyer, F., "Two Point Boundary Problems by Quasilinearization," *J. SIAM* **13,** 586–602 (1965).
6. Keller, H. B., "Accurate Difference Methods for Two-Point Boundary-Value Problems," *SIAM J. Number. Anal.* **11,** 305–320, (1974).
7. Mellor, G. L., and Herring, H. J., "A Method of Calculating Compressible Turbulent Boundary Layers," NASA Report CR-1144 (1968).
8. Keller, H. B., and Cebeci, T., "Accurate Numerical Methods for Boundary-Layer Flows," Part 2, "Two-Dimensional Turbulent Flows," *AIAA J.* **10,** 1193 (1972).
9. Bradshaw, P., Cebeci, T., and Whitelaw, J. H., *Engineering Calculation Methods for Turbulent Flow.* Academic, New York (1981).
10. Prandtl, L., "Turbulent Flow," NACA TM 435 (1926).

[8] Often it is found that the magnitude of u is relatively small in the reversed-flow region, which motivates what is called the FLARE approximation (after Flügge-Lotz And REyhner, who first tried it [17]), namely, to set $u\,\partial u/\partial x = 0$ whenever calculations indicate $u < 0$.

11. Van Driest, E. R., "On Turbulent Flow Near a Wall," *J. Aeronaut. Sci.* **23,** 1007 (1956).
12. Pai, S.-I, *Viscous Flow Theory. I-Laminar Flow.* Van Nostrand, Princeton, N. J. (1956), p. 62.
13. Cebeci, T., and Smith, A. M. O., *Analysis of Turbulent Boundary Layers.* Academic, New York (1974).
14. Briley, W. R., and McDonald, H., "Numerical Prediction of Incompressible Separation Bubbles," *J. Fluid Mech.* **69,** 631–656 (1975).
15. Pletcher, R. H., "Calculation of Separated Flows by Viscous-Inviscid Interaction," *Recent Adv. Numer. Methods Fluids.* **3,** 383–414 (1984).
16. Catherall, D., and Mangler, K. W., "The Integration of the Two-Dimensional Laminar Boundary-Layer Equations Past a Point of Vanishing Skin Friction," *J. Fluid Mech.* **26,** 163–182 (1966).
17. Flügge-Lotz, I., and Reynher, T. A., "The Interaction of a Shock Wave with a Laminar Boundary Layer," *Int. J. Nonlinear Mech.* **3,** 173–199 (1968).

10.7. PROBLEMS

1. For the similar solutions of the laminar boundary layer, show that the skin-friction coefficient can be written

$$c_f = \frac{2}{\sqrt{\mathrm{Re}_x}} f''(0)$$

where

$$\mathrm{Re}_x \equiv \rho \frac{V_e x}{\mu}$$

and that the displacement thickness is

$$\frac{\delta^*}{x} = \frac{1}{\sqrt{\mathrm{Re}_x}} \lim_{\eta \to \infty} [\eta - f(\eta)]$$

2. Show that, for the similar solutions of the laminar boundary layer, the momentum thickness is given by

$$\frac{\theta}{\delta} = \frac{2}{3m+1} [f''(0) - m \frac{\delta^*}{\delta}]$$

Hint: You will need some trickery to evaluate

$$I \equiv \int f'^2 \, d\eta$$

Integration by parts gives you an integral of ff''. Use the Falkner–Skan equation to put that integral in terms of I. Then collect the terms proportional to I, and so show

$$\frac{3m+1}{m+1} \int_0^\infty f'(\eta)^2 \, d\eta = f + \frac{2m}{m+1} \eta + \frac{2}{m+1} f'' \Bigg|_0^\infty$$

3. Use FALKSK or FSKBOX to find, for $-0.09 < m < 1.0$, the values of $\Delta\eta$ and η_∞ required to obtain δ^*/δ and $\sqrt{\mathrm{Re}_x}\, c_f$ accurate to four decimal places. To calculate c_f, some modification of FALKSK is required. From problem 1, c_f is proportional to $f'' = u^{*''}$ at $\eta = 0$. Derive a second-order-accurate finite difference scheme to calculate $u^{*'}$ at $\eta = 0$ given (numerically) the values of u^* at η_1, η_2, and η_3 (see equations 10-74 and 10-75 for guidance).

4. Examine the rate of convergence of the method implemented in program FALKSK for the similar solutions by repeating the calculations with the number of intervals doubled each time. Study both evenly spaced and unevenly spaced cases. For the uneven mesh spacing defined by equation 10-41, you can double the number of intervals by changing ρ and η_2 (RATIO and DETA in the program) to ρ' and η_2', where

$$\rho' = \sqrt{\rho}, \qquad \eta_2' = \frac{\eta_2}{1+\rho'}$$

5. Repeat problem 4 for the box method implemented in program FSKBOX. Note that FSKBOX routinely outputs four measures of the convergence of its trials: the maximum and root-mean-squared values of the changes in the solution from one trial to the next (CHGMAX and CHGRMS, respectively) and the maximum and RMS values of the residuals of the difference equations (RESMAX and RESRMS). The iterations are said to converge when CHGMAX $< 10^{-6}$. Would it make any substantial difference in the "converged" results were some other criterion to be used?

6. Revise FALKSK (or FSKBOX) to repeat the calculations on a mesh that is twice as fine and then to extrapolate the two results to high-order accuracy; see Section 9.2.

7. Compare the Thwaites method (as implemented in the laminar part of program INTGRL) with the finite-difference method of program BDYLAY as to accuracy and speed for the following cases:
   (a) The Howarth problem, $V_e = 1 - x/L$.
   (b) Flow past an ellipse.
   (c) Flow past a symmetric airfoil, using data output by DUBLET or PANEL.

8. Check whether the solution output by program BDYLAY satisfies the momentum integral equation 7.34. This is a very useful check on any finite-difference method for the solution of the boundary-layer equations.

9. Modify BDYLAY to use the box method for the difference equation in the η direction. Compare its performance (with regards to accuracy and speed) with the program furnished.

10. **Very instructive problem:** Merge a panel method with BDYLAY by constructing an intermediate program that will (a) locate the stagnation point according to the panel-method solution and (b) create two files suitable for input into BDYLAY, one for the flow on the upper side of the stagnation point and one for the flow on the lower side.

11. **A major project:** Now take the output of BDYLAY and compute the effective value of V_n at the points at which V_e is prescribed from equation 7-29. Then modify the

panel method that generated the data on V_e to compute the displacement effect of the boundary layer on the pressure distribution.

12. **A major project:** Revise BDYLAY to treat a boundary layer that starts out laminar from a stagnation point but undergoes transition to turbulence.

10.8. COMPUTER PROGRAMS

```
C      PROGRAM FALKSK
C
C         FINITE-DIFFERENCE METHOD FOR SIMILAR SOLUTIONS OF
C         LAMINAR INCOMPRESSIBLE BOUNDARY LAYER
C
       COMMON /TRID/ A(251),B(251),C(251),D(251),V(251)
       DIMENSION     U(251),ETA(251)
       REAL          M
C
C                    INPUT EXPONENT IN EXTERNAL VELOCITY DISTRIBUTION (M),
C                    APPROXIMATION TO INFINITY (ETAINF), FIRST
C                    INPUT FIRST INTEGRATION INTERVAL (DETA), AND
C                    RATIO OF SUCCESSIVE INTERVALS (RATIO)
C
       PRINT,        ' INPUT M,ETAINF,DETA,RATIO',
       READ,        M,ETAINF,DETA,RATIO
       NINF         =  2 + ALOG(1. - (1.-RATIO)*ETAINF/DETA)/ALOG(RATIO)
C
C                    SET UP INITIAL GUESS
C
       ETA(1)       =  0.0
       U(1)         =  0.0
       ETA(2)       =  DETA
       DO 100       I = 2,NINF
       U(I)         =  1.0
       DETA         =  DETA*RATIO
       IF (ETA(I) .GE. 1.0)      GO TO 100
       U(I)         =  ETA(I)
   100 ETA(I+1)     =  ETA(I) + DETA
C
C                    ITERATE FOR VELOCITY AND STREAM FUNCTION
C
       IPRINT       =  0
       PRINT 1000
       DO 220       N = 1,25
C
C                    --  SET UP TRIDIAGONAL SYSTEM
C
       B(1)         =  1.0
       C(1)         =  0.0
       D(1)         =  0.0
       F            =  0.0
       IUP          =  NINF - 1
       DO 200       I = 2,IUP
       F            =  F + (ETA(I) - ETA(I-1))*.5*(U(I) + U(I-1))
       FACT1        =  2./(ETA(I+1)-ETA(I))/(ETA(I+1)-ETA(I-1))
       FACT2        =  2./(ETA(I)-ETA(I-1))/(ETA(I+1)-ETA(I-1))
       P            =  .5*F*(M + 1)/(ETA(I+1) - ETA(I-1))
       Q            =  - 2.*M*U(I)
```

```
      R            =  - M*(1. + U(I)**2)
      A(I)         =  FACT2 - P
      B(I)         =  - FACT2 - FACT1 + Q
      C(I)         =  FACT1 + P
  200 D(I)         =  R
      A(NINF)      =  0.0
      B(NINF)      =  1.0
      D(NINF)      =  1.0
C
C                    --  UPDATE SOLUTION FOR VELOCITY
C
      CALL TRIDIG(NINF)
      DUMAX        =  0.0
      F            =  0.0
      DO 210       I = 2,NINF
      DU           =  V(I) - U(I)
      IF (ABS(DU) .GT. DUMAX)DUMAX = ABS(DU)
      U(I)         =  V(I)
      F            =  F + (ETA(I) - ETA(I-1))*.5*(U(I) + U(I-1))
      IF (IPRINT .EQ. 1)  PRINT 1010, ETA(I),F,U(I)
  210 DETA         =  DETA*RATIO
      IF (IPRINT .EQ. 1)     GO TO 300
      IF (DUMAX .LT. 1.E-6)  IPRINT = 1
  220 CONTINUE
  300 DELSTR       =  ETA(NINF) - F
      PRINT 1020, DELSTR
      STOP
 1000 FORMAT(/////,6X,'ETA',8X,'F',9X,'U',/)
 1010 FORMAT(3F10.5)
 1020 FORMAT(///,' DISPLACEMENT THICKNESS =',F10.5)
      END
C=====================================================
      SUBROUTINE TRIDIG(NEQNS)

      (SEE PROGRAM LINBVP)
```

```
C     PROGRAM FSKBOX
C
C        KELLER'S BOX METHOD FOR SIMILAR SOLUTIONS OF
C        LAMINAR INCOMPRESSIBLE BOUNDARY LAYER
C
      COMMON /COEF/ A(3,3),B(3,3),C(3,3),D(3)
      COMMON /DATA/ M,NINF
      COMMON /SOLN/ ETA(50),F(50,3),DF(50,3)
      REAL           M
C
C                    INPUT EXPONENT IN EXTERNAL VELOCITY DISTRIBUTION (M)
C                    APPROXIMATION TO INFINITY (ETAINF),
C                    FIRST INTEGRATION INTERVAL (DETA), AND
C                    RATIO OF SUCCESSIVE INTERVALS (RATIO)
C
      PRINT,       ' INPUT M,ETAINF,DETA,RATIO',
      READ,        M,ETAINF,DETA,RATIO
      RATLOG       =  ALOG(RATIO)
      NINF         =  2 + ALOG(1. - (1.-RATIO)*ETAINF/DETA)/RATLOG
      ETA(1)       =  0.0
      ETA(2)       =  DETA
      DO 100       I = 2,NINF
      DETA         =  DETA*RATIO
```

```
  100 ETA(I+1)    =  ETA(I) + DETA
      CALL INITAL(NINF)
C
C                   ITERATE FOR VELOCITY AND STREAM FUNCTION
C
      IPRINT      =  0
      PRINT 1030
      DO 240      ITRY = 1,25
      RESMAX      =  0.0
      RESRMS      =  0.0
      DO 210      I = 1,NINF
      CALL COFISH(I)
      CALL REDUCE(I)
      DO 200      J = 1,3
      IF (ABS(D(J)) .GT. RESMAX)          RESMAX = ABS(D(J))
  200 RESRMS      =  RESRMS + D(J)**2
  210 CONTINUE
      RESRMS      =  SQRT(RESRMS/(3.*NINF))
      CALL BACKSUB(NINF)
      CHGMAX      =  0.0
      CHGRMS      =  0.0
      DO 230      I = 1,NINF
      DO 220      J = 1,3
      IF (ABS(DF(I,J)) .GT. CHGMAX)       CHGMAX = ABS(DF(I,J))
      CHGRMS      =  CHGRMS + DF(I,J)**2
  220 F(I,J)      =  F(I,J) + DF(I,J)
      IF (IPRINT .EQ. 0)      GO TO 230
      PRINT 1010, ETA(I),(F(I,J),J=1,3)
  230 CONTINUE
      IF (IPRINT .EQ. 1)      GO TO 300
      CHGRMS      =  SQRT(CHGRMS/(3.*NINF))
      PRINT 1040, ITRY,RESMAX,RESRMS,CHGMAX,CHGRMS
      IF (CHGMAX .GT. 1.E-6) GO TO 240
      IPRINT      =  1
      PRINT 1000
  240 CONTINUE
  300 DELSTR      =  ETA(NINF) - F(NINF,1)
      PRINT 1020, DELSTR
      STOP
 1000 FORMAT(/////,6X,'ETA',8X,'F',9X,'U',8X,'TAU',/)
 1010 FORMAT(4F10.5)
 1020 FORMAT(///,' DISPLACEMENT THICKNESS =',F10.5)
 1030 FORMAT(//,'                  CONVERGENCE HISTORY',
     +       //,' TRIAL    RESMAX     RESRMS     CHGMAX     CHGRMS',/)
 1040 FORMAT(I4,4E10.3)
      END
C==========================================================
      SUBROUTINE INITAL(NINF)
C
C                 SET INITIAL GUESSES
C
      COMMON /SOLN/ ETA(50),F(50,3),DF(50,3)
      REAL          M
      DO 110      I = 1,NINF
      IF (ETA(I) .GE. 1.)        GO TO 100
      F(I,1)      =  ETA(I)*ETA(I)*0.5
      F(I,2)      =  ETA(I)
      F(I,3)      =  1.0
```

```
      GO TO 110
  100 F(I,1)      =  ETA(I) - 0.5
      F(I,2)      =  1.
      F(I,3)      =  0.
  110 CONTINUE
      RETURN
      END
C==============================================================
      SUBROUTINE REDUCE(I)
C
C                    REDUCTION PHASE OF DIRECT SOLUTION OF
C                    BLOCK TRIDIAGONAL SYSTEM
C
      COMMON /COEF/ A(3,3),B(3,3),C(3,3),D(3)
      COMMON        COFSAV(50,3,7)
      COMMON /COF/  AA(3,7),MUP
      MUP         =  3
      NUP         =  3
      DO 60       M = 1,MUP
      COFSAV(I,M,1+NUP)       =  D(M)
      DO 50       N= 1,NUP
   50 COFSAV(I,M,N)           =  C(M,N)
   60 CONTINUE
      NUP         =  4
      IF (I .EQ. 1)           GO TO 200
C
C                    ELIMINATE (I-1)TH UNKNOWN FROM ITH EQUATION
C
      LUP         =  3
      DO 120      L = 1,LUP
      DO 120      M = 1,MUP
      AML         =  A(M,L)
      DO 100      N = 1,MUP
  100 B(M,N)      =  B(M,N) - AML*COFSAV(I-1,L,N)
      COFSAV(I,M,1+MUP) = COFSAV(I,M,1+MUP) - AML*COFSAV(I-1,L,1+MUP)
  120 CONTINUE
C
C
C                    FACTOR OUT B FROM ITH EQUATION
C
  200 DO 230      M = 1,MUP
      DO 210      N = 1,MUP
  210 AA(M,N)     =  B(M,N)
      DO 220      N = 1,NUP
  220 AA(M,N+MUP) = COFSAV(I,M,N)
  230 CONTINUE
      CALL GAUSS(4)
      DO 240      M = 1,MUP
      DO 240      N = 1,NUP
  240 COFSAV(I,M,N) = AA(M,N+MUP)
      RETURN
      END
C==============================================================
      SUBROUTINE BACKSUB(IMAX)
C
C                    BACK-SUBSTITUTION PHASE OF DIRECT SOLUTION OF
C                    BLOCK TRIDIAGONAL SYSTEM
C
```

```
      COMMON         COFSAV(50,3,7)
      COMMON /SOLN/  ETA(50),F(50,3),U(50,3)
      MUP          =  3
      DO 100       M = 1,MUP
  100 U(IMAX,M)    =  COFSAV(IMAX,M,1+MUP)
      DO 130       L = 2,IMAX
      I            =  IMAX + 1 - L
      NUP          =  MUP
      DO 120       M = 1,MUP
      TEMP         =  COFSAV(I,M,1+NUP)
      DO 110       N = 1,NUP
  110 TEMP         =  TEMP - COFSAV(I,M,N)*U(I+1,N)
  120 U(I,M)       =  TEMP
  130 CONTINUE
      RETURN
      END
C===========================================================
      SUBROUTINE COFISH(I)
C
C                    SET BLOCKS ON ITH ROW OF BLOCK-TRIDIAGONAL MATRIX
C
      COMMON /COEF/A(3,3),B(3,3),C(3,3),D(3)
      COMMON /DATA/ M,NINF
      COMMON /SOLN/ ETA(50),F(50,3),DF(50,3)
      REAL           M
C
C                    INITIALIZE BLOCKS
C
      DO 110       J = 1,3
      DO 100       K = 1,3
      A(J,K)       =  0.0
      B(J,K)       =  0.0
  100 C(J,K)       =  0.0
  110 D(J)         =  0.0
      IF (I .GT. 1)           GO TO 200
C
C                    BOUNDARY CONDITIONS AT WALL
C
      B(1,1)       =  1.0
      B(2,2)       =  1.0
      GO TO 220
C
C                    FINITE-DIFFERENCE EQUATIONS
C
  200 DETA         =  ETA(I) - ETA(I-1)
      DO 210       J = 1,2
      A(J,J)       =  -1.
      A(J,J+1)     =  - .5*DETA
      B(J,J)       =  1.
      B(J,J+1)     =  - .5*DETA
  210 D(J)         =  .5*DETA*(F(I,J+1) + F(I-1,J+1)) - F(I,J) + F(I-1,J)
      IF (I .EQ. NINF)        GO TO 230
  220 DETA         =  ETA(I+1) - ETA(I)
      B(3,1)       =  .25*DETA*(M + 1.)*F(I,3)
      B(3,2)       =  - DETA*M*F(I,2)
      B(3,3)       =  -1. + .25*DETA*(M + 1.)*F(I,1)
      C(3,1)       =  .25*DETA*(M + 1.)*F(I+1,3)
      C(3,2)       =  - DETA*M*F(I+1,2)
      C(3,3)       =  1. + .25*DETA*(M + 1.)*F(I+1,1)
```

```
      D(3)        =  F(I,3)-F(I+1,3)+.5*DETA*(M*(F(I,2)**2+F(I+1,2)**2)
     +               - M*2.
     +               - .5*(M + 1.)*(F(I,1)*F(I,3) + F(I+1,1)*F(I+1,3)))
      GO TO 300
C
C                    BOUNDARY CONDITION AT INFINITY
C
  230 B(3,2)      =  1.0
  300 RETURN
      END
C=======================================================================
      SUBROUTINE GAUSS(NRHS)

      (SEE PROGRAM DUBLET)
```

```
C     PROGRAM BDYLAY
C
C        SOLUTION OF LAMINAR INCOMPRESSIBLE BOUNDARY LAYER
C        FOR ARBITRARY PRESSURE GRADIENT
C
      COMMON /UETA/ U(100),ETA(100),UM1(100),UM2(100)
      COMMON /NUM/ PI,NX,TAU
      COMMON /REY/ RE
      COMMON /MISC/ M,DELTA,FACT,FACT2,PRINTU,DX,NINF
      DIMENSION     XX(100)
      LOGICAL       PRINTU
      REAL         M
      PI          =  3.1415926535
C
C                    INPUT THICKNESS RATIO OF ELLIPSE
C
      PRINT,      ' INPUT TAU
      READ,       TAU
C
C                    INPUT NUMBER OF STATIONS ALONG AIRFOIL
C
      PRINT,      ' INPUT NX
      READ,       NX
C
C                    FIND DISTANCES BETWEEN NODES ALONG SURFACE
C
      XX(1)       =  0.0
      DO 100      I = 2,NX
      DX          =  X(I) - X(I-1)
      DY          =  Y(I) - Y(I-1)
  100 XX(I)       =  XX(I-1) + SQRT(DX*DX + DY*DY)
      XX(NX+1)    =  XX(NX-2)
C     VE(NX+1)    =  VE(NX-2)
C
C                    INPUT REYNOLDS NUMBER BASED ON REFERENCE V AND L
C
      PRINT,      ' INPUT RE
      READ,       RE
C
C
C                    INPUT APPROXIMATION TO INFINITY (ETAINF),
C                    FIRST INTEGRATION INTERVAL (DETA),
C                    RATIO OF SUCCESSIVE INTERVALS (RATIO)
C
```

```
      PRINT,      ' INPUT ETAINF,DETA,RATIO'
      READ,       ETAINF,DETA,RATIO
      PRINT 1010,
      READ  1020, IANS
      PRINTU      =  (IANS .EQ. 1HY)
C
C                   SET UP MESH AND INITIAL GUESS
C
      ETA(1)      =  0.0
      U(1)        =  0.0
      NINF        =  2 + ALOG(1. - (1.-RATIO)*ETAINF/DETA)/ALOG(RATIO)
      DO 200      I = 2,NINF
      ETA(I)      =  ETA(I-1) + DETA
      DETA        =  DETA*RATIO
      IF (ETA(I) .GE. 1.0) THEN
         U(I)     =  1.0
      ELSE
         U(I)     =  ETA(I)
      ENDIF
  200 CONTINUE
      PRINT 1000
C
C                   FIND INITIAL BOUNDARY LAYER
C
      DO 300      I = 1,NINF
      UM1(I)      =  0.0
  300 UM2(I)      =  0.0
      CALL VDOT(VE(1),VE(2),VE(3),XX(1),XX(2),XX(3),VGRAD)
      M           =  1.0
      DELTA       =  SQRT(1./VGRAD/RE)
      DX          =  XX(2)
      FACT        =  1.0
      FACT2       =  1.0
      CALL INTEG(0.0,0.0)
C
C                   FIND BOUNDARY LAYER PROFILE AT ITH STATION
C
      DO 410      I = 2,NX
      DO 400      J = 1,NINF
      UM2(J)      =  UM1(J)
  400 UM1(J)      =  U(J)
      CALL VDOT(VE(I),VE(I-1),VE(I+1),XX(I),XX(I-1),XX(I+1),VGRAD)
      M           =  XX(I)*VGRAD/VE(I)
      DELTA       =  SQRT(XX(I)/VE(I)/RE)
      IF (I .GT. 2) THEN
         DX       =  XX(I-1) - XX(I-2)
         FACT     =  (XX(I) - XX(I-2))/(XX(I) - XX(I-1))
         FACT2    =  FACT - 1./FACT
      ENDIF
      CALL INTEG(XX(I),VE(I))
  410 CONTINUE
 1000 FORMAT(////,5X,'X',9X,'UE',9X,'M',9X,'THETA',8X,'H',
     +       9X,'CF',/)
 1010 FORMAT(* PRINT OUT DETAILS OF VELOCITY PROFILE (Y/N) *)
 1020 FORMAT(A1)
      END
C==================================================
```

```
      SUBROUTINE INTEG(X,UE)
C
C                  INTEGRATE FOR VELOCITY PROFILE AT X
C
      COMMON /UETA/ U(100),ETA(100),UM1(100),UM2(100)
      COMMON /TRID/ A(251),B(251),C(251),D(251),V(251)
      COMMON /REY/ RE
      COMMON /MISC/ M,DELTA,FACT,FACT2,PRINTU,DX,NINF
      DIMENSION      USS(100)
      REAL           M
      LOGICAL        PRINTU,CONVRG
C
C                  SET UP FOR ITERATIONS
C
      CONVRG     =  .FALSE.
      USS(1)     =  0.0
      DO 100     I = 2,NINF
      USS(I)     =  FACT*UM1(I) - UM2(I)/FACT

  100 CONTINUE
C
C                  ITERATE FOR VELOCITY AND STREAM FUNCTION
C
      DO 230     KOUNT = 1,100
C
C                  --  SET UP TRIDIAGONAL SYSTEM
C
      B(1)       =  1.0
      C(1)       =  0.0
      D(1)       =  0.0
      F          =  0.0
      FSS        =  0.0
      IUP        =  NINF - 1
      DO 210     I = 2,IUP
      HMINUS     =  ETA(I) - ETA(I-1)
      HPLUS      =  ETA(I+1) - ETA(I)
      F          =  F + HMINUS*.5*(U(I) + U(I-1))
      FSS        =  FSS + .5*HMINUS*(USS(I) + USS(I-1))
      P          =  .5*F*(M + 1) + X*(FACT2*F - FSS)/DX
      A(I)       =  (2./HMINUS - P)/(HPLUS + HMINUS)
      C(I)       =  (2./HPLUS + P)/(HPLUS + HMINUS)
      B(I)       =  - A(I) - C(I) - 2.*M*U(I)
     +              - X*(2.*FACT2*U(I) - USS(I))/DX
  210 D(I)       =  - M*(1. + U(I)**2)- X*FACT2*U(I)**2/DX
      A(NINF)    =  0.0
      B(NINF)    =  1.0
      D(NINF)    =  1.0
C
C                  --  UPDATE SOLUTION FOR VELOCITY
C
      CALL TRIDIG(NINF)
      DUMAX      =  0.0
      F          =  0.0
      THETA      =  0.0
      DO 220     I = 2,NINF
      DU         =  V(I) - U(I)
      IF (ABS(DU) .GT. DUMAX)  DUMAX = ABS(DU)
      U(I)       =  V(I)
      IF (U(I) .LT. 0.0)       GO TO 310
      IF (.NOT. CONVRG)        GO TO 220
      H          =  .5*(ETA(I) - ETA(I-1))
```

```
      THETA       =  THETA + H*(U(I)*(1.-U(I)) + U(I-1)*(1.-U(I-1)))
      F           =  F + H*(U(I) + U(I-1))
      IF (PRINTU) PRINT 1020, ETA(I),F,U(I)
  220 CONTINUE
      IF (CONVRG)               GO TO 300
      CONVRG      =  (DUMAX .LT. 1.E-6)
  230 CONTINUE
      PRINT,      ' *** ITERATIONS FAILED TO CONVERGE'
      RETURN
C
C                   FIND DISPLACEMENT AND MOMENTUM THICKNESSES
C                   AND SKIN-FRICTION COEFFICIENT
C
  300 DELSTR      =  (ETA(NINF) - F)*DELTA
      THETA       =  THETA*DELTA
      H           =  DELSTR/THETA
      CF          =  2.*(ETA(3)*U(2)/ETA(2) - ETA(2)*U(3)/ETA(3))
     +               /(ETA(3) - ETA(2))/DELTA/UE/RE
      PRINT 1020, X,UE,M,THETA,H,CF
      IF (ABS(1. - U(NINF-1)) .GT. .01) PRINT 1030
      RETURN
  310 PRINT, ' **** LAMINAR SEPARATION'
      STOP
 1020 FORMAT(3F10.5,F12.7,F10.5,F11.6)
 1030 FORMAT(' **** WARNING: ETAINF PROBABLY TOO SMALL')
      END
C=========================================================
      SUBROUTINE TRIDIG(NEQNS)

      (SEE PROGRAM LINBVP)

C================================================================
      SUBROUTINE VDOT(V1,V2,V3,X1,X2,X3,VGRAD)
C
C                   FIND VELOCITY GRADIENT AT NODES
C
C                   VGRAD = DV/DX AT X1
C
      FACT        =  (X3 - X1)/(X2 - X1)
      VGRAD       =  ((V2 - V1)*FACT - (V3 - V1)/FACT)/(X3 - X2)
      RETURN
      END
C============================================================
      FUNCTION X(I)
C
C                   X FOR ELLIPSE
C
      COMMON /NUM/ PI,NX,TAU
      X           =  - COS(PI*(I - 1)/FLOAT(NX - 1))
      RETURN
      END
C=============================================================
      FUNCTION Y(I)
C
C                   Y FOR ELLIPSE
C
      COMMON /NUM/ PI,NX,TAU
      Y           =  SIN(PI*(I - 1)/FLOAT(NX - 1))*TAU
      RETURN
      END
C============================================================
```

```
      FUNCTION VE(I)
C
C                   VE FOR ELLIPSE
C
      COMMON /NUM/ PI,NX,TAU
      VE           =  (1. + TAU)*SQRT((1. - X(I)**2)
     +                  /(1. - (1.-TAU**2)*X(I)**2))
      RETURN
      END
```

CHAPTER 11

COMPRESSIBLE POTENTIAL FLOW PAST AIRFOILS

It may have surprised you to find air treated as an incompressible fluid, as we have done so far in this book. In fact, of course, air is quite easily compressed. However, as we shall see shortly, air in steady flow has approximately constant density if the flow speed is everywhere small compared to that of a sound wave, which is usually about 300 m/sec. This condition is met in many flows of aerodynamic interest, but not in those related to flight at speeds economical for air transport, to say nothing of flight at the even higher speeds that are technologically feasible. Flight at or above the speed of sound—in the *transonic* and *supersonic* ranges—respectively, introduces phenomena not observed at lower speeds. This chapter will discuss some of those phenomena and methods for analyzing them.

11.1. SHOCK WAVES AND SOUND WAVES

One of the most striking features of inviscid compressible fluid flow is its ability to support discontinuities in the fluid properties.[1] These discontinuities propagate through the flow at a definite, finite speed and are called *shock waves*. Examples include waves created by explosions and, more to the point of this book, waves created by the motion of aircraft.

[1] Such discontinuities dissolve into regions of rapid changes if the viscosity of the fluid is taken into account. However, as is shown in Appendix I, the thickness of a shock wave in a real fluid is extremely small even compared to the thickness of the typical boundary layer. Since, as is also shown in Appendix I, the change in flow properties across the shock is not affected by viscosity, the internal structure of a shock wave is generally an object of only academic interest.

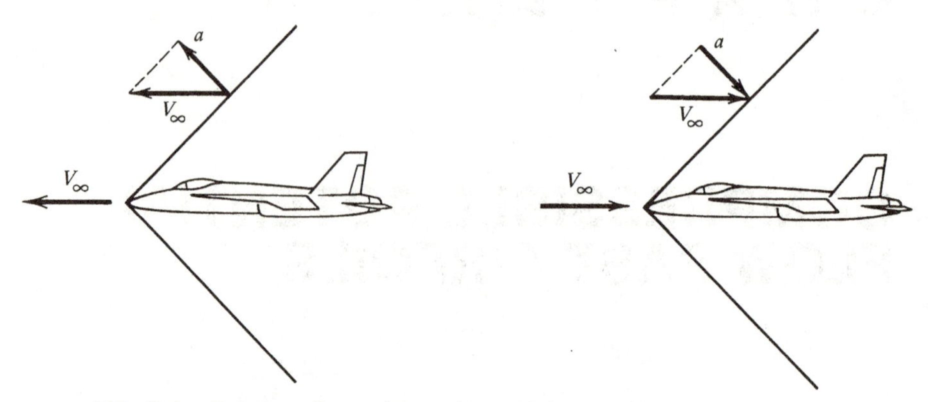

FIG. 11.1. Relation of sound speed to vehicle speed in supersonic flight.

A *sound wave* is a weak shock wave, one that creates relatively small discontinuities. The *speed of sound* is the speed at which a sound wave moves normal to itself. As we shall see, there are remarkable differences between steady flows in which the flow speed is everywhere less than the speed of sound (*subsonic*) and flows in which the speed is at least partly *supersonic*.

We will now compute the speed of sound in a perfect gas. For later convenience, we shall do so in the context of a wave created by a body moving at a constant supersonic speed through still air. As shown on the left side of Fig. 11.1, the sound speed—which, if the wave is weak enough to be called a sound wave, equals the speed of the wave normal to itself, relative to the undisturbed air ahead—equals the component of the vehicle speed normal to the wave. Then, if we adopt a frame of reference fixed to the aircraft, in which the flow appears steady, the sound speed is the normal component of the onset flow velocity, as shown on the right side of Fig. 11.1.

Let us consider, then, the steady flow through a discontinuity that is inclined at an angle θ_w to a uniform onset flow. Our analysis will be applicable even to curved waves in nonuniform flows, since we can concentrate our attention on a part of the wave that is small enough that the flow is locally uniform and the wave locally plane.

As indicated in Fig. 11.2, we shall use the subscript "1" for flow properties upstream of the discontinuity and "2" for properties downstream. The angle

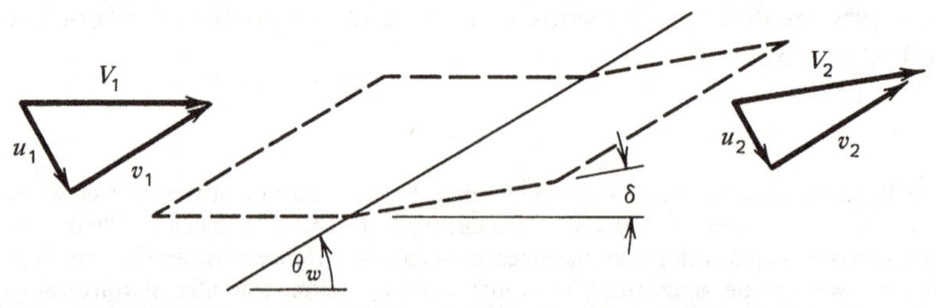

FIG. 11.2. Control volume for analysis of steady flow through a shock (or sound) wave.

through which the flow is deflected as it goes through the wave is δ, whereas u, v are the velocity components normal and tangential to the wave, respectively.

We shall analyze the flow in a control volume that straddles the wave and whose surfaces are parallel either to the wave or to the local flow direction, again as shown in Fig. 11.2. Let the area of the intersection of the control volume with the wave be A. Note that the only control surfaces through which mass, momentum, and energy flow are those parallel to the discontinuity and that both inlet and outlet surfaces have area A.

Now applying mass conservation to the control volume, we get, from equation 2-7,

$$\rho_1 u_1 A = \rho_2 u_2 A \equiv \dot{m} \qquad (11\text{-}1)$$

There is a net force on the control volume in the direction normal to the wave, so that the normal component of Newton's law, equation 2-8, gives

$$p_1 A - p_2 A = \dot{m}(u_2 - u_1)$$

or, using equation 11-1,

$$p_2 - p_1 = \rho_1 u_1 (u_1 - u_2) \qquad (11\text{-}2)$$

However, there is no net force in the direction along the wave, and so

$$0 = \dot{m}(v_2 - v_1) \qquad (11\text{-}3)$$

The work done by the pressure forces on the ends of the control volume produces a net energy flux through the system. Thus, from equation 2-9,

$$p_1 u_1 A - p_2 u_2 A = \dot{m}[e_2 + \tfrac{1}{2}u_2^2 + \tfrac{1}{2}v_2^2 - (e_1 + \tfrac{1}{2}u_1^2 + \tfrac{1}{2}v_1^2)] \qquad (11\text{-}4)$$

For a perfect gas with constant specific heats, thermodynamics and equation 2-11 show that

$$\begin{aligned} e = c_v T &= c_v \frac{p}{\rho R} \\ &= \frac{c_v}{c_p - c_v}\frac{p}{\rho} \\ &= \frac{1}{\gamma - 1}\frac{p}{\rho} \end{aligned} \qquad (11\text{-}5)$$

in which T is the absolute temperature, R the gas constant, c_p and c_v the specific heats at constant pressure and volume, respectively, and γ their ratio:

$$\gamma \equiv \frac{c_p}{c_v} \qquad (11\text{-}6)$$

For air at ordinary temperatures (under about 400 °F), γ is close to 1.4. Then, from equations 11-1 and 11-5,

$$p_1 u_1 A + \dot{m} e_1 = A\left(p_1 u_1 + \rho_1 u_1 \frac{1}{\gamma - 1} \frac{p_1}{\rho_1} \right)$$

$$= A \frac{\gamma}{\gamma - 1} p_1 u_1$$

The same equation holds with the subscript "1" changed to "2." Thus, using equation 11-3 as well, we can put equation 11-4 in the form

$$\frac{\gamma}{\gamma - 1}(p_2 u_2 - p_1 u_1) = \tfrac{1}{2}\rho_1 u_1 (u_1^2 - u_2^2) \tag{11-7}$$

Thus far we have not restricted the wave as to its strength. Let us now assume, as is appropriate for a sound wave, that

$$\Delta u \equiv u_2 - u_1 \ll V_1$$

$$\Delta p \equiv p_2 - p_1 \ll p_1$$

$$\Delta \rho \equiv \rho_2 - \rho_1 \ll \rho_1$$

Then, as argued above, u_1, the component of the onset flow velocity normal to the sound wave, becomes the sound speed a.

Equation 11-2 may now be written

$$\Delta p \cong -\rho_1 a \Delta u \tag{11-8}$$

whereas equation 11-7 is approximated as follows:

$$\frac{\gamma}{\gamma - 1}[(p_1 + \Delta p)(a + \Delta u) - p_1 a] \cong \tfrac{1}{2}\rho_1 a[a^2 - (a + \Delta u)^2]$$

or

$$\frac{\gamma}{\gamma - 1}(p_1 \Delta u + a \Delta p) \cong -\rho_1 a^2 \Delta u$$

Eliminating Δp with equation 11-8, we get

$$\frac{\gamma}{\gamma - 1}(p_1 - \rho_1 a^2)\Delta u \cong -\rho_1 a^2 \Delta u$$

or

$$\gamma p_1 \cong [1 + (\gamma - 1)]\rho_1 a^2$$

and we have the formula we sought for the speed of sound in a perfect gas:

$$a^2 = \frac{\gamma p}{\rho} \tag{11-9}$$

11.2. EQUATIONS OF COMPRESSIBLE STEADY POTENTIAL FLOW

Although this chapter is not intended to be a primer on compressible flow, we will at least take the time to derive the equations that will be found most useful. We consider only inviscid irrotational flows. To reiterate the argument made in Chapter 2, a fluid particle whose angular velocity is initially zero will not start to rotate if subjected only to normal forces; you'd need shear forces, which are zero in the nonviscous approximation.[2]

Starting with the irrotational flow relation

$$\text{curl}\,\mathbf{V} = 0 \tag{11-10}$$

we can derive a compressible version of Bernoulli's equation from the Euler equations. For a steady flow, the x component of equation 2-32 is

$$\rho\left(u\frac{\partial u}{\partial x} + v\frac{\partial u}{\partial y} + w\frac{\partial u}{\partial z}\right) + \frac{\partial p}{\partial x} = 0 \tag{11-11}$$

From the y and z components of equation 11-10,

$$\frac{\partial u}{\partial y} = \frac{\partial v}{\partial x}, \qquad \frac{\partial u}{\partial z} = \frac{\partial w}{\partial x}$$

so equation 11-11 can be written

$$\rho\left(u\frac{\partial u}{\partial x} + v\frac{\partial v}{\partial x} + w\frac{\partial w}{\partial x}\right) + \frac{\partial p}{\partial x} = \frac{1}{2}\rho\frac{\partial V^2}{\partial x} + \frac{\partial p}{\partial x} = 0$$

Similar equations follow from the y and z components of equation 2-32, and so we write

$$\frac{\rho}{2}dV^2 + dp = 0 \tag{11-12}$$

From problem 2 of Chapter 2, for steady flow,

$$\mathbf{V}\cdot\nabla\left(e + \frac{1}{2}V^2 + \frac{p}{\rho}\right) = 0$$

Thus

$$h + \tfrac{1}{2}V^2$$

is constant along streamlines, where

$$h \equiv e + \frac{p}{\rho} \tag{11-13}$$

[2] An initially irrotational flow can become rotational if it passes through a strong, curved shock wave. The shock waves with which this chapter deals will be supposed to be too weak to rotate the flow. This condition is satisfied if the local Mach number of the flow relative to the shock is less than about 1.3.

is the *enthalpy* of the fluid. Then, for flows that are uniform far from the body,

$$h + \tfrac{1}{2}V^2 = \text{constant} \tag{11-14}$$

throughout the fluid. This quantity is called the *stagnation enthalpy*, since it is the value of the enthalpy at stagnation points (where $V = 0$).

For a perfect gas, from equations 11-13 and 11-5,

$$h = \frac{\gamma}{\gamma - 1}\frac{p}{\rho} \tag{11-15}$$

Combining equations 11-12, 11-14, and 11-15, we get

$$\begin{aligned} dp &= -\tfrac{1}{2}\rho\, dV^2 \\ &= \rho\, dh \\ &= \rho\frac{\gamma}{\gamma - 1}\left(\frac{dp}{\rho} - \frac{p}{\rho^2}\, d\rho\right) \end{aligned}$$

which simplifies to

$$\frac{dp}{p} = \gamma\frac{d\rho}{\rho}$$

whose solution is the *isentropic relationship*

$$p/\rho^\gamma = \text{constant} \tag{11-16}$$

It is convenient to evaluate the constants in equations 11-14 and 11-16 far upstream, where conditions will be indicated by the subscript ∞. Thus, from equations 11-14 and 11-15, we get

$$\frac{\gamma}{\gamma - 1}\frac{p}{\rho} + \frac{1}{2}V^2 = \frac{\gamma}{\gamma - 1}\frac{p_\infty}{\rho_\infty} + \frac{1}{2}V_\infty^2 \tag{11-17}$$

while equation 11-16 becomes

$$\frac{p}{\rho^\gamma} = \frac{p_\infty}{\rho_\infty^\gamma} \tag{11-18}$$

Introducing the sound speed a from equation 11-9, we can rewrite equation 11-17 as

$$\frac{1}{\gamma - 1}a^2 + \frac{1}{2}V^2 = \frac{1}{\gamma - 1}a_\infty^2 + \frac{1}{2}V_\infty^2 \tag{11-19}$$

Also, we can eliminate p from equation 11-9 with the aid of equation 11-18:

$$\begin{aligned} a^2 &= \frac{\gamma p_\infty}{\rho_\infty}\cdot\frac{p}{p_\infty}\frac{\rho_\infty}{\rho} \\ &= a_\infty^2(\rho/\rho_\infty)^{\gamma-1} \end{aligned} \tag{11-20}$$

A pressure formula similar to Bernoulli's equation can now be derived, as follows. From equations 11-18 and 11-20,

$$\frac{p}{p_\infty} = \left(\frac{a}{a_\infty}\right)^{2\gamma/(\gamma-1)}$$

Then, using equation 11-19 to eliminate a^2, we get

$$\frac{p}{p_\infty} = \left[1 + \frac{\gamma - 1}{2}\frac{V_\infty^2 - V^2}{a_\infty^2}\right]^{\gamma/(\gamma-1)} \tag{11-21}$$

which may be regarded as the analog for a perfect gas of the Bernoulli equation 2-37.

For a steady two-dimensional compressible flow, the differential continuity equation 2-26 can be written

$$\begin{aligned} 0 &= \frac{\partial}{\partial x}\rho u + \frac{\partial}{\partial y}\rho v \\ &= \rho\left(\frac{\partial u}{\partial x} + \frac{\partial v}{\partial y}\right) + u\frac{\partial \rho}{\partial x} + v\frac{\partial \rho}{\partial y} \end{aligned}$$

so

$$\begin{aligned} \frac{\partial u}{\partial x} + \frac{\partial v}{\partial y} &= -\frac{u}{\rho}\frac{\partial \rho}{\partial x} - \frac{v}{\rho}\frac{\partial \rho}{\partial y} \\ &= -u\frac{\partial \ln \rho}{\partial x} - v\frac{\partial \ln \rho}{\partial y} \end{aligned} \tag{11-22}$$

But, from equation 11-20,

$$\ln a^2 = \ln a_\infty^2 + (\gamma - 1)\ln \rho - (\gamma - 1)\ln \rho_\infty$$

so

$$d \ln \rho = \frac{1}{\gamma - 1} d \ln a^2 \tag{11-23}$$

and, from equation 11-19,

$$\begin{aligned} d \ln a^2 &= d \ln\left[a_\infty^2 + \frac{\gamma - 1}{2}(V_\infty^2 - V^2)\right] \\ &= -\left(\frac{\gamma - 1}{2} dV^2 \Big/ a^2\right) \end{aligned} \tag{11-24}$$

Thus, from equations 11-22 to 11-24,

$$\begin{aligned} \frac{\partial u}{\partial x} + \frac{\partial v}{\partial y} &= \frac{u}{2a^2}\frac{\partial V^2}{\partial x} + \frac{v}{2a^2}\frac{\partial V^2}{\partial y} \\ &= \frac{u}{a^2}\left(u\frac{\partial u}{\partial x} + v\frac{\partial v}{\partial x}\right) + \frac{v}{a^2}\left(u\frac{\partial u}{\partial y} + v\frac{\partial v}{\partial y}\right) \end{aligned}$$

since $V^2 = u^2 + v^2$. This simplifies to

$$\left(1 - \frac{u^2}{a^2}\right)\frac{\partial u}{\partial x} + \left(1 - \frac{v^2}{a^2}\right)\frac{\partial v}{\partial y} = \frac{uv}{a^2}\left(\frac{\partial v}{\partial x} + \frac{\partial u}{\partial y}\right) \tag{11-25}$$

Note that, if u and v are small compared to a, equation 11-25 reduces to

$$\frac{\partial u}{\partial x} + \frac{\partial v}{\partial y} = 0$$

the form of the continuity equation for incompressible flow.

11.3. THE PRANDTL–GLAUERT EQUATION

Consider now the uniform steady flow of a perfect gas past a thin airfoil. If, as we assume, the flow is irrotational, the velocity can be expressed as the gradient of a velocity potential Φ, so

$$u = \frac{\partial \Phi}{\partial x}, \qquad v = \frac{\partial \Phi}{\partial y} \tag{11-26}$$

Since a^2 is given in terms of $V^2 = u^2 + v^2$ by equation 11-19, we can reduce equation 11-25 to a partial differential equation (too horrendous to write down here) for Φ. Boundary conditions are, as in Chapters 3 and 4, that

$$\nabla\Phi \to \mathbf{V}_\infty \tag{11-27}$$

far from the body, flow tangency

$$\frac{v}{u} = \frac{dY}{dx} \tag{11-28}$$

on the body, and (perhaps) a Kutta condition to fix the circulation.

For flow past a thin airfoil, we can, as in Chapter 4, simplify the flow tangency condition to

$$\frac{v}{V_\infty} \cong \frac{dY}{dx} \qquad \text{at } y = 0\pm \tag{11-29}$$

Also, it is convenient to introduce a *perturbation potential* ϕ by writing

$$\Phi = \Phi_\infty + \varepsilon\phi \tag{11-30}$$

ε being some dimensionless measure of the airfoil's thickness, camber, and/or angle of attack, and Φ_∞ being the potential of the onset flow. Then

$$\nabla\Phi_\infty = \mathbf{V}_\infty = V_\infty(\cos\alpha\hat{\mathbf{i}} + \sin\alpha\hat{\mathbf{j}})$$

where α is the angle of attack of the flow far upstream, see Fig. 11.3, so

$$\begin{aligned}\Phi_\infty &= V_\infty(x\cos\alpha + y\sin\alpha)\\ &\cong V_\infty(x + \alpha y)\end{aligned}$$

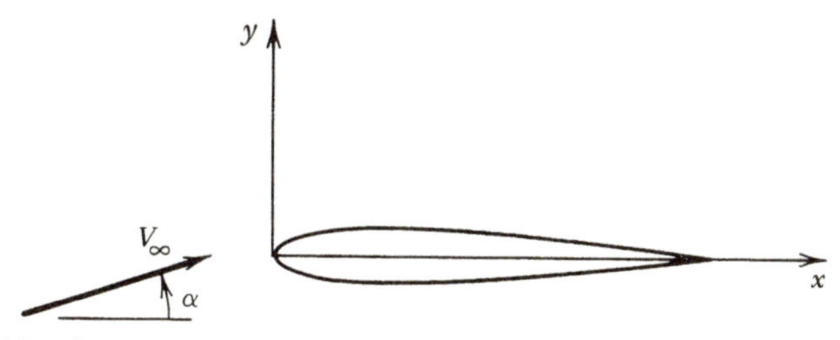

FIG. 11.3. Nomenclature for analysis of compressible flow past an airfoil.

if α is small, as we shall suppose. Then

$$u \simeq V_\infty + \varepsilon \frac{\partial \phi}{\partial x}$$

$$v \simeq V_\infty \alpha + \varepsilon \frac{\partial \phi}{\partial y} \tag{11-31}$$

Substituting equations 11-31 into 11-19 and dropping terms quadratic in the small quantities α and ε, we get

$$\begin{aligned} a^2 &\simeq a_\infty^2 + \frac{\gamma - 1}{2} V_\infty^2 - \frac{\gamma - 1}{2}\left[V_\infty^2 + 2\varepsilon V_\infty \frac{\partial \phi}{\partial x} + \varepsilon^2 \left(\frac{\partial \phi}{\partial x}\right)^2\right] \\ &\quad - \frac{\gamma - 1}{2}\left[V_\infty^2 \alpha^2 + 2V_\infty \alpha\varepsilon \frac{\partial \phi}{\partial y} + \varepsilon^2 \left(\frac{\partial \phi}{\partial y}\right)^2\right] \\ &\simeq a_\infty^2 - (\gamma - 1)\varepsilon V_\infty \frac{\partial \phi}{\partial x} \end{aligned}$$

The differential equation 11-25 then becomes

$$\begin{aligned} &\left[a_\infty^2 - (\gamma - 1)\varepsilon V_\infty \frac{\partial \phi}{\partial x} - V_\infty^2 - 2\varepsilon V_\infty \frac{\partial \phi}{\partial x} + O(\alpha^2, \varepsilon^2)\right]\varepsilon \frac{\partial^2 \phi}{\partial x^2} \\ &\qquad + \left[a_\infty^2 - (\gamma - 1)\varepsilon V_\infty \frac{\partial \phi}{\partial x} + O(\varepsilon^2, \alpha^2)\right]\varepsilon \frac{\partial^2 \phi}{\partial y^2} \\ &\qquad\qquad = -\left(V_\infty + \varepsilon \frac{\partial \phi}{\partial x}\right)\left(V_\infty \alpha + \varepsilon \frac{\partial \phi}{\partial y}\right) 2\varepsilon \frac{\partial^2 \phi}{\partial x \partial y} \end{aligned}$$

or, to a first approximation,

$$(a_\infty^2 - V_\infty^2)\frac{\partial^2 \phi}{\partial x^2} + a_\infty^2 \frac{\partial^2 \phi}{\partial y^2} = 0$$

Dividing through by a_∞^2, we get

$$(1 - M_\infty^2)\frac{\partial^2 \phi}{\partial x^2} + \frac{\partial^2 \phi}{\partial y^2} = 0 \tag{11-32}$$

where

$$M_\infty \equiv V_\infty / a_\infty \tag{11-33}$$

is the *Mach number* of the onset flow.

Equation 11-32 is called the *Prandtl–Glauert equation*. From equations 11-26, 11-27, 11-29, and 11-31, the boundary conditions on its solution are

$$\nabla\phi \to 0 \qquad \text{as } x^2 + y^2 \to \infty$$

$$\varepsilon\frac{\partial\phi}{\partial y} = V_\infty\left(\frac{dY}{dx} - \alpha\right) \qquad \text{at } y = 0\pm \tag{11-34}$$

It must also satisfy a Kutta condition.

Note that, as $M_\infty \to 0$, the Prandtl–Glauert equation reduces to the familiar Laplace equation. However, unlike the Laplace equation, it is valid only for flows that are nearly uniform and directed along the x axis.[3]

When $V \cong V_\infty$, the pressure formula (11-21) may be simplified to

$$\frac{p}{p_\infty} \cong 1 + \frac{\gamma}{2}\frac{V_\infty^2 - V^2}{a_\infty^2}$$

$$\cong 1 + \frac{\gamma}{2}\left[V_\infty^2 - \left(V_\infty^2 + 2\varepsilon V_\infty\frac{\partial\phi}{\partial x}\right)\right]\Big/ a_\infty^2 = 1 - \gamma\varepsilon\frac{V_\infty}{a_\infty^2}\frac{\partial\phi}{\partial x}$$

in which we have used equation 11-31 and dropped terms quadratic in α and/or ε. Then the pressure coefficient is

$$C_p \equiv \frac{p - p_\infty}{\frac{1}{2}\rho_\infty V_\infty^2}$$

$$\simeq -\frac{\gamma\varepsilon p_\infty V_\infty}{\frac{1}{2}\rho_\infty a_\infty^2 V_\infty^2}\frac{\partial\phi}{\partial x} \tag{11-35}$$

from which a_∞^2 may be eliminated via equation 11-9 to yield

$$C_p \simeq -\frac{2}{V_\infty}\varepsilon\frac{\partial\phi}{\partial x} \tag{11-36}$$

just as in incompressible flow (compare equation 3-35).

11.4. SUBSONIC FLOW PAST THIN AIRFOILS

For a subsonic flow, the Prandtl–Glauert equation 11-32 is so close to Laplace's equation that its solutions can be related to those of Laplace's equation. However, the related solutions are about airfoils that differ in size by a scale factor.

[3] Also in contrast with the Laplace equation, the Prandtl–Glauert equation attaches special significance to the x direction; it is anisotropic. Generally, the small-perturbation assumptions are most accurate if the x axis is parallel to the onset flow. In this chapter, however, we linearize the boundary conditions and apply them on the chord line of the airfoil, in which case no accuracy is lost by choosing the x axis to be the chord line, which we do, for convenience.

To show this, we first remove the Mach number from the partial differential equation 11-32 by introducing a change of coordinates, namely,

$$\xi \equiv x/\beta$$

$$\eta \equiv y$$

$$\tilde{\phi}(\xi, \eta) \equiv \phi(x, y) \tag{11-37}$$

where

$$\beta \equiv \sqrt{1 - M_\infty^2} \tag{11-38}$$

Then, from the chain rule, we find

$$\frac{\partial^2 \tilde{\phi}}{\partial \xi^2} + \frac{\partial^2 \tilde{\phi}}{\partial \eta^2} = 0 \tag{11-39}$$

Letting

$$Y(x) = \beta \tilde{Y}(\xi) \tag{11-40}$$

we can also remove M_∞ from the boundary conditions (11-34):

$$\nabla \tilde{\phi} \to 0 \qquad \text{as } \xi^2 + \eta^2 \to \infty$$

$$\varepsilon \frac{\partial \tilde{\phi}}{\partial \eta} = V_\infty \left(\frac{d\tilde{Y}}{d\xi} - \alpha \right) \qquad \text{at } \eta = 0\pm \tag{11-41}$$

Comparing equation 11-39 with 11-32, and equations 11-41 with 11-34, we see that $\tilde{\phi}$ is the perturbation potential of an incompressible flow (flow at $M_\infty = 0$) about a body whose shape is, from equations 11-37 and 11-40, the same as the body of interest, but which is blown up in both longitudinal and lateral dimensions by a factor of $1/\beta$, as shown in Fig. 11.4.

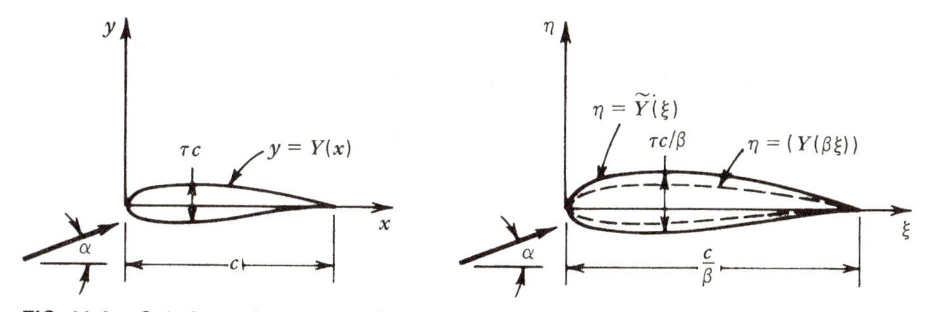

FIG. 11.4. Solution of subsonic-flow problems according to the Prandtl–Glauert equation.

Example 1: Ellipse at Zero Angle of Attack Consider the subsonic flow past the ellipse shown in Fig. 11.5, whose shape is described by

$$Y(x) = \pm \tau \sqrt{1 - x^2}$$

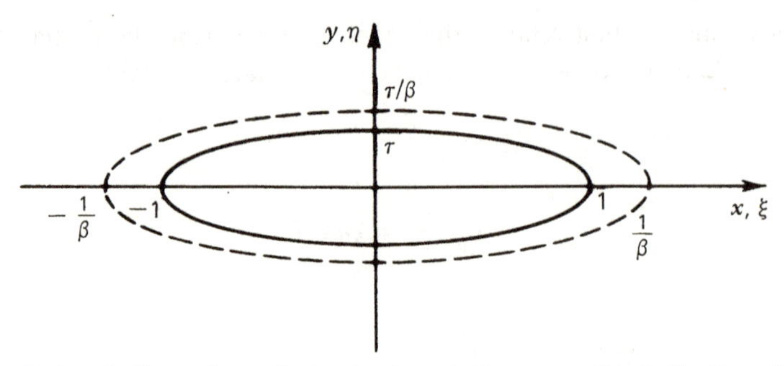

FIG. 11.5. Related ellipses in analysis of subsonic flow according to the Prandtl–Glauert equation.

According to equation 11-37, the perturbation potential of this flow equals that of the incompressible flow about the scaled-up ellipse (also shown in Fig. 11.5)

$$\tilde{Y}(\xi) = \pm\frac{\tau}{\beta}\sqrt{1-\beta^2\xi^2}$$

whose thickness ratio is also τ. From equation 3-41, the perturbation surface velocity on such an ellipse in such a flow is (in the present nomenclature)

$$\varepsilon\frac{\partial\tilde{\phi}}{\partial\xi}\bigg|_{\eta=0\pm} = V_\infty\tau$$

From the chain rule and equations 11-37, we get

$$\frac{\partial\phi}{\partial x} = \frac{\partial\tilde{\phi}}{\partial\xi}\frac{\partial\xi}{\partial x} = \frac{1}{\beta}\frac{\partial\tilde{\phi}}{\partial\xi} \tag{11-42}$$

Thus the perturbation velocity on the ellipse in a subsonic flow is

$$\varepsilon\frac{\partial\phi}{\partial x}\bigg|_{y=0} = \frac{1}{\beta}V_\infty\tau$$

and the corresponding pressure coefficient is, from equation 11-36

$$C_p\bigg|_{y=0\pm} = -2\frac{\tau}{\beta}$$

Equation 11-42 is a general mathematical result, not restricted to flows past ellipses. Substituting it into equation 11-36 yields

$$C_p(x, y) \simeq \frac{1}{\beta}\left(-2\varepsilon\frac{\partial\tilde{\phi}}{\partial\xi}\right) \equiv \frac{1}{\beta}\tilde{C}_p(\xi, \eta) \tag{11-43}$$

Now, because the body shape in the ξ, η plane $\tilde{Y}(\xi)$ changes with the Mach number of the onset flow only by a scale factor, the transformed perturbation velocity components

$$\varepsilon \frac{\partial \tilde{\phi}}{\partial \xi}, \qquad \varepsilon \frac{\partial \tilde{\phi}}{\partial \eta}$$

are, at corresponding points in the transformed plane (that is, at points with coordinates $\xi = x/\beta, \eta = y/\beta$), independent of Mach number. Strictly speaking, this does not mean that those velocity components (or, from equation 11-43, the transformed pressure coefficient $\tilde{C}_p$) are independent of Mach number at points corresponding to the body surface, because a point on the body surface in the x–y plane $[x, Y(x)]$ maps through equations 11-37 onto $[x/\beta, Y(x)]$, which is not on the body contour $\tilde{Y}(\xi)$ to which the solution of equations 11-39 and 11-41 refer; see Fig. 11.4. However, in thin-airfoil theory, one can approximate the velocity and pressure on the body surface by their values on either side of the x axis. Since the point $(x, 0\pm)$ does map onto $(x/\beta, 0\pm)$, we see from equation 11-43 that the pressure distribution on a thin airfoil in subsonic flow is $1/\beta$ times what can be recognized as the pressure distribution on an identically shaped body in an incompressible flow. The jumps in pressure across the bodies are related by the same factor, and so, then, are their lift coefficients:

$$c_l \approx \frac{1}{\beta} \tilde{c}_l \tag{11-44}$$

Here $\tilde{c}_l$ is the lift coefficient of the same airfoil at the same angle of attack in an incompressible flow. Thus, the lift coefficient of an airfoil increases with the Mach number of the onset flow and is proportional to the lift at zero Mach number.[4] This is called the *Prandtl–Glauert rule* [1].

11.5. SUPERSONIC FLOW PAST THIN AIRFOILS

If $M_\infty > 1$, the β defined by equation 11-38 is imaginary, and the transformations (11-37) and (11-40) do not make any sense. Also, the first of the boundary conditions (11-41) is troublesome, since $\xi^2 + \eta^2$ can be small—even zero—very far from the airfoil. Thus the result of equation 11-44 for the dependence of an airfoil's lift on the Mach number of the onset flow is valid only if the Mach number is subsonic.

However, as we shall see, the lift of an airfoil in supersonic flow is even easier to calculate than its value in incompressible flow, and supersonic aerodynamics is

[4] In slender-body theory, the counterpart of thin-airfoil theory that applies to fuselage-like bodies, one must evaluate the pressure on the body surface, because

$$\lim_{r \to 0} \frac{\partial \phi}{\partial x}$$

does not exist; see problem 15 of Chapter 3. However, it is possible to relate the pressure on a slender body in subsonic flow to the incompressible-flow result for a body whose shape differs from the physical body by being stretched only in the x direction. That is, the β is dropped from equation 11-40. Thus the thickness ratio of the body dealt with in the transformed space is β times that of the real body. This is called *Gothert's rule* [1].

actually remarkably simple. Let

$$B \equiv \sqrt{M_\infty^2 - 1} \tag{11-45}$$

so that equation 11-32 becomes

$$B^2 \frac{\partial^2 \phi}{\partial x^2} - \frac{\partial^2 \phi}{\partial y^2} = 0 \tag{11-46}$$

If $f(t)$ is any twice-differentiable function of t, application of the chain rule shows that

$$f(x - By)$$

solves equation 11-46. Similarly, if $g(t)$ is twice differentiable,

$$g(x + By)$$

also satisfies equation 11-46. We will show that any solution of equation 11-46 can be decomposed into two functions of one variable each:

$$\phi(x, y) = f(x - By) + g(x + By) \tag{11-47}$$

The proof will consist of constructing the functions f and g that meet the boundary conditions on ϕ for supersonic flow past an airfoil, equations 11-34.

To facilitate this task, Fig. 11.6 shows some of the lines on which $x \pm By$ is constant. Note that f is constant on the lines whose slope dy/dx is $+1/B$, whereas g is constant on those of slope $-1/B$.

The lines sketched in Fig. 11.6 divide the flow into zones, which have been numbered. Except in zones 5 and 8, a line on which $x - By$ is constant can be drawn that connects any given point in the flow field with points infinitely far upstream of

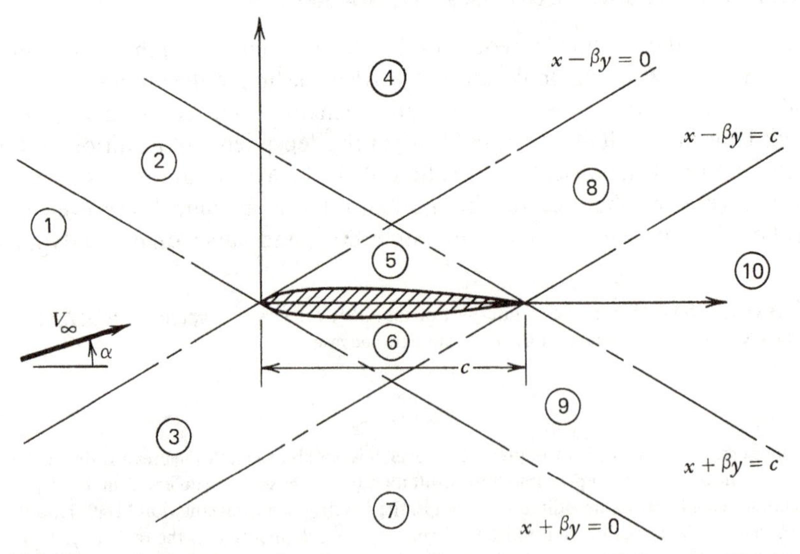

FIG. 11.6. Important zones of supersonic flow past an airfoil according to the Prandtl–Glauert equation.

the airfoil. Thus the value of $f(x - By)$ in most of the flow field can be deduced by examining its value far upstream. The problem in zones 5 and 8 is that the airfoil cuts through such a line. Since equation 11-46 does not hold "inside" the airfoil, the value of $f(x - By)$ could change across the airfoil and so need not be the same in zones 5 and 6. Similarly, the value of $g(x + By)$ is the same as its value far upstream of the airfoil, except in zones 6 and 9.

From equation 11-47,

$$\frac{\partial \phi}{\partial x} = f'(x - By) + g'(x + By)$$

$$\frac{\partial \phi}{\partial y} = -Bf'(x - By) + Bg'(x + By) \qquad (11\text{-}48)$$

where the primes denote differentiation with respect to the indicated arguments. Since the perturbation velocity components must vanish far upstream of the airfoil, we may solve equation 11-48 for the values of f' and g' there, namely, zero. Thus f and g are constant far upstream. Since the perturbation potential is the sum of f and g, and since only derivatives of ϕ are of interest, the values of these constants may be set arbitrarily. For convenience, we take

$$f = g = 0$$

far upstream. Then, since f is constant on lines $x - By = \text{constant}$, f is zero everywhere that can be reached by a line with slope

$$\frac{dy}{dx} = \frac{1}{B}$$

from a point far upstream; that is, everywhere except in zones 5 and 8 of Fig. 11.6. Similarly, g is nonzero only in zones 6 and 9, so, from equation 11-47,

$$\begin{aligned} \phi &= f(x - By) \qquad && \text{in zones 5 and 8} \\ &= g(x + By) && \text{in zones 6 and 9} \\ &= 0 && \text{elsewhere} \end{aligned} \qquad (11\text{-}49)$$

Where they are not zero, the values of f and g can be found from the flow tangency condition (11-34). Since, from equation 11-49,

$$\frac{\partial \phi}{\partial y} = -Bf'(x - By)$$

in zone 5, we have, at $y = 0+$ and $0 < x < c$,

$$-\varepsilon Bf'(x) = V_\infty \left[\frac{dY_u}{dx}(x) - \alpha \right] \qquad (11\text{-}50)$$

where $y = Y_u(x)$ locates the upper surface of the airfoil. Similarly,

$$\frac{\partial \phi}{\partial y} = Bg'(x + By)$$

in zone 6, and so, from equation 11-34,

$$\varepsilon B g'(x) = V_\infty \left[\frac{dY_l}{dx}(x) - \alpha \right] \tag{11-51}$$

where $y = Y_l(x)$ locates the airfoil's lower surface. Then, from equations 11-31 and 11-48–11-51, the fluid velocity components are

$$\begin{aligned} u &= V_\infty \left[1 - \frac{1}{B}\{ Y_u'(x - By) - \alpha \} \right] && \text{in zones 5 and 8} \\ &= V_\infty \left[1 + \frac{1}{B}\{ Y_l'(x + By) - \alpha \} \right] && \text{in zones 6 and 9} \\ &= V_\infty && \text{elsewhere} \\ v &= V_\infty Y_u'(x - By) && \text{in zones 5 and 8} \\ &= V_\infty Y_l'(x + By) && \text{in zones 6 and 9} \\ &= V_\infty \alpha && \text{elsewhere} \end{aligned} \tag{11-52}$$

As shown in Fig. 11.7, the streamlines of this flow are either parallel to the onset flow direction or to the airfoil's upper or lower surface. The onset flow is undisturbed until hit by waves that emanate from the airfoil's leading edge and have slope $\pm 1/B$. Its velocity then changes abruptly to satisfy the flow-tangency condition and later reverts to its initial state when hit by waves that emanate from the body's trailing edge.

Note that the flows above and below the airfoil are completely independent. This is, in part, a result of approximations made in our analysis. If the airfoil's leading edge is blunt, the waves detach from the airfoil, as shown in Fig. 11.8, and the flow is

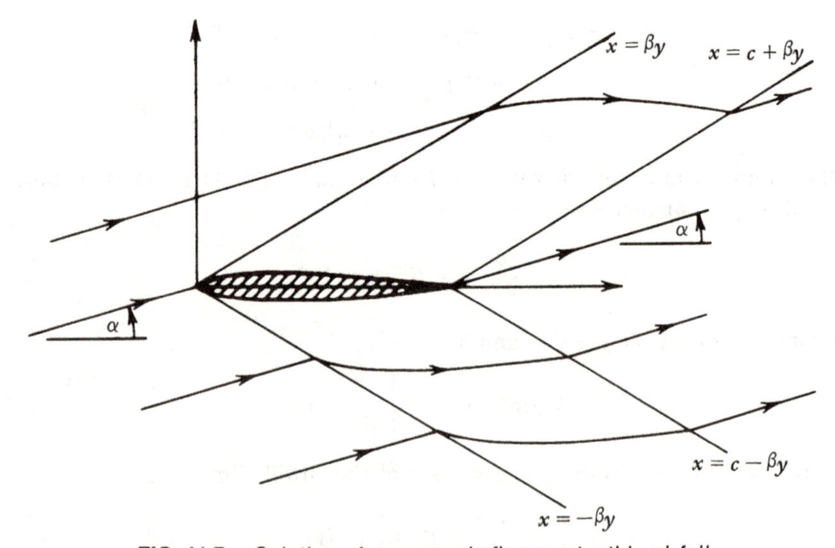

FIG. 11.7. Solution of supersonic flow past a thin airfoil.

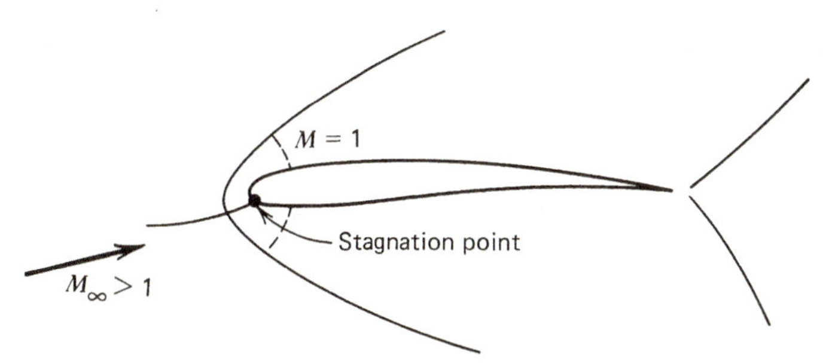

FIG. 11.8. Supersonic flow past a blunt-nosed airfoil.

subsonic near the nose and includes a stagnation point. As usual, this invalidates the small-perturbation theory, at least locally. However, if the leading edge is sufficiently sharp, the flow is everywhere supersonic, and the independence of the flows on either side of the airfoil is realized. In a purely supersonic flow, what is done at a particular point influences the flow elsewhere only in a wedge- (or, in three dimensions, cone-) shaped region behind the point that is bounded by the sound waves that are (or could be) sent out from that point. Conversely, the flow properties at such a point are influenced only by what happens in an upstream-facing wedge or cone with vertex at the point. One speaks of *zones of influence, silence*, and *dependence*; see Fig. 11.9.

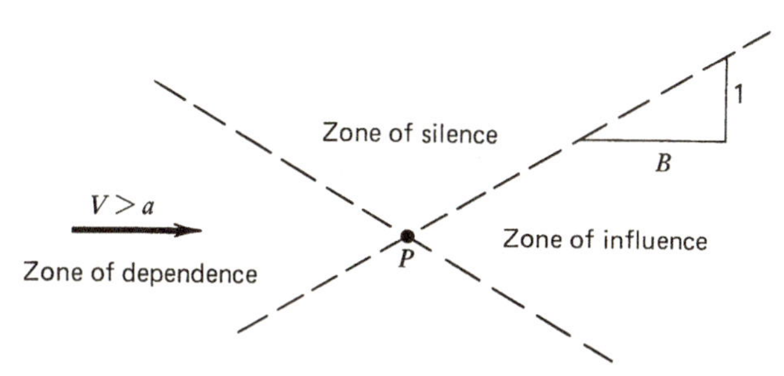

FIG. 11.9. Compartmentalization of a supersonic flow.

When the waves are attached to the airfoil's leading edge, there can be no flow around the edge. Then there is no "leading-edge suction," as in the incompressible flows discussed in Chapter 4, which allowed airfoils to have zero drag (in accordance with the Kutta–Joukowski theorem) even when the net force was "obviously" perpendicular to the chord line. Similarly, when the waves are attached to the trailing edge, the independence of the flow on either side makes a Kutta condition untenable, and the inability of the flow to go around the edge makes it unnecessary, too.

Example 2: Flow Past a Flat Plate at Angle of Attack Suppose that the airfoil under consideration has no thickness and no camber:

$$Y_u(x) = Y_l(x) = 0$$

Then, from equations 11-52, the velocity is constant along the airfoil's surfaces,

$$\frac{u}{V_\infty} = 1 \pm \frac{\alpha}{B} \qquad \text{at } y = 0\pm$$

as is then the pressure distribution. From equations 11-31 and 11-36,

$$C_p = \mp \frac{2\alpha}{B} \qquad \text{at } y = 0\pm$$

Thus the net force on the plate per unit span is, as can be seen from Fig. 11.10,

$$\begin{aligned} F' &= c(p|_{y=0-} - p|_{y=0+}) \\ &= c \cdot \tfrac{1}{2}\rho V_\infty^2 (C_p|_{y=0-} - C_p|_{y=0+}) \\ &= 2\rho V_\infty^2 \frac{c\alpha}{B} \end{aligned}$$

Since this force is directed normal to the plate, the airfoil experiences drag as well as lift, the respective coefficients being

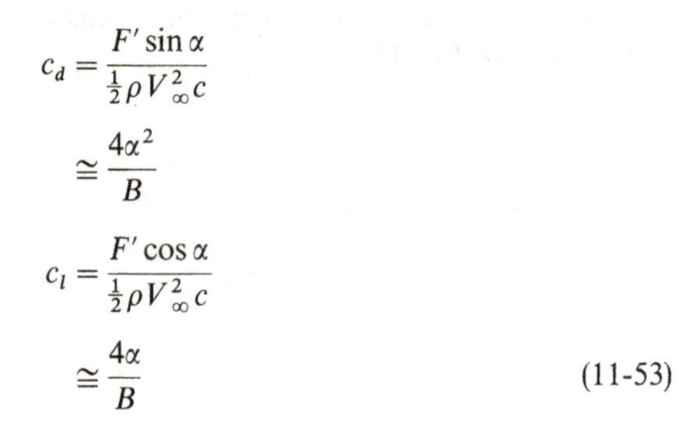

$$\begin{aligned} c_d &= \frac{F' \sin\alpha}{\frac{1}{2}\rho V_\infty^2 c} \\ &\cong \frac{4\alpha^2}{B} \\ c_l &= \frac{F' \cos\alpha}{\frac{1}{2}\rho V_\infty^2 c} \\ &\cong \frac{4\alpha}{B} \end{aligned} \tag{11-53}$$

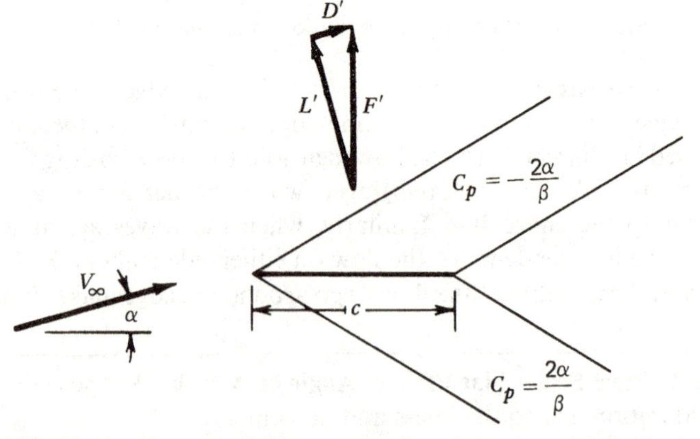

FIG. 11.10. Supersonic flow past a flat plate at angle of attack.

11.6. TRANSONIC FLOW PAST THIN AIRFOILS

Equations 11-44 and 11-53 both predict that the lift of a thin airfoil becomes infinite when the onset-flow Mach number goes to 1.0, since then $B = \beta = 0$. This is not correct, of course, nor is it true that the drag is zero subsonically but blows up as M_∞ approaches 1.0 from above, as equation 11-53 would have you believe.

Such anomalies indicate that the Prandtl–Glauert equation 11-32 is invalid when M_∞ is near 1.0. This difficulty should be expected, in view of the vastly different characters of subsonic and supersonic flow. The transition range is called *transonic flow*.

To obtain an equation that is valid for thin airfoils in transonic flow, it is necessary to return to equations 11-19 and 11-25 and to consider in their approximation the possibility that $M_\infty \cong 1.0$. While we will not accept the prediction of the Prandtl–Glauert equation that, at $M_\infty = 1.0$, $\partial^2\phi/\partial y^2 = 0$, we will trust its implication that, in the limit as $M_\infty \to 1.0$, ϕ changes much more slowly with y than with x. This is not unlike the situation we encountered in Chapter 7, in which the properties of a viscous flow were determined to vary much more rapidly in the direction normal to a solid surface than along the surface. As we did then, we will introduce different characteristic lengths for flow variations in the x and y directions, L and Δ, respectively, and allow for a disparity between these lengths as we use them to estimate the sizes of the different terms in equations 11-19 and 11-25. Specifically, we expect that

$$\Delta \gg L \tag{11-54}$$

By definition of a characteristic length, the various derivatives of ϕ can be estimated as follows:

$$\frac{\partial\phi}{\partial x} \sim \frac{\delta\phi}{L} \sim \frac{\phi}{L}$$

$$\frac{\partial\phi}{\partial y} \sim \frac{\delta\phi}{\Delta} \sim \frac{\phi}{\Delta} \tag{11-55}$$

That $\delta\phi$ (the change in ϕ) is of the order of ϕ itself follows from the boundary condition that ϕ vanish far upstream of the airfoil.

It will simplify our analysis if we introduce a complete set of dimensionless variables:

$$\xi \equiv x/L$$

$$\eta \equiv y/\Delta$$

$$\phi' \equiv \phi/LV_\infty \tag{11-56}$$

Then, from the estimates (11-55) and the chain rule,

$$\frac{\partial\phi'}{\partial\xi} = \frac{1}{LV_\infty}\frac{\partial\phi}{\partial x}\cdot L \sim \frac{1}{V_\infty}\frac{\phi}{L} = \phi'$$

$$\frac{\partial\phi'}{\partial\eta} = \frac{1}{LV_\infty}\frac{\partial\phi}{\partial y}\cdot\Delta \sim \frac{1}{V_\infty}\frac{\phi}{L} = \phi' \tag{11-57}$$

while equations 11-31 become

$$u = V_\infty + \varepsilon V_\infty \frac{\partial \phi'}{\partial \xi}$$

$$v = V_\infty \alpha + \varepsilon V_\infty \frac{L}{\Delta} \frac{\partial \phi'}{\partial \eta} \tag{11-58}$$

Since ϕ' will be the center of our attention for the rest of this chapter, we will, for convenience, drop the primes in what follows. Thus equations 11-57 and 11-58 will henceforth be used without the prime on ϕ.

Now we can start our approximations:

$$\begin{aligned} V^2 &= u^2 + v^2 \\ &= V_\infty^2 \left\{ 1 + 2\varepsilon \frac{\partial \phi}{\partial \xi} + \varepsilon^2 \left(\frac{\partial \phi}{\partial \xi} \right)^2 + \alpha^2 + 2\alpha\varepsilon \frac{L}{\Delta} \frac{\partial \phi}{\partial \eta} + \varepsilon^2 \left(\frac{L}{\Delta} \right)^2 \left(\frac{\partial \phi}{\partial \eta} \right)^2 \right\} \\ &\simeq V_\infty^2 \left\{ 1 + 2\varepsilon \frac{\partial \phi}{\partial \xi} \right\} \end{aligned}$$

The neglected terms are clearly much smaller than those retained, in view of equations 11-54 and 11-57. Thus, from equation 11-19,

$$a^2 \simeq a_\infty^2 - \varepsilon(\gamma - 1) V_\infty^2 \frac{\partial \phi}{\partial \xi}$$

as in the analysis that led to the Prandtl–Glauert equation, and so

$$a^2 - u^2 \simeq a_\infty^2 - V_\infty^2 - \varepsilon(\gamma + 1) V_\infty^2 \frac{\partial \phi}{\partial \xi}$$

also much as before. Now, however, we shall not neglect the last term compared to $a_\infty^2 - V_\infty^2$, which is also small.

The leading terms of an approximation to equation 11-25 appropriate in transonic flow are thus

$$\begin{aligned} (a_\infty^2 - V_\infty^2) \frac{\varepsilon V_\infty}{L} \frac{\partial^2 \phi}{\partial \xi^2} - \varepsilon(\gamma + 1) V_\infty^2 \frac{\partial \phi}{\partial \xi} \cdot \frac{\varepsilon V_\infty}{L} \frac{\partial^2 \phi}{\partial \xi^2} \\ + a_\infty^2 \varepsilon \frac{V_\infty L}{\Delta^2} \frac{\partial^2 \phi}{\partial \eta^2} = V_\infty^2 \alpha \cdot 2 \frac{\varepsilon V_\infty}{\Delta} \frac{\partial^2 \phi}{\partial \xi \, \partial \eta} \end{aligned} \tag{11-59}$$

From equations 11-57, we expect the derivatives of ϕ to be all of the same size. When a_∞ is not too close to V_∞, the characteristic lengths L and Δ are of about the same size, and the first and third terms on the left side of equation 11-59 clearly dominate the others. This leads us back to the Prandtl–Glauert equation 11-32, once we return to dimensional variables via equation 11-56.

As M_∞ approaches 1.0, however, the first term of equation 11-59 vanishes. To get an equation different from the Prandtl–Glauert equation in this limit—which we must do, since that equation is invalid in the transonic range—one of the other terms of equation 11-59 must become as big as the third term. The ratio of the term on the right side of equation 11-59 to the third term on the left contains a factor $\alpha\Delta/L$, whereas the ratio of the second term on the left to the third term is proportional to $\varepsilon(\Delta/L)^2$. Since $\Delta \gg L$, either one (or both) of these ratios could become comparable to unity, even though α and ε are both small.

To decide which of these possibilities is correct, we must have some knowledge of the relative sizes of α and ε. This we can get from the flow tangency condition (11-29). Let τ be a measure of the airfoil's thickness and camber, so that, if

$$Y(x) \equiv \tau L F(\xi) \tag{11-60}$$

then

$$\frac{dY}{dx} = \tau \frac{dF}{d\xi}$$

and $dF/d\xi$ is of order one. Then equations 11-29 and 11-58 give us

$$\varepsilon \frac{V_\infty L}{\Delta} \frac{\partial \phi}{\partial \eta} = V_\infty \left(\tau \frac{dF}{d\xi} - \alpha \right) \qquad \text{at } y = 0 \pm \tag{11-61}$$

Assuming α is of the same order of magnitude as τ, we deduce from equation 11-61 that

$$\frac{\varepsilon L}{\Delta} \sim \tau \tag{11-62}$$

Then the ratio of the term on the right side of equation 11-59 to the second term on the left is

$$-\frac{\alpha L}{\varepsilon \Delta} \sim \frac{\alpha}{\tau} \left(\frac{L}{\Delta} \right)^2 \sim \left(\frac{L}{\Delta} \right)^2$$

which we decided in equation 11-54 is quite small.

Therefore, the right side of equation 11-59 may be neglected even in the transonic range; it is the second term on the left that must balance the third term when M_∞ is near 1.0. In order to achieve this balance, their ratio

$$\varepsilon \left(\frac{\Delta}{L} \right)^2 \sim 1$$

so that equation 11-62 becomes

$$\left(\frac{L}{\Delta} \right)^3 \sim \tau$$

This gives us

$$\frac{L}{\Delta} \sim \tau^{1/3}$$

which, when substituted into equation 11-62, yields

$$\varepsilon \sim \tau^{2/3}$$

We have now introduced a number of parameters whose definitions need to be tightened up: ε, in equation 11-30; L and Δ, in equation 11-55; and τ, in equation 11-60. The characteristic length for flow property variations in the x direction, L, can clearly be taken to be the airfoil chord. Then τL can be set equal to the maximum y coordinate of the airfoil. As for ε and Δ, with τ and L now fixed we can simply turn the preceding estimates into definitions; that is,

$$\Delta \equiv \tau^{-1/3} L \tag{11-63}$$

$$\varepsilon \equiv \tau^{2/3} \tag{11-64}$$

Substituting these definitions into equation 11-59, and dividing through by $a_\infty^2 V_\infty \varepsilon L/\Delta^2$, we then get

$$\left[\frac{1 - M_\infty^2}{\tau^{2/3}} - (\gamma + 1)M_\infty^2 \frac{\partial \phi}{\partial \xi}\right]\frac{\partial^2 \phi}{\partial \xi^2} + \frac{\partial^2 \phi}{\partial \eta^2} = 0 \tag{11-65}$$

the right side of equation 11-59 having been discarded in accordance with the size estimates made above. The flow tangency condition now becomes, from equations 11-61, 11-63, and 11-64,

$$\frac{\partial \phi}{\partial \eta} = \frac{dF}{d\xi} - \frac{\alpha}{\tau} \qquad \text{at } y = 0\pm \tag{11-66}$$

Since the second term in the brackets of equation 11-65 is important only when M_∞ is near 1.0, it is sometimes replaced with

$$(\gamma + 1)\frac{\partial \phi}{\partial \xi}$$

Then the Mach number enters the problem only in combination with τ in the form

$$\frac{1 - M_\infty^2}{\tau^{2/3}} \equiv K \tag{11-67}$$

which is called the *transonic similarity parameter*. In this approximation, solutions of equation 11-65 subject to the boundary condition (11-66) can be used for flows at different transonic Mach numbers past bodies of the same shape but different thickness parameter τ. All one must do is to keep the K defined by equation 11-67 constant, along with the body-shape function F defined by equation 11-60 and, as is demanded by equation 11-66, the ratio α/τ. We shall retain the M_∞^2 factor in equation 11-65, whose solutions will therefore only approximately have this similarity property, but we will use the definition 11-67 for the first term in the brackets of equation 11-65.

11.6.1. Aerodynamics in the Transonic Range

Equation 11-65 is called the *transonic small-disturbance equation*. It differs crucially from the Prandtl–Glauert equation 11-32, which governs compressible

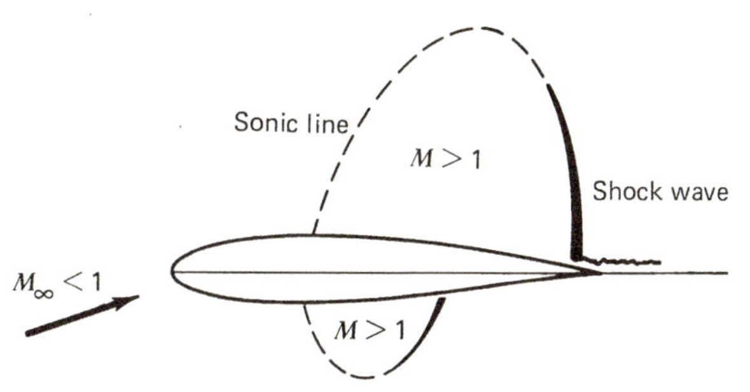

FIG. 11.11. Transonic flow past an airfoil.

flows at Mach numbers not too close[5] to 1.0 by being nonlinear in the unknown ϕ. Therefore, the superposition techniques that are used to solve the Prandtl–Glauert equation in subsonic and supersonic flow are not available in the transonic range.

Because of their relative difficulty (of which the nonlinearity of the governing equations will be seen to be just a small part), practical problems in transonic flow were not solved until the 1970s, decades after workable methods were developed for sub- and supersonic flow problems. The question may arise, how important are such problems? Did the size of the effort expended on transonic flow reflect some necessity or was it due just to the challenge of a relatively difficult problem?

In fact, it is not really a very good idea to fly at speeds very close to that of sound. Since a subsonic flow accelerates as its streamlines are squeezed together in making way for the body, zones of locally supersonic flow appear when the upstream Mach number is still subsonic, as shown in Fig. 11.11. As the streamlines diverge near the aft end of the body, the flow outside the supersonic zone decelerates, but, because streamline divergence accelerates a supersonic flow,[6] the transition back to subsonic speeds is rarely smooth. Thus the supersonic zone usually terminates abruptly, with a shock wave. As does the compression wave at the forward end of a body in supersonic flow, this creates a pressure drag that is called *wave drag*. What is worse, it imposes a sharp positive pressure gradient on the boundary layer, which may separate, further increasing the drag. The separated flow is often unstable, leading to a loss of control surface effectiveness (e.g., "aileron buzz").

Thus it may seem like the best thing to do is not to fly so fast that the flow becomes supersonic. The onset flow Mach number at which the local Mach number somewhere in the flow first reaches 1.0 is called the *critical Mach number*; flying "subcritically" ensures that the flow speed is everywhere subsonic.[7] The thicker the body, the lower the critical Mach number. For a typical transport aircraft, M_{crit} is about 0.75. See problem 2 for an approximate calculation of M_{crit} for a thin airfoil.

However, the drag that accompanies the appearance of shock waves is a function of the shock strength. By careful design, the shock strength can be kept acceptably

[5] Nor too high, either; some neglected terms become too large to ignore in the hypersonic flow regime.

[6] See, for example, Kuethe and Chow [2], p. 210.

[7] The terms "subcritical" and "supercritical" are also used to distinguish flows whose Reynolds numbers are below and above the Reynolds number at which transition is expected to take place. They have yet another meaning in open-channel flow.

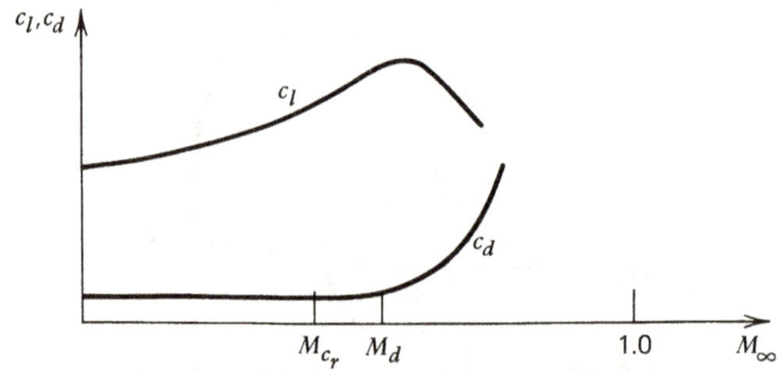

FIG. 11.12. Dependence on Mach number of lift and drag coefficients of typical airfoil.

low well past the critical Mach number, as indicated in Fig. 11.12. Further, the Prandtl–Glauert-based result of equation 11-44 is correct in predicting the lift coefficient to increase as the Mach number increases toward 1.0. As shown in Fig. 11.12, this increase continues until the Mach number is somewhat higher than the drag-rise Mach number M_d (the value of M_∞ at which the drag begins to increase markedly). To understand the subsequent decrease in c_l, note that the shock waves that induce the drag rise move toward the trailing edge as M_∞ increases. Since, for a lifting airfoil, the flow speed is higher on the upper surface than on the lower, the size of the supersonic zone is generally higher above the airfoil than below, as shown in Fig. 11.11, and the upper-surface shock reaches the trailing edge first. With further increases in M_∞ the upper-surface pressure distribution is relatively unchanged, but the rearward motion of the shock on the lower surface decreases the average pressure on that surface, and the lift coefficient reaches a maximum at some subsonic value of M_∞, as shown in Fig. 11.12.

Thus, the lift/drag ratio reaches a maximum just above the drag/rise Mach number M_d, which is in turn in the "supercritical" Mach number range. Aerodynamic analysis of supercritical flows is therefore of vital importance in commercial aircraft design. To increase utilization of the aircraft and/or its passengers, it is necessary to fly as fast as possible, but without consuming too much fuel (i.e., without incurring too much drag).

11.6.2. Solution of the Transonic Small-Disturbance Equation: Subcritical Flow

Although the case of real practical interest is that of supercritical flow, we will first develop a method that will turn out to be useful only for subcritical flow. This limitation is not obvious *a priori*, but, once we have the method down on paper, it will be possible to identify and then to remove its deficiency.

The techniques available for the solution of equation 11-65 being severely limited by its nonlinearity, we shall adopt a numerical approach based on finite-difference approximations to the governing partial differential equations. Since the algebraic equations that result from those approximations must be expected to be nonlinear, we will plan on an iterative method for their solution. Therefore, our method will resemble the one developed in Chapter 9 for the incompressible thin-airfoil problem.

This time our motivation for iteration is not just to save storage, but to deal with the nonlinearity of the problem.

Our first step is to set up the finite-difference equations. As in Chapter 9, we seek approximations ϕ_{ij} to the potential at ξ_i, η_j, the intersections of mesh lines aligned with the ξ and η axes, as shown in Fig. 11.13. Let h_E, h_N, h_W, and h_S be the distances from a typical mesh point ξ_i, η_j to its nearest neighbors, as shown in Fig. 11.14. Using ϕ_O for the potential at the central point, and ϕ_E, ϕ_N, ϕ_W, and ϕ_S for the potentials at its neighbors, we can write the finite-difference approximation to equation 11-65 as

$$\left[K - (\gamma + 1)M_\infty^2 \frac{\phi_E - \phi_W}{h_E + h_W}\right]\left(\frac{\phi_E - \phi_O}{h_E} - \frac{\phi_O - \phi_W}{h_W}\right)\frac{2}{h_E + h_W} + \left(\frac{\phi_N - \phi_O}{h_N} - \frac{\phi_O - \phi_S}{h_S}\right)\frac{2}{h_N + h_S} = 0 \tag{11-68}$$

To solve a system of equations like (11-68) by one of the iterative methods introduced in Chapter 9, we must be able to solve each equation for the potential at

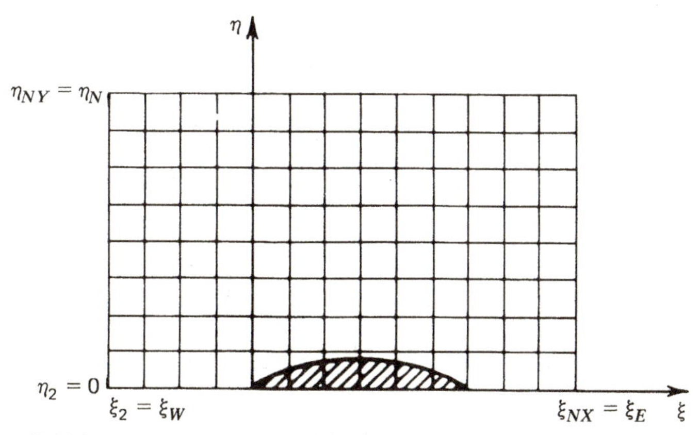

FIG. 11.13. Grid for finite-difference analysis of transonic small-disturbance equation.

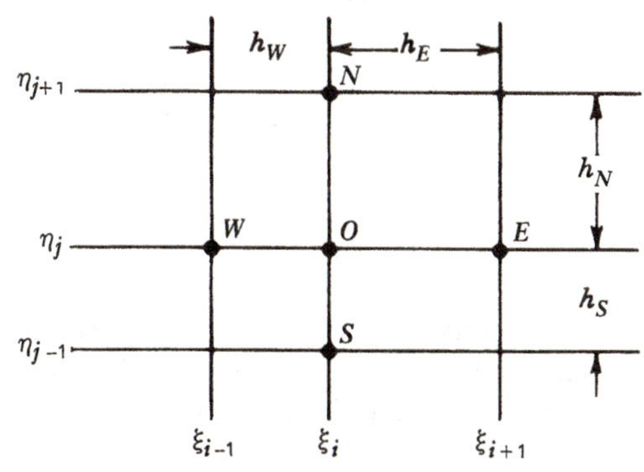

FIG. 11.14. Nomenclature for finite-difference approximation of transonic small-disturbance equation.

the central point in terms of the potentials at the neighboring points. This is surprisingly easy in the case of equation 11-68. Although the partial differential equation that it approximates is nonlinear, equation 11-68 is linear in the unknown ϕ_O. In the nomenclature of Chapter 9, a Gauss–Seidel method can then be set up by replacing ϕ_E and ϕ_N in equation 11-68 by their tilded counterparts, that is by "previous" approximations to those unknowns, and by starring the other ϕ's to indicate that they are "current" approximations to those variables. The ϕ_O^*'s are then updated in lexicographical order, that is, in order of increasing i and j. Similarly, to perform a successive overrelaxation method, we set

$$\phi_O^* = \tilde{\phi}_O + \omega(\phi'_O - \tilde{\phi}_O) \tag{11-69}$$

in which ϕ'_O is the quantity computed by the Gauss–Seidel method just described.

11.6.3. Conservative versus Nonconservative Difference Schemes

Unfortunately, it turns out that, before discretizing equation 11-65, it is advisable to rewrite it as

$$\frac{\partial}{\partial \xi}\left[K\frac{\partial \phi}{\partial \xi} - \frac{\gamma+1}{2}M_\infty^2\left(\frac{\partial \phi}{\partial \xi}\right)^2\right] + \frac{\partial^2 \phi}{\partial \eta^2} = 0 \tag{11-70}$$

The straightforward finite-difference approximation to equation 11-70 is

$$\begin{aligned}\frac{2}{h_E + h_W}&\left\{\left[K\frac{\phi_E - \phi_O}{h_E} - \frac{\gamma+1}{2}M_\infty^2\left(\frac{\phi_E - \phi_O}{h_E}\right)^2\right]\right.\\ &\left.- \left[K\frac{\phi_O - \phi_W}{h_W} - \frac{\gamma+1}{2}M_\infty^2\left(\frac{\phi_O - \phi_W}{h_W}\right)^2\right]\right\}\\ &+ \frac{2}{h_N + h_S}\left(\frac{\phi_N - \phi_O}{h_N} - \frac{\phi_O - \phi_S}{h_S}\right) = 0\end{aligned} \tag{11-71}$$

which, unlike equation 11-68, is nonlinear in the potential at the central point.

The differential equations 11-65 and 11-70 are entirely equivalent; the latter is simply the *divergence* or *conservation* form of the former. That is, equation 11-70 is of the form

$$\frac{\partial f}{\partial \xi} + \frac{\partial g}{\partial \eta} \equiv \operatorname{div}\mathbf{F} = 0 \tag{11-72}$$

in which the "vector" $\mathbf{F}$ is

$$\mathbf{F} = f\hat{\mathbf{e}}_\xi + g\hat{\mathbf{e}}_\eta$$

Such an equation can be converted into what is called a *conservation law* by application of the divergence theorem, equation 2-17; if $\mathscr{V}$ is a region in which equation 11-72 is valid, S its boundary and $\hat{\mathbf{n}}$ a unit vector normal to S, then

$$\int_{\mathscr{V}} \operatorname{div}\mathbf{F}\, d\mathscr{V} = 0 = \int_S \mathbf{F}\cdot\hat{\mathbf{n}}\, dS \tag{11-73}$$

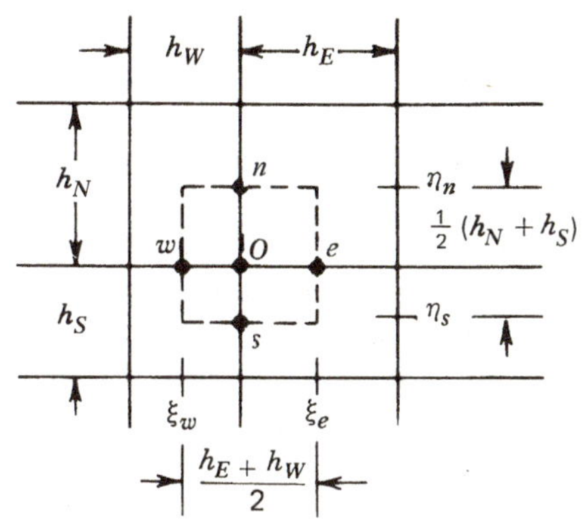

FIG. 11.15. Approximation of conservation law.

The conservation laws of fluid mechanics are basically expressed in terms of surface integrals that describe the flux of mass, momentum, or energy through the surface of a control volume; see equations 2-7 through 2-9. Differential equations that express the same laws, equations 2-26 to 2-28, are derived by using the divergence theorem as in equation 11-73. However, this requires that the vector **F** be continuous in $\mathscr{V}$. We are interested here in flows containing shock waves, for which the integral expressions of the conservation laws are still valid, but the differential versions are not.[8] The differential equations can be used on either side of such discontinuities, but not on them.

The virtue of the conservation form of a differential equation is that its finite-difference approximation is consistent with a numerical approximation to the associated conservation law. Consider the control volume shown dashed in Fig. 11.15. A finite-difference approximation to equation 11-72 at the central point O is

$$\frac{2}{h_E + h_W}(f_e - f_w) + \frac{2}{h_N + h_S}(g_n - g_s) = 0 \tag{11-74}$$

in which the subscripts e, n, w, and s indicate evaluation at the mesh-line midpoints shown in Fig. 11.15. Applying the divergence theorem to this control volume, we get, from equation 11-73,

$$\int_{\xi=\xi_e} f\,d\eta + \int_{\eta=\eta_n} g\,d\xi - \int_{\xi=\xi_w} f\,d\eta - \int_{\eta=\eta_s} g\,d\xi = 0$$

to which a numerical approximation (specifically, the "midpoint rule") is

$$f_e\frac{h_N + h_S}{2} + g_n\frac{h_E + h_W}{2} - f_w\frac{h_N + h_S}{2} - g_s\frac{h_E + h_W}{2} = 0$$

which is just a multiple of equation 11-74.

[8] Thus it was that we used an integral formulation in our analysis of shock waves, equations 11-1 to 11-4.

The finite-difference approximation (11-71) to the differential equation 11-70 is, therefore, consistent with the conservation law on which the equation is based, namely, conservation of mass. The approximation (11-68) to equation 11-65 differs from equation 11-71 by terms that are insignificant if the potential varies smoothly from one point to the next, but not in the presence of discontinuities.

It should be noted that this argument has a hidden flaw. The differential equations 11-65 and 11-70 are based primarily on the continuity equation, plus certain relations derived under the assumption of irrotational flow, which was noted to be a valid approximation if the shocks are sufficiently weak. Because of this assumption, these differential equations are not consistent with the integral law of conservation of momentum. Therefore, there is room for argument on the question of whether one should work with the "conservative" or nonconservative form of the equations of transonic potential flow.[9]

It is generally agreed that one should work with the conservative form of the transonic small-disturbance equation; from a practical viewpoint, the iterations then seem to converge faster. However, if the small-disturbance approximation is not made—if we work with the so-called full-potential equations 11-19 and 11-25—it is relatively hard to get the finite-difference method based on the conservative form of the equations to converge. Also, the results obtained from the nonconservative form often agree better with experiment. We shall avoid controversy by working only with the small-disturbance equation, in conservation form.

In this case, the difference between the conservative and nonconservative difference equations 11-71 and 11-68 is very minor. The first term of equation 11-71 may be factored, with the result

$$\left[K - \frac{\gamma+1}{2} M_\infty^2 \left(\frac{\phi_E - \phi_O}{h_E} + \frac{\phi_O - \phi_W}{h_W}\right)\right]\left(\frac{\phi_E - \phi_O}{h_E} - \frac{\phi_O - \phi_W}{h_W}\right)\frac{2}{h_E + h_W}$$

$$+\left(\frac{\phi_N - \phi_O}{h_N} - \frac{\phi_O - \phi_S}{h_S}\right)\frac{2}{h_N + h_S} = 0 \qquad \text{(11-71a)}$$

which is exactly the same as equation 11-68 in the case of even mesh spacing, and is virtually the same in any case. However, we will find important differences between the two approaches once we identify and remove an error in our approach thus far, which is our next step.

11.6.4. Supercritical Flow and Upwind Differencing

To solve a system of nonlinear equations like (11-71) for the potential at the central point, it is first necessary to linearize them around a guess for the ϕ_0's. Then an iterative scheme can be set up for the solution of the linearized equations and boundary conditions that would have roughly the same structure as the program THINAIR developed in Chapter 9.

[9] No such question should arise if we do not assume the flow to be irrotational, since the appropriate differential equations are then consistent with all the conservation laws. Thus methods that attempt to solve the Euler equations are very likely to be conservative.

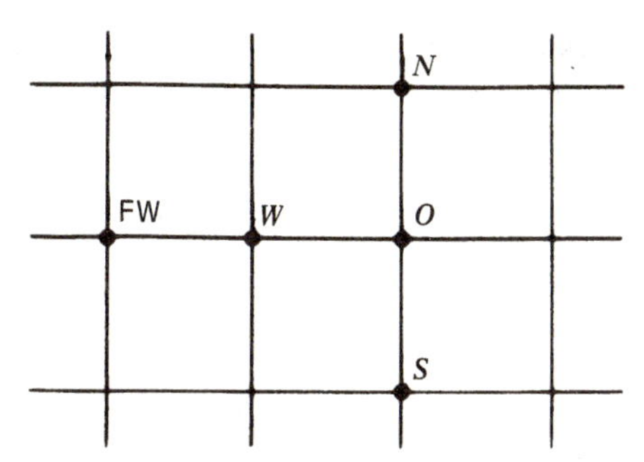

FIG. 11.16. Mesh points involved in upwind differencing.

However, such a procedure works only if the flow is subcritical. If the local Mach number is supersonic anywhere in the flow field, the iterations diverge.

The source of this problem is the use of difference formulas centered at the point at which the differential equation is being approximated. Although central differences are generally more accurate than one-sided differences, their use here implies that the solution at the central point depends on what happens at all of its neighbors. However, as we noted earlier, the flow properties at a point in a supersonic flow depend only on what happens upstream. Thus, if the flow is supersonic, we should be approximating the ξ derivatives in equation 11-70 with difference formulas that involve only points upstream of the point at which the equation is being approximated. In particular, we could center the approximations to the ξ derivatives at a point just upstream of the one at which the η derivatives are being approximated, as is illustrated in Fig. 11.16.[10]

The realization that "upwind differencing" was necessary in the supersonic part of the flow was the crucial step taken by Murman and Cole [4], who thereby paved the way for the accurate computation of practical transonic flows. Few papers have had as much impact on the science of aerodynamics.

The implementation of their method is best described in terms of a "switching function"

$$\begin{aligned} \mu_{ij} &\equiv 1 \qquad \text{if } M > 1 \qquad \text{at } \xi_i, \eta_j \\ &\equiv 0 \qquad \text{if } M \leq 1 \qquad \text{at } \xi_i, \eta_j \end{aligned} \tag{11-75}$$

Let the two terms of equation 11-70 be called P and Q,

$$\begin{aligned} P &\equiv \frac{\partial}{\partial \xi}\left[K\frac{\partial \phi}{\partial \xi} - \frac{\gamma+1}{2}M_\infty^2\left(\frac{\partial \phi}{\partial \xi}\right)^2\right] \\ Q &\equiv \frac{\partial}{\partial \eta}\left[\frac{\partial \phi}{\partial \eta}\right] \end{aligned} \tag{11-76}$$

[10] An alternative way of ensuring the proper dependence of the finite-difference formula is to evaluate the density at an upwind point; see, for example, Ref. [3]. Such an approach simplifies the development of methods for solving the full-potential equation in three dimensions.

and P_{ij} and Q_{ij} be their respective central-difference approximations at ξ_i, η_j. After a factoring like that which led to equation 11-71a, we find

$$P_{ij} = [K - \frac{(\gamma+1)}{2h} M_\infty^2 (\phi_{i+1j} - \phi_{i+1j})](\phi_{i+1j} - 2\phi_{ij} + \phi_{ij-1})\frac{1}{h^2}$$

$$Q_{ij} = (\phi_{ij+1} - 2\phi_{ij} + \phi_{ij-1})\frac{1}{k^2} \tag{11-77}$$

For simplicity, we are now specializing our analysis to the case of even mesh spacing:

$$\begin{aligned} \xi_{i+1j} - \xi_{ij} &= h \\ \eta_{ij+1} - \eta_{ij} &= k \end{aligned} \qquad \text{for all } i, j \tag{11-78}$$

Then equation 11-71a can be written

$$P_{ij} + Q_{ij} = 0$$

while the Murman–Cole idea is to use instead

$$P_{ij} + Q_{ij} = \mu_{ij}(P_{ij} - P_{i-1j}) \tag{11-79}$$

However, as was later recognized by Murman [5], simply biasing the differences upstream of a supersonic point yields a difference equation that is consistent with a differential equation that is not in conservation form. The right side of equation 11-79 can be interpreted as proportional to an approximation to the ξ derivative of P. Thus, identifying the quantities in the square brackets of equation 11-76 with the functions f and g introduced in equation 11-72, we regard equation 11-79 as an approximation to a differential equation of the form

$$\frac{\partial f}{\partial \xi} + \frac{\partial g}{\partial \eta} = \mu \Delta\xi \frac{\partial P}{\partial \xi} \tag{11-80}$$

rather than equation 11-72. This is not in divergence form, because of the variability of μ. An almost-equivalent equation (from equation 11-75, μ is constant except at transitions through Mach 1.0) that *is* in divergence form is

$$\frac{\partial}{\partial \xi}(f - \mu \Delta\xi P) + \frac{\partial g}{\partial \eta} = 0 \tag{11-81}$$

whose finite-difference approximation is, rather than equation 11-79,

$$P_{ij} + Q_{ij} = \mu_{ij} P_{ij} - \mu_{i-1j} P_{i-1j} \tag{11-82}$$

To use equation 11-82, we must be able to calculate μ_{ij}, and so, from equation 11-75, to decide whether the flow at ξ_i, η_j is sub or supersonic. In problem 5, you will show that the local flow speed V is greater than the local speed of sound if

$$(\gamma + 1)\frac{\partial \phi}{\partial \xi} > K$$

Thus, in view of the form of equation 11-77 for P_{ij}, we set

$$\begin{aligned} \mu_{ij} &= 1 \qquad \text{if } A_{ij} < 0 \\ &= 0 \qquad \text{if not} \end{aligned} \tag{11-83}$$

where

$$A_{ij} \equiv K - \frac{(\gamma + 1)}{2h} M_{\infty}^2 (\phi_{i+1j} - \phi_{i-1j}) \tag{11-84}$$

It is useful to pursue the above interpretation of upwind differencing as an approximation to an equation somewhat different from the one with which we started. Recall from equation 11-76 that P is the ξ derivative of a quantity that, in turn, involves first derivatives of the ξ component of the velocity $\varepsilon\, \partial\phi/\partial\xi$ (see equation 11-58). Thus we see that equation 11-80 has some of the structure of the Navier–Stokes equations, with the right side of the equation playing the role of a viscous term. Without the upwind differencing, equation 11-80 is more like the nonviscous Euler equations from which, of course, it was derived. Thus upwind differencing has the effect of introducing an artificially large viscous term into the differential equation. This argument can be made rather solid, since the quantities P_{ij} and Q_{ij} are second-order accurate approximations to P and Q, whereas the term on the right side of equation 11-80 is of first order in $\Delta\xi$.

In a viscous flow, a shock wave develops a structure, with rapid changes in the fluid properties taking place over a short but finite distance—the shock-wave thickness—and the flow properties being essentially constant outside this structure; see Fig. 11.17. As is shown in Appendix I, the thickness of a weak shock wave is proportional to the viscosity of the fluid.

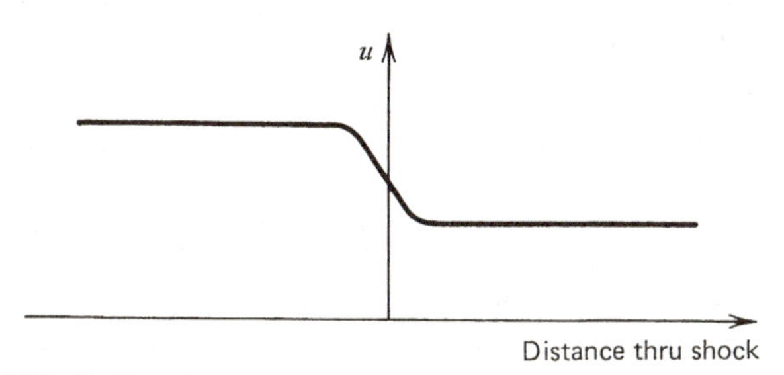

FIG. 11.17. Structure of a shock wave in a viscous fluid.

In the case of equation 11-80, the "viscosity" role is played by the mesh spacing $\Delta\xi$. This "explains" the behavior of solutions obtained with the algorithm we are about to develop in detail. Flows containing imbedded shocks, like the one pictured in Fig. 11.11, are approximated by a solution that exhibits rapid changes over a few mesh points, as shown in Fig. 11.18. One of the most exciting challenges in computational fluid dynamics has been to develop methods that would produce as sharp an approximation to the shock wave as possible and also to get its strength and location right. Many methods—see especially the work of Jameson [6, 7]—

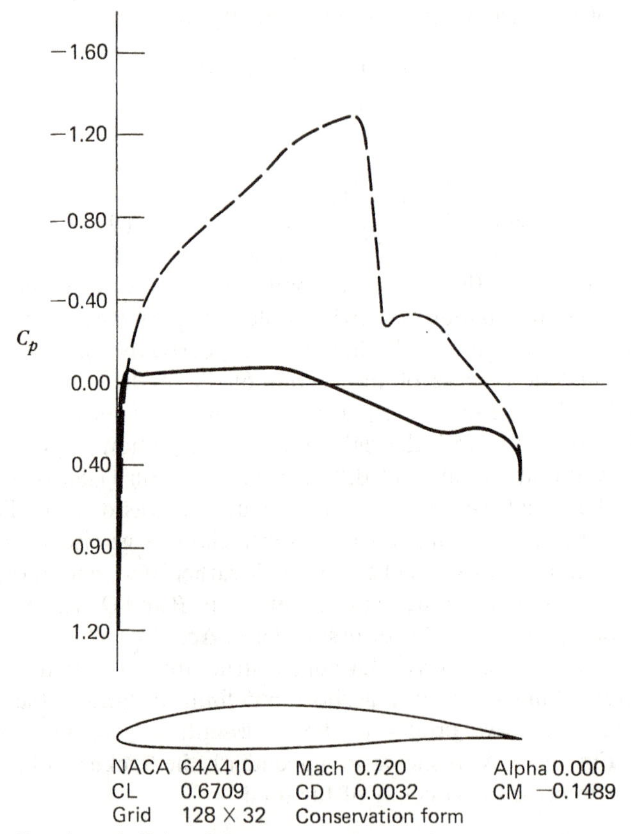

FIG. 11.18. Results of finite-difference analysis of transonic flow past an airfoil. Reprinted with permission from Ref. [7].

add artificial viscous terms of convenient form to the differential equations *before* they start differencing them.[11]

Another effect of this pseudoviscosity is to make the solution irreversible. The Euler equations describe a flow that is reversible; if $\{V, p, \rho\}$ describe one solution, another is given by $\{-V, p, \rho\}$. If the first solution satisfies a flow-tangency condition $\mathbf{V} \cdot \hat{\mathbf{n}} = 0$ on a certain body in the flow, so does the second. Thus. according to the Euler equations, if the onset flow in turned around ($\mathbf{V}_\infty \to -\mathbf{V}_\infty$), the body feels exactly the same pressure distribution. If there are any shock waves in the first flow,

[11] Methods that rely on the artificial viscosity of the finite-difference equations to detect shock waves automatically are called *shock-capturing methods.* It would be more accurate to treat shocks as discontinuities across which the jump conditions of Section 11.1 must be satisfied, and to use numerical approximations to the differential equations away from the shock. Methods that attempt to do so are called *shock-fitting methods.* Unless the shock location is known *a priori,* which is not so in the case illustrated depicted in Fig. 11.11, shock fitting is obviously more cumbersome than shock capturing. Also, unless shock-fitted coordinates are used, the inherent advantage in accuracy of shock fitting is negated by the difficulty of setting boundary conditions on the shock.

they appear in the reversed flow as "expansion shocks," with the fluid's density and pressure decreasing as it passes through the shock, since the shock jump conditions derived in Section 11.1 are also reversible.

In actuality, flows are not reversible, because of viscous action. Neither the Navier–Stokes equations of Chapter 6 nor the boundary-layer equations of Chapters 7 and 10 admit reversed solutions. In particular, since shock waves are inviscid approximations to solutions of the Navier–Stokes equations in which the flow properties vary extremely rapidly between the same values allowed by inviscid theory on either side of a discontinuity, the irreversibility of the viscous solution rules out expansion shocks.[12]

Were we to solve the problem depicted in Fig. 11.11 with a finite-difference method based on central differences, the solution would be reversible. The use of upwind differencing brings the method into conformance with the reality of irreversibility. In fact, however, we would not be able to find a solution using central differences; such a method would be numerically unstable.

11.6.5. The Relaxation Iteration

We now have, in equations 11-82, an acceptable finite-difference approximation to the transonic small-disturbance equation 11-70. It is time to solve them, subject to discretized versions of the boundary conditions (11-34) and (11-66). We shall do so for the same simple problem studied in Chapter 9, steady flow past a thin symmetric airfoil at zero angle of attack, except that the flow will now be transonic rather than incompressible.

As we did in Chapter 10, we will set up a rectangular grid in which the axis of symmetry is the line $\eta = \eta_2$ and "upstream infinity" the line $\xi = \xi_2$, the lines $\eta = \eta_1$ and $\xi = \xi_1$ both lying outside the computational domain; see Fig. 11.13. Then the boundary conditions are (compare equations 9-28 and 9-32)

$$\left.\begin{aligned} \frac{\phi_{3j} - \phi_{1j}}{2h} &= 0 \\ \frac{\phi_{NX+1,j} - \phi_{NX-1,j}}{2h} &= 0 \end{aligned}\right\} \quad \text{for } j = 2, \ldots, NY$$

$$\frac{\phi_{i3} - \phi_{i1}}{2k} = 0 \qquad \text{off the airfoil}$$

$$= \left.\frac{dF}{d\xi}\right|_{\xi_i} \qquad \text{on the airfoil } (0 \le \xi \le 1)$$

$$\frac{\phi_{i,NY+1} - \phi_{i,NY-1}}{2k} = 0 \qquad \text{for } i = 2, \ldots, NX \tag{11-85}$$

[12] The irreversibility of a compression shock wave can also be proved by a nonviscous argument based on the second law of thermodynamics; an expansion shock would imply an adiabatic decrease of entropy. See, for example, Ref. [1], p. 60.

Here $NX - 1$ is the number of constant-ξ mesh lines and $NY - 1$ the number of constant-η lines in the computational domain; again, see Fig. 11.13.

In the program THINAIR developed in Section 9.7, the finite-difference equations are solved by the successive overrelaxation (SOR) method recalled above in equation 11-69. Here we shall employ a variation on that scheme, known as successive *line* overrelaxation (SLOR). Rather than update our approximations for the potential one point at a time, we shall solve simultaneously for new approximations to the potential on each constant-ξ line. As is especially appropriate if the flow contains a supersonic zone, in which the properties at any point depend only on what happens upstream, we will sweep through the mesh one line at a time, from upstream to downstream. Since equations 11-77 and 11-85 contain at most three ϕ's on the line $\xi = \xi_i$, and since those ϕ's have consecutive j indexes, the system of equations to be solved for the unknowns on a line will be tridiagonal, and so easily solved with the Thomas algorithm described in Chapter 9, once they are linearized.

In keeping with the nomenclature discussed earlier, we shall replace $\phi_{i+1,j}$ in equation 11-77 by $\tilde{\phi}_E$, the result obtained at the point to the east of the central point, see Fig. 11.14, in the previous sweep through the mesh (or the initial assumption, if this is our first sweep). The potentials with the subscripts $(i-1,j)$ and $(i-2,j)$ are replaced with ϕ_W^* and ϕ_{FW}^*, respectively, see Fig. 11.16, whereas the unknowns at points N, O, and S are given primes, indicating that they are to be solved for. Thus, from equations 11-77, the terms of equation 11-82 are now written

$$P_{ij} = \left[K - \frac{\gamma+1}{2h} M_\infty^2(\tilde{\phi}_E - \phi_W^*)\right](\tilde{\phi}_E - 2\phi'_O + \phi_W^*)\frac{1}{h^2}$$

$$Q_{ij} = (\phi'_N - 2\phi'_O + \phi'_S)\frac{1}{k^2}$$

$$P_{i-1,j} = \left[K - \frac{\gamma+1}{2h} M_\infty^2(\phi'_O - \phi_{FW}^*)\right](\phi'_O - 2\phi_W^* + \phi_{FW}^*)\frac{1}{h^2} \qquad \text{(11-86)}$$

As in any iterative process, it is convenient to rewrite these equations in terms of the changes in the unknowns from the results of the previous sweep. Let

$$\delta\phi_O \equiv \phi'_O - \tilde{\phi}_O$$

for example. Also, let $\tilde{P}_{ij}$, $\tilde{Q}_{ij}$, and $\tilde{P}_{i-1,j}$ be the quantities obtained if the primed variables in equations 11-86 are replaced by their tilded counterparts. Then equations 11-86 may be written

$$P_{ij} = \tilde{P}_{ij} - \frac{2}{h^2}\tilde{A}_{ij}\,\delta\phi_O$$

$$Q_{ij} = \tilde{Q}_{ij} + \frac{1}{k^2}(\delta\phi_N - 2\delta\phi_O + \delta\phi_S)$$

$$P_{i-1j} = \tilde{P}_{i-1j} + \frac{1}{h^2}\tilde{A}_{i-1j}\,\delta\phi_O - \frac{\gamma+1}{2h} M_\infty^2(\tilde{\phi}_O - 2\phi_W^* + \phi_{FW}^*)\frac{\delta\phi_O}{h^2}$$

in which

$$\tilde{A}_{ij} = K - \frac{\gamma + 1}{2h} M_\infty^2(\tilde{\phi}_E - \phi_W^*)$$

$$\tilde{A}_{i-1j} = K - \frac{\gamma + 1}{2h} M_\infty^2(\tilde{\phi}_O - \phi_{FW}^*)$$

and we have linearized $P_{i-1,j}$ by neglecting the term quadratic in $\delta\phi_O$. Then equation 11-82 may be written

$$-(1 - \mu_{ij})\frac{2}{h^2}\tilde{A}_{ij}\,\delta\phi_O + \frac{1}{k^2}(\delta\phi_N - 2\delta\phi_O + \delta\phi_S)$$

$$+ \mu_{i-1j}\frac{1}{h^2}\left[\tilde{A}_{i-1j} - \frac{\gamma + 1}{2h} M_\infty^2(\tilde{\phi}_O - 2\phi_W^* + \phi_{FW}^*)\right]\delta\phi_O = -R_{ij} \quad (11\text{-}87)$$

where R_{ij} is the *residual*

$$R_{ij} = (1 - \mu_{ij})\,\tilde{P}_{ij} + \tilde{Q}_{ij} + \mu_{i-1j}\,\tilde{P}_{i-1j}$$

that is, the result of substituting into equation 11-82 the current approximations—tilded or starred, as the case may be—for all the unknowns. Its magnitude can be examined from one iteration to the next to help decide whether or not to continue the iterations, since its vanishing indicates convergence.

It remains to discuss the overrelaxation parameter ω of equation 11-69. Overrelaxation is desirable in the subsonic part of the flow. The properties of a subsonic flow depend on and influence the solution everywhere else in the flow. Since the relaxation process fixes the value of the potential to the east of the central point at its previously known value, the change computed at the central point is generally underestimated. Hence the use of a multiplicative factor greater than one to hasten convergence.

In the supersonic zone, however, the solution at a point depends *only* on what happens upstream. Since the upstream points are all accounted for in equation 11-87, overrelaxation is unnecessary in the supersonic zone and, in fact, it is also undesirable. While Murman and Cole [4] used a value of $\omega < 1$ in the supersonic zone, so as to avoid divergence of their particular iteration scheme, the recommended value of ω when $A_{ij} < 0$ is 1.0.

Program TSDE, which implements the method described above, thus asks you for two relaxation factors, one for subsonic points (RLXSUB) and one for supersonic points (RLXSUP). Suggested values are 1.5 to 1.9 in the subsonic zone and 1.0 in the supersonic zone. Unlike the program THINAIR that TSDE resembles in general structure, TSDE will not try to optimize these choices for you. You will have to experiment to find the best value of RLXSUB for your particular mesh.

11.6.6. The Poisson Iteration

No matter what you choose for the relaxation factors, convergence of the relaxation iterations described above is painfully slow. The problem is in the subsonic zone. As the local flow speed increases toward the sonic velocity, the coefficient of $\partial^2\phi/\partial\xi^2$ in equation 11-65 decreases, making it difficult to communicate information in the ξ direction by the relaxation process. One way around this is to alternate the directions of the line relaxation process; that is, to collect unknowns on lines parallel to the ξ and η axes in alternate sweeps through the mesh. Many of the more efficient methods currently used for solving transonic flows employ a variation on this idea called *approximate factorization*; see Baker [8] for a recent review.

A simpler (though somewhat less effective) remedy is the linearization of equation 11-65 by transposition of its nonlinear term to the right side,

$$K\frac{\partial^2\phi}{\partial\xi^2} + \frac{\partial^2\phi}{\partial\eta^2} = (\gamma + 1)\, M_\infty^2 \frac{\partial\phi}{\partial\xi}\frac{\partial^2\phi}{\partial\xi^2} \tag{11-88}$$

taking values for the nonlinear term from the previous sweep through the mesh. The operator on the left is essentially the Laplacian, so that the equation being solved is of the form

$$\nabla^2\phi = f$$

where f is a known function of the independent variables. This is called the *Poisson equation*, and the use of equation 11-88 as the basis for a sweep through the mesh is called a *Poisson sweep*.

In most programs that use Poisson sweeps, they are performed by a *fast direct Poisson solver*, which takes advantage of the special symmetries of the Laplacian to solve equation 11-88 without iteration but also without the storage and time consumption of the direct methods discussed in Chapter 9. See Jameson [7] for references to the various approaches that have been used. Program TSDE, however, uses a line relaxation method to solve equation 11-88, so as to simplify the programming (and avoid a discussion of the intricacies of the admittedly more efficient fast direct method). That is, equation 11-88 is approximated by

$$K\frac{\tilde{\phi}_E - 2\phi'_O + \phi^*_W}{h^2} + \frac{\phi'_N - 2\phi'_O + \phi'_S}{k^2} = (\gamma + 1)\, M_\infty^2 \frac{\tilde{\phi}_E - \phi^*_W}{2h}\frac{\tilde{\phi}_E - 2\tilde{\phi}_O + \phi^*_W}{h^2}$$

so that equation 11-87 is replaced by

$$-\frac{2}{h^2}K\delta\phi_O + \frac{1}{h^2}(\delta\phi_N - 2\delta\phi_O + \delta\phi_S) = -R_{ij} \tag{11-89}$$

In the supersonic zone, the right side of equation 11-88 is bigger than the $\partial^2\phi/\partial\xi^2$ term on the left. Thus, if the Poisson process of approximating the right side of equation 11-88 in terms of the results of the previous sweep were the sole basis of the solution, the iterations would diverge for any supercritical flow. A fast direct solver is nevertheless used throughout the flow field, even in the supersonic zone, since each Poisson sweep is followed by several relaxation sweeps. The Poisson sweep quickly improves the solution in the subsonic zone, where the relaxation method is slow, whereas the relaxation sweeps are efficient in correcting the errors of the Poisson approach in the supersonic part of the field.

When, as in program TSDE, an iterative approach is used for the Poisson process, several Poisson sweeps are necessary to have any effect on the solution in the subsonic zone. To avoid a serious buildup of errors in the supersonic zone during these sweeps, the supersonic relaxation factor RLXSUP should be set to zero. This, then, is how TSDE distinguishes between relaxation and Poisson sweeps, by the value specified for RLXSUP. If it is < 0.5, a Poisson sweep is assumed; if not, a relaxation sweep (again, the recommended values are 0.0 and 1.0, respectively).

Even with alternate series of Poisson and relaxation sweeps, convergence in program TSDE requires hundreds of iterations. Therefore, on exiting the program, the solution is saved on a file (TSDATA) that can be used for the initial guess when the program is restarted. The file also contains data on the mesh spacing ($\text{DX} = h$, $\text{DY} = k$) and size ($\text{XE} = \xi_E$, $\text{XW} = \xi_W$, $\text{YN} = \eta_N$, see Fig. 11.13). However, the Mach number and airfoil thickness ratio are input iteractively, so that converged data for one pair of values of M_∞ and τ can be used for the initial guess for another pair of input values.

Program TSDE also prints out (optionally) a map of the supersonic zone. Otherwise, its structure will be found basically similar to the program THINAIR introduced in Section 9.7 for the solution of the symmetric incompressible airfoil problem.

11.7. REFERENCES

1. Liepmann, H. W., and Roshko, A., *Elements of Gasdynamics*. Wiley, New York (1957).
2. Kuethe, A. M. and Chow, C.-Y., *Foundations of Aerodynamics*, 3rd ed. Wiley, New York (1976).
3. Hafez, M., South, J. C., and Murman, E. M., "Artificial Compressibility Method for Numerical Solutions of Transonic Full Potential Equation," *AIAA J.* **17**, 145–152 (1979).
4. Murman, E., and Cole, J. D., "Calculation of Plane Steady Transonic Flows," *AIAA J.* **9**, 114–121 (1971).
5. Murman, E., "Analysis of Embedded Shock Waves Calculated by Relaxation Methods," *AIAA J.* **12**, 626–633 (1974).
6. Jameson, A., "Iterative Solution of Transonic Flows over Airfoils and Wings, Including Flows at Mach 1," *Commun. Pure Appl. Math.* **27**, 283–309 (1974).
7. Jameson, A., "Transonic Flow Calculations", in *Numerical Methods in Fluid Dynamics*, H. J. Wirz and J. J. Smolderen, eds. Hemisphere, Washington, D. C. (1978).
8. Baker, T. J., "Approximate Factorization Methods", *Proceedings, IMA Conference on Numerical Methods in Aeronautical Fluid Dynamics*, P. L. Roe, ed. Academic, Reading, (1982), pp. 115–141.
9. Jameson, A., "The Evolution of Computational Methods in Aerodynamics," *J. Appl. Mech.* **50**, 1052–1070 (1983).

11.8. PROBLEMS

1. Solve equations 11-2 and 11-7 for the pressure jump across a shock wave in terms of the Mach number of the normal component of the upstream velocity, u_1/a_1, without making the weak-wave approximation.

2. A certain thin airfoil feels a minimum pressure coefficient of -0.2 at Mach 0.0. Estimate the critical Mach number for this airfoil.
3. Find the moment coefficient, aerodynamic center, and center of pressure of a flat-plate airfoil in supersonic flow, as a function of the onset flow Mach number and angle of attack.
4. Calculate the drag of the symmetric parabolic-arc airfoil whose surfaces are located at

$$y = \pm 0.2\, x(1 - x/c)$$

The angle of attack is 2° and the Mach number is 2.0.

5. Show that the transonic flow past a thin airfoil is locally sonic at points for which

$$K = (\gamma + 1) M_\infty^2 \partial\phi/\partial\xi$$

6. The convergence of program TSDE is slow in any case, but it gets worse as the mesh is refined. Modify TSDE so that it performs some iterations on a coarser mesh than the one of interest, so as to get a better first approximation to the desired solution than just a wild guess.[13] That is, revise the program so that you can change the mesh spacing and use the results already computed as a trial solution with the new data. You will probably want to halve the mesh spacing, so that the old results are directly usable. Of course, you will still have to interpolate in those results for data at those new mesh points that lie between the old mesh points. Test to see if this strategy saves you any computing time, compared to starting with the desired mesh. Note that you can write a separate program that just operates on the data stored in file TSDATA.
7. Compare the pressure distributions yielded by program TSDE for a thin airfoil at various subsonic Mach numbers (including some in the supercritical range) with those you obtain by applying the rule of equation 11-43. Use program THINAIR to generate the incompressible-flow results.

11.9. COMPUTER PROGRAM

```
      PROGRAM TSDE(INPUT,OUTPUT,TAPE1)
C
C        FINITE-DIFFERENCE SOLUTION OF COMPRESSIBLE POTENTIAL FLOW
C        PAST A THIN SYMMETRIC AIRFOIL AT ZERO ANGLE OF ATTACK
C
C        OVERRELAXATION BY LINES
C
      COMMON /UXYN/ PHI(100,50),X(100),Y(50),NX,NY
      COMMON /TRID/ A(50),B(50),C(50),D(50),DPHI(50)
      COMMON /COEF/ K,UFAC,TAU23
      COMMON /TRAN/ ACOF(50),ACOFUP(50),RLXSUP
      REAL          MACH,K
C
```

[13] A very powerful variation on this theme is the *multigrid method*, in which one uses corrections computed on a sequence of increasingly coarse meshes to improve the solution on the mesh of interest; see Jameson [9] for a review of this and other work in progress.

```
C                  INPUT SPACING OF MESH LINES IN X,Y DIRECTIONS (DX,DY),
C                  MACH NUMBER (MACH), THICKNESS RATIO (TAU),
C                  AND FIELD DIMENSIONS (XW < X < XE, 0 < Y < YN)
C
      PRINT, ' INITIAL GUESS FROM FILE (1) OR NOT (0) ',
      READ,       IANS
      IF (IANS .EQ. 1) THEN
         CALL GETPF('TAPE1','TSDATA',0,0)
         READ (1,1000)  DX,DY,XW,XE,YN
      ELSE
         PRINT,' INPUT DX,DY',
         READ,       DX,DY
         PRINT, ' INPUT XW,XE,YN',
         READ, XW,XE,YN
      ENDIF
      ILE         =  (2.5*DX - XW)/DX
      ITE         =  (1.0 + 2.5*DX - XW)/DX
      NX          =  (XE + 2.5*DX - XW)/DX
      NY          =  (YN + 2.5*DY)/DY
      GAMMA       =  1.4
      PRINT, ' INPUT MACH,TAU',
      READ,       MACH,TAU
      TAU23       =  TAU**(2./3.)
      K           =  (1. - MACH**2)/TAU23
      UFAC        =  (GAMMA + 1.)*MACH**2/(2.*DX)
C
C                  MAKE INITIAL GUESS AND SET UP MESH
C
      DO 70       I = 1,NX
      X(I)        =  (I - ILE)*DX
      DO 70       J = 1,NY
      Y(J)        =  (J - 2)*DY
   70 PHI(I,J)      =  0.0
      IF (IANS .EQ. 1) THEN
         DO 80  I = 1,NX
         READ (1,1000)  (PHI(I,J), J = 1,NY)
   80    CONTINUE
      ENDIF
 1000 FORMAT(4E18.8)
      KOUNT       =  0
      CHGOLD      =  1.0
C
C                  INPUT MAXIMUM NUMBER OF TRIES (MAXTRY)
C                  AND OVERRELAXATION FACTORS (RLXSUB AND
C                  RLXSUP IN SUB- AND SUPERSONIC ZONES, RESPECTIVELY)
C
C                  RLXSUP = 1 FOR RELAXATION ITERATION
C                         = 0 FOR POISSON ITERATION
C
  100 PRINT,' INPUT MAXTRY,RLXSUB,RLXSUP',
      READ,       MAXTRY,RLXSUB,RLXSUP
      IF (MAXTRY .EQ. 0) THEN
         REWIND 1
         WRITE (1,1000) DX,DY,XW,XE,YN
         DO 110 I = 1,NX
         WRITE (1,1000) (PHI(I,J), J = 1,NY)
  110    CONTINUE
         CALL REPLACE('TAPE1','TSDATA',0,0)
         STOP
```

```
      ENDIF
      PRINT 2000
      DO 240      ITRY = 1,MAXTRY
      KOUNT       =  KOUNT + 1
      CHGRMS      =  0.0
      CHGMAX      =  0.0
      NUMSUP      =  0
      DO 170      I = 2,NX
      CALL COFISH(I)
      CALL TRIDIG(2,NY)
      DO 160      J = 2,NY
      IF (ACOF(J) .LT. 0.0) THEN
          NUMSUP  = NUMSUP + 1
          DPHI(J) = DPHI(J)*RLXSUP
      ELSE
          DPHI(J) = DPHI(J)*RLXSUB
      ENDIF
      PHI(I,J)    =  PHI(I,J) + DPHI(J)
      IF (ABS(DPHI(J)) .LT. CHGMAX)  GO TO 160
      CHGMAX      =  ABS(DPHI(J))
      IMAX        =  I
      JMAX        =  J
  160 CHGRMS      =  CHGRMS + DPHI(J)**2
  170 CONTINUE
C
C                   -- EVALUATE TRIAL
C
      CHGRMS      =  SQRT(CHGRMS/FLOAT(NX*NY))
      IF (CHGRMS .LT. 1.E-6) GO TO 300
      RATIO       =  CHGRMS/CHGOLD
      CHGOLD      =  CHGRMS
      XMAX        =  X(IMAX)
      YMAX        =  Y(JMAX)
      PRINT 2010,        KOUNT,CHGRMS,RATIO,XMAX,YMAX,CHGMAX,NUMSUP
  240 CONTINUE
  300 PRINT,      ' DO YOU WANT CONTOUR MAPS (Y/N)',
      READ 2020, IANS
      IF (IANS .EQ. 1HN)   GO TO 310
      CALL CONTOUR(40)
      CALL SONIC(40)
  310 CALL CPOUT(ILE,ITE)
      GO TO 100
 2000 FORMAT(///,
     +   ' TRIAL  RMS CHANGE  CHANGE RATIO      XMAX       YMAX
     +   'MAX CHANGE NUMSUP')
 2010 FORMAT(I5,E11.3,2F13.4,F10.4,E11.3,I5)
 2020 FORMAT(A1)
      END
C=================================================================
      SUBROUTINE COFISH(I)
C
C                   SET COEFFICIENTS OF TRIDIAGONAL SYSTEM
C                   FOR CHANGE OF POTENTIAL IN ITH COLUMN
C
      COMMON /UXYN/ PHI(100,50),X(100),Y(50),NX,NY
      COMMON /TRID/ A(50),B(50),C(50),D(50),DPHI(50)
      COMMON /COEF/ K,UFAC,TAU23
      COMMON /TRAN/ ACOF(50),ACOFUP(50),RLXSUP
      REAL          K
C
```

```
C                   SET POTENTIAL OUTSIDE COMPUTATIONAL SPACE
C
      IF (I .GT. 2)            GO TO 120
      DO 110      J = 2,NY
      ACOFUP(J)     =  K
  110 PHI(1,J)    =  0.0
  120 IF (I .LT. NX)           GO TO 140
      DO 130      J = 2,NY
  130 PHI(NX+1,J)=  PHI(NX-1,J)
  140 PHI(I,1)    =  PHI(I,3) - (Y(3) - Y(1))*FP(X(I))
      PHI(I,NY+1)=  PHI(I,NY-1)
C
C                   SET GLOBAL COEFFICIENTS OF FINITE-DIFFERENCE EQUATIONS
C
      DX          =  X(3) - X(2)
      DY          =  Y(3) - Y(2)
      DO 160      J = 2,NY
      ACOF(J)     =  K - UFAC*(PHI(I+1,J) - PHI(I-1,J))
      CN          =  DX/DY
      CS          =  CN
      CO          =  - CN - CS
      A(J)        =  CS
      B(J)        =  CO
      C(J)        =  CN
      D(J)        =  - CN*PHI(I,J+1) - CO*PHI(I,J) - CS*PHI(I,J-1)
      IF ((ACOF(J) .LT. 0.0) .AND. (RLXSUP .GT. 0.5)) GO TO 150
C
C                   -- SUBSONIC AT CURRENT POINT
C
      CE          =  DY/DX*ACOF(J)
      CW          =  CE
      CO          =  - CW - CE
      D(J)        =  D(J) - CE*PHI(I+1,J) - CO*PHI(I,J) - CW*PHI(I-1,J)
      IF (RLXSUP .LT. 0.5)    CO = -2.*DY/DX*K
      B(J)        =  B(J) + CO
  150 IF ((ACOFUP(J) .GT. 0.0) .OR. (RLXSUP .LT. 0.5)) GO TO 160
C
C                   -- SUPERSONIC AT UPSTREAM POINT
C
      CO          =  DY/DX*ACOFUP(J)
      CW          =  - 2.*CO
      CFW         =  CO
      B(J)        =  B(J) + CO
     +            - UFAC*DY/DX*(PHI(I,J) - 2.*PHI(I-1,J) + PHI(I-2,J))
      D(J)        =  D(J) - CO*PHI(I,J) - CW*PHI(I-1,J) - CFW*PHI(I-2,J)
  160 CONTINUE
C
C                   ADJUST EQUATIONS AT UPPER AND LOWER BOUNDARIES
C
      C(2)        =  C(2) + A(2)
      A(NY)       =  A(NY) + C(NY)
C
C                   STORE COEFFICIENTS OF D2PHI/DX2 TERM
C
      DO 170      J = 2,NY
  170 ACOFUP(J)     =  ACOF(J)
      RETURN
      END
C==========================================================
```

```
      FUNCTION FP(X)
C
C                    SLOPE OF AIRFOIL SURFACE
C
      FP          =  0.0
      IF ((X .LT. 0.0) .OR. (X .GT. 1.0)) RETURN
      FP          =  2. - 4.*X
      RETURN
      END
C=======================================================
      SUBROUTINE TRIDIG(ILO,NEQNS)
C
C                    TRIDIAGONAL SYSTEM SOLVER
C
      COMMON /TRID/ A(50),B(50),C(50),D(50),V(50)
      ILOP        =  ILO + 1
      DO 100      I = ILOP,NEQNS
      FACTOR      =  A(I)/B(I-1)
      B(I)        =  B(I) - FACTOR*C(I-1)
  100 D(I)        =  D(I) - FACTORjD(I-1)
      V(NEQNS)    =  D(NEQNS)/B(NEQNS)
      DO 200      L = ILOP,NEQNS
      I           =  NEQNS + ILO - L
  200 V(I)        =  (D(I) -C(I)*V(I+1))/B(I)
      RETURN
      END
C========================================================
      SUBROUTINE CONTOUR(NWIDTH)

      (SEE PROGRAM THINAIR)

C=============================================================
      SUBROUTINE TERPLT(NWIDTH,DY)

      (SEE PROGRAM THINAIR)

C=====================================================
      SUBROUTINE CPOUT(ILE,ITE)
C
C                    PRINT AND PLOT PRESSURE DISTRIBUTION ON AIRFOIL
C
      COMMON /UXYN/ PHI(100,50),X(100),Y(50),NX,NY
      COMMON /SKAL/ NZERO,YMULT,YSLOP
      COMMON /COEF/ K,UFAC,TAU23
      REAL          K
      NZERO       =  31
      YMULT       =  20.
      YSLOP       =  .5/YMULT
      PRINT 1000
      ILEP        =  ILE + 1
      DO 100      I = ILEP,ITE
      XP          =  .5*(X(I) + X(I-1))
      U           =  (PHI(I,2) - PHI(I-1,2))/(X(I) - X(I-1))*TAU23
      CP          =  - 2.*U
      CALL PLOTXY(XP,CP)
  100 CONTINUE
 1000 FORMAT(////,' PRESSURE DISTRIBUTION ON AIRFOIL',//,
     +              4X,'X',8X,'CP',/)
      RETURN
      END
```

```
C==========================================================
      SUBROUTINE PLOTXY(X,Y)

      (SEE PROGRAM DUBLET)

C===============================================================
      SUBROUTINE SONIC(NWIDTH)
C
C                    OUTLINE OF SUPERSONIC REGION
C
      COMMON /UXYN/ PHI(100,50),X(100),Y(50),NX,NY
      COMMON /COEF/ K,UFAC,TAU23
      COMMON /CNTR/ UPLOT(50),ISYM(100)
      COMMON /UPLT/ U(100,50)
      REAL           K
      JGAP            =  NWIDTH - 8
      DO  100       I = 2,NX
      DO  100       J = 2,NY
      U(I,J)        =  (K - UFAC*(PHI(I+1,J) - PHI(I-1,J)))
  100 CONTINUE
      YMAX          =  Y(NY) - Y(2)
      XMAX          =  X(NX) - X(2)
      DY            =  YMAX/FLOAT(NWIDTH)
      DX            =  DY*1.67
      LINES         =  (XMAX + .5*DX)/DX +1
      PRINT 1000,              Y(2),JGAP,Y(NY),X(2)
      XLINE         =  X(2)
      DO 200        J = 2,NY
  200 UPLOT(J)      =   1.0
      CALL TERPLT(NWIDTH,DY)
      DO 260        L = 2,LINES
      XLINE         =  XLINE + DX
      DO 210        I = 2,NX
      IF (XLINE .LT. X(I))   GO TO 220
  210 CONTINUE
  220 DO 230        J = 2,NY
      TEST          =  U(I,J) + (XLINE - X(I))*(U(I-1,J) - U(I,J))
     +                         /(X(I-1) - X(I))
      UPLOT(J)      =  1.0
      IF (TEST .LT. 0.0)        UPLOT(J) = 0.0
  230 CONTINUE
      CALL TERPLT(NWIDTH,DY)
  260 CONTINUE
      PRINT 1010,               X(NX)
 1000 FORMAT(/////,' SUPERSONIC ZONE',////,
     +              //,5X,'Y =',F6.3,=X,'Y =',F6.3,/,' X =',F6.3)
 1010 FORMAT(' X =',F6.3)
      RETURN
      END
```

APPENDIX A

AN IMPORTANT INTEGRAL

Let

$$I_n(\theta_0) \equiv \mathrm{P}\int_0^\pi \frac{\cos n\theta}{\cos\theta - \cos\theta_0}\, d\theta \tag{A-1}$$

We will find the following identity useful

$$\cot\frac{\theta_0 - \theta}{2} + \cot\frac{\theta_0 + \theta}{2} = \frac{2\sin\theta_0}{\cos\theta - \cos\theta_0}$$

Proof of identity: the left side is

$$\frac{\cos\dfrac{\theta_0-\theta}{2}}{\sin\dfrac{\theta_0-\theta}{2}} + \frac{\cos\dfrac{\theta_0+\theta}{2}}{\sin\dfrac{\theta_0+\theta}{2}} = \frac{\cos\dfrac{\theta_0-\theta}{2}\sin\dfrac{\theta_0+\theta}{2} + \cos\dfrac{\theta_0+\theta}{2}\sin\dfrac{\theta_0-\theta}{2}}{\sin\dfrac{\theta_0-\theta}{2}\sin\dfrac{\theta_0+\theta}{2}}$$

$$= \frac{\sin\left(\dfrac{\theta_0+\theta}{2} + \dfrac{\theta_0-\theta}{2}\right)}{\frac{1}{2}\left[\cos\left(\dfrac{\theta_0+\theta}{2} - \dfrac{\theta_0-\theta}{2}\right) - \cos\left(\dfrac{\theta_0+\theta}{2} + \dfrac{\theta_0-\theta}{2}\right)\right]}$$

$$= \frac{2\sin\theta_0}{\cos\theta - \cos\theta_0}$$

Therefore, equation A-1 can be written

$$I_n(\theta_0) = \frac{1}{2\sin\theta_0}\,\mathrm{P}\int_0^\pi \cos n\theta\left(\cot\frac{\theta_0-\theta}{2} + \cot\frac{\theta_0+\theta}{2}\right) d\theta \tag{A-2}$$

But, if $z = -\theta$

$$\mathrm{P}\!\!\int_0^{\pi} \cos n\theta \cot \frac{\theta_0 - \theta}{2}\, d\theta = -\mathrm{P}\!\!\int_0^{-\pi} \cos(-nz) \cot \frac{\theta_0 + z}{2}\, dz$$

$$= +\mathrm{P}\!\!\int_{-\pi}^{0} \cos nz \cot \frac{\theta_0 + z}{2}\, dz$$

$$= \mathrm{P}\!\!\int_{-\pi}^{0} \cos n\theta \cot \frac{\theta_0 + \theta}{2}\, d\theta$$

where we let $z = \theta$ in the last line. Thus equation A-2 simplifies to

$$I_n(\theta_0) = \frac{1}{2 \sin \theta_0} \mathrm{P}\!\!\int_{-\pi}^{\pi} \cos n\theta \cot \frac{\theta_0 + \theta}{2}\, d\theta \tag{A-3}$$

Let $x \equiv \theta_0 + \theta$. Then $\theta = x - \theta_0$, and equation A-3 can be written

$$I_n = \frac{1}{2 \sin \theta_0} \mathrm{P}\!\!\int_{-\pi+\theta_0}^{\pi+\theta_0} (\cos nx \cos n\theta_0 + \sin nx \sin n\theta_0) \cdot \cot \frac{x}{2}\, dx \tag{A-4}$$

But every term in the integrand has a period 2π, and since, if $f(x) = f(x + 2\pi)$

$$\int_{-\pi+\theta_0}^{-\pi} f(x)\, dx = \int_{-\pi+\theta_0}^{-\pi} f(x + 2\pi)\, dx = \int_{\pi+\theta_0}^{\pi} f(z)\, dz = -\int_{\pi}^{\pi+\theta_0} f(z)\, dz$$

where now $z = x + 2\pi$, the limits on the integral in equation A-4 can be replaced by $(-\pi, \pi)$. Then

$$2 \sin \theta_0\, I_n(\theta_0) = \cos n\theta_0\, J_n + \sin n\theta_0\, K_n \tag{A-5}$$

where

$$J_n \equiv \mathrm{P}\!\!\int_{-\pi}^{\pi} \cos nx \cot \frac{x}{2}\, dx$$

$$K_n \equiv \int_{-\pi}^{\pi} \sin nx \cot \frac{x}{2}\, dx \tag{A-6}$$

By symmetry

$$J_n = 0 \tag{A-7}$$

As for K_n, first note

$$\cot \frac{x}{2} = \frac{\cos \frac{x}{2}}{\sin \frac{x}{2}} = \frac{\sqrt{1 + \cos x}}{\sqrt{1 - \cos x}} \cdot \frac{\sqrt{1 + \cos x}}{\sqrt{1 + \cos x}}$$

$$= \frac{1 + \cos x}{\sin x}$$

Then, for the case $n = 1$,

$$K_1 \equiv \int_{-\pi}^{\pi} \sin x \cot \frac{x}{2}\, dx = \int_{-\pi}^{\pi} \sin x \frac{1 + \cos x}{\sin x}\, dx$$

$$= 2\pi \tag{A-8}$$

Also, since

$$\sin nx = \sin(n-1)x \cos x + \cos(n-1)x \sin x$$

we can write

$$\sin nx \cot \frac{x}{2} = \sin(n-1)x \frac{\cos x + \cos^2 x}{\sin x} + \cos(n-1)x(1 + \cos x)$$

$$= \sin(n-1)x \frac{\cos x + 1 - \sin^2 x}{\sin x} + \cos(n-1)x(1 + \cos x)$$

$$= \sin(n-1)x \frac{1 + \cos x}{\sin x} + \cos(n-1)x$$

$$- \sin x \sin(n-1)x + \cos(n-1)x \cos x$$

$$= \sin(n-1)x \cot \frac{x}{2} + \cos(n-1)x + \cos nx$$

Then, from equation A-6, we get,

$$K_n = \int_{-\pi}^{\pi} \sin(n-1)x \cot \frac{x}{2}\, dx + \int_{-\pi}^{\pi} [\cos(n-1)x + \cos nx]\, dx$$

$$= K_{n-1} + 0$$

This recursion formula shows all the K_n's to be equal. From equation A-8, their common value is 2π. Thus, from equations A-6 and A-7,

$$I_n(\theta_0) = \frac{\sin n\theta_0\, K_n}{2 \sin \theta_0}$$

$$= \frac{2\pi \sin n\theta_0}{2 \sin \theta_0}$$

so, from equation A-1,

$$⨍_0^{\pi} \frac{\cos n\theta}{\cos \theta - \cos \theta_0}\, d\theta = \pi \frac{\sin n\theta_0}{\sin \theta_0}$$

APPENDIX B

THE INTEGRAL $I_n = ⨍_0^c \frac{t^n\,dt}{x-t}$

Consider first the case $n = 0$:

$$
\begin{aligned}
I_0 &\equiv ⨍_0^c \frac{dt}{x-t} \equiv \lim_{\varepsilon\to 0}\int_0^{x-\varepsilon}\frac{dt}{x-t} + \int_{x+\varepsilon}^c \frac{dt}{x-t} \\
&= \lim_{\varepsilon\to 0}\left[-\ln|x-t|\Big|_0^{x-\varepsilon} - \ln|x-t|\Big|_{x+\varepsilon}^c\right] \\
&= \lim_{\varepsilon\to 0}\left[-\ln\frac{\varepsilon}{x} - \ln\frac{c-x}{\varepsilon}\right] \\
&= \ln\frac{x}{c-x} \qquad \text{(B-1)}
\end{aligned}
$$

Note that this same result could be obtained by more-or-less ignoring the Cauchy principal value formalism:

$$
⨍_0^c \frac{dt}{x-t} = -\ln|x-t|\Big|_0^c = \ln\frac{x}{c-x}
$$

This is generally the case; for a smooth numerator of the integrand, correct results are obtained simply by being careful to include absolute-value signs in the argument of any logarithmic terms.

Now look at

$$
I_{-1/2} \equiv \int_0^c \frac{t^{-1/2}\,dt}{x-t}
$$

Let

$$
r \equiv \sqrt{x}, \qquad s \equiv \sqrt{t}
$$

Then

$$\begin{aligned} I_{-1/2} &= \int_0^{\sqrt{c}} \frac{2\,ds}{r^2 - s^2} \\ &= \frac{1}{r} \, \mathrm{P}\!\!\int_0^{\sqrt{c}} \left[\frac{1}{r-s} + \frac{1}{r+s} \right] ds \\ &= -\frac{1}{r} \ln|r-s| \Big|_0^{\sqrt{c}} + \frac{1}{r} \ln|r+s| \Big|_0^{\sqrt{c}} \\ &= \frac{1}{\sqrt{x}} \ln \frac{\sqrt{x} + \sqrt{c}}{\sqrt{c} - \sqrt{x}} \end{aligned} \qquad \text{(B-2)}$$

The values of I_n for other integral and half-integral values of n can be obtained from a recursion formula, which is derived as follows. Write

$$\begin{aligned} I_n &= \mathrm{P}\!\!\int_0^c \frac{t^{n-1}(t - x + x)\,dt}{x - t} \\ &= x I_{n-1} - \int_0^c t^{n-1}\,dt \end{aligned}$$

Thus,

$$I_n = x I_{n-1} - \frac{c^n}{n} \qquad \text{(B-3)}$$

and so

$$I_{-1/2} = \frac{1}{\sqrt{x}} \ln \frac{\sqrt{c} + \sqrt{x}}{\sqrt{c} - \sqrt{x}} \qquad \text{(B-2)}$$

$$I_0 = \ln \frac{x}{c - x} \qquad \text{(B-1)}$$

$$I_{1/2} = \sqrt{x} \ln \frac{\sqrt{c} + \sqrt{x}}{\sqrt{c} - \sqrt{x}} - 2\sqrt{c}$$

$$I_1 = x \ln \frac{x}{c - x} - c$$

$$I_{3/2} = x^{3/2} \ln \frac{\sqrt{c} + \sqrt{x}}{\sqrt{c} - \sqrt{x}} - 2x\sqrt{c} - \frac{2}{3} c^{3/2}$$

and so on.

APPENDIX C

POTENTIAL FLOW PAST A CORNER

Consider the flow in the vicinity of a corner that, as shown in Fig. C.1, may be concave or convex. In terms of polar coordinates centered at the corner, we assume the velocity potential to be of the form

$$\phi = f(r)g(\theta) \tag{C-1}$$

Plugging into Laplace's equation, which for polar coordinates is

$$\frac{\partial^2 \phi}{\partial r^2} + \frac{1}{r}\frac{\partial \phi}{\partial r} + \frac{1}{r^2}\frac{\partial^2 \phi}{\partial \theta^2} = 0 \tag{C-2}$$

we get

$$f''g + \frac{1}{r}f'g + \frac{1}{r^2}fg'' = 0$$

which can be rearranged into

$$\frac{r^2 f'' + rf'}{f} = -\frac{g''}{g}$$

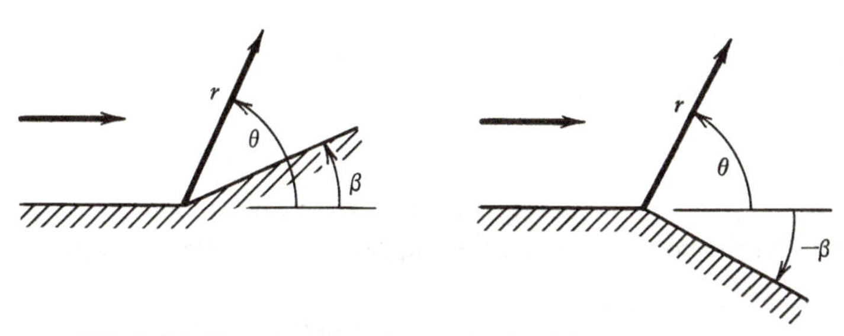

FIG. C.1. Nomenclature for analysis of flow past a corner.

Since $f = f(r)$, the left side of this equation is independent of θ. But, since $g = g(\theta)$, the right side is independent of r. Therefore, each side is constant, say k^2:

$$-\frac{g''}{g} = k^2 = r^2\frac{f''}{f} + r\frac{f'}{f}$$

or

$$g'' + k^2 g = 0 \tag{C-3}$$

$$r^2 f'' + rf' - k^2 f = 0 \tag{C-4}$$

This process is called *separation of variables.* It is often useful in solving partial differential equations. See Chapter 9 for an application of the method to the heat equation.

Equation C-3 is linear and homogeneous with constant coefficients. Its solution can therefore be expressed in terms of exponentials, or, better yet, trigonometric functions:

$$g(\theta) = A\sin k\theta + B\cos k\theta \tag{C-5}$$

Equation C-4 is also easy to solve. It is "homogeneous in r" and so is known to have a solution of the form

$$f(r) = Cr^\lambda$$

Substituting this into equation C-4, we get

$$C\lambda(\lambda - 1)r^\lambda + C\lambda r^\lambda - k^2 Cr^\lambda = 0$$

so

$$k^2 = \lambda(\lambda - 1) + \lambda = \lambda^2$$

and

$$\lambda = \pm k$$

Since, for a given k, both λ's solve equation C-4, its general solution must be of the form

$$f(r) = Cr^k + Dr^{-k} \tag{C-6}$$

Thus, from equations C-1, C-5, and C-6, a solution of the Laplace equation in polar coordinates C-2 can be written

$$\phi = (Cr^k + Dr^{-k})(A\sin k\theta + B\cos k\theta) \tag{C-7}$$

To be definite, suppose $k > 0$ but do not assume it to be an integer.

Now bring in the boundary conditions appropriate for the ramp problem:

$$\frac{\partial\phi}{\partial\theta} = 0 \qquad \text{at } \theta = \beta \text{ and } \theta = \pi \tag{C-8}$$

Substituting equation C-7 into C-8, we find

$$0 = (Cr^k + Dr^{-k})k(A\cos k\theta - B\sin k\theta) \qquad \text{at } \theta = \beta \qquad \text{and } \theta = \pi$$

$$0 = A\cos k\beta - B\sin k\beta$$

$$0 = A\cos k\pi - B\sin k\pi$$

Solve for A/B:

$$\frac{A}{B} = \frac{\sin k\beta}{\cos k\beta} = \frac{\sin k\pi}{\cos k\pi} \tag{C-9}$$

Then

$$\sin k\pi \cos k\beta - \cos k\pi \sin k\beta = 0$$

or

$$\sin k(\pi - \beta) = 0$$

so

$$k(\pi - \beta) = n\pi \qquad \text{for } n = 0, \pm 1, \pm 2, \pm 3, \ldots$$

Since k was supposed to be positive, and, from Fig. C.1, $|\beta| < \pi$, n must be nonnegative:

$$k = \frac{n\pi}{\pi - \beta} \qquad \text{for } n = 0, 1, 2, 3, \ldots \tag{C-10}$$

Then, as $\beta \to 0$, $k \to n$ and

$$\phi \to (Cr^n + Dr^{-n})\, B\cos n\theta$$

But the flow should then be approaching one that is uniform in the x direction, that is,

$$\phi \to V_\infty x = V_\infty r\cos\theta \qquad \text{as } \beta \to 0$$

Thus $D = 0$ and the only n that makes any sense is

$$n = 1$$

Equation C-7 then reduces to

$$\phi = r^k(A\sin k\theta + B\cos k\theta) \tag{C-11}$$

with, from equation C-10,

$$k = \frac{\pi}{\pi - \beta} \tag{C-12}$$

Now look at the flow velocity on the ramp surface:

$$V = \frac{\partial\phi}{\partial r} = kr^{k-1}(A\sin k\theta + B\cos k\theta) \tag{C-13}$$

We see that, at the corner itself ($r = 0$),

$$V = 0 \qquad \text{if } k > 1$$

and

$$V \to \infty \qquad \text{if } k < 1$$

But, from equation C-12, whether $k > 1$ or <1 depends on whether $\beta > 0$ or <0. Thus, a convex corner ($\beta < 0$) is a singular point for the velocity field, whereas a concave corner ($\beta > 0$) is a stagnation point.

In view of equations C-9 and C-10, equation C-13 also shows that, for a given distance r from the corner, the magnitude of V is the same on both sides of the corner ($\theta = \beta$ and $\theta = \pi$). Specifically, we find

$$\frac{V|_{\theta=\beta}}{V|_{\theta=\pi}} = \frac{A \sin k\beta + B \cos k\beta}{A \sin k\pi + B \cos k\pi}$$

$$= \frac{B \cos k\beta}{B \cos k\pi} \cdot \frac{\dfrac{A}{B}\dfrac{\sin k\beta}{\cos k\beta} + 1}{\dfrac{A}{B}\dfrac{\sin k\pi}{\cos k\pi} + 1} = \frac{\cos k\pi}{\cos k\beta}$$

$$= \frac{\cos \dfrac{\pi^2}{\pi - \beta}}{\cos \dfrac{\pi\beta}{\pi - \beta}}$$

But

$$\cos \frac{\pi^2}{\pi - \beta} = \cos\left(\pi + \frac{\beta\pi}{\pi - \beta}\right)$$

$$= -\cos \frac{\beta\pi}{\pi - \beta}$$

Thus

$$V|_{\theta=\beta} = -V|_{\theta=\pi}$$

APPENDIX D

UNIQUENESS OF SOLUTIONS OF LAPLACE EQUATION

In Section 4.2, we found that an ostensible reasonable problem—find ϕ so that

$$\nabla^2\phi = 0 \qquad \text{in flow field} \tag{D-1}$$

$$\nabla\phi \to \mathbf{V}_\infty \qquad \text{far from the body} \tag{D-2}$$

$$\nabla\phi \cdot \hat{\mathbf{n}} = 0 \qquad \text{on the body surface} \tag{D-3}$$

—does not have a unique solution. Our first try at a solution (the symmetric airfoil problem) involved an arbitrary constant, to which the lift turned out to be proportional. To fix the constant, we invented the "Kutta condition," whose justification was left to Chapter 7. You may be left worrying whether this is enough; whether even with the Kutta condition, our final solution is still not unique. When "reasonable" problems don't have unique solutions, what exactly *do* you have to do to specify the solution uniquely?

The question is strictly mathematical; physical reasoning failed us once and so is clearly inadequate. Reference to the real flow is also useless; we have made too many idealizations (incompressibility, negligible viscosity) for such references to be reliable.

In the case of flow past a two-dimensional airfoil, the answer is that, besides having ϕ satisfy the equations given above, we must specify the circulation

$$\Gamma = -\oint_C \nabla\phi \cdot \mathbf{dl} \tag{D-4}$$

around the airfoil. This is what the Kutta condition accomplishes; there is just one value of Γ for which the Kutta condition is satisfied.

We'll now prove that, for a given $\mathbf{V}_\infty$ and body shape, there is only one potential function that meets all of equations D-1 through D-4. First we will show that the

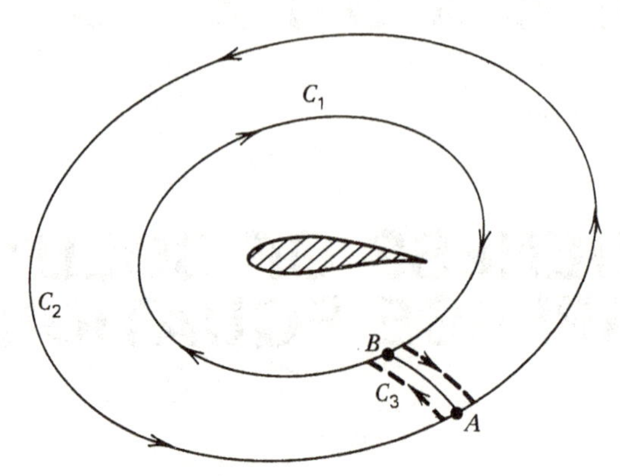

FIG. D.1. Contours needed to show independence of circulation of path of integration around the airfoil.

integral

$$\oint \nabla\phi \cdot \mathbf{dl}$$

has the same value, no matter what contour we pick for the integration, so long as that contour encloses the airfoil. Let C_1 and C_2 be two contours enclosing the wing, and C_3 a third line that connects the other two, as shown in Fig. D.1. Integrate $\nabla\phi \cdot \mathbf{dl}$ along a closed contour (call it C) which starts at A, goes counterclockwise on C_2 back to A, then goes to B along one side of C_3, clockwise on C_1 back to B, and finally back to A along the other side of C_3. The two integrations along C_3 cancel, since, at any point along the two paths, $\nabla\phi$ is the same, but $\mathbf{dl}$ is oppositely directed. Thus the result of the integral is simply

$$\oint_C \nabla\phi \cdot \mathbf{dl} = \oint_{C_2} \nabla\phi \cdot \mathbf{dl} + \oint_{C_1} \nabla\phi \cdot \mathbf{dl}$$

$$= \oint_{C_1} \nabla\phi \cdot \mathbf{dl} - \oint_{C_2} \nabla\phi \cdot \mathbf{dl} \tag{D-5}$$

(note the reversal of the direction of integration in the last integral). But, by Stokes's theorem, equation 2-25,

$$\oint_C \mathbf{V} \cdot \mathbf{dl} = \iint_S \hat{\mathbf{n}} \cdot \operatorname{curl} \mathbf{V}\, dS \tag{D-6}$$

where S is the area enclosed by the closed contour C, and $\hat{\mathbf{n}}$ is a unit normal to S. In equation D-5, $\mathbf{V} = \nabla\phi$, and S is the area between C_1 and C_2. Since the curl of any gradient is zero so is the right side of equation D-6, and hence of D-5 also. Thus, from equation D-5,

$$\oint_{C_2} \nabla\phi \cdot \mathbf{dl} = \oint_{C_1} \nabla\phi \cdot \mathbf{dl} \tag{D-7}$$

which is what we wanted to prove.

Note that Stokes's theorem does *not* show

$$\oint_{C_1} \nabla\phi \cdot \mathbf{dl} = \int_{S_1} \hat{\mathbf{n}} \cdot \text{curl } \nabla\phi \, dS = 0$$

since the surface S_1 enclosed by C_1 includes, not only the flow field in which ϕ is defined, but also the cross section of the wing, within which ϕ need not exist.

Another fact we will need is that a function ϕ is not single valued in a certain region unless, for every closed contour C in that region

$$\oint_C \nabla\phi \cdot \mathbf{dl} = 0 \tag{D-8}$$

That is, the change in ϕ between any two points P_1 and P_2 is

$$\phi(P_2) - \phi(P_1) = \int_{P_1}^{P_2} \nabla\phi \cdot \mathbf{dl}$$

If we get the same result for this integral regardless of path, then ϕ is single valued. But if that is so, we could integrate from P_1 to P_2 along some path C_1, then back to P_1 along another path C_2, as shown in Fig. D.2, and expect to get zero for

$$\oint_{C_1+C_2} \nabla\phi \cdot \mathbf{dl}$$

If this integral is not zero, the integral from P_1 to P_2 depends on the path, and ϕ is not single valued, QED.

So much for the preliminaries. Now let ϕ_1 and ϕ_2 be two functions of x and y that each satisfy equations D-1, D-2, D-3, and D-4. Note that the integral in (D-4) is now known to be independent of what path C we take around the airfoil. Then, if

$$g \equiv \phi_1 - \phi_2 \tag{D-9}$$

we see that

$$\nabla^2 g = 0 \tag{D-10}$$

$$\nabla g \to 0 \quad \text{far from body} \tag{D-11}$$

$$\nabla g \cdot \hat{\mathbf{n}} = 0 \quad \text{on body} \tag{D-12}$$

$$\oint_{\text{body}} \nabla g \cdot \mathbf{dl} = 0 \tag{D-13}$$

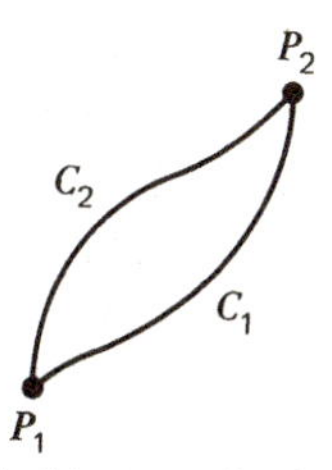

FIG. D.2. Contours needed to test single valuedness of potential.

We will show that these conditions make $\nabla g = 0$, which would imply equations D-1 to D-4 are sufficient to determine a unique solution. For then, if ϕ_1 is any one solution, any other solution ϕ_2 would have the same gradient ($\nabla\phi_1 = \nabla\phi_2$). Then ϕ_1 and ϕ_2 differ by at most a constant, which won't affect the result for the velocity field.

Let C_B be the body contour, C_∞ a surface far from the body, and C_C a connecting line, as shown in Fig. D.3. Since $\nabla g \cdot \hat{\mathbf{n}} = 0$ on C_B and C_∞,

$$\oint_{C_B} \hat{\mathbf{n}} \cdot g \, \nabla g \, dl = \oint_{C_\infty} \hat{\mathbf{n}} \cdot g \nabla g \, dl = 0 \tag{D-14}$$

where $\hat{\mathbf{n}}$ is a unit vector normal to the contour, directed out of the flow.

We now look at the integral

$$\oint_C \hat{\mathbf{n}} \cdot g \nabla g \, dl$$

where the contour C encloses the entire flow field: it starts at point A in Fig. D.3, goes counterclockwise on C_∞ back to A, goes to B on one side of C_C, goes around C_B clockwise back to B, and then back to A on the other side of C_C. Because of equation D-13 (here is where the specification of the circulation is necessary), g is single valued. Then (and *only* then) the two integrals along C_C cancel, and so, in view of equation D-14,

$$\oint_C \hat{\mathbf{n}} \cdot g \nabla g \, dl = 0$$

But, from the divergence theorem, equation 2-17,

$$\oint_C \hat{\mathbf{n}} \cdot \mathbf{f} \, dl = \int_S \operatorname{div} \mathbf{f} \, dS$$

where S is the area enclosed by C. Therefore,

$$\int_S \operatorname{div}(g \nabla g) \, dS = 0$$

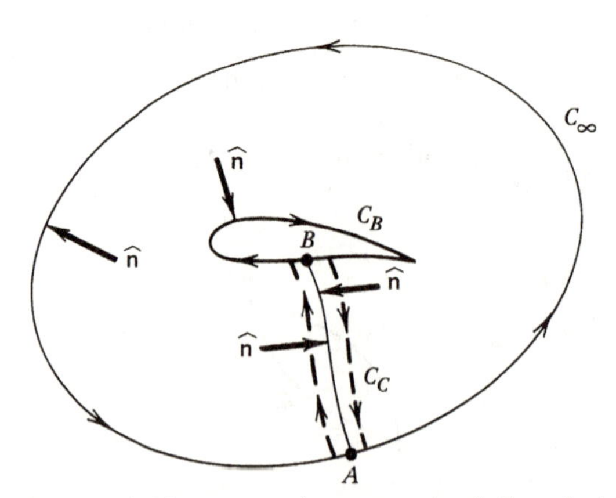

FIG. D.3. Contours needed to prove uniqueness of solution of airfoil problem.

Now, from equations 2-14 and 2-16,

$$\operatorname{div}(g\nabla g) = \frac{\partial}{\partial x}\left(g\frac{\partial g}{\partial x}\right) + \frac{\partial}{\partial y}\left(g\frac{\partial g}{\partial y}\right)$$
$$= \left(\frac{\partial g}{\partial x}\right)^2 + \left(\frac{\partial g}{\partial y}\right)^2 + g\left(\frac{\partial^2 g}{\partial x^2} + \frac{\partial^2 g}{\partial y^2}\right)$$

Applying equation D-10, we see that the last two terms drop out, leaving

$$\operatorname{div}(g\nabla g) = \left(\frac{\partial g}{\partial x}\right)^2 + \left(\frac{\partial g}{\partial y}\right)^2 = |\nabla g|^2$$

so that equation D-15 may be written

$$\int_S |\nabla g|^2 \, dS = 0$$

This can only be true if

$$\nabla g = 0$$

everywhere in the flow field. Therefore, any two solutions of equations D-1 to D-4 have the same gradient (velocity field), which is what we wanted to show.

APPENDIX E

FOURIER SERIES EXPANSIONS

Series expansions are often used to simplify operations on certain functions; for example, integration. Expansions in powers of the independent variable (i.e., *power series*) are just one example:

$$f(x) = a_0 + a_1 x + a_2 x^2 + \cdots \tag{E-1}$$

Fourier series are expansions in trigonometric functions:

$$f(\theta) = a_0 + \sum_{n=1} a_n \cos n\theta + b_n \sin n\theta \tag{E-2}$$

They are obviously especially useful for periodic functions. However, when we need to deal with a function over a finite interval, we can always imagine that the interval is the period of the function, and "continue" the function beyond the interval of interest so that it is periodic, as indicated in Fig. E.1. Since the "period" of the function sketched in Fig. E.1 is c, while that of the series (E-2) is 2π, we set

$$\frac{x}{c} = \frac{\theta}{2\pi} \tag{E-3a}$$

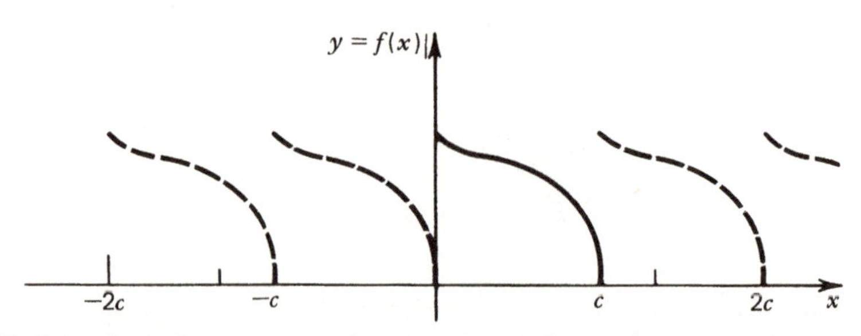

FIG. E.1. Periodic extension of a function defined over the interval $(0, c)$.

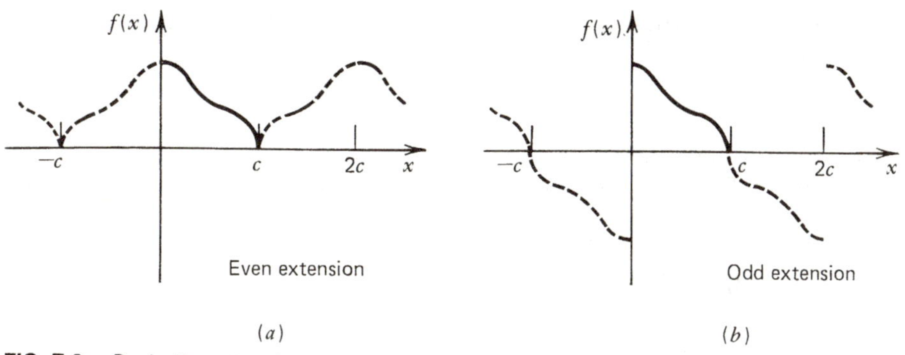

FIG. E.2. Periodic extensions of a function defined over the interval (0, *c*). (*a*) Even extension. (*b*) Odd extension.

We can also regard the interval of interest as half the period of the function, defining it over the other half of the period so the extended function is "even"

$$f(x) = f(-x)$$

or "odd"

$$f(x) = -f(-x)$$

see Fig. E.2. In either case, the period is $2c$, so the conversion from x to θ takes the form

$$\frac{x}{2c} = \frac{\theta}{2\pi} \tag{E-3b}$$

Since the sine is an odd function, while the cosine (including the special case $\cos 0\theta$, or constant term) is even, different Fourier expansions would be used for the different extensions:

$$f(x) = a_0 + \sum_{n=1} a_n \cos \frac{n\pi x}{c} \qquad \text{(even extension)} \tag{E-4}$$

$$= \sum_{n=1} b_n \sin \frac{n\pi x}{c} \qquad \text{(odd extension)} \tag{E-5}$$

In order for a series expansion to be useful, it is necessary to define the coefficients in terms of properties of the function. For a power series, the easiest way is to match derivatives of the function and its expansion at the origin. Then

$$a_n = \frac{1}{n!} \frac{d^n f}{dx^n}(0)$$

For the Fourier series, the coefficients are defined so that *the mean-squared difference between the function and its expansion over the period is minimized.* Consider, for example, a cosine series. By "mean-squared difference (or error)" we mean

$$E = \frac{1}{\pi}\int_0^\pi [f(\theta) - a_0 - \sum_{n=1} a_n \cos n\theta]^2 \, d\theta$$

For a given function $f(\theta)$, the error E depends on how we choose $a_0, a_1, a_2, \ldots$. Thus, to minimize E, set to zero the partial derivatives of E with respect to those coefficients:

$$0 = \frac{\partial E}{\partial a_0} = \frac{1}{\pi}\int_0^\pi 2[f(\theta) - a_0 - \sum_{n=1} a_n \cos n\theta] \, d\theta \tag{E-6}$$

$$0 = \frac{\partial E}{\partial a_m} = \frac{1}{\pi}\int_0^\pi 2[f(\theta) - a_0 - \sum_{n=1} a_n \cos n\theta]\cos m\theta \, d\theta$$

$$\text{for } m = 1, 2, 3, \ldots. \tag{E-7}$$

Since

$$\int_0^\pi \cos n\theta \, d\theta = \frac{1}{n} \sin n\theta \Big|_0^\pi = 0$$

the integral in equation E-6 becomes

$$0 = \frac{2}{\pi}\int_0^\pi f(\theta) \, d\theta - \frac{2}{\pi} a_0 \int_0^\pi d\theta$$

so

$$a_0 = \frac{1}{\pi}\int_0^\pi f(\theta) \, d\theta \tag{E-8}$$

whereas equation E-7 reduces to

$$\int_0^\pi f(\theta) \cos m\theta \, d\theta = \sum_{n=1} a_n \int_0^\pi \cos n\theta \cos m\theta \, d\theta \qquad \text{for } m = 1, 2, \ldots \tag{E-9}$$

Equation E-9 looks like a set of simultaneous equations for the coefficients a_n. However, it is much simpler than that, because of a property of trigonometric functions called *orthogonality*:

$$\begin{aligned} \int_0^\pi \cos n\theta \cos m\theta \, d\theta &= 0 \qquad \text{if } n \neq m \\ &= \frac{\pi}{2} \qquad \text{if } n = m \neq 0 \\ &= \pi \qquad \text{if } n = m = 0 \end{aligned} \tag{E-10}$$

This makes every term of the sum on the right side of (E-9) vanish except the mth, so

$$\int_0^\pi f(\theta) \cos m\theta \, d\theta = \frac{\pi}{2} a_m$$

or

$$a_m = \frac{2}{\pi}\int_0^\pi f(\theta)\cos m\theta\, d\theta \qquad \text{for } m = 1, 2, \ldots \tag{E-11}$$

It remains to discuss the *convergence* of the series; that is, whether we can make difference between the series and the associated function arbitrarily small by computing enough terms of the series. Power series generally converge only for limited values of the independent variable; for example,

$$\frac{1}{1-x} = 1 + x + x^2 + \cdots$$

converges only for $|x| < 1$. Fourier series converge under very weak conditions on the function. The function may even have finite jump discontinuities (in fact, it may have an infinite number of such jumps, so long as the sum of the jumps is finite) and still converge in the mean-squared sense. That is, for example,

$$E_N \equiv \frac{1}{\pi}\int_0^\pi [f(x) - a_0 - \sum_{n=1}^{N} a_n \cos n\theta]^2\, d\theta$$

vanishes as $N \to \infty$. Where it does converge, a Fourier series may be integrated term by term, but differentiation is possible only under restrictions on the function.

REFERENCE

Any good book on "mathematical methods for engineers and/or physicists" will do. I like especially H. Sagan's *Boundary and Eigenvalue Problems in Mathematical Physics*, Wiley, New York (1961).

PROBLEMS

Show that the coefficients of the Fourier sine series

$$f(\theta) = \sum_{n=1} b_n \sin n\theta$$

are given by

$$b_n = \frac{2}{\pi}\int_0^\pi f(\theta)\sin n\theta\, d\theta$$

if we want the mean-squared error of the expansion to be minimized. You will need to evaluate

$$\int_0^\pi \sin n\theta \sin m\theta\, d\theta$$

It may help to note that

$$\sin n\theta \sin m\theta = \tfrac{1}{2}\cos(n-m)\theta - \tfrac{1}{2}\cos(n+m)\theta$$

APPENDIX F

DOWNWASH DUE TO A HORSESHOE VORTEX

Consider a horseshoe vortex of strength Γ whose corners are at (x_a, y_a) and (x_a, y_b) and which is semi-infinite in the x direction, as shown in Fig. F.1. The velocity field may be calculated in one of two ways, neither of which is easy to explain. One rests on the *Biot–Savart Law*, according to which the velocity $\mathbf{V}$ induced by a vortex line is given by

$$\mathbf{V} = \frac{\Gamma}{4\pi} \oint \frac{\mathbf{dl} \times \mathbf{r}}{r^3}$$

in which (r, θ) are polar coordinates of the field point (x, y) relative to the segment $\mathbf{dl}$ of the (closed) vortex line, as shown in Fig. F.2. Note the right-hand rule relationship between $\mathbf{dl}$ and the circulation Γ. I don't know of any easy way to derive this formula; see, for example, Ref. [1].

In applying the Biot–Savart Law to the horseshoe vortex, let us be content to compute the downwash on the x–y plane. According to the right-hand rule

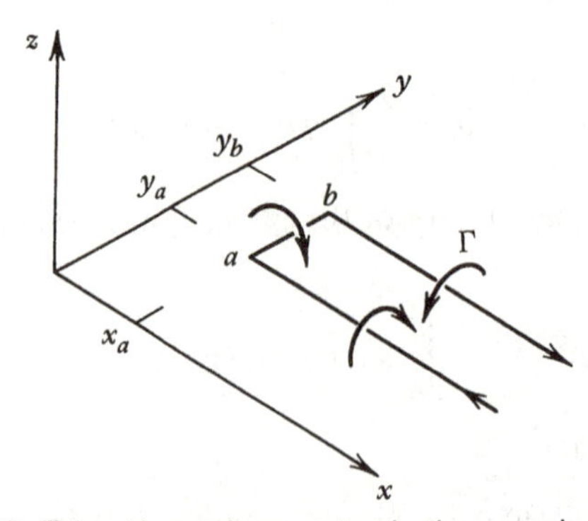

FIG. F.1. Horseshoe vortex in the x–y plane.

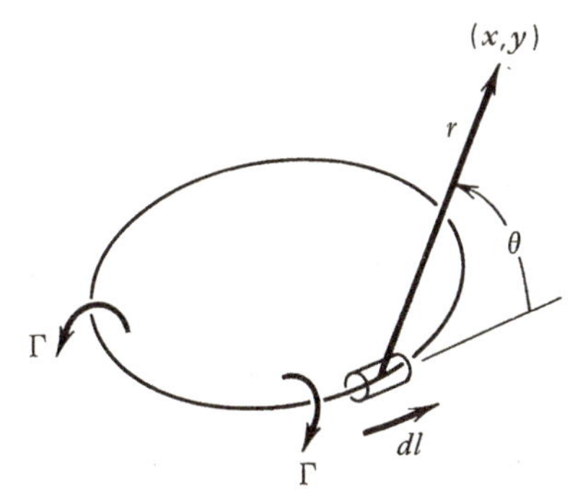

FIG. F.2. Nomenclature for Biot–Savart law.

described above, the integration should be from ∞ to a to b to ∞, as shown in Fig. F.3. Let (ξ, η) be the coordinates of **dl**, so

$$\mathbf{r} = (x - \xi)\hat{\mathbf{i}} + (y - \eta)\hat{\mathbf{j}}$$

and

$$r = \{(x - \xi)^2 + (y - \eta)^2\}^{1/2}$$

along the whole path. Also, note

$$\begin{aligned} \mathbf{dl} &= d\xi\,\hat{\mathbf{i}} \qquad \text{on } \infty a \text{ and } b\infty \\ &= d\eta\,\hat{\mathbf{j}} \qquad \text{on } ba \end{aligned}$$

Then, with the help of integral tables, we get

$$\begin{aligned} 4\pi w(x, y) &= -4\pi \mathbf{V} \cdot \mathbf{k} \\ &= -\hat{\mathbf{k}} \cdot \left\{ \int_{\infty}^{x_a} \frac{d\xi(y - y_a)\hat{\mathbf{k}}}{[(x-\xi)^2 + (y-y_a)^2]^{3/2}} - \int_{y_a}^{y_b} \frac{d\eta(x - x_a)\hat{\mathbf{k}}}{[(x-x_a)^2 + (y-\eta)^2]^{3/2}} \right. \\ &\quad \left. + \int_{x_a}^{\infty} \frac{d\xi(y - y_b)\hat{\mathbf{k}}}{[(x-\xi) + (y-y_b)^2]^{3/2}} \right\} \\ &= -\frac{(\xi - x)}{(y - y_a)[(x-\xi)^2 + (y-y_a)^2]^{1/2}}\Bigg|_{\xi=\infty}^{\xi=x_a} \\ &\quad + \frac{(\eta - y)}{(x - x_a)[(x-x_a)^2 + (y-\eta)^2]^{1/2}}\Bigg|_{\eta=y_a}^{\eta=y_b} \\ &\quad - \frac{(\xi - x)}{(y - y_b)[(x-\xi)^2 + (y-y_b)^2]^{1/2}}\Bigg|_{\xi=x_a}^{\xi=\infty} \end{aligned}$$

When $x = x_a$, the part of the vortex parallel to the y axis is ignored, since it contributes nothing to the downwash in the plane $x = x_a$. A little algebra and you recover the result of equation 5-9.

A second way of getting that result depends on the assertion that, for any distribution of vortices, there is an equivalent distribution of doublets whose axes

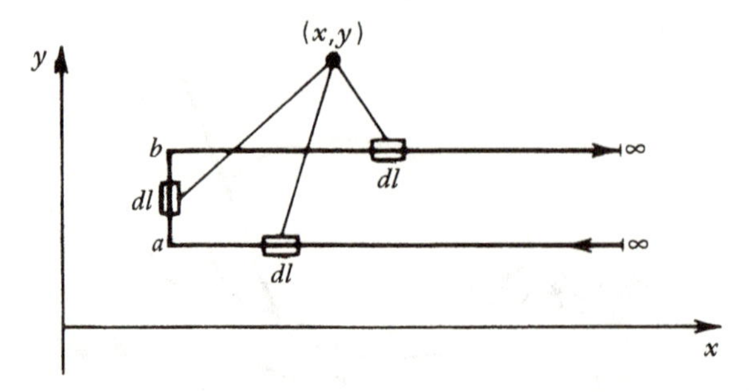

FIG. F.3. Nomenclature for application of Biot–Savart law to downwash induced by horseshoe vortex of Fig. F.1.

are normal to the surface on which they are distributed. You have already seen this to be true in the two-dimensional case. In problem 12 of Chapter 3, you showed

$$\phi = \frac{-1}{2\pi}\int_{x_s}^{x_f} \gamma(t)\tan^{-1}\frac{y}{x-t}\,dt = \frac{-1}{2\pi}\int_{x_s}^{\infty} \mu(t)\frac{y}{(x-t)^2+y^2}\,dt$$

in which the strength of the doublet distribution is just the negative of the integral of the strength of the vortex distribution γ:

$$\mu(x) = -\int_{x_s}^{x} \gamma(t)\,dt$$

If the vortex distribution is concentrated into a single vortex of strength Γ at $x = x_0$, that is, if the vortex strength is a *delta function*

$$\gamma(x) = \Gamma\delta(x - x_0)$$

the doublet strength becomes a *step function*

$$\mu(x) = -\Gamma \qquad \text{if } x > x_0$$
$$= 0 \qquad \text{if } x < x_0$$

and so the formulas given above for the potential become

$$\phi = -\frac{1}{2\pi}\Gamma\tan^{-1}\frac{y}{x-x_0}$$
$$= \frac{1}{2\pi}\Gamma\int_{x_0}^{\infty}\frac{y}{(x-t)^2+y^2}\,dt \qquad \text{(F-1)}$$

That is, a concentrated vortex is equivalent to a constant strength distribution of doublets from the location of the vortex to infinity.[1]

[1] If you don't follow this delta-function argument, you can still at least verify equation F-1 simply by evaluating the integral.

With the justification being results proved for the two-dimensional case, we now assert that the horseshoe vortex under discussion is equivalent to a constant-strength distribution of doublets over the semi-infinite rectangle within the horseshoe shown in Fig. F.1, $x_a < x < \infty$, $y_a < y < y_b$. The potential of a three-dimensional doublet with axis in the z direction (i.e., the direction normal to the rectangle) is, as in the two-dimensional case, obtained as the negative of the derivative of the potential of a point source

$$\phi_D = -\frac{\partial}{\partial z}\left(-\frac{1}{4\pi r}\right)$$
$$= -\frac{1}{4\pi}\frac{z}{r^3}$$

where r is the distance from the source. Letting the strength of the distribution be $-\Gamma$, we thus get for the potential of a horseshoe vortex of strength Γ,

$$\phi = \frac{\Gamma}{4\pi}\int_{y_a}^{y_b} d\eta \int_{x_a}^{\infty} d\xi \frac{z}{[(x-\xi)^2 + (y-\eta)^2 + z^2]^{3/2}}$$

This integral may be evaluated with the help of integral tables (e.g., in Dwight's tables, see entries 200.02, 120.1, and 387.):

$$\phi = \frac{\Gamma}{4\pi}\int_{y_a}^{y_b} d\eta \frac{z(\xi - x)}{[(\eta - y)^2 + z^2][(\xi - x)^2 + (\eta - y)^2 + z]^{1/2}}\Bigg|_{x_a}^{\infty}$$
$$= \frac{\Gamma}{4\pi}\int_{y_a}^{y_b} d\eta \frac{z}{[(\eta - y)^2 + z^2]}\left\{1 + \frac{x - x_a}{[(x - x_a)^2 + (\eta - y)^2 + z^2]^{1/2}}\right\}$$
$$= \frac{\Gamma}{4\pi}\left\{\tan^{-1}\left(\frac{\eta - y}{z}\right) + \tan^{-1}\left[\frac{(\eta - y)(x - x_a)}{z[(x - x_a)^2 + (\eta - y)^2 + z^2]^{1/2}}\right]\right\}\Bigg|_{y_a}^{y_b}$$

To get the formula 5-9 for the downwash of the horseshoe vortex, just differentiate this result with respect to z and set $z = 0$:

$$w(x, y) = -\frac{\partial \phi}{\partial z}(x, y, 0)$$

REFERENCE

1. Batchelor, G. K., *An Introduction to Fluid Dynamics.* Cambridge University Press, Cambridge/New York (1967), Section 2.4 and 2.6.

APPENDIX G

GEOMETRICAL DEMONSTRATION THAT STRAIN IS A TENSOR

The proof given in the text that the rate of strain defined by equation 6-27 satisfies the requirements for a second-order tensor, equation 6-46, is rather dry, being based on the chain rule of partial differentiation. To provide a feel for the physical content of the claim that "strain is a tensor," we present here a geometrical proof of the same result, for the special case in which the deformation takes place solely in the x_1–x_2 plane.

Specifically, we will explore the effect of a rotation of coordinates through an angle θ about the x_3 axis, see Fig. 6.25, on the strain components in the primed (rotated) and unprimed (original) coordinate systems. For this purpose, consider the deformation of two fluid particles, whose initial shapes are, as shown in Fig. G.1, triangular prisms, with faces parallel to the x_1, x_2, and x_3 axes and to either the x'_1 or x'_2 axis. Let the lengths of the faces along the primed axes be l.

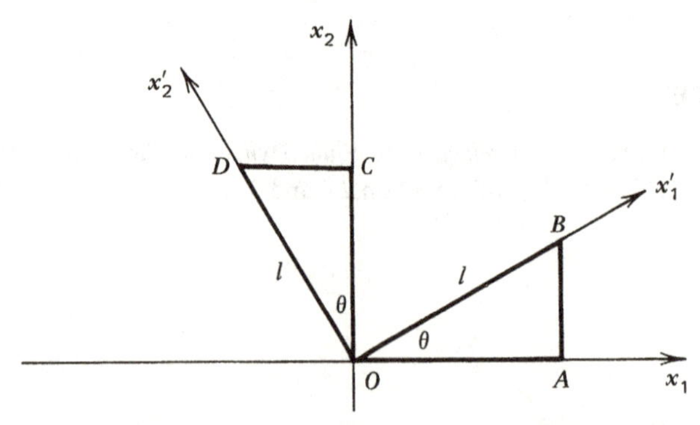

FIG. G.1. Test elements for definition of strain components: undeformed positions.

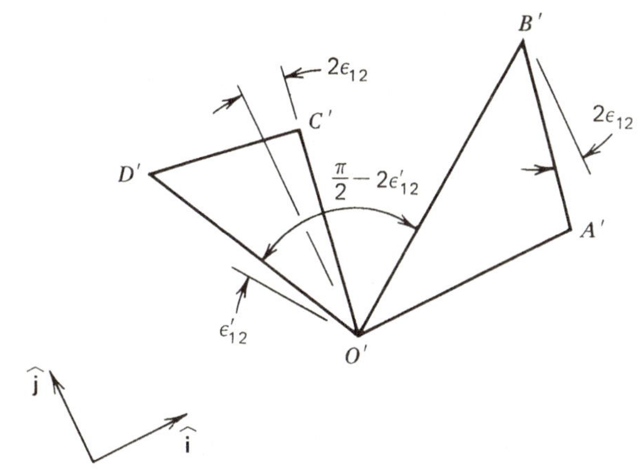

FIG. G.2. Test elements for definition of strain components: deformed positions.

In time Δt, the particles may move into the deformed positions shown in Fig. G.2. There O', A',... are the new positions of the edges whose projections on the x_1–x_2 plane are called O, A,... in Fig. G.1. Since the normal strain ε_{ii} is the percentage increase in length of a line originally in the x_i direction, the deformed lengths of edges OA, OB, AB,... may be expressed in terms of their original lengths (for which see Fig. G.1) and the appropriate normal strain components as follows:

$$\begin{aligned}
\overline{O'A'} &= l\cos\theta(1+\varepsilon_{11}) \\
\overline{O'B'} &= l(1+\varepsilon'_{11}) \\
\overline{A'B'} &= l\sin\theta(1+\varepsilon_{22}) \\
\overline{O'C'} &= l\cos\theta(1+\varepsilon_{22}) \\
\overline{O'D'} &= l(1+\varepsilon'_{22}) \\
\overline{C'D'} &= l\sin\theta(1+\varepsilon_{11})
\end{aligned} \tag{G-1}$$

The shear strain ε_{ij} also has a geometric interpretation; it is half of the decrease in the angle between lines originally in the x_i and x_j directions. This leads to the labeling of certain angles in Fig. G.2.

Now the vector $\mathbf{O'B'}$ can be written

$$\mathbf{O'B'} = \mathbf{O'A'} + \mathbf{A'B'}$$

Introducing orthogonal unit vectors $\hat{\mathbf{i}}$ and $\hat{\mathbf{j}}$, with $\hat{\mathbf{i}}$ along the deformed edge $O'A'$, as shown in Fig. G.2, we obtain

$$\begin{aligned}
\mathbf{O'B'} &= \overline{O'A'}\,\hat{\mathbf{i}} + \overline{A'B'}(\hat{\mathbf{j}}\cos 2\varepsilon_{12} + \hat{\mathbf{i}}\sin 2\varepsilon_{12}) \\
&= [(1+\varepsilon_{11})l\cos\theta + (1+\varepsilon_{22})l\sin\theta\sin 2\varepsilon_{12}]\hat{\mathbf{i}} + (1+\varepsilon_{22})l\sin\theta\cos 2\varepsilon_{12}\hat{\mathbf{j}} \\
&\approx l(\cos\theta\,\hat{\mathbf{i}} + \sin\theta\,\hat{\mathbf{j}}) + l([\varepsilon_{11}\cos\theta + 2\varepsilon_{12}\sin\theta]\hat{\mathbf{i}} + \varepsilon_{22}\sin\theta\,\hat{\mathbf{j}})
\end{aligned} \tag{G-2}$$

in which we have assumed the time interval Δt to be so small that all the strain components are also small, and so have neglected their products. Similarly,

$$\mathbf{O'D'} = l(-\sin\theta\,\hat{\mathbf{i}} + \cos\theta\,\hat{\mathbf{j}}) + l([-\varepsilon_{11}\sin\theta + 2\varepsilon_{12}\cos\theta]\hat{\mathbf{i}} + \varepsilon_{22}\cos\theta\,\hat{\mathbf{j}}) \quad \text{(G-3)}$$

Then, since

$$\mathbf{O'B'}\cdot\mathbf{O'B'} = |\overline{O'B'}|^2$$

we obtain, from equation G-2 and the second of equations G-1,

$$l^2(1 + 2\varepsilon'_{11}) \approx l^2(1 + 2\varepsilon_{11}\cos^2\theta + 4\varepsilon_{12}\sin\theta\cos\theta + 2\varepsilon_{22}\sin^2\theta)$$

or

$$\varepsilon'_{11} = \varepsilon_{11}\cos^2\theta + 2\varepsilon_{12}\sin\theta\cos\theta + \varepsilon_{22}\sin^2\theta \quad \text{(G-4)}$$

Since, for the rotation of coordinates pictured in Fig. 6.25,

$$\hat{\mathbf{e}}_1\cdot\hat{\mathbf{e}}'_1 = \cos\theta \qquad \hat{\mathbf{e}}_1\cdot\hat{\mathbf{e}}'_2 = -\sin\theta$$

$$\hat{\mathbf{e}}_2\cdot\hat{\mathbf{e}}'_1 = \sin\theta \qquad \hat{\mathbf{e}}_2\cdot\hat{\mathbf{e}}'_2 = \cos\theta$$

equation G-4 meets the requirement of equation 6-46, with $i = j = 1$. Similarly, from

$$\mathbf{O'D'}\cdot\mathbf{O'D'} = \overline{O'D'}^2$$

we get

$$\varepsilon'_{22} = \varepsilon_{11}\sin^2\theta - \varepsilon_{12}\sin\theta\cos\theta + \varepsilon_{22}\cos^2\theta \quad \text{(G-5)}$$

whereas, since

$$\mathbf{O'D'}\cdot\mathbf{O'B'} = \overline{O'D'}\,\overline{O'B'}\cos\left(-\frac{\pi}{2} - 2\varepsilon'_{12}\right)$$

$$\approx \overline{O'D'}\,\overline{O'B'}\,2\varepsilon'_{12}$$

see Fig. G.2, we get

$$-\varepsilon_{11}\sin\theta\cos\theta + \varepsilon_{12}(\cos^2\theta - \sin^2\theta) + \varepsilon_{22}\sin\theta\cos\theta = \varepsilon'_{12} \quad \text{(G-6)}$$

Equations G-5 and G-6 also agree with equation 6-46, the first with $i = j = 2$ and the second with $i = 1, j = 2$. Since these are the only nonzero strain components for the plane motion under consideration, we have thus verified that the geometrical definitions of the strain components, which were derived in Chapter 6 from their mathematical definition (6-27), meet the requirements of second-order tensors.

APPENDIX H

OPTIMIZATION OF THE SOR METHOD FOR THE LAPLACE EQUATION

Suppose, for convenience, that the finite-difference mesh consists of evenly spaced lines, as in equation 9-37. Then the finite-difference approximation to Laplace's equation is equation 9-38. If we try to solve the finite-difference equations by the iterative method known as successive overrelaxation, trial values for the Φ_{ij}'s are updated at each mesh point by the formula

$$\Phi_O^* = (1 - \omega)\tilde{\Phi}_O + \frac{\omega}{4}(\tilde{\Phi}_N + \Phi_S^* + \tilde{\Phi}_E + \Phi_W^*) \tag{H-1}$$

Here tildes indicate values taken from the preceding sweep through the mesh and stars data updated in the current sweep. It is supposed that points are swept from west to east and from south to north. The constant ω is the overrelaxation factor. The objective of this appendix is to determine the value of ω that yields the most rapid rate of convergence of the iterations.

Following Garabedian [1], we correlate successive sweeps through the mesh with an *artificial time* t, so that the nth sweep yields data at $t = nk$, say. Further, we suppose the quantities $\tilde{\Phi}_N$, $\tilde{\Phi}_O$, Φ_S^*, and so on to represent values of a continuous function of position and (artificial) time, and make appropriate Taylor-series expansions of these values about the central point at the tilded (previous) "time level":

$$\tilde{\Phi}_N = \tilde{\Phi} + h\frac{\partial\tilde{\Phi}}{\partial y} + \frac{h^2}{2}\frac{\partial^2\tilde{\Phi}}{\partial y^2} + \cdots$$

$$\tilde{\Phi}_E = \tilde{\Phi} + h\frac{\partial\tilde{\Phi}}{\partial x} + \frac{h^2}{2}\frac{\partial^2\tilde{\Phi}}{\partial x^2} + \cdots$$

$$\Phi_O^* = \tilde{\Phi} + k\frac{\partial\tilde{\Phi}}{\partial t} + \frac{k^2}{2}\frac{\partial^2\tilde{\Phi}}{\partial t^2} + \cdots$$

$$\Phi_S^* = \tilde{\Phi} + k\frac{\partial\tilde{\Phi}}{\partial t} + \frac{k^2}{2}\frac{\partial^2\tilde{\Phi}}{\partial t^2} - h\frac{\partial\tilde{\Phi}}{\partial y} - hk\frac{\partial^2\tilde{\Phi}}{\partial y\partial t} + \frac{h^2}{2}\frac{\partial^2\tilde{\Phi}}{\partial y^2} + \cdots$$

$$\Phi_W^* = \tilde{\Phi} + k\frac{\partial\tilde{\Phi}}{\partial t} + \frac{k^2}{2}\frac{\partial^2\tilde{\Phi}}{\partial t^2} - h\frac{\partial\tilde{\Phi}}{\partial x} - hk\frac{\partial^2\tilde{\Phi}}{\partial x\partial t} + \frac{h^2}{2}\frac{\partial^2\tilde{\Phi}}{\partial x^2} + \cdots$$

Here it is understood that quantities on the right side are evaluated at the central point and at the tilded time level. Substituting these series into equation H-1, we find, after some cancellation, that $\tilde{\Phi}^{\cdot}$ satisfies the partial differential equation[1]

$$\left(1 - \frac{\omega}{2}\right)\left(k\frac{\partial\tilde{\Phi}}{\partial t} + \frac{k^2}{2}\frac{\partial^2\tilde{\Phi}}{\partial t^2}\right) = \frac{\omega h^2}{4}\left(\frac{\partial^2\tilde{\Phi}}{\partial x^2} + \frac{\partial^2\tilde{\Phi}}{\partial y^2}\right) - \frac{\omega hk}{4}\left(\frac{\partial^2\tilde{\Phi}}{\partial x\partial t} + \frac{\partial^2\tilde{\Phi}}{\partial y\partial t}\right) \tag{H-2}$$

terms of third and higher order in h and k having been dropped.

As the artificial time $t \to \infty$—that is, as we perform more and more sweeps through the mesh—the solution of the partial differential equation (H-2) ought to become independent of time, in which case it reduces to the Laplace equation. Our object, then, is to pick ω so that the decay of the solution $\tilde{\Phi}$ to its steady state $\tilde{\Phi}$ is as rapid as possible. The analysis will be by the separation-of-variables method illustrated in Chapter 9 in connection with the heat equation.

Because of the mixed partial derivatives $\partial^2\tilde{\Phi}/\partial x\partial t$ and $\partial^2\tilde{\Phi}/\partial y\partial t$, equation H-2 does not admit a product solution of the form $X(x)\,Y(y)\,T(t)$. However, we can change variables so as to eliminate those derivatives, thus paving the way for a product solution in the new variables.

Let

$$\begin{aligned} \xi &\equiv x \\ \eta &\equiv y \\ \tau &\equiv t + \alpha x + \beta y \\ \phi(\xi, \eta, \tau) &\equiv \tilde{\Phi}(x, y, t) \end{aligned} \tag{H-3}$$

Then, from the chain rule,

$$\frac{\partial\tilde{\Phi}}{\partial x} = \frac{\partial\phi}{\partial\xi} + \alpha\frac{\partial\phi}{\partial\tau}$$

$$\frac{\partial\tilde{\Phi}}{\partial y} = \frac{\partial\phi}{\partial\eta} + \beta\frac{\partial\phi}{\partial\tau}$$

$$\frac{\partial\tilde{\Phi}}{\partial t} = \frac{\partial\phi}{\partial\tau}$$

[1] Since equation H-1 represents a method for solving finite-difference equations that approximate Laplace's partial differential equation, this may not seem like progress!

and so on, so that equation H-2 becomes

$$\left(1-\frac{\omega}{2}\right)\left(k\frac{\partial\phi}{\partial\tau}+\frac{k^2}{2}\frac{\partial^2\phi}{\partial\tau^2}\right)$$
$$=\frac{\omega h^2}{4}\left(\frac{\partial^2\phi}{\partial\xi^2}+2\alpha\frac{\partial^2\phi}{\partial\xi\partial\tau}+\alpha^2\frac{\partial^2\phi}{\partial\tau^2}+\frac{\partial^2\phi}{\partial\eta^2}+2\beta\frac{\partial^2\phi}{\partial\eta\partial\tau}+\beta^2\frac{\partial^2\phi}{\partial\tau^2}\right)$$
$$-\frac{\omega hk}{4}\left(\frac{\partial^2\phi}{\partial\xi\partial\tau}+\alpha\frac{\partial^2\phi}{\partial\tau^2}+\frac{\partial^2\phi}{\partial\eta\partial\tau}+\beta\frac{\partial^2\phi}{\partial\tau^2}\right) \quad \text{(H-4)}$$

The troublesome mixed derivatives now disappear if we set

$$\alpha=\beta=\frac{k}{2h} \quad \text{(H-5)}$$

Then equation (H-4) becomes

$$\left(1-\frac{\omega}{2}\right)k\frac{\partial\phi}{\partial\tau}+\left(1-\frac{\omega}{4}\right)\frac{k^2}{2}\frac{\partial^2\phi}{\partial\tau^2}=\frac{\omega h^2}{4}\left(\frac{\partial^2\phi}{\partial\xi^2}+\frac{\partial^2\phi}{\partial\eta^2}\right) \quad \text{(H-6)}$$

Assume a solution of product form

$$\phi=F(\xi,\eta)T(\tau) \quad \text{(H-7)}$$

Substituting equation H-7 into H-6 and dividing by ϕ, we get

$$\frac{\left(1-\frac{\omega}{2}\right)kT'+\left(1-\frac{\omega}{4}\right)\frac{k^2}{2}T''}{T}=\frac{\omega h^2}{4}\frac{\nabla^2F}{F} \quad \text{(H-8)}$$

Since, according to equation H-7, the right side of equation H-8 is independent of τ and the left of ξ and η, both sides are constant, which we choose to call $\mu\omega h^2/4$. Then

$$\nabla^2F=\mu F \quad \text{(H-9)}$$

$$\left(1-\frac{\omega}{4}\right)\frac{k^2}{2}T''+\left(1-\frac{\omega}{2}\right)kT'-\frac{\mu\omega h^2}{4}T=0 \quad \text{(H-10)}$$

Since equation H-10 is a linear, homogeneous, ordinary differential equation with constant coefficients, its general solution is of exponential form, say

$$T=Ce^{-\gamma\tau} \quad \text{(H-11)}$$

Substituting equation H-11 into H-10, we obtain a quadratic for $k\gamma$

$$\left(1-\frac{\omega}{4}\right)\frac{k^2\gamma^2}{2}-\left(1-\frac{\omega}{2}\right)k\gamma-\frac{\mu\omega h^2}{4}=0 \quad \text{(H-12)}$$

whose solutions are, for a given μ,

$$k\gamma = \frac{1 - \frac{\omega}{2} \pm \left[\left(1 - \frac{\omega}{2}\right)^2 + \frac{\mu\omega h^2}{2}\left(1 - \frac{\omega}{4}\right) \right]^{1/2}}{1 - \frac{\omega}{4}} \tag{H-13}$$

Thus, for each μ and F that satisfy equation H-9, equation H-6 has two solutions,

$$e^{-\gamma_+(\mu)}F(\xi, \eta), \qquad e^{-\gamma_-(\mu)}F(\xi, \eta)$$

in which the subscripts on the γ's indicate the choice of sign in front of the radical in equation H-13.

When $\mu = 0$, one of the γ's of equation H-13 is zero, and equation H-9 reduces to the Laplace equation satisfied by the solution $\Phi(\xi, \eta)$, say, toward which the iterative solution $\phi(\xi, \eta, \tau)$ is supposed to converge as $\tau \to \infty$ (and $h \to 0$). Therefore, let the F corresponding to $\gamma = 0$ meet whatever boundary conditions are imposed on Φ. That F will then *be* Φ and will be written as such in equation H-17 below. Further, the F's that correspond to other values of μ may then be required to satisfy homogeneous versions of the boundary conditions imposed on Φ. A function that is required to satisfy only homogeneous equations could be just zero. The μ's that allow satisfaction of equation H-9 and homogeneous boundary conditions by functions F that are not identically zero are called *eigenvalues* of the Laplacian, and the corresponding F's are *eigenfunctions* belonging to those eigenvalues.

In general, there are an infinite number of eigenvalues of equation H-9; compare equation 9-53 and also equation C-10. Let them be labeled $\mu_1, \mu_2, \mu_3, \ldots$. Also, let the corresponding γ's given by equation H-13 in terms of the μ's be $\gamma_{1+}, \gamma_{1-}, \gamma_{2+}, \gamma_{2-}, \ldots$, and the corresponding eigenfunctions be $F_1, F_2, \ldots$.

$$\nabla^2 F_n = \mu_n F_n \tag{H-14}$$

By applying the divergence theorem to $F_n \nabla F_n$ in the region $\mathscr{V}$ of interest, it is easy to show that, if F_n satisfies equation H-14 in $\mathscr{V}$ and homogeneous conditions on its boundary,

$$\mu_n = -\frac{\int_{\mathscr{V}} |\nabla F_n|^2 \, d\mathscr{V}}{\int_{\mathscr{V}} F_n^2 \, d\mathscr{V}} \tag{H-15}$$

which is real and less than zero for any nonzero F_n. Let the μ_n's be ordered so that

$$0 > \mu_1 \geq \mu_2 \geq \mu_3 \geq \cdots \tag{H-16}$$

Since no one of the eigenfunctions is liable to be proportional to the difference between the converged solution $\Phi(\xi, \eta)$ and the initial guess $\phi(\xi, \eta, 0)$, we trust that they may be superposed to do so, and write

$$\phi(\xi, \eta, \tau) = \Phi(\xi, \eta) + \sum_{n=1} F_n(\xi, \eta)(a_n e^{-\gamma_{n+}\tau} + b_n e^{-\gamma_{n-}\tau}) \tag{H-17}$$

Our object, it may be recalled, is to choose ω so that $\phi \to \Phi$ as τ increases, as rapidly as possible. Since τ enters equation H-17 only in the arguments of the exponentials, we thus want the $\gamma_{n\pm}$'s to be as large as possible. Rather, since equation H-13 allows γ to be complex if the radicand is negative, we must maximize the real parts of $\gamma_{n\pm}$. For if

$$\gamma_{n\pm} = \alpha_{n\pm} + i\beta_{n\pm}$$

then

$$|e^{-\gamma_{n\pm}\tau}| = e^{-\alpha_{n\pm}\tau}$$

Note that, for the iterations to converge at all, we need $\alpha_{n\pm} > 0$ for every n.

Since, from equation H-16, $\mu_n < 0$, equation H-13 shows $\gamma_{n\pm}$ to be purely imaginary ($\alpha_{n\pm} = 0$) when $\omega = 2$, in which case the iterations therefore do not converge. For ω close enough to 2 that the radicand of equation H-13 is still negative, the real part of $\gamma_{n\pm}$ is

$$\alpha_{n\pm} = \frac{1 - \dfrac{\omega}{2}}{1 - \dfrac{\omega}{4}} \tag{H-18}$$

whose derivative with respect to ω is

$$\frac{d\alpha_{n\pm}}{d\omega} = \frac{-\frac{1}{4}}{\left(1 - \dfrac{\omega}{4}\right)^2} < 0$$

Since we need $\alpha_{n\pm} > 0$ and as large as possible, we thus require $\omega < 2$.

When the radicand of equation H-13 is positive, $\gamma_{n\pm}$ is real, so

$$\alpha_{n\pm} = \gamma_{n\pm}$$

and, if $\omega < 2$,

$$\alpha_{n+} > \alpha_{n-}$$

It can be shown that the graphs of $\alpha_{n\pm}$ versus ω form a single-parameter family of curves that intersect the line described by equation H-18 with infinite slope, as shown in Fig. H.1. As the figure also shows, the smallest $\alpha_{n\pm}$ at any ω, and so the one associated with the slowest-decaying term of equation H-17, is α_{1-}, since, from equation H-16, μ_1 is the eigenvalue of equation H-14 smallest in magnitude. Thus the best possible value of ω is the one that maximizes α_{1-}. That value is seen from Fig. H.1 to be the one at which the curve of α_{1-} intersects the line described by equation H-18, namely, that which zeroes the radicand of equation H-13 when $\mu = \mu_1$. Calling this optimum value ω_c, we find

$$\omega_c = 2\left\{1 - \left[1 - \frac{1}{1 - (\mu_1 h^2/2)}\right]^{1/2}\right\}$$

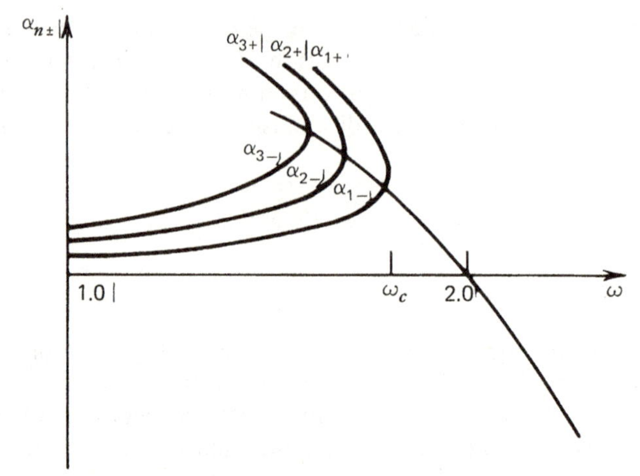

FIG. H.1. Variation with overrelaxation factor of time constant for convergence of SOR method.

which may be approximated for small $h^2\mu_1$ by

$$\omega_c \approx 2\left(1 - \sqrt{-\frac{\mu_1 h^2}{2}}\right) \tag{H-19}$$

Although this result seems to require a knowledge of μ_1, the smallest (in magnitude) eigenvalue of the Laplacian, it is possible to determine ω_c experimentally; that is, by executing the SOR method and manipulating its output. To do so, it is necessary that the ω used in the trial runs be less than ω_c, so that the smallest $\alpha_{n\pm}$ is α_{1-} (otherwise the $\alpha_{n\pm}$'s all coalesce into the value given by equation H-18). Since the terms of equation H-17 associated with other $\gamma_{n\pm}$'s then decay faster, after enough trials the solution should be well approximated by

$$\phi \cong \Phi + F_1 e^{-\alpha_{1-}\tau} \tag{H-20}$$

Since k is the "time step" from one iteration to the next, the change in the solution in one sweep through the mesh is

$$\delta(\tau) \cong F_1[e^{-\alpha_{1-}(\tau+k)} - e^{-\alpha_{1-}\tau}] = F_1 e^{-\alpha_{1-}\tau}(e^{-\alpha_{1-}k} - 1)$$

Similarly, the change in the solution in the next sweep is

$$\delta(\tau + k) \cong F_1 e^{-\alpha_{1-}(\tau+k)}(e^{-\alpha_{1-}k} - 1)$$

and so

$$-\alpha_{1-}k \cong \ln \frac{\|\delta(\tau + k)\|}{\|\delta(\tau)\|} \tag{H-21}$$

in which the double bars indicate some measure of the changes in the solution in one iteration through the whole mesh; for example, the square root of the sum of the squares of the changes at each mesh point. Then the value of μh^2 may be obtained from equation H-12, with (because $\omega < \omega_c$)γ set equal to α_{1-}:

$$\mu_1 h^2 = \frac{4}{\omega}\left[\left(1 - \frac{\omega}{4}\right)\frac{k^2\alpha_{1-}^2}{2} - \left(1 - \frac{\omega}{2}\right)k\alpha_{1-}\right] \qquad \text{(H-22)}$$

Taking $k\alpha_{1-}$ from equation H-20, we substitute equation H-22 into equation H-19 to get our estimate for ω_c.

Note, from equation H-19, that

$$2 - \omega_c = O(h)$$

and so, from equation H-18, that the best value of α_{1-} is

$$\alpha_{1-} = O(h)$$

Thus, as the mesh is refined, the optimum value of ω approaches 2, and, from equation H-20, the number of iterations to reduce the error to satisfactory values increases.

REFERENCE

1. Garabedian, P. R., "Estimation of the Relaxation Factor for Small Mesh Size," *Math. Tables Other Aids Comput.* **10**, 183–185 (1956).

APPENDIX I

STRUCTURE OF A WEAK SHOCK WAVE

Consider a steady, one-dimensional, compressible, viscous flow in the x direction. The x component of the Navier–Stokes equations can be found by combining equations 6-3, 6-27, and 6-56

$$\frac{d\rho u^2}{dx} = -\frac{dp}{dx} + \frac{d}{dx}\mu'\frac{du}{dx} \tag{I-1}$$

in which μ' is the sum of the absolute and bulk viscosities, whereas the continuity equation is

$$\frac{d\rho u}{dx} = 0 \tag{I-2}$$

These equations may both be integrated with respect to x, with the results

$$p + \rho u^2 = \mu'\frac{du}{dx} + \text{constant} \tag{I-3}$$

$$\rho u = \text{constant} \tag{I-4}$$

Equations I-3 and I-4 (along with the energy equation, which we will not look at in detail in this appendix) admit a solution in which p, ρ, and u vary with x but achieve constant values outside the region in which they vary, as shown in Fig. 11.17. Such a situation may exist under the same conditions that create a shock wave in a nonviscous flow. Thus we shall use the above equations to discover the effects of viscosity on a shock wave.

Indicating the values of the flow properties far upstream and downstream of the region of variation with the subscripts 1 and 2, respectively, we see from equations I-3 and I-4 that the changes from one side of the region to the other are connected by

$$p_1 + \rho_1 u_1^2 = p_2 + \rho_2 u_2^2$$

$$\rho_1 u_1 = \rho_2 u_2$$

which are the same as in a nonviscous flow; see equations 11-1 and 11-2. Again, the same thing is true for the energy equation, so that the "jump conditions" derived from the equations of Chapter 11 are valid for the changes in flow properties across a shock wave in a viscous fluid, too.

In the viscous case, however, the changes are not instantaneous, but take place over a finite length, which may be called the *shock-wave thickness.* We shall estimate this thickness for the case of a weak shock wave. Since the change in entropy through a weak shock is negligible, the isentropic pressure-density relation (11-16) is approximately valid. Then

$$\frac{dp}{dx} = \frac{\gamma p}{\rho}\frac{d\rho}{dx} = a^2 \frac{d\rho}{dx} = -\frac{\rho a^2}{u}\frac{du}{dx}$$

in which we have introduced the sound speed from equation 11-9 and used the continuity equation I-2. Substituting this result into equation I-1, we get

$$\rho(u^2 - a^2)\frac{du}{dx} = u\frac{d}{dx}\mu'\frac{du}{dx} \tag{I-5}$$

Now, in a weak shock wave, the flow speed normal to the shock u is close to the sound speed a, the difference being proportional to the change in u (or p, or ρ) across the shock, Δu, say. Thus

$$\begin{aligned} u^2 - a^2 &= (u-a)(u+a) \\ &\cong 2u(u-a) \\ &\sim u\Delta u \end{aligned}$$

The characteristic length of the flow is the shock-wave thickness L, which may be used to estimate derivatives of u,

$$\frac{du}{dx} \sim \frac{\Delta u}{L}$$

so that the left side of equation I-5 is of the order of

$$\sim \rho u \Delta u \frac{\Delta u}{L}$$

Similarly, the right side of equation I-5 may be estimated by

$$u\mu'\frac{\Delta u}{L^2}$$

Equating the two estimates, we can solve for L, the shock-wave thickness:

$$L \sim \frac{\mu'}{\rho \Delta u}$$

Thus the Reynolds number based on the shock-wave thickness and the change in velocity across the shock is of the order of 1. The thickness of a shock wave of measurable strength is, therefore, extremely small. For comparison, recall from equation 7-15 that the Reynolds number based on the thickness of a laminar boundary layer is of the order of the square root of the Reynolds number based on distance along the boundary layer, which is usually in the millions. Thus the thickness of a typical shock wave is much smaller than that of the typical boundary layer.

INDEX